# SEMICONDUCTOR PHYSICS AND DEVICES

**BASIC PRINCIPLES**

# SEMICONDUCTOR PHYSICS AND DEVICES

## BASIC PRINCIPLES

DONALD A. NEAMEN
University of New Mexico

Homewood, IL 60430
Boston, MA 02116

To my wife, Mimi,
for her patience, understanding,
and support during this project.

 This symbol indicates that the paper in this book is made from recycled paper. Its fiber content exceeds the recommended minimum of 50% waste paper fibers as specified by the EPA.

Cover photo: Courtesy of AT&T Archives

© RICHARD D. IRWIN, INC., 1992

Sponsoring editor: Bill Stenquist
Project editor: Paula M. Buschman
Production manager: Bette K. Ittersagen
Art manager: Kim Meriwether
Compositor: Bi-Comp, Inc.
Typeface: 10/12 Times Roman
Printer: R. R. Donnelley & Sons Company

**Library of Congress Cataloging-in-Publication Data**

Neamen, Donald A.
    Semiconductor physics and devices : basic principles / Donald A. Neamen.
        p.    cm.
    Includes bibliographical references and index.
    ISBN 0-256-08405-X
    1. Semiconductors.   I. Title.
QC611.N39 1992
537.6'22—dc20                                           91–31944

*Printed in the United States of America*
1 2 3 4 5 6 7 8 9 0 DOC 9 8 7 6 5 4 3 2

# PREFACE

## PHILOSOPHY AND GOALS

The purpose of this book is to provide a basis for understanding the characteristics, operation, and limitations of semiconductor devices. In order to gain this understanding, it is essential to have a thorough knowledge of the physics of the semiconductor material. The goal of this book is to bring together quantum mechanics, the quantum theory of solids, semiconductor material physics, and semiconductor device physics. All of these components are vital to the understanding of both the operation of present day devices and any future developments in the field.

The amount of physics presented in this text is greater than what is covered in most introductory semiconductor device books. Although this coverage is more extensive, the author has found that once the basic introductory and material physics have been thoroughly covered, the physics of the semiconductor device follows quite naturally and can be covered fairly quickly and efficiently.

Since the objective of this text is to provide an introduction to the theory of semiconductor devices, there is a great deal of advanced theory that is not considered. In addition, fabrication processes are not described in detail. There are a few references and general discussions about processing techniques such as diffusion and ion implantation, but only where the results of this processing have direct impact on device characteristics.

## PREREQUISITES

This book is intended for junior and senior undergraduates. The prerequisites for understanding the material are college mathematics, up to and including differential equations, and college physics, including an introduction to modern physics and electrostatics. Prior completion of an introductory course in electronic circuits is helpful, but not essential.

## ORGANIZATION

The text begins with the introductory physics, moves on to the semiconductor material physics, and then covers the physics of semiconductor devices. Chapter 1 presents an introduction to the crystal structure of solids, leading to the ideal single-crystal semiconductor material. Chapters 2 and 3 introduce quantum mechanics and the quantum theory of solids, which together provide the necessary basic physics.

Chapters 4 through 6 cover the semiconductor material physics. Chapter 4 presents the physics of the semiconductor in thermal equilibrium; Chapter 5 treats the transport phenomena of the charge carriers in a semiconductor. The nonequilibrium excess carrier characteristics are then developed in Chapter 6. Understanding the behavior of excess carriers in a semiconductor is vital to the goal of understanding the device physics.

The physics of semiconductor devices is developed in Chapters 7 through 15. Chapter 7 treats the electrostatics of the basic pn junction and Chapter 8 covers the current-voltage characteristics of the pn junction. Metal–semiconductor junctions, both rectifying and nonrectifying, and semiconductor heterojunctions are considered in Chapter 9. Chapter 10 treats the bipolar transistor and Chapter 11 covers the junction field-effect transistor. The physics of the metal-oxide-semiconductor field-effect transistor is presented in Chapters 12 and 13. Once the physics of the pn junction is developed, the chapters dealing with the three basic transistors may be covered in any order—these chapters are written so as not to depend on each other. Optical devices such as solar cells, photodetectors, LEDs, and LASER diodes are considered in Chapter 14. Finally, Chapter 15 treats the physics of a few other semiconductor devices, including the charge-coupled device and microwave generation devices.

## FEATURES

- ☐ A Preview section introduces each chapter, generally beginning with a discussion of how the material from the previous chapter will be used in the present chapter. The preview section then gives the topics to be covered in the chapter, elucidates the importance of these topics, and explains how they fit into the overall picture.
- ☐ An extensive number of examples are used throughout the text to reinforce the theoretical concepts being developed.
- ☐ A Summary and Review section follows the text of each chapter. This section summarizes the results that were derived in the chapter and reviews the basic concepts that were developed. The section ends by discussing the significance of the material to the remaining chapters.
- ☐ A Glossary of Important Terms follows the Summary and Review section of each chapter.
- ☐ A large number of problems are given at the end of each chapter, organized according to the subject of each section in the chapter body.

In many chapters, problems are included at the end in a Summary and Review section. These are design and open-ended problems.

☐ A Reading List finishes up each chapter. The references that are at an advanced level compared to this text are indicated by an asterisk.

☐ Answers to selected problems are given in the last appendix. Knowing the answer to a problem is an aid and a reinforcement in problem solving.

☐ A solutions manual is available to instructors.

## USE OF THE BOOK

The text is intended for a one semester course at the junior or senior level. As with most textbooks, there is more material than can be conveniently covered in one semester; this allows each instructor some flexibility in designing the course to his or her own specific needs. However, the text is not an encyclopedia. Sections in each chapter that can be skipped without loss of continuity are identified by an asterisk in both the table of contents and in the chapter itself. These sections, although important to the development of semiconductor device physics, can be postponed to a later time.

The material in the text has been used extensively in a course that is required for junior-level electrical engineering students at the University of New Mexico. Slightly less than half of the semester is devoted to the first six chapters; the remainder of the semester is devoted to the pn junction, the bipolar transistor, and the metal-oxide-semiconductor field-effect transistor. A few other special topics may be briefly considered near the end of the semester.

The first part of Chapter 3 develops qualitatively the idea of the band theory of semiconductors. An instructor choosing this approach need not, then, delve further into the more rigorous mathematics that follows later in Chapter 3.

Although the bipolar transistor is discussed in Chapter 10 before the JFET or MOSFET, each chapter, dealing with one of the three basic types of transistors, is written to stand alone. Any one of the transistor types may be covered first. Optical and other semiconductor devices are considered in the last two chapters. Although several device types are discussed in only two chapters, they are covered in sufficient detail for the reader to gain a basic understanding of their operation and characteristics.

## NOTES TO THE READER

This book introduces the physics of semiconductor materials and devices. Although many electrical engineering students are more comfortable building electronic circuits or writing computer programs than studying the underlying principles of semiconductor devices, the material presented here is vital to an understanding of the limitations of electronic devices, such as the microprocessor.

Mathematics is used extensively throughout the book. This may at times seem tedious, but the end result is an understanding that will not otherwise occur. Although some of the mathematical models used to describe physical processes may seem abstract, they have withstood the test of time in their ability to describe and predict these physical processes.

The reader is encouraged to continually refer to the preview sections so that the objective of the chapter and the purpose of each topic can be kept in mind. This constant review is especially important in the first six chapters, dealing with basic physics.

The reader must keep in mind that, although some sections may be skipped without loss of continuity, many instructors will choose to cover these topics. The fact that sections are marked with an asterisk does not minimize the importance of these subjects.

It is also important that the reader keep in mind that there may be questions still unanswered at the end of a course. Although the author dislikes the phrase, "it can be shown that. . . ," there are some concepts used here that rely on derivations beyond the scope of the text. This book is intended as an introduction to the subject. Those questions remaining unanswered at the end of the course, the reader is encouraged to keep "in a desk drawer." Then, during the next course in this area of concentration, he or she can take out these questions and search for the answers.

## ACKNOWLEDGMENTS

I am indebted to the many students I have had over the years who have helped in the evolution of this text. I am grateful for their enthusiasm and constructive criticism.

The University of New Mexico has my appreciation for providing an atmosphere conducive to writing this book, and Sandia National Laboratories has my gratitude for allowing me time for writing during a sabbatical.

I want to thank the many people at Richard D. Irwin, Inc., for their support. In particular, I am grateful to Bill Stenquist, the sponsoring editor, who supported and encouraged me throughout this project. His attention to the vast number of details is appreciated. I also appreciate the efforts of Paula Buschman, the project editor, who guided the work through its final phase, including gently, but firmly, pushing me through proofreading.

The following reviewers deserve thanks for reading the manuscript in its various phases:

Timothy J. Drummond, *Sandia Laboratories;* J. L. Davidson, *Vanderbilt University;* Robert Jackson, *University of Massachusetts, Amherst;* C. H. Wu, *University of Missouri, Rolla;* Bruce P. Johnson, *University of Nevada, Reno;* William Wilson, *Rice University;* Dennis Polla, *University of Minnesota;* G. E. Stillman, *University of Illinois, Urbana-Champaign;* D. K. Reinhard, *Michigan State University;* Len Trombetta, *University of Hous-*

*ton;* Dan Moore, *Virginia Polytechnic Institute and State University;* Richard C. Jaeger, *Auburn University;* Anand Kulkarni, *Michigan Technological University;* and Ronald D. Schrimpf, *University of Arizona.*

I appreciate the many fine and thorough reviews—your suggestions have made this a better book.

**Donald A. Neamen**

## READING LIST

Colclaser, R. A., and S. Diehl-Nagle. *Materials and Devices for Electrical Engineers and Physicists.* New York: McGraw-Hill, 1985.

Rhoderick, E. H. *Metal-Semiconductor Contacts.* Oxford: Clarendon Press, 1978.

Runyan, W. R., and K. E. Bean. *Semiconductor Integrated Circuit Processing Technology.* Reading, Mass.: Addison-Wesley, 1990.

Torrey, H. C., and C. A. Whitmer. *Crystal Rectifiers.* New York: McGraw-Hill, 1948.

**D.A.N.**

# CONTENTS

---

\* Indicates sections which may be skipped without loss of continuity.

x

*CHAPTER 15
# OTHER SEMICONDUCTOR DEVICES   665

# APPENDICES   710

# INDEX   731

# SEMICONDUCTORS AND THE INTEGRATED CIRCUIT

The development of the transistor and the integrated circuit (IC) has led to remarkable electronic capabilities. The IC permeates almost every facet of our daily lives. Perhaps the most dramatic example of IC technology is the digital computer—a relatively small desktop computer today has more capability than the equipment that just a few years ago would fill a whole room. The semiconductor electronics field continues to be a fast-changing one, with thousands of papers published each year.

The semiconductor device has a fairly long history, although the greatest explosion of IC technology has occurred during the last two decades.[1] The metal–semiconductor contact dates back to the early work of Braun in 1874, who discovered the asymmetric nature of electrical conduction between metal contacts and semiconductors, such as copper, iron, and lead sulphide. These devices were used as detectors in early experiments on radio. In 1906, Pickard took out a patent for a point contact detector using silicon and, in 1907, Pierce published rectification characteristics of diodes made by sputtering metals onto a variety of semiconductors.

By 1935, selenium rectifiers and silicon point contact diodes were available for use as radio detectors. With the development of radar, the need for detector diodes and mixers increased. Methods of achieving high-purity silicon and germanium were developed during this time. A significant advance in our understanding of the metal–semiconductor contact was aided by developments in the semiconductor physics. Perhaps most important during this period was Bethe's thermionic-emission theory in 1942, according to which the current is determined by the process of emission of electrons into the metal rather than by drift or diffusion.

---

[1] This brief introduction is intended to give a flavor of the history of the semiconductor device and integrated circuit. Thousands of engineers and scientists have made significant contributions to the development of semiconductor electronics—the few events and names mentioned here are not meant to imply that these are the only significant events or people involved in the semiconductor history.

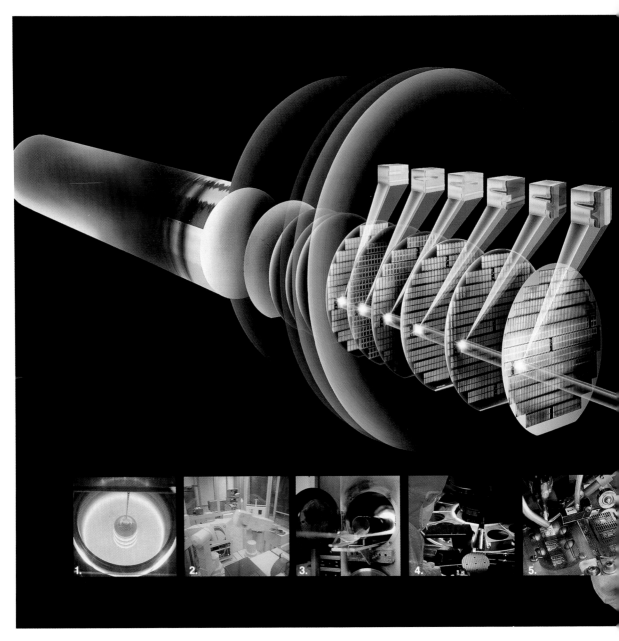

# THE CREATION OF AN INTEGRATED CIRCUIT

Placing 1 million transistors in a space this size [    ] is one of the most complex manufacturing operations imaginable. Yet Texas Instruments daily manufactures millions of integrated circuits—chips—of astonishing complexity and intricacy.

The precision involved is measured in microns or smaller; a micron being one-millionth of a meter or about one-fiftieth the diameter of a human hair; fabrication facilities are 1,000 times cleaner than the cleanest operating rooms; and billions of dollars have been invested in robots and other highly automated machinery.

Shown here are five major steps in the process of building an integrated circuit, plus an overview of an advanced technique called "trenching." It was developed by TI for making high-density memory chips.

### 1. Crystal Growth and Slicing

A chip begins with drawing a silicon ingot in a fiery furnace containing molten silicon. Special high-speed saws slice the ingots into wafers about the thickness of a dime, and they are then ground thinner and polished mirror smooth.

### 2. Photolithography Patterning

Creating the actual circuitry of each chip is done by photolithography. In this process, masks are used to expose a chemical coating called photoresist to ultraviolet light. This causes the photoresist to harden in desired patterns when developed.

### 3. Diffusion

An early step in building a chip is baking (diffusing) impurities into the silicon wafers in a diffusion furnace. The impurities alter the electrical characteristics of the silicon, creating separate regions with excess negative or positive charges. To determine the flow of these charges in any circuit, the impurities are diffused along a pattern defined by previous photolithography and etching processes. The example shown is the beginning of a high-density memory chip.

### 4. Etching

In this operation, the wafer moves to a plasma reactor where electrically excited gases etch the surface into the pattern defined by the photolithography process. In this manner, trenches—about three microns deep—are etched into the wafer. The trenches conserve valuable surface area and allow TI to place a million capacitors and a million transistors on a single chip.

Repeating this basic cycle—laying down materials, patterning, and etching—builds up the layers of the circuit. One of the final steps is to lay down an aluminum interconnect pattern that joins (integrates) the circuit components. In the case of the example, the connection to the aluminum interconnect pattern occurs at the edges of the chips.

### 5. Assembly, Bonding, and Packaging

The finished chips (there can be hundreds on each wafer) are tested, separated from each other, mounted in a frame, and wired to fit product specifications. The mounted chips are then sealed in packages of either ceramic or plastic to prevent damage and contamination and to facilitate handling. A series of stress tests simulating operating environments and a final electrical test assure reliability.

Another big breakthrough came in December 1947 when the first transistor was constructed and tested at Bell Telephone Laboratories by William Shockley, John Bardeen, and Walter Brattain. This first transistor was a point contact device and used polycrystalline germanium. The transistor effect was soon demonstrated in silicon as well. A significant improvement occurred at the end of 1949 when single-crystal material was used rather than the polycrystalline material. The single crystal yields uniform and improved properties throughout the whole semiconductor material.

The next significant step in the development of the transistor was the use of the diffusion process to form the necessary junctions. This process allowed better control of the transistor characteristics and yielded higher-frequency devices. The diffused mesa transistor was commercially available in germanium in 1957 and in silicon in 1958. The diffusion process also allowed many transistors to be fabricated on a single silicon slice, so the cost of these devices decreased.

Up to this point, each component in an electronic circuit had to be individually connected by wires. In September 1958, Jack Kilby of Texas Instruments demonstrated the first integrated circuit, which was fabricated in germanium. At about the same time, Robert Noyce of Fairchild Semiconductor introduced the integrated circuit in silicon using a planar technology. These first circuits used bipolar transistors. Practical MOS transistors were then developed in the mid-60s. The MOS technologies, especially CMOS, have become a major focus for IC design and development. Silicon is the main semiconductor material. Gallium arsenide and other compound semiconductors are used for special applications requiring very high frequency devices and for optical devices.

Since that first IC, circuit design has become more sophisticated, and the integrated circuit more complex. A single silicon chip may be on the order of 1 square centimeter and contain over a million transistors. Some ICs may have more than a hundred terminals, while an individual transistor has only three. An IC can contain the arithmetic, logic, and memory functions on a single semiconductor chip—the primary example of this type of IC is the microprocessor. Intense research on silicon processing and increased automation in design and manufacturing have led to lower costs and higher fabrication yields.

Understanding the semiconductor device and its characteristics must start with the semiconductor material. A specific goal of this text is to introduce the basic physics of the semiconductor material. This basic knowledge will lead to a sound understanding of existing devices, and it is also fundamental for understanding or developing new devices. The physics of semiconductor materials and devices will be covered in this text. Although semiconductor processing is an extremely important part of the IC technology, it is a topic left for other texts.

# THE CRYSTAL STRUCTURE OF SOLIDS

## CHAPTER 1

**PREVIEW**

This text deals with the electrical properties and characteristics of semiconductor materials and devices. The electrical properties of solids are therefore of primary interest. The semiconductor is in general a single-crystal material. The electrical properties of a single-crystal material are determined not only by the chemical composition but also by the arrangement of atoms in the solid; this being true, a brief study of the crystal structure of solids is warranted. The formation, or growth, of the single-crystal material is an important part of semiconductor technology. A short discussion of several growth techniques is included in this chapter to provide the reader with some of the terminology that describes semiconductor device structures.

We begin by presenting some of the common semiconductor materials. We then discuss some simple types of single-crystal lattice structures, including the simple cubic, body-centered cubic, and face-centered cubic. The single-crystal structures of the most common semiconductor materials are then described. These structures are referred to as the diamond or zincblende crystals. The bonding between atoms is closely related to the lattice structure and, in particular, the relationship between the diamond lattice and covalent bonding is considered. The perfect single-crystal lattice is an idealized structure; imperfections and impurities, which may affect the electrical properties of a solid, do exist in real materials. Several types of defects and impurities that occur in crystals are discussed.

Finally, several semiconductor material growth techniques are presented. The growth of semiconductor materials involves a phase change from a liquid or gas to a solid crystalline phase. One common growth technique is the Czochralski method. The substrate, the basic starting material for the fabrication of integrated circuits, is usually derived from this method. The epitaxial process is also of fundamental importance in the fabrication of integrated circuits. An epitaxial layer is formed on top of a single-crystal substrate; this is the region in which active semiconductor devices are fabricated.

The study of single-crystal lattices can become complex and lengthy, and the study of semiconductor growth techniques can involve a vast amount of theoretical as well as empirical detail. Whole books are devoted to each of these topics. This introductory chapter, however, should provide the necessary background in single crystals and crystal growth for the basic understanding of the electrical properties of semiconductor materials and devices.

## 1.1 SEMICONDUCTORS

Semiconductors are a group of materials having conductivities between those of metals and insulators. Two general classifications of semiconductors are the elemental semiconductor materials, found in group IV of the periodic table, and the compound semiconductor materials, most of which are formed from special combinations of group III and group V elements. Table 1–1 shows a portion of the periodic table in which the more common semiconductors are found and Table 1–2 lists a few of the semiconductor materials. (Semiconductors can also be formed from combinations of group II and group VI elements, but in general these will not be considered in this text.)

The elemental materials, those that are composed of single species of atoms, are silicon and germanium. Silicon is by far the most common semiconductor used in integrated circuits and will be emphasized to a great extent.

**Table 1–1**  A portion of the periodic table

| III | IV | V |
|-----|-----|-----|
| B | C | |
| Al | Si | P |
| Ga | Ge | As |
| In | | Sb |

**Table 1–2**  A list of some semiconductor materials

| Elemental semiconductors | |
|---|---|
| Si | Silicon |
| Ge | Germanium |

| Compound semiconductors | |
|---|---|
| AlP | Aluminum phosphide |
| AlAs | Aluminum arsenide |
| GaP | Gallium phosphide |
| GaAs | Gallium arsenide |
| InP | Indium phosphide |

The two-element, or *binary*, compounds such as gallium arsenide or gallium phosphide are formed by combining one group III and one group V element. Gallium arsenide is the most common of the compound semiconductors. Its good optical properties make it useful in optical devices. GaAs is also used in specialized applications in which, for example, high speed is required.

We can also form a three-element, or *ternary*, compound semiconductor. An example is $Al_xGa_{1-x}As$, in which the subscript $x$ indicates the fraction of the lower atomic number element component. More complex semiconductors can also be formed that provide flexibility when choosing material properties.

## 1.2 TYPES OF SOLIDS

Amorphous, polycrystalline, and single crystal are the three general types of solids. Each type is characterized by the size of an ordered region within the material. An ordered region is a spatial volume in which atoms or molecules have a regular geometric arrangement or periodicity. Amorphous materials have order only within a few atomic or molecular dimensions, while polycrystalline materials have a high degree of order over many atomic or molecular dimensions. These ordered regions, or single-crystal regions, vary in size and orientation with respect to one another. The single-crystal regions are called grains and are separated from one another by grain boundaries. Single-crystal materials, ideally, have a high degree of order, or regular geometric periodicity, throughout the entire volume of the material. The advantage of a single-crystal material is that, in general, its electrical properties are superior to those of a nonsingle-crystal material, since grain boundaries tend to degrade the electrical characteristics. Two-dimensional representations of amorphous, polycrystalline, and single-crystal materials are shown in Figure 1–1.

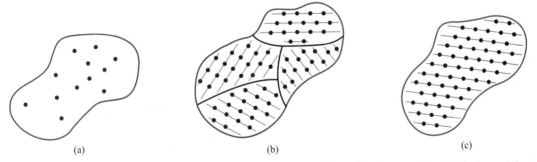

(a)      (b)      (c)

**Figure 1–1** Schematics of three general types of crystals: (a) amorphous, (b) polycrystalline, (c) single crystal.

## 1.3 SPACE LATTICES

Our primary concern will be the single crystal with its regular geometric periodicity in the atomic arrangement. A representative unit, or group of atoms, is repeated at regular intervals in each of the three dimensions to form the single crystal. The periodic arrangement of atoms in the crystal is called the *lattice*.

### 1.3.1 Primitive and Unit Cell

We can represent a particular atomic array by a dot which is called a lattice point. Figure 1–2 shows an infinite two-dimensional array of lattice points. The simplest means of repeating an atomic array is by translation. Each lattice point in Figure 1–2 can be translated a distance $a_1$ in one direction and a distance $b_1$ in a second noncolinear direction to generate the two-dimensional lattice. A third noncolinear translation will produce the three-dimensional lattice. The translation directions need not be perpendicular.

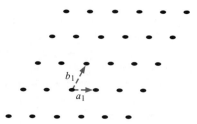

**Figure 1–2**   Two-dimensional representation of a single-crystal lattice.

Since the three-dimensional lattice is a periodic repetition of a group of atoms, we do not need to consider the entire lattice, but only a fundamental unit that is being repeated. A *unit cell* is a small volume of the crystal that can be used to reproduce the entire crystal. A unit cell is not a unique entity. Figure 1–3 shows several possible unit cells in a two-dimensional lattice.

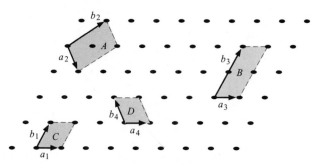

**Figure 1–3**   Two-dimensional representation of a single-crystal lattice showing various possible unit cells.

The unit cell A can be translated in directions $a_2$ and $b_2$, the unit cell B can be translated in directions $a_3$ and $b_3$, and the entire two-dimensional lattice can be constructed by the translations of either of these unit cells. The unit cells C and D in Figure 1–3 can also be used to construct the entire lattice by using the appropriate translations. This discussion of two-dimensional unit cells can easily be extended to three dimensions to describe a real single crystal material.

A *primitive cell* is the smallest unit cell that can be repeated to form the lattice. In many cases, it is more convenient to use a unit cell that is not a primitive cell. Unit cells may be chosen that have orthogonal sides, for example, whereas the sides of a primitive cell may be nonorthogonal.

A generalized three-dimensional unit cell is shown in Figure 1–4. The relationship between this cell and the lattice is characterized by three vectors **a**, **b**, and **c**, which need not be perpendicular and which may or may not be equal in length. Every equivalent lattice point in the three-dimensional crystal can be found using the vector

$$\mathbf{r} = p\mathbf{a} + q\mathbf{b} + s\mathbf{c} \qquad\qquad (1\text{--}1)$$

where p, q, and s are integers. Since the location of the origin is arbitrary, we will let p, q, and s be positive integers for simplicity.

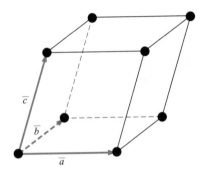

**Figure 1–4**  A generalized primitive unit cell.

### 1.3.2 Basic Crystal Structures

Before we discuss the semiconductor crystal, let us consider three crystal structures and determine some of the basic characteristics of these crystals. Figure 1–5 shows the simple cubic, body-centered cubic, and face-centered cubic structures. For these simple structures, we may choose unit cells such that the general vectors **a**, **b**, and **c** are perpendicular to each other and the lengths are equal. The *simple cubic* (sc) structure has an atom located at each corner; the *body-centered cubic* (bcc) structure has an additional atom at the center of the cube; and the *face-centered cubic* (fcc) structure has additional atoms on each face plane.

By knowing the crystal structure of a material and its lattice dimensions, we can determine several characteristics of the crystal. For example, we can determine the volume density of atoms.

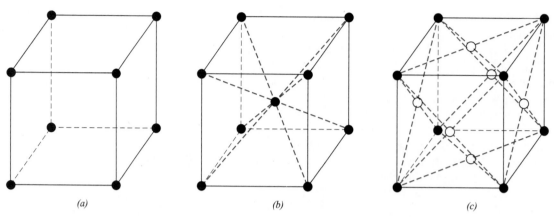

(a)                    (b)                    (c)

**Figure 1–5**   Three lattice types: (a) simple cubic, (b) body-centered cubic, (c) face-centered cubic.

---

**Example 1–1**

**Objective:** To find the volume density of atoms in a crystal.

Consider a single-crystal material that is a body-centered cubic with a lattice constant $a = 5$ Å $= 5 \times 10^{-8}$ cm. A corner atom is shared by eight unit cells which meet at each corner so that each corner atom effectively contributes one eighth of its volume to each unit cell. The eight corner atoms then contribute an equivalent of one atom to the unit cell. If we add the body-centered atom to the corner atoms, each unit cell contains an equivalent of two atoms.

**Solution:** The volume density of atoms is then found as

$$\text{Density} = \frac{2 \text{ atoms}}{(5 \times 10^{-8})^3} = 1.6 \times 10^{22} \text{ atoms per cm}^3$$

**Comment:** The volume density of atoms just calculated represents the order of magnitude of density for most materials. The actual density is a function of the crystal type and crystal structure since the packing density, number of atoms per unit cell, depends on crystal structure.

---

### 1.3.3 Crystal Planes and Miller Indices

Since real crystals are not infinitely large, they eventually terminate at a surface. Semiconductor devices are fabricated at or near a surface, so the surface properties may influence the device characteristics. We would like to be able to describe these surfaces in terms of the lattice. Surfaces, or planes through the crystal, can be described by first considering the intercepts of the plane along the **a**, **b**, and **c** axes used to describe the lattice.

## Example 1–2

**Objective:** To describe the plane shown in Figure 1–6.
  The lattice points in Figure 1–6 are shown along the **a**, **b**, and **c** axes only.

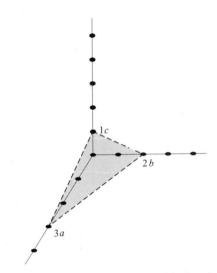

**Figure 1–6**  A representative crystal-lattice plane.

**Solution:** Using Equation (1–1), the intercepts of the plane correspond to p = 3, q = 2, and s = 1. Now write the reciprocals of the intercepts, which gives

$$\left(\frac{1}{3}, \frac{1}{2}, \frac{1}{1}\right)$$

Multiply by the lowest common denominator, which in this case is 6, to obtain (2, 3, 6). The plane in Figure 1–6 is then referred to as the (236) plane. The integers are referred to as the *Miller indices*. We will refer to a general plane as the (hkl) plane.

**Comment:** We can show that the same three Miller indices are obtained for any plane which is parallel to the one shown in Figure 1–6. Any parallel plane is entirely equivalent to any other.

---

  Three planes that are commonly considered in a cubic crystal are shown in Figure 1–7. The plane in Figure 1–7a is parallel to the **b** and **c** axes so the intercepts are given as p = 1, q = ∞ and s = ∞. Taking the reciprocal, we obtain the Miller indices as (1,0,0) so the plane shown in Figure 1–7a is referred to as the (100) plane. Again, any plane parallel to the one shown in Figure 1–7a and separated by an integral number of lattice constants is

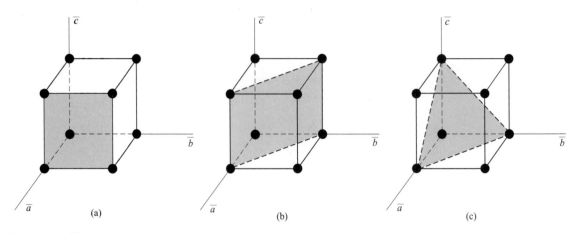

**Figure 1-7** Three lattice planes: (a) (100) plane, (b) (110) plane, (c) (111) plane.

equivalent and is referred to as the (100) plane. One advantage to taking the reciprocal of the intercepts to obtain the Miller indices is that the use of infinity is avoided when describing a plane that is parallel to an axis. If we were to describe a plane passing through the origin of our system, we would obtain infinity as one or more of the Miller indices after taking the reciprocal of the intercepts. However, the location of the origin of our system is entirely arbitrary and so, by translating the origin to another equivalent lattice point, we can avoid the use of infinity in the set of Miller indices.

For the simple cubic structure, the body-centered cubic, and the face-centered cubic, there is a high degree of symmetry. The axes can be rotated by 90 degrees in each of the three dimensions and each lattice point can again be described by Equation (1–1) as

$$\mathbf{r} = p\mathbf{a} + q\mathbf{b} + s\mathbf{c} \qquad (1\text{--}1)$$

Each face plane of the cubic structure shown in Figure 1–7a is entirely equivalent. These planes are grouped together and are referred to as the {100} set of planes.

We may also consider the planes shown in Figures 1–7b and 1–7c. The intercepts of the plane shown in Figure 1–7b are $p = 1$, $q = 1$, and $s = \infty$. The Miller indices are found by taking the reciprocal of these intercepts and, as a result, this plane is referred to as the (110) plane. In a similar way, the plane shown in Figure 1–7c is referred to as the (111) plane.

One characteristic of a crystal that can be determined is the distance between nearest equivalent parallel planes. Another characteristic is the surface concentration of atoms, # per cm$^2$, that are cut by a particular plane. Again, a single-crystal semiconductor is not infinitely large and must terminate at some surface. The surface density of atoms may be important, for example, in determining how another material, such as an insulator, will "fit" on the surface of a semiconductor material.

**Example 1–3**

**Objective:** To calculate the surface density of atoms on a particular plane in a crystal.

Consider the body-centered cubic structure and the (110) plane shown in Figure 1–8a. Assume the atoms can be represented as hard spheres with the closest atoms touching each other. Assume the lattice constant is $a_1 = 5$ Å. Figure 1–8b shows how the atoms are cut by the (110) plane.

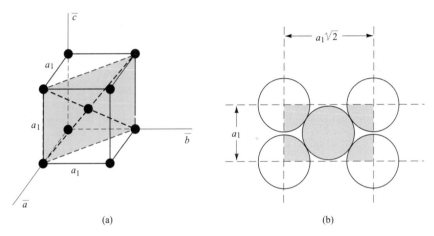

(a)          (b)

**Figure 1–8** (a) The (110) plane in a body-centered cubic and (b) the atoms cut by the (110) plane in a body-centered cubic.

The atom at each corner is shared by four similar equivalent lattice planes, so each corner atom effectively contributes one fourth of its area to this lattice plane as indicated in the figure. The four corner atoms then effectively contribute one atom to this lattice plane. The atom in the center is completely enclosed in the lattice plane. There is no other equivalent plane which cuts the center atom and the corner atoms, so the entire center atom is included in the number of atoms in the crystal plane. The lattice plane in Figure 1–8b then contains two atoms.

**Solution:** The surface density is found by dividing the number of lattice atoms by the surface area, or in this case

$$\text{Surface density} = \frac{2 \text{ atoms}}{(a_1)(a_1\sqrt{2})} = \frac{2}{(5 \times 10^{-8})^2(\sqrt{2})}$$

which is

$$5.66 \times 10^{14} \text{ atoms per cm}^2$$

**Comment:** The surface density of atoms is a function of the particular crystal plane in the lattice and generally varies from one crystal plane to another.

In addition to describing crystal planes in a lattice, we may want to describe a particular direction in the crystal. The direction can be expressed as a set of three integers which are the components of a vector in that direction. For example, the body diagonal in a simple cubic lattice is composed of vector components 1,1,1. The body diagonal is then described as the [111] direction. The brackets are used to designate direction as distinct from the parentheses used for the crystal planes. Note that in the simple cubic lattices, the [hkl] direction is perpendicular to the (hkl) plane. This perpendicularity may not be true in noncubic lattices.

### 1.3.4 The Diamond Structure

As already stated, silicon is the most common semiconductor material. Silicon is referred to as a group IV element and has a diamond crystal structure. Germanium is also a group IV element and has the same diamond structure. A unit cell of the diamond structure, shown in Figure 1–9, is more complicated than the simple cubic structures that we have considered up to this point.

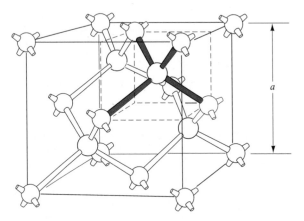

**Figure 1–9**   The diamond structure.

We may begin to understand the diamond lattice by considering the tetrahedral structure shown in Figure 1–10. This structure is basically a body-centered cubic with four of the corner atoms missing. Every atom in the tetrahedral structure has four nearest neighbors and it is this structure which is the basic building block of the diamond lattice.

There are several ways to visualize the diamond structure. One way in which we can gain a further understanding of the diamond lattice is by considering Figure 1–11. Figure 1–11a shows two body-centered cubic, or tetrahedral, structures diagonally adjacent to each other. The shaded circles

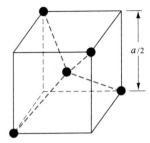

**Figure 1–10** The tetrahedral structure of closest neighbors in the diamond lattice.

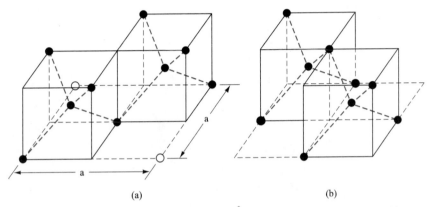

(a)                                         (b)

**Figure 1–11** Portions of the diamond lattice: (a) bottom half and (b) top half.

represent atoms in the lattice that are generated when the structure is translated to the right or left, one lattice constant, $a$. Figure 1–11b represents the top half of the diamond structure. The top half again consists of two tetrahedral structures joined diagonally, but which are at 90 degrees with respect to the bottom-half diagonal. An important characteristic of the diamond lattice is that any atom within the diamond structure will have four nearest neighboring atoms. We will note this characteristic again in our discussion of atomic bonding in the next section.

The diamond structure refers to the particular lattice in which all atoms are of the same species, such as silicon or germanium. The zincblende (sphalerite) structure differs from the diamond structure only in that there are two different types of atoms in the lattice. Compound semiconductors, such as GaAs, have the zincblende structure shown in Figure 1–12. The important feature of both the diamond and the zincblende structures is that the atoms are joined together to form a tetrahedron. Figure 1–13 shows the basic tetrahedral structure of GaAs in which each Ga atom has four nearest As neighbors and each As atom has four nearest Ga neighbors. This figure also begins to show the interpenetration of two sublattices that can be used to generate the diamond or zincblende lattice.

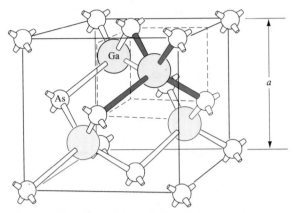

**Figure 1–12**   The zincblende (sphalerite) lattice of GaAs.

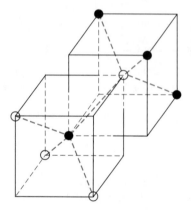

**Figure 1–13**   The tetrahedral structure of closest neighbors in the zincblende lattice.

## 1.4 ATOMIC BONDING

We have been considering various single-crystal structures. The question arises as to why one particular crystal structure is favored over another for a particular assembly of atoms. A fundamental law of nature is that the total energy of a system in thermal equilibrium tends to reach a minimum value. The interaction that occurs between atoms to form a solid and to reach the minimum total energy depends on the type of atom or atoms involved. The type of bond, or interaction, between atoms, then, depends on the particular atom or atoms in the crystal. If there is not a strong bond between atoms, they will not "stick together" to create a solid.

The interaction between atoms can be described by quantum mechanics. Although an introduction to quantum mechanics is presented in the next chapter, the quantum-mechanical description of the atomic bonding interaction is still beyond the scope of this text. We can nevertheless obtain a qualitative understanding of how various atoms interact by considering the valence, or outermost, electrons of an atom.

The atoms at the two extremes of the periodic table (excepting the inert elements) tend to lose or gain valence electrons, thus forming ions. These ions then essentially have complete outer energy shells. The elements in group I of the periodic table tend to lose their one electron and become positively charged, while the elements in group VII tend to gain an electron and become negatively charged. These oppositely charged ions then experience a coulomb attraction and form a bond referred to as an *ionic bond*. If the ions were to get too close, a repulsive force would become dominant, so an equilibrium distance results between these two ions. In a crystal, negatively charged ions tend to be surrounded by positively charged ions and positively charged ions tend to be surrounded by negatively charged ions, so a periodic array of the atoms is formed to create the lattice. A classic example of ionic bonding is sodium chloride.

The interaction of atoms tends to form closed valence shells such as we see in ionic bonding. Another atomic bond that tends to achieve closed-valence energy shells is covalent bonding, an example of which is found in the hydrogen molecule. A hydrogen atom has one electron and needs one more electron to complete the lowest energy shell. A schematic of two noninteracting hydrogen atoms, and the hydrogen molecule with the covalent bonding, are shown in Figure 1–14. Covalent bonding results in electrons being shared between atoms, so that in effect the valence energy shell of each atom is full.

H•        •H

(a)

H⦂H

(b)

Figure 1–14  Representation of (a) hydrogen valence electrons and (b) covalent bonding in a hydrogen molecule.

Atoms in group IV of the periodic table, such as silicon and germanium, also tend to form covalent bonds. Each of these elements has four valence electrons and needs four more electrons to complete the valence energy shell. If a silicon atom, for example, has four nearest neighbors, with each neighbor atom contributing one valence electron to be shared, then the center atom will in effect have eight electrons in its outer shell. Figure 1–15a schematically shows five noninteracting silicon atoms with the four valence electrons around each atom. A two-dimensional representation of the covalent bonding in silicon is shown in Figure 1–15b. The center atom has eight shared valence electrons.

**Figure 1–15**   Representation of (a) silicon valence electrons and (b) covalent bonding in the silicon crystal.

A significant difference between the covalent bonding of hydrogen and of silicon is that, when the hydrogen molecule is formed, it has no additional electrons to form additional covalent bonds, while the outer silicon atoms always have valence electrons available for additional covalent bonding. The silicon array may then be formed into an infinite crystal, with each silicon atom having four nearest neighbors and eight shared electrons. The four nearest neighbors in silicon forming the covalent bond correspond to the tetrahedral structure and the diamond lattice, which were shown in Figures 1–10 and 1–9, respectively. Atomic bonding and crystal structure are obviously directly related.

The third major atomic bonding scheme is referred to as metallic bonding. Group I elements have one valence electron. If two sodium atoms (Z = 11), for example, are brought into close proximity, the valence electrons interact in a way similar to that in covalent bonding. When a third sodium atom is brought into close proximity with the first two, the valence electrons can also interact and continue to form a bond. Solid sodium has a body-centered cubic structure, so each atom has eight nearest neighbors with each atom sharing many valence electrons. We may think of the positive metallic ions as being surrounded by a sea of negative electrons, the solid being held together by the electrostatic forces. This description gives a qualitative picture of the metallic bond.

A fourth type of atomic bond, called the Van der Waals bond, is the weakest of the chemical bonds. A hydrogen fluoride (HF) molecule, for example, is formed by an ionic bond. The effective center of the positive charge of the molecule is not the same as the effective center of the negative charge. This nonsymmetry in the charge distribution results in a small electric dipole which can interact with the dipoles of other HF molecules. With these weak dipole interactions, solids formed by the Van der Waals bonds have a relatively low melting temperature—in fact, most of these materials are in gaseous form at room temperature.

## *1.5   IMPERFECTIONS AND IMPURITIES IN SOLIDS

Up to this point, we have been considering an ideal single-crystal structure. In a real crystal, the lattice is not perfect, but contains imperfections or defects; that is, the perfect geometric periodicity is disrupted in some manner. Imperfections tend to alter the electrical properties of a material and, in some cases, electrical parameters can be dominated by these defects or impurities.

### 1.5.1   Imperfections in Solids

One type of imperfection that all crystals have in common is atomic thermal vibration. A perfect single crystal contains atoms at particular lattice sites, the atoms separated from each other by a distance we have assumed to be constant. The atoms in a crystal, however, have a certain thermal energy, which is a function of temperature. The thermal energy causes the atoms to vibrate in a random manner about an equilibrium lattice point. This random thermal motion causes the distance between atoms to randomly fluctuate, slightly disrupting the perfect geometric arrangement of atoms. This imperfection, called lattice vibrations, affects some electrical parameters, as we will see later in our discussion of semiconductor material characteristics.

Another type of defect is called a point defect. There are several of this type that we need to consider. Again, in an ideal single-crystal lattice, the atoms are arranged in a perfect periodic arrangement. However, in a real crystal, an atom may be missing from a particular lattice site. This defect is referred to as a vacancy and is schematically shown in Figure 1–16a. In another situation, an atom may be located between lattice sites. This defect is referred to as an interstitial and is schematically shown in Figure 1–16b. In

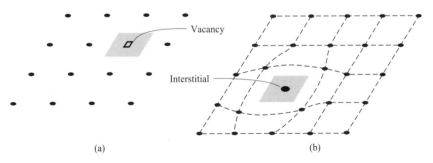

(a)                                                   (b)

**Figure 1–16**   Two-dimensional representation of a single-crystal lattice showing (a) a vacancy defect and (b) an interstitial defect.

the case of vacancy and interstitial defects, not only is the perfect geometric arrangement of atoms broken, but also the ideal chemical bonding between atoms is disrupted, which tends to change the electrical properties of the material. A vacancy and interstitial may be in close enough proximity to exhibit an interaction between the two point defects. This vacancy-interstitial defect, also known as a Frenkel defect, produces different effects than the simple vacancy or interstitial.

The point defects involve single atoms or single-atom locations. In forming single-crystal materials, more complex defects may occur. A line defect, for example, occurs when an entire row of atoms is missing from its normal lattice site. This defect is referred to as a line dislocation and is shown in Figure 1–17. As with a point defect, a line dislocation disrupts both the normal geometric periodicity of the lattice and the ideal atomic bonds in the crystal. This dislocation can also alter the electrical properties of the material, usually in a more unpredictable manner than the simple point defects.

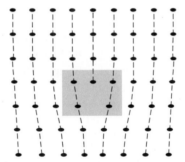

**Figure 1–17**   A two-dimensional representation of a line dislocation.

Other complex dislocations can also occur in a crystal lattice. However, this introductory discussion is intended only to present a few of the basic types of defect, and to show that a real crystal is necessarily not a perfect lattice structure. The effect of these imperfections on the electrical properties of a semiconductor will be considered in later chapters.

### 1.5.2 Impurities in Solids

Foreign atoms, or impurity atoms, may be present in a crystal lattice. Impurity atoms may be located at normal lattice sites, in which case they are called substitutional impurities. Impurity atoms may also be located between normal sites, in which case they are called interstitial impurities. Both these impurities are lattice defects and are schematically shown in Figure 1–18. Some impurities, such as oxygen in silicon, tend to be essentially inert;

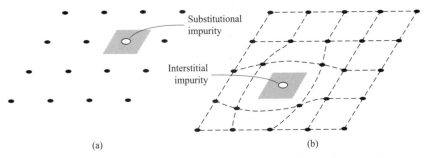

**Figure 1–18**    Two-dimensional representation of a single-crystal lattice showing (a) a substitutional impurity and (b) an interstitial impurity.

however, other impurities, such as gold or phosphorus in silicon, can drastically alter the electrical properties of the material.

In Chapter 4 we will see that, by adding controlled amounts of particular impurity atoms, the electrical characteristics of a semiconductor material can be favorably altered. The technique of adding impurity atoms to a semiconductor material in order to change its conductivity is called doping. There are two general methods of doping: impurity diffusion and ion implantation.

The actual diffusion process depends to some extent on the material but, in general, impurity diffusion occurs when a semiconductor crystal is placed in a high-temperature ($\approx 1000°C$) gaseous atmosphere containing the desired impurity atom. At this high temperature, many of the crystal atoms can randomly move in and out of their single-crystal lattice sites. Vacancies may be created by this random motion so that impurity atoms can move through the lattice by hopping from one vacancy to another. Impurity diffusion is the process by which impurity particles move from a region of high concentration near the surface, to a region of lower concentration within the crystal. When the temperature decreases, the impurity atoms become permanently frozen into the substitutional lattice sites. Diffusion of various impurities into selected regions of a semiconductor allows us to fabricate complex electronic circuits in a single semiconductor crystal.

Ion implantation generally takes place at a lower temperature than diffusion. A beam of impurity ions is accelerated to kinetic energies in the range of 50 KeV or greater and then directed to the surface of the semiconductor. The high-energy impurity ions enter the crystal and come to rest at some average depth from the surface. One advantage of ion implantation is that controlled numbers of impurity atoms can be introduced into specific regions of the crystal. A disadvantage of this technique is that the incident impurity atoms collide with the crystal atoms, causing lattice-displacement damage. However, most of the lattice damage can be removed by thermal annealing, in which the temperature of the crystal is raised for a short time. Thermal annealing is a required step after implantation.

## *1.6 GROWTH OF SEMICONDUCTOR MATERIALS

The success in fabricating very large scale integrated (VLSI) circuits is a result, to a large extent, of the development of and improvement in the formation or growth of pure single-crystal semiconductor materials. The high purity requirement means that extreme care is necessary in the growth and the treatment of the material at each step of the fabrication process. The mechanics and kinetics of crystal growth are extremely complex and will be described in only very general terms in this text. However, a general knowledge of the growth techniques and terminology is valuable.

### 1.6.1 Growth from a Melt

A common technique for growing single-crystal materials is called the Czochralski method. In this technique, a small piece of single-crystal material, known as a seed, is brought into contact with the surface of the same material in liquid phase, and then slowly pulled from the melt. As the seed is slowly pulled, solidification occurs along the plane between the solid–liquid interface. Usually the crystal is also rotated slowly as it is being pulled, to provide a slight stirring action to the melt, resulting in a more uniform temperature. Controlled amounts of specific impurity atoms, such as boron or phosphorus, may be added to the melt so that the grown semiconductor crystal is intentionally doped with the impurity atom. A schematic of the Czochralski growth process and a silicon ingot or boule grown by this process are shown in Figure 1–19.

Some impurities may be present in the ingot that are undesirable. Zone refining is a common technique for purifying material. A high-temperature coil, or r-f induction coil, is slowly passed along the length of the boule. The temperature induced by the coil is high enough so that a thin layer of liquid is formed. At the solid–liquid interface, there is a distribution of impurities between the two phases. The parameter which describes this distribution is called the segregation coefficient: the ratio of the concentration of impurities in the solid to the concentration in the liquid. If the segregation coefficient is 0.1, for example, the concentration of impurities in the liquid is a factor of 10 greater than that in the solid. As the liquid zone moves through the material, the impurities are driven along with the liquid. After several passes of the r-f coil, most impurities are at the end of the bar, which can then be cut off. The moving molten zone, or the zone-refining technique, can result in considerable purification.

After the semiconductor material is grown, the ingot is cut to a uniform diameter and then wafers are formed by slicing the ingot. The wafer must be thick enough to mechanically support itself. A mechanical two-sided lapping operation produces a flat wafer of uniform thickness. Since the lapping procedure can leave a surface damaged and contaminated by the mechanical operation, the surface must be removed by chemical etching. The final step

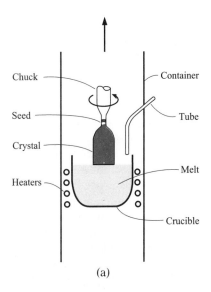

(a)

Photo courtesy of Texas Instruments Incorporated.

(b)

**Figure 1–19**   (a) Model of a crystal puller and (b) photograph of silicon granules, silicon ingot, slices, wafers, and integrated circuits.

is polishing. This provides a smooth surface on which devices may be fabricated or further growth processes may be carried out. This final semiconductor wafer is called the substrate material.

### 1.6.2 Epitaxial Growth

Epitaxial growth is a process whereby a thin, single-crystal layer of material is grown on the surface of a single-crystal substrate. In the epitaxial process, the single-crystal substrate acts as the seed, although the process takes place far below the melting temperature. When an epitaxial layer is grown on a substrate of the same material, the process is termed homoepitaxy. Growing silicon on a silicon substrate is one example of a homoepitaxy process. At present, a great deal of work is being done with heteroepitaxy. In a heteroepitaxy process, although the substrate and epitaxial materials are not the same, the two crystal structures should be very similar if single-crystal growth is to be obtained and if a large number of defects are to be avoided at the epitaxial–substrate interface. Growing epitaxial layers of the ternary alloy AlGaAs on a GaAs substrate is one example of a heteroepitaxy process.

One epitaxial growth technique that has been used extensively is called chemical vapor phase deposition (CVD). Silicon epitaxial layers, for example, are grown on silicon substrates by the controlled deposition of silicon atoms onto the surface from a chemical vapor containing silicon. In one method, silicon tetrachloride reacts with hydrogen at the surface of a heated substrate. The silicon atoms are released in the reaction and can be deposited onto the substrate, while the other chemical reactant, HCl, is in gaseous form and is swept out of the reactor. A sharp demarcation between the impurity doping in the substrate and in the epitaxial layer can be achieved using the CVD process. This technique allows great flexibility in the fabrication of semiconductor devices.

Liquid-phase epitaxy is another epitaxial growth technique. A compound of the semiconductor with another element may have a melting temperature lower than that of the semiconductor itself. The semiconductor substrate is held in the liquid compound and, since the temperature of the melt is lower than the melting temperature of the substrate, the substrate does not melt. As the solution is slowly cooled, a single-crystal semiconductor layer grows on the seed crystal. This technique, which occurs at a lower temperature than the Czochralski method, is useful in growing group III-V compound semiconductors.

A versatile technique for growing epitaxial layers is the molecular beam epitaxy (MBE) process. A substrate is held in vacuum at a temperature normally in the range of 400–800°C, a relatively low temperature compared to many semiconductor-processing steps. Semiconductor and dopant atoms are then evaporated onto the surface of the substrate. In this technique, the doping can be precisely controlled resulting in very complex doping profiles. Complex ternary compounds, such as AlGaAs, can be grown on substrates, such as GaAs, where abrupt changes in the crystal composition are desired. Many layers of various types of epitaxial compositions can be grown on a substrate in this manner. These structures are extremely beneficial in optical devices such as LASER diodes, which will be considered in later chapters of the text.

## 1.7 SUMMARY AND REVIEW

To begin this chapter, we listed a few common semiconductor materials. The properties of semiconductors are determined to a large extent by the single-crystal lattice structure, which can be described by a unit cell. A unit cell is a small volume of crystal that is used to reproduce the entire crystal. We considered three simple lattice structures—the simple cubic, body-centered cubic, and face-centered cubic—and considered the unit cells commonly used for these lattices. Since semiconductor devices are fabricated at or near a surface, it is convenient to describe these surfaces in terms of the lattice, using the set of integers called the Miller indices.

Silicon, the most common semiconductor, has the diamond lattice structure; that is, atoms are formed in a tetrahedral configuration with four nearest neighbor atoms. Germanium, the other elemental semiconductor, also has the diamond lattice. The binary semiconductors have a zincblende lattice, which is basically the same as the diamond lattice. The tetrahedral configuration in the diamond lattice is directly related to the atomic bonding in the semiconductor. The elemental semiconductors have four valence electrons forming covalent bonds with the four nearest neighbor atoms. Covalent bonding refers to the sharing of valence electrons between atoms.

Ideally, we would like to have perfect single-crystal semiconductors. However, imperfections and impurities do exist in real materials. We discussed various types of defects, including the simple point defects and the more complex line defect. By doping, that is, adding controlled amounts of specific types of impurity atoms, the electrical characteristics of a semiconductor can be altered in a favorable manner, as we will see in later chapters. These impurity atoms may be added during the growth of the semiconductor material, or later, by a diffusion or ion implantation process.

The growth of semiconductor materials is a very complex process. A brief discussion of growth methods introduced some of the terminology. The concepts of a substrate and an epitaxial layer are particularly important. The substrate is the starting point for fabricating integrated circuits. An active semiconductor device is usually fabricated in a thin epitaxial layer of semiconductor material grown on top of the substrate.

An introduction to quantum mechanics is given in the next chapter, which will then be applied, in Chapter 3, to describe the behavior and characteristics of the electrons involved in covalent bonding. Covalent bonding and the valence electrons will be referred to extensively in the next few chapters as we describe the electrical properties of semiconductor materials. In the chapters dealing with semiconductor devices, various semiconductor surfaces will be described by the Miller indices. The terminology we have introduced, such as impurity, ion implantation, and epitaxial layer, will be used extensively throughout the remainder of the text.

## GLOSSARY OF IMPORTANT TERMS

**Binary semiconductor:** A two-element compound semiconductor, such as gallium arsenide (GaAs).

**Covalent bonding:** The bonding between atoms in which valence electrons are shared.

**Diamond lattice:** The atomic crystal structure of silicon, for example, in which each atom has four nearest neighbors in a tetrahedral configuration.

**Elemental semiconductor:** A semiconductor composed of a single species of atom, such as silicon or germanium.

**Epitaxial layer:** A thin, single-crystal layer of material formed on the surface of a substrate.

**Lattice:** The periodic arrangement of atoms in a crystal.

**Miller indices:** The set of integers used to describe a crystal plane.

**Primitive cell:** The smallest unit cell that can be repeated to form a lattice.

**Substrate:** A semiconductor wafer or other material used as the starting material for further semiconductor processing, such as epitaxial growth or diffusion.

**Ternary semiconductor:** A three-element compound semiconductor, such as aluminum gallium arsenide (AlGaAs).

**Unit cell:** A small volume of a crystal that can be used to reproduce the entire crystal.

**Zincblende lattice:** A lattice structure identical to the diamond lattice except that there are two types of atoms instead of one.

## PROBLEMS

### Section 1.2

1. Determine the number of atoms per unit cell in a (*a*) face-centered cubic, (*b*) body-centered cubic, and (*c*) diamond lattice.

2. A crystal is composed of two elements, A and B. The basic crystal structure is a body-centered cubic with elements A at each of the corners and element B in the center. The effective radius of element A is 1.02 Å. Assume the elements are hard spheres with the surface of each A-type atom in contact with the surface of its nearest A-type neighbor. Calculate (*a*) the maximum radius of the B-type atom that

will fit into this structure, and (b) the volume density (# per cm$^3$) of A-type atoms and the volume density of B-type atoms.

3.  Assume that each atom is a hard sphere with the surface of each atom in contact with the surface of its nearest neighbor. Determine what percentage of the total unit cell volume is occupied in a (a) face-centered cubic, (b) body-centered cubic, and (c) diamond lattice.

4.  If the lattice constant of silicon is 5.43 Å, calculate (a) the distance from the center of one silicon atom to the center of its nearest neighbor, (b) the number density of silicon atoms (# per cm$^3$), and (c) the mass density (grams per cm$^3$) of silicon.

5.  Consider a three dimensional cubic lattice with a lattice constant equal to a. (a) Sketch the following planes: (i) (100), (ii) (110), (iii) (310), and (iv) (230). (b) Sketch the following directions: (i) [100], (ii) [110], (iii) [310], and (iv) [230].

6.  Sketch the following crystal planes for a simple cubic lattice: (a) (100), (b) (011), (c) (111), and (d) (432).

7.  The lattice constant of a simple cubic cell is 4.50 Å. Calculate the distance between the nearest parallel (a) (100), (b) (110), and (c) (111) planes.

8.  The lattice constant of a single crystal is 5.63 Å. Calculate the surface density of atoms (# per cm$^2$) on the following planes: (i) (100), (ii) (110), (iii) (111) for each of the following lattice structures: (a) simple cubic, (b) body-centered cubic, and (c) face-centered cubic.

9.  Determine the surface density of atoms for silicon on the (a) (100) plane, (b) (110) plane, and (c) (111) plane.

10.  Consider a face-centered cubic lattice. Assume the atoms are hard spheres with the surfaces of the nearest neighbors touching. Assume the radius of the atom is 1.34 Å. (a) Calculate the volume density of atoms in the crystal. (b) Calculate the distance between nearest (110) planes. (c) Calculate the surface density of atoms on the (100) plane.

## Section 1.3

11.  Calculate the density of valence electrons in silicon.

12.  The structure of GaAs is the zincblende lattice. The lattice constant is 5.65 Å. Calculate the density of valence electrons in GaAs.

## Section 1.4

13.  If $1 \times 10^{15}$ boron atoms per cm$^3$ are added to silicon as a substitutional impurity, determine what percentage of the silicon atoms are displaced in the single crystal lattice.

14.  If $1 \times 10^{15}$ gold atoms per cm$^3$ are added to silicon as a substitu-
     tional impurity and are distributed uniformly throughout the semi-
     conductor, determine the distance between gold atoms in terms of
     silicon lattice constants. (Assume the gold atoms are distributed in a
     rectangular or cubic array.)

## READING LIST

1.  Azaroff, L. V., and J. J. Brophy. *Electronic Processes in Materials*. New
    York: McGraw-Hill, 1963.

2.  Kittel, C. *Introduction to Solid State Physics*. 6th ed. New York: Wiley,
    1986.

3.  McKelvey, J. P. *Solid State and Semiconductor Physics*. New York: Harper
    & Row, 1966.

4.  Pierret, R. F. *Semiconductor Fundamentals*. Vol. 1 of *Modular Series on
    Solid State Devices*. 2nd ed. Reading, Mass.: Addison-Wesley, 1988.

5.  Pierret, R. F. *Advanced Semiconductor Fundamentals*. Vol. 6 of *Modular
    Series on Solid State Devices*. Reading, Mass.: Addison-Wesley, 1987.

6.  Streetman, B. G. *Solid State Electronic Devices*. 3rd ed. Englewood Cliffs,
    N.J.: Prentice Hall, 1990.

7.  Sze, S. M. *VLSI Technology*. New York: McGraw-Hill, 1983.

*8.  Wolfe, C. M.; N. Holonyak, Jr.; and G. E. Stillman. *Physical Properties of
    Semiconductors*. Englewood Cliffs, N.J.: Prentice Hall, 1989.

\* Indicates references that are at an advanced level compared to this text.

# INTRODUCTION TO QUANTUM MECHANICS

The motion of large objects, such as planets and satellites, can be predicted to a high degree of accuracy using classical theoretical physics based on Newton's laws of motion. But certain experimental results, involving electrons and high-frequency electromagnetic waves, appear to be inconsistent with classical physics. However, these experimental results can be predicted using the principles of quantum mechanics. The quantum mechanical wave theory is the basis for the theory of semiconductor physics.

We are ultimately interested in semiconductor materials whose electrical properties are directly related to the behavior of electrons in the crystal lattice. The behavior and characteristics of these electrons can be described by the formulation of quantum mechanics called wave mechanics. The essential elements of this wave mechanics are presented in this chapter. Three basic principles of quantum mechanics are initially discussed, and then Schrodinger's wave equation is introduced.

Schrodinger's wave equation will be applied to particular potential functions; the analysis will allow the reader to become familiar with the application of the wave equation as well as to observe some basic differences between the results of quantum mechanics and those of classical mechanics. The wave equation will be applied to the one-electron, or hydrogen, atom. The results of this analysis will be used to determine how the initial portion of the periodic table of elements can be developed.

This chapter is intended to provide an introduction to quantum mechanics. Some basic properties of electron behavior are developed using wave mechanics so that the results can be generalized in the next chapter to qualitatively and mathematically describe electron behavior in the single-crystal lattice.

## 2.1 PRINCIPLES OF QUANTUM MECHANICS

Before we delve into the mathematics of quantum mechanics, there are three principles we need to consider: the principle of energy quanta, the wave–particle duality principle, and the uncertainty principle.

### 2.1.1 Energy Quanta

One experiment that demonstrates an inconsistency between experimental results and the classical theory of light is called the photoelectric effect. If monochromatic light is incident on a clean surface of a material, then under certain conditions, electrons (photoelectrons) are emitted from the surface. According to classical physics, if the intensity of the light is large enough, the work function of the material will be overcome and an electron will be emitted from the surface independent of the incident frequency. This result is not observed. The observed effect is that, at a constant incident intensity, the maximum kinetic energy of the photoelectron varies linearly with frequency with a limiting frequency $\nu = \nu_o$, below which no photoelectron is produced. This result is shown in Figure 2–1. If the incident intensity varies at a constant frequency, the rate of photoelectron emission changes, but the maximum kinetic energy remains the same.

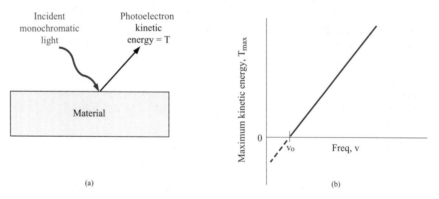

**Figure 2–1**   (a) The photoelectric effect and (b) the maximum kinetic energy of the photoelectron as a function of incident frequency.

Planck postulated in 1900 that thermal radiation is emitted from a heated surface in discrete packets of energy called *quanta*. The energy of these quanta is given by $E = h\nu$, where $\nu$ is the frequency of the radiation and $h$ is a constant now known as Planck's constant ($h = 6.625 \times 10^{-34}$ J − sec). Then in 1905, Einstein interpreted the photoelectric results by suggesting that the energy in a light wave is also contained in discrete packets or bundles. The particle-like packet of energy is called a *photon*, whose energy is also given by $E = h\nu$. A photon with sufficient energy, then, can knock an electron from the surface of the material. The minimum energy required to remove an electron is called the *work function* of the material and any excess photon energy goes into the kinetic energy of the photoelectron. This result was confirmed experimentally as demonstrated in Figure 2–1. The photoelectric effect shows the discrete nature of the photon and demonstrates the particle-like behavior of the photon.

The maximum kinetic energy of the photoelectron can be written as

$$T_{max} = \tfrac{1}{2} mv^2 = h\nu - h\nu_o \qquad (\nu \geq \nu_o) \qquad (2\text{–}1)$$

where $h\nu$ is the incident photon energy and $h\nu_o$ is the minimum energy, or work function, required to remove an electron from the surface.

---

**Example 2–1**

**Objective:** To calculate the photon energy corresponding to a particular wavelength.

Consider an X-ray with a wavelength of $\lambda = 0.708 \times 10^{-8}$ cm.

**Solution:** The energy is

$$E = h\nu = \frac{hc}{\lambda} = \frac{(6.625 \times 10^{-34})(3 \times 10^{10})}{0.708 \times 10^{-8}} = 2.81 \times 10^{-15} \text{ J}$$

This value of energy may be given in the more common unit of electron-volt (see appendix F). We have

$$E = \frac{2.81 \times 10^{-15}}{1.6 \times 10^{-19}} = 1.75 \times 10^4 \text{ eV}$$

**Comment:** The reciprocal relation between photon energy and wavelength is demonstrated: a large energy corresponds to a short wavelength.

---

## 2.1.2 Wave–Particle Duality

We have seen in the last section that light waves, in the photoelectric effect, behave as if they are particles. The particle-like behavior of electromagnetic waves was also instrumental in the explanation of the Compton effect. In this experiment, an X-ray beam was incident on a solid. A portion of the X-ray beam was deflected and the frequency of the deflected wave had shifted compared to the incident wave. The observed change in frequency and the deflected angle corresponded exactly to the expected results of a "billiard ball" collision between an X-ray quanta, or photon, and an electron in which both energy and momentum are conserved.

In 1924, de Broglie postulated the existence of matter waves. He suggested that since waves exhibit particle-like behavior, then particles should be expected to show wave-like properties. The hypothesis of de Broglie was the existence of a *wave–particle duality principle*. The momentum of a photon is given by

$$p = \frac{h}{\lambda} \tag{2-2}$$

where $\lambda$ is the wavelength of the light wave. Then, de Broglie hypothesized that the wavelength of a particle can be expressed as

$$\lambda = \frac{h}{p} \tag{2-3}$$

where $p$ is the momentum of the particle and $\lambda$ is known as the *de Broglie wavelength* of the matter wave.

The wave nature of electrons has been tested in several ways. One experiment by Davisson and Germer in 1927 involved electrons from a heated filament being accelerated at normal incidence onto a single crystal of nickel. A detector measured the scattered electrons as a function of angle. Figure 2–2 shows the experimental setup and Figure 2–3 shows the results. The

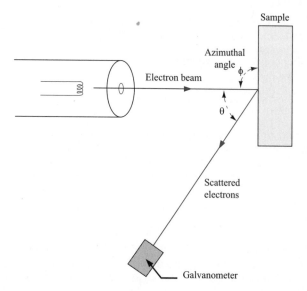

**Figure 2–2**   Experimental arrangement of the Davisson-Germer experiment.

existence of a peak in the density of scattered electrons can be explained as a constructive interference of waves scattered by the periodic atoms in the planes of the nickel crystal. The angular distribution is very similar to an interference pattern produced by light diffracted from a grating.

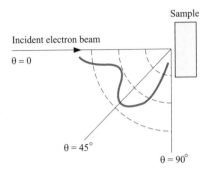

**Figure 2–3** Scattered electron flux as a function of scattering angle for the Davisson-Germer experiment.

---

**Example 2–2**

**Objective:** To calculate the de Broglie wavelength of a particle.

Consider an electron traveling at a velocity of $10^7$ cm/sec $= 10^5$ m/sec.

**Solution:** The momentum is given by

$$p = mv = (9.11 \times 10^{-31})(10^5) = 9.11 \times 10^{-26}$$

Then, the de Broglie wavelength is

$$\lambda = \frac{h}{p} = \frac{6.625 \times 10^{-34}}{9.11 \times 10^{-26}} = 7.27 \times 10^{-9} \text{ m}$$

or

$$\lambda = 72.7 \text{ Å}$$

**Comment:** This calculation shows the order of magnitude of the de Broglie wavelength for a "typical" electron.

---

In some cases electromagnetic waves behave as if they are particles (photons) and sometimes particles behave as if they are waves. This wave–particle duality principle of quantum mechanics applies primarily to small particles such as electrons, but it has also been shown to apply to protons and neutrons. For very large particles, we can show that the relevant equations reduce to those of classical mechanics. The wave–particle duality principle is the basis on which we will use wave theory to describe the motion and behavior of electrons in a crystal.

### 2.1.3 The Uncertainty Principle

The Heisenberg uncertainty principle, given in 1927, also applies primarily to very small particles, and states that we cannot describe with absolute accuracy the behavior of these subatomic particles. The uncertainty principle describes a fundamental relationship between conjugate variables, including position and momentum and also energy and time.

The first statement of the uncertainty principle is that it is impossible to simultaneously describe with absolute accuracy the position and momentum of a particle. If the uncertainty in the momentum is $\Delta p$ and the uncertainty in the position is $\Delta x$, then the uncertainty principle is stated as[1]

$$\Delta p \Delta x \geq \hbar \qquad (2\text{-}4)$$

where $\hbar$ is defined as $\hbar = h/2\pi = 1.054 \times 10^{-34}$ J − sec and is called a modified Planck's constant. This statement may be generalized to include angular position and angular momentum.

The second statement of the uncertainty principle is that it is impossible to simultaneously describe with absolute accuracy the energy of a particle and the instant of time the particle has this energy. Again, if the uncertainty in the energy is given by $\Delta E$ and the uncertainty in the time is given by $\Delta t$, then the uncertainty principle is stated as

$$\Delta E \Delta t \geq \hbar \qquad (2\text{-}5)$$

One way to visualize the uncertainty principle is to consider the simultaneous measurement of position and momentum, and the simultaneous measurement of energy and time. The uncertainty principle implies that these simultaneous measurements are in error to a certain extent. However, the modified Planck's constant $\hbar$ is very small; the uncertainty principle is only significant for subatomic particles. We must keep in mind nevertheless that the uncertainty principle is a fundamental statement and does not deal only with measurements. While we may not be able to know the exact position or momentum of a particle, it is still possible to predict the most probable position or the most probable momentum. In dealing with large numbers of electrons in a semiconductor, for example, the predictions from quantum mechanics will yield statistical results. We will discuss the uncertainty principle again as we consider various aspects of quantum mechanics in later sections of the chapter.

---

[1] In some texts, the uncertainty principle is stated as $\Delta p \Delta x \geq \hbar/2$. We are interested here in the order of magnitude and will not be concerned with small differences.

## 2.2 SCHRODINGER'S WAVE EQUATION

The various experimental results involving electromagnetic waves and particles, which could not be explained by classical laws of physics, showed that a revised formulation of mechanics was required. Schrodinger, in 1926, provided a formulation called wave mechanics, which incorporated the principles of quanta introduced by Planck, and the wave–particle duality principle introduced by de Broglie. Based on the wave–particle duality principle, we will describe the motion of electrons in a crystal by wave theory. This wave theory is described by Schrodinger's wave equation.

### 2.2.1 The Wave Equation

The one-dimensional, nonrelativistic Schrodinger's wave equation is given by

$$\frac{-\hbar^2}{2m} \cdot \frac{\partial^2 \Psi(x,t)}{\partial x^2} + V(x)\Psi(x,t) = j\hbar \frac{\partial \Psi(x,t)}{\partial t} \qquad (2-6)$$

where $\Psi(x,t)$ is the wave function, $V(x)$ is the potential function assumed to be independent of time, $m$ is the mass of the particle, and $j$ is the imaginary constant $\sqrt{-1}$. There are theoretical arguments that justify the form of Schrodinger's wave equation, but the equation is a basic postulate of quantum mechanics. The wave function $\Psi(x,t)$ will be used to describe the behavior of the system and, mathematically, $\Psi(x,t)$ can be a complex quantity.

We may determine the time-dependent portion of the wave function and the position-dependent, or time-independent, portion of the wave function by using the technique of separation of variables. Assume that the wave function can be written in the form

$$\Psi(x,t) = \psi(x)\phi(t) \qquad (2-7)$$

where $\psi(x)$ is a function of the position $x$ only and $\phi(t)$ is a function of time $t$ only. Substituting this form of the solution into Schrodinger's wave equation, we obtain

$$\frac{-\hbar^2}{2m} \phi(t) \frac{\partial^2 \psi(x)}{\partial x^2} + V(x)\psi(x)\phi(t) = j\hbar\psi(x) \frac{\partial \phi(t)}{\partial t} \qquad (2-8)$$

If we divide by the total wave function, Equation (2–8) becomes

$$\frac{-\hbar^2}{2m} \cdot \frac{1}{\psi(x)} \cdot \frac{\partial^2 \psi(x)}{\partial x^2} + V(x) = j\hbar \cdot \frac{1}{\phi(t)} \cdot \frac{\partial \phi(t)}{\partial t} \qquad (2-9)$$

Since the left side of Equation (2–9) is a function of position $x$ only and the right side of the equation is a function of time $t$ only, each side of this equation must be equal to a constant. We will denote this separation of variables constant by $\eta$.

The time-dependent portion of Equation (2–9) is then written as

$$\eta = j\hbar \cdot \frac{1}{\phi(t)} \cdot \frac{\partial \phi(t)}{\partial t} \qquad (2\text{–}10)$$

where again the parameter $\eta$ is called a separation constant. The solution of Equation (2–10) can be written in the form

$$\phi(t) = e^{-j(\eta/\hbar)t} \qquad (2\text{–}11)$$

The form of this solution is the classical exponential form of a sinusoidal wave where $\eta/\hbar$ is the radian frequency $\omega$. We have that $E = h\nu$ or $E = h\omega/2\pi$. Then $\omega = \eta/\hbar = E/\hbar$ so that the separation constant is equal to the total energy $E$ of the particle.

The time-independent portion of Schrodinger's wave equation can now be written from Equation (2–9) as

$$\frac{-\hbar^2}{2m} \cdot \frac{1}{\psi(x)} \cdot \frac{\partial^2 \psi(x)}{\partial x^2} + V(x) = E \qquad (2\text{–}12)$$

where the separation constant is the total energy $E$ of the particle. Equation (2–12) may be written as

$$\frac{\partial^2 \psi(x)}{\partial x^2} + \frac{2m}{\hbar^2}(E - V(x))\psi(x) = 0 \qquad (2\text{–}13)$$

where again $m$ is the mass of the particle, $V(x)$ is the potential experienced by the particle, and $E$ is the total energy of the particle. This time-independent Schrodinger's wave equation can also be justified on the basis of the classical wave equation as shown in Appendix E. The pseudo-derivation in the appendix is a simple approach but shows the plausibility of the time-independent Schrodinger's equation.

### 2.2.2 Physical Meaning of the Wave Function

We are ultimately trying to use the wave function $\Psi(x,t)$ to describe the behavior of an electron in a crystal. The function $\Psi(x,t)$ is a wave function, so it is reasonable to ask what the relation is between the function and the electron. The total wave function is the product of the position-dependent, or time-independent, function and the time-dependent function. We have from Equation (2–7) that

$$\Psi(x,t) = \psi(x)\phi(t) = \psi(x)e^{-j(E/\hbar)t} \qquad (2\text{–}14)$$

Since the total wave function $\Psi(x,t)$ is a complex function, it cannot by itself represent a real physical quantity.

Max Born postulated in 1926 that the function $|\Psi(x,t)|^2$ is the probability of finding the particle between $x$ and $x + dx$ at a given time, or that $|\Psi(x,t)|^2$ is a probability density function. We have that

$$|\Psi(x,t)|^2 = \Psi(x,t) \cdot \Psi^*(x,t) \qquad (2-15)$$

where $\Psi^*(x,t)$ is the complex conjugate function. Therefore

$$\Psi^*(x,t) = \psi^*(x) \cdot e^{+j(E/\hbar)t}$$

Then the product of the total wave function and its complex conjugate is given by

$$\Psi(x,t)\Psi^*(x,t) = [\psi(x)e^{-j(E/\hbar)t}][\psi^*(x)e^{+j(E/\hbar)t}] = \psi(x)\psi^*(x) \qquad (2-16)$$

Therefore, we have that

$$|\Psi(x,t)|^2 = \psi(x)\psi^*(x) = |\psi(x)|^2 \qquad (2-17)$$

is the probability density function and is independent of time. We will determine the probability density function for several examples, and, since this property is independent of time, we will, in general, only be concerned with the time-independent wave function.

### 2.2.3 Boundary Conditions

Since the function $|\psi(x)|^2$ represents the probability density function, then for a single particle, we must have that

$$\int_{-\infty}^{\infty} |\psi(x)|^2 \, dx = 1 \qquad (2-18)$$

The probability of finding the particle somewhere is certain. Equation (2–18) allows us to normalize the wave function and is one boundary condition that is used to determine some wave function coefficients.

The remaining boundary conditions imposed on the wave function and its derivative are postulates. However, we may state the boundary conditions and present arguments which justify why they must be imposed. The wave function and its first derivative must have the following properties if the total energy $E$ and the potential $V(x)$ are finite everywhere.

Condition 1.   $\psi(x)$ must be finite, single-valued, and continuous.

Condition 2.   $\partial\psi(x)/\partial x$ must be finite, single-valued, and continuous.

Since $|\psi(x)|^2$ is a probability density, then $\psi(x)$ must be finite and single-valued. If the probability density were to become infinite at some point in space, then the probability of finding the particle at this position would be certain and the uncertainty principle would be violated. If the total energy $E$ and the potential $V(x)$ are finite everywhere, then from Equation (2–13), the second derivative must be finite, which implies that the first derivative must be continuous. The first derivative is related to the particle momentum, which must be finite and single-valued. Finally, a finite first derivative implies that the function itself must be continuous. In some of the specific examples that we will consider, the potential function will become infinite in particular regions of space. For these cases, the first derivative will not necessarily be continuous, but the remaining boundary conditions will still hold.

## 2.3 APPLICATIONS OF SCHRODINGER'S WAVE EQUATION

The following examples demonstrate the techniques used in wave mechanics and also provide results and concepts that will be utilized later in the discussion of the semiconductor physics.

### 2.3.1 Electron in Free Space

As a first example of applying the Schrodinger's wave equation, consider the motion of an electron in free space. If there is no force acting on the particle, then the potential function $V(x)$ will be constant and we must have $E > V(x)$. Assume, for simplicity, that the potential function $V(x) = 0$ for all $x$. Then, the time-independent wave equation can be written from Equation (2–13) as

$$\frac{\partial^2 \psi(x)}{\partial x^2} + \frac{2mE}{\hbar^2} \psi(x) = 0 \qquad (2\text{–}19)$$

The solution to this differential equation can be written in the form

$$\psi(x) = A \exp\left[\frac{jx\sqrt{2mE}}{\hbar}\right] + B \exp\left[\frac{-jx\sqrt{2mE}}{\hbar}\right] \qquad (2\text{–}20)$$

We can differentiate this solution, substitute back into the differential equation, and indeed show that this function is a solution of the wave equation.

Recall that the time-dependent portion of the solution is

$$\phi(t) = e^{-j(E/\hbar)t} \qquad (2\text{–}21)$$

Then the total solution for the wave function is given by

$$\Psi(x,t) = A \exp\left[\frac{j}{\hbar}(x\sqrt{2mE} - Et)\right] + B \exp\left[\frac{-j}{\hbar}(x\sqrt{2mE} + Et)\right] \qquad (2\text{–}22)$$

This wave function solution is a traveling wave which means that a particle moving in free space is represented by a traveling wave. The first term, with the coefficient $A$, is a wave traveling in the $+x$ direction, while the second term, with the coefficient $B$, is a wave traveling in the $-x$ direction. The value of these coefficients will be determined from boundary conditions. We will again see the traveling-wave solution for an electron in a crystal or semiconductor material.

Assume, for the moment, that we have a particle traveling in the $+x$ direction, which will be described by the $+x$ traveling wave. The coefficient $B \equiv 0$. We can write the traveling-wave solution in the form

$$\Psi(x,t) = A \exp [j(kx - \omega t)] \qquad (2-23)$$

where $k$ is a wave number and is

$$k = \frac{2\pi}{\lambda} \qquad (2-24)$$

The parameter $\lambda$ is the wavelength and, comparing Equation (2–23) to Equation (2–22), the wavelength is given by

$$\lambda = \frac{h}{\sqrt{2mE}} \qquad (2-25)$$

From de Broglie's wave-particle duality principle, the wavelength is also given by

$$\lambda = \frac{h}{p} \qquad (2-26)$$

A free particle with a well-defined energy will also have a well-defined wavelength and momentum.

The probability density function is

$$\Psi(x,t)\Psi^*(x,t) = AA^* \qquad (2-27)$$

which is a constant independent of position. A free particle with a well-defined momentum can be found anywhere with equal probability. This result is in agreement with the Heisenberg uncertainty principle in that a precise momentum implies an undefined position.

A localized free particle must be defined by a wave packet, formed by a superposition of wave functions with different momentum or $k$ values. Consider adding two waves in which the wave number differs by a differential factor. Let

$$\Psi(x,t) = \Psi_1(x,t) + \Psi_2(x,t) \qquad (2-28)$$

where 
$$\Psi_1(x,t) = \sin [kx - \omega t] \qquad (2-29a)$$

and 
$$\Psi_2(x,t) = \sin [(k + dk)x - (\omega + d\omega)t] \qquad (2-29b)$$

We have that

$$\sin A + \sin B = 2 \cos [\tfrac{1}{2}(A - B)] \sin [\tfrac{1}{2}(A + B)] \qquad (2-30)$$

Applying this trigonometric identity to the wave functions, we have

$$\Psi(x,t) = 2 \cos [\tfrac{1}{2}(dk\, x - d\omega\, t)] \sin [\tfrac{1}{2}(2k + dk)x - (2\omega + d\omega)t] \qquad (2-31)$$

Since $dk \ll 2k$ and $d\omega \ll 2\omega$, Equation (2–31) can be written as

$$\Psi(x,t) \cong 2 \cos [\tfrac{1}{2}(dk\, x - d\omega\, t)] \cdot \sin [kx - \omega t] \qquad (2-32)$$

A plot of $\Psi(x,t)$ as a function of $x$ at a fixed time is shown in Figure 2–4. The oscillations of $\Psi(x,t)$ are contained within an envelope of periodic vary-

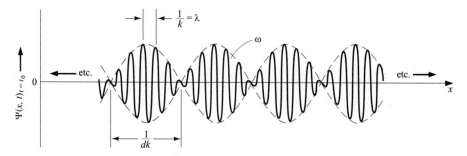

**Figure 2–4**   The sum of two sinusoidal waves of slightly different frequencies and wave numbers.

ing amplitude. If we add more waves with different $k$ values, we will begin to localize the particle within a single wave packet such as shown in Figure 2–5. The particle may be localized to a higher degree of accuracy by adding an increasing number of waves with different $k$ values, or momentum, so the uncertainty principle is still valid.

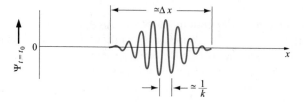

**Figure 2–5**   A wave packet.

### 2.3.2 The Infinite Potential Well

The problem of a particle in the infinite potential well is a classic example of a bound particle. The potential $V(x)$ as a function of position for this problem is shown in Figure 2–6. The particle is assumed to exist in region II so the particle is contained within a finite region of space. The time-independent Schrodinger's wave equation is again given by Equation (2–13) as

$$\frac{\partial^2 \psi(x)}{\partial x^2} + \frac{2m}{\hbar^2}(E - V(x))\psi(x) = 0 \tag{2–13}$$

where $E$ is the total energy of the particle. If $E$ is finite, the wave function must be zero, or $\psi(x) = 0$, in both regions I and III. A particle cannot penetrate these infinite potential barriers, so the probability of finding the particle in regions I and III is zero.

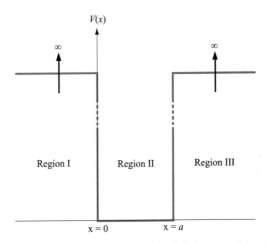

V(x)

Region I     Region II     Region III

x = 0          x = a

**Figure 2-6** Potential function of the infinite potential well.

The time-independent Schrodinger's wave equation in region II, where $V = 0$, becomes

$$\frac{\partial^2 \psi(x)}{\partial x^2} + \frac{2mE}{\hbar^2} \psi(x) = 0 \tag{2-33}$$

A particular form of solution to this equation is given by

$$\psi(x) = A_1 \cos Kx + A_2 \sin Kx \tag{2-34}$$

where

$$K = \sqrt{\frac{2mE}{\hbar^2}} \tag{2-35}$$

One boundary condition is that the wave function $\psi(x)$ must be continuous so that

$$\psi(x = 0) = \psi(x = a) = 0 \tag{2-36}$$

Applying the boundary condition at $x = 0$, we must have that $A_1 \equiv 0$. At $x = a$, we have

$$\psi(x = a) = 0 = A_2 \sin Ka \tag{2-37}$$

This equation is valid if $Ka = n\pi$, where the parameter $n$ is a positive integer, or $n = 1, 2, 3, \ldots$. The parameter $n$ is referred to as a quantum number. We can write

$$K = \frac{n\pi}{a} \tag{2-38}$$

Negative values of $n$ simply introduce a negative sign in the wave function and yield redundant solutions for the probability density function. We cannot physically distinguish any difference between $+n$ and $-n$ solutions. Because of this redundancy, negative values of $n$ are not considered.

The coefficient $A_2$ can be found from the normalization boundary condition that was given by Equation (2–18) as $\int_{-\infty}^{\infty} \psi(x)\psi^*(x)dx = 1$. If we assume that the wave function solution $\psi(x)$ is a real function, then $\psi(x) = \psi^*(x)$. Substituting the wave function into Equation (2–18), we have

$$\int_0^a A_2^2 \sin^2 Kx\,dx = 1 \tag{2–39}$$

Evaluating this integral gives[2]

$$A_2 = \sqrt{\frac{2}{a}} \tag{2–40}$$

Finally, the time-independent wave solution is given by

$$\psi(x) = \sqrt{\frac{2}{a}} \sin\left(\frac{n\pi x}{a}\right) \qquad \text{where} \qquad n = 1, 2, 3, \ldots \tag{2–41}$$

This solution represents the electron in the infinite potential well and is a standing wave solution. The free electron was represented by a traveling wave and now the bound particle is represented by a standing wave.

The parameter $K$ in the wave solution was defined by Equations (2–35) and (2–38). Equating these two expressions for $K$, we obtain

$$\frac{2mE}{\hbar^2} = \frac{n^2\pi^2}{a^2} \tag{2–42}$$

The total energy can then be written as

$$E = E_n = \frac{\hbar^2 n^2 \pi^2}{2ma^2} \qquad \text{where} \qquad n = 1, 2, 3, \ldots \tag{2–43}$$

For the particle in the infinite potential well, the wave function is now given by

$$\psi(x) = \sqrt{\frac{2}{a}} \sin Kx \tag{2–44}$$

where the constant $K$ must have discrete values, implying that the total energy of the particle can only have discrete values. This result means that

---

[2] A more thorough analysis would show that the coefficient $A_2$ could be $+\sqrt{2/a}$, $-\sqrt{2/a}$, $+j\sqrt{2/a}$, or $-j\sqrt{2/a}$. Since the wave function itself has no physical meaning, the choice of which coefficient to use is immaterial: they all produce the same probability density function.

the energy of the particle is *quantized*. That is, the energy of the particle can only have particular discrete values. The quantization of the particle energy is contrary to results from classical physics, which would allow the particle to have continuous energy values. The discrete energies lead to quantum states which will be considered in more detail in this and later chapters. The quantization of the energy of a bound particle is an extremely important result.

---

**Example 2–3**

**Objective:** To calculate the first three energy levels of an electron in an infinite potential well.

Consider an electron in an infinite potential well of width 5 Å.

**Solution:** From Equation (2–43) we have

$$E_n = \frac{\hbar^2 n^2 \pi^2}{2ma^2} = \frac{n^2(1.054 \times 10^{-34})^2 \, \pi^2}{2(9.11 \times 10^{-31})(5 \times 10^{-10})^2} = n^2(2.41 \times 10^{-19}) \quad \text{J}$$

or
$$E_n = \frac{n^2(2.41 \times 10^{-19})}{1.6 \times 10^{-19}} = n^2(1.51) \text{ eV}$$

Then,
$$E_1 = 1.51 \text{ eV}, \, E_2 = 6.04 \text{ eV}, \, E_3 = 13.59 \text{ eV}$$

**Comment:** This calculation shows the order of magnitude of the energy levels of a bound electron.

---

Figure 2–7a shows the first four allowed energies for the particle in the infinite potential well, and Figures 2–7b and 2–7c show the corresponding wave functions and probability functions. We may note that as the energy increases, the probability of finding the particle at any given value of $x$ becomes more uniform.

## 2.3.3 The Step Potential Function

Consider now a step potential function as shown in Figure 2–8. In the previous section, we considered a particle being confined between two potential barriers. In this example, we will assume that a flux of particles is incident on the potential barrier. We will assume that the particles are traveling in the $+x$ direction and that they originated at $x = -\infty$. A particularly interesting result is obtained for the case when the total energy of the particle is less than the barrier height, or $E < V_0$.

We again need to consider the time-independent wave equation in each of the two regions. This general equation was given in Equation (2–13) as $\partial^2\psi(x)/\partial x^2 + 2m/\hbar^2 (E - V(x))\psi(x) = 0$. The wave equation in region I, in which $V = 0$, is

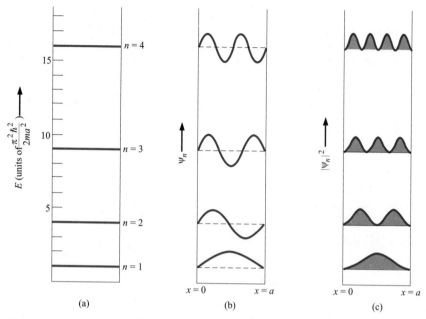

**Figure 2–7**   Particle in an infinite potential well. (a) Four lowest discrete energy levels. (b) Corresponding wave functions. (c) Corresponding probability functions. (From Pierret [7]).

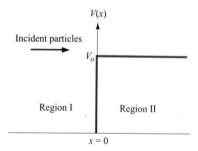

**Figure 2–8**   The step potential function.

$$\frac{\partial^2 \psi_1(x)}{\partial x^2} + \frac{2mE}{\hbar^2} \psi_1(x) = 0 \tag{2–45}$$

The general solution to this equation can be written in the form

$$\psi_1(x) = A_1 e^{jK_1 x} + B_1 e^{-jK_1 x} \quad (x \le 0) \tag{2–46}$$

where the constant $K_1$ is

$$K_1 = \sqrt{\frac{2mE}{\hbar^2}} \tag{2–47}$$

The first term in Equation (2–46) is a traveling wave in the $+x$ direction that represents the incident wave, and the second term is a traveling wave in the $-x$ direction that represents a reflected wave. As in the case of a free particle, the incident and reflected particles are represented by traveling waves.

For the incident wave, $A_1 \cdot A_1^*$ is the probability density function of the incident particles. If we multiply this probability density function by the incident velocity, then $v_i \cdot A_1 \cdot A_1^*$ is the flux of incident particles in units of #/cm²-sec. Likewise the quantity $v_r \cdot B_1 \cdot B_1^*$ is the flux of the reflected particles, where $v_r$ is the velocity of the reflected wave. (The parameters $v_i$ and $v_r$ in these terms are actually the magnitudes of the velocity only.)

In region II, the potential is $V = V_0$. If we assume that $E < V_0$, then the differential equation describing the wave function in region II can be written as

$$\frac{\partial^2 \psi_2(x)}{\partial x^2} - \frac{2m}{\hbar^2}(V_0 - E)\psi_2(x) = 0 \tag{2–48}$$

The general solution may then be written in the form

$$\psi_2(x) = A_2 e^{-K_2 x} + B_2 e^{+K_2 x} \quad (x \geq 0) \tag{2–49}$$

where

$$K_2 = \sqrt{\frac{2m(V_0 - E)}{\hbar^2}} \tag{2–50}$$

One boundary condition is that the wave function $\psi_2(x)$ must remain finite which means that the coefficient $B_2 \equiv 0$. The wave function is now given by

$$\psi_2(x) = A_2 e^{-K_2 x} \quad (x \geq 0) \tag{2–51}$$

The wave function at $x = 0$ must be continuous so that

$$\psi_1(0) = \psi_2(0) \tag{2–52}$$

Then from Equations (2–46), (2–51), and (2–52), we obtain

$$A_1 + B_1 = A_2 \tag{2–53}$$

Since the potential function is everywhere finite, the first derivative of the wave function must also be continuous so that

$$\frac{\partial \psi_1}{\partial x}\bigg|_{x=0} = \frac{\partial \psi_2}{\partial x}\bigg|_{x=0} \tag{2–54}$$

Using Equations (2–46), (2–51), and (2–54), we obtain

$$jK_1 A_1 - jK_1 B_1 = -K_2 A_2 \tag{2–55}$$

We can solve Equations (2–53) and (2–55) to determine the coefficients $B_1$ and $A_2$ in terms of the incident wave coefficient $A_1$. The results are

$$B_1 = \frac{-(K_2^2 + 2jK_1K_2 - K_1^2)A_1}{(K_2^2 + K_1^2)} \qquad (2\text{--}56)$$

and
$$A_2 = \frac{2K_1(K_1 - jK_2)A_1}{(K_2^2 + K_1^2)} \qquad (2\text{--}57)$$

The reflected probability density function is given by

$$B_1 \cdot B_1^* = \frac{(K_2^2 - K_1^2 + 2jK_1K_2)(K_2^2 - K_1^2 - 2jK_1K_2)A_1 \cdot A_1^*}{(K_2^2 + K_1^2)^2} \qquad (2\text{--}58)$$

We can define a reflection coefficient, $R$, as the ratio of the reflected flux to the incident flux, which is written as

$$R = \frac{v_r \cdot B_1 \cdot B_1^*}{v_i \cdot A_1 \cdot A_1^*} \qquad (2\text{--}59)$$

where $v_i$ and $v_r$ are the incident and reflected velocities, respectively, of the particles. In region I, $V = 0$ so that $E = T$, where $T$ is the kinetic energy of the particle. The kinetic energy is given by

$$T = \tfrac{1}{2}mv^2 \qquad (2\text{--}60)$$

so that the constant $K_1$, from Equation (2–47), may be written as

$$K_1 = \sqrt{\frac{2m}{\hbar^2}\left(\frac{1}{2}mv^2\right)} = \sqrt{m^2\frac{v^2}{\hbar^2}} = \frac{mv}{\hbar} \qquad (2\text{--}61)$$

The incident velocity can then be written as

$$v_i = \frac{\hbar}{m} \cdot K_1 \qquad (2\text{--}62)$$

Since the reflected particle also exists in region I, the reflected velocity (magnitude) is given by

$$v_r = \frac{\hbar}{m} \cdot K_1 \qquad (2\text{--}63)$$

The incident and reflected velocities (magnitudes) are equal. The reflection coefficient is then

$$R = \frac{v_r \cdot B_1 \cdot B_1^*}{v_i \cdot A_1 \cdot A_1^*} = \frac{B_1 \cdot B_1^*}{A_1 \cdot A_1^*} \qquad (2\text{--}64)$$

Substituting the expression from Equation (2–58) into Equation (2–64), we obtain

$$R = \frac{B_1 \cdot B_1^*}{A_1 \cdot A_1^*} = \frac{(K_2^2 - K_1^2)^2 + 4K_1^2K_2^2}{(K_2^2 + K_1^2)^2} = 1.0 \qquad (2\text{--}65)$$

The result of $R = 1$ implies that all of the particles incident on the potential barrier for $E < V_0$ are eventually reflected. Particles are not absorbed or transmitted through the potential barrier. This result is entirely consistent with classical physics and one might ask why we should consider this problem in terms of quantum mechanics. The interesting result is in terms of what happens in region II.

The wave solution in region II was given by Equation (2–51) as $\psi_2(x) = A_2 e^{-K_2 x}$. The coefficient $A_2$ from Equation (2–53) is $A_2 = A_1 + B_1$, which we derived from the boundary conditions. For the case of $E < V_0$, the coefficient $A_2$ is not zero. If $A_2$ is not zero, then the probability density function $\psi_2(x) \cdot \psi_2^*(x)$ of the particle being found in region II is not equal to zero. This result implies that there is a finite probability that the incident particle will penetrate the potential barrier and exist in region II. The probability of a particle penetrating the potential barrier is another difference between classical and quantum mechanics: the quantum mechanical penetration is classically not allowed. Although there is a finite probability that the particle may penetrate the barrier, since the reflection coefficient in region I is unity, the particle in region II must eventually turn around and move back into region I.

---

**Example 2–4**

**Objective:** To calculate the penetration depth of a particle impinging on a potential barrier.

Consider an incident electron which is traveling at a velocity of $1 \times 10^5$ m/sec in region I.

**Solution:** With $V(x) = 0$, the total energy is also equal to the kinetic energy so that

$$E = T = \tfrac{1}{2}mv^2 = 4.56 \times 10^{-21} \text{ J} = 2.85 \times 10^{-2} \text{ eV}$$

Now, assume that the potential barrier at $x = 0$ is twice as large as the total energy of the incident particle, or that $V_0 = 2E$. The wave function solution in Region II is $\psi_2(x) = A_2 e^{-K_2 x}$, where the constant $K_2$ is given by $K_2 = \sqrt{2m(V_0 - E)/\hbar^2}$.

In this example, we want to determine the distance $x = d$ at which the wave function magnitude has decayed to $e^{-1}$ of its value at $x = 0$. Then, for this case, we have $K_2 d = 1$ or

$$1 = d \sqrt{\frac{2m(2E - E)}{\hbar^2}} = d\sqrt{\frac{2mE}{\hbar^2}}$$

The distance is then given by

$$d = \sqrt{\frac{\hbar^2}{2mE}} = \frac{1.054 \times 10^{-34}}{\sqrt{2(9.11 \times 10^{-31})(4.56 \times 10^{-21})}} = 11.6 \times 10^{-10} \text{ m}$$

or

$$d = 11.6 \text{ Å}$$

**Comment:** This penetration distance corresponds to approximately two lattice constants of silicon. The numbers used in this example are rather arbitrary. We used a distance at which the wave function decayed to $e^{-1}$ of its initial value. We could have arbitrarily used $e^{-2}$, for example, but the results give an indication of the magnitude of penetration depth.

The case when the total energy of a particle, which is incident on the potential barrier, is greater than the barrier height, or $E > V_0$, is left as an exercise at the end of the chapter.

### 2.3.4 The Potential Barrier

We now want to consider the potential barrier function which is shown in Figure 2–9. The more interesting problem, again, is in the case when the total energy of an incident particle is $E < V_0$. Again assume that we have a flux of incident particles originating on the negative $x$ axis traveling in the $+x$

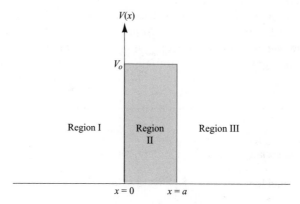

**Figure 2–9**   The potential barrier function.

direction. As before, we need to solve Schrodinger's time-independent wave equation in each of the three regions. The solutions of the wave equation in regions I, II, and III are given, respectively, as

$$\psi_1(x) = A_1 e^{jK_1 x} + B_1 e^{-jK_1 x} \tag{2–66a}$$

$$\psi_2(x) = A_2 e^{K_2 x} + B_2 e^{-K_2 x} \tag{2–66b}$$

$$\psi_3(x) = A_3 e^{jK_1 x} + B_3 e^{-jK_1 x} \tag{2–66c}$$

where

$$K_1 = \sqrt{\frac{2mE}{\hbar^2}} \tag{2–67a}$$

and

$$K_2 = \sqrt{\frac{2m}{\hbar^2}(V_0 - E)} \tag{2–67b}$$

The coefficient $B_3$ in Equation (2–66c) represents a negative traveling wave in region III. However, once a particle gets into region III, there are no potential changes to cause a reflection, therefore the coefficient $B_3$ must be zero. We must keep both exponential terms in Equation (2–66b) since the potential barrier width is finite; that is, neither term will become unbounded.

We have four boundary relations for the boundaries at $x = 0$ and $x = a$ corresponding to the wave function and its first derivative being continuous. We can solve for the four coefficients $B_1$, $A_2$, $B_2$, and $A_3$ in terms of $A_1$. The wave solutions in the three regions are shown in Figure 2–10.

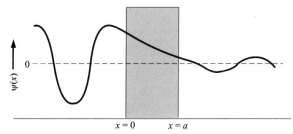

**Figure 2–10**   The wave functions through the potential barrier.

One particular parameter of interest is the transmission coefficient, in this case defined as the ratio of the transmitted flux in region III to the incident flux in region I. Then the transmission coefficient $T$ is

$$T = \frac{v_t \cdot A_3 \cdot A_3^*}{v_i \cdot A_1 \cdot A_1^*} = \frac{A_3 \cdot A_3^*}{A_1 \cdot A_1^*} \qquad (2\text{–}68)$$

where $v_t$ and $v_i$ are the velocities of the transmitted and incident particles, respectively. Since the potential $V = 0$ in both regions I and III, the incident and transmitted velocities are equal. The transmission coefficient may be determined by solving the boundary condition equations. For the special case when $E \ll V_0$, we find that

$$T \cong 16 \left(\frac{E}{V_0}\right) \left(1 - \frac{E}{V_0}\right) \exp\left(-2K_2 a\right) \qquad (2\text{–}69)$$

Equation (2–69) implies that there is a finite probability that a particle impinging a potential barrier will penetrate the barrier and will appear in region III. This phenomenon is called tunneling and it, too, violates classical mechanics. We will see later how this quantum mechanical tunneling phenomenon can be applied to semiconductor device characteristics, such as the tunnel diode.

---

**Example 2–5**

**Objective:** To calculate the probability of an electron tunneling through a potential barrier.

Consider an electron with an energy of 2 $eV$ impinging on a potential barrier with $V_0 = 20$ $eV$ and a width of 3 Å.

**Solution:** Equation (2–69) is the tunneling probability. The factor $K_2$ is

$$K_2 = \sqrt{\frac{2m(V_0 - E)}{\hbar^2}} = \sqrt{\frac{2(9.11 \times 10^{-31})(20 - 2)(1.6 \times 10^{-19})}{(1.054 \times 10^{-34})^2}}$$

or

$$K_2 = 2.17 \times 10^{10} \quad m^{-1}$$

Then

$$T = 16(0.1)(1 - 0.1) \exp\left[-2(2.17 \times 10^{10})(3 \times 10^{-10})\right]$$

and finally

$$T = 3.17 \times 10^{-6}$$

**Comment:** The tunneling probability may appear to be a small value, but the value is not zero. If a large number of particles impinge on a potential barrier, a significant number can penetrate the barrier.

---

## *2.4 EXTENSIONS OF THE WAVE THEORY TO ATOMS

So far in this chapter, we have considered several one-dimensional potential energy functions, and solved Schrodinger's time-independent wave equation to obtain the probability function of finding a particle at various positions. Consider now the one-electron, or hydrogen, atom potential function. We will only briefly consider the mathematical details and wave function solutions, but the results are extremely interesting and important.

### 2.4.1 The One-Electron Atom

The nucleus is a heavy, positively charged proton and the electron is a light, negatively charged particle which, in the classical sense, is revolving around the nucleus. The potential function is due to the coulomb attraction between the proton and electron and is given by

$$V(r) = \frac{-e^2}{4\pi\varepsilon_0 r} \tag{2–70}$$

where $e$ is the magnitude of the electronic charge and $\varepsilon_0$ is the permittivity of free space. This potential function, although spherically symmetric, leads to a three-dimensional problem in spherical coordinates.

We may generalize the time-independent Schrodinger's wave equation to three dimensions by writing

$$\nabla^2\psi(r, \theta, \phi) + \frac{2m_0}{\hbar^2}(E - V(r))\psi(r, \theta, \phi) = 0 \tag{2–71}$$

where $\nabla^2$ is called the Laplacian operator, or "del squared." The parameter $m_0$ is the rest mass of the electron.[3]

In spherical coordinates, Schrodinger's wave equation may be written as

$$\frac{1}{r^2} \cdot \frac{\partial}{\partial r}\left(r^2 \frac{\partial \psi}{\partial r}\right) + \frac{1}{r^2 \sin^2 \theta} \cdot \frac{\partial^2 \psi}{\partial \phi^2} + \frac{1}{r^2 \sin \theta} \cdot \frac{\partial}{\partial \theta}\left(\sin \theta \cdot \frac{\partial \psi}{\partial \theta}\right)$$
$$+ \frac{2m_0}{\hbar^2}(E - V(r))\psi = 0 \tag{2-72}$$

The solution to Equation (2–72) can be determined using the separation of variables technique. We will assume that the solution to the time-independent wave equation can be written in the form

$$\psi(r, \theta, \phi) = R(r) \cdot \Theta(\theta) \cdot \Phi(\phi) \tag{2-73}$$

where $R$, $\Theta$, and $\Phi$ are functions only of $r$, $\theta$, and $\phi$, respectively. Substituting this form of solution into Equation (2–72), we will obtain

$$\frac{\sin^2 \theta}{R} \cdot \frac{\partial}{\partial r}\left(r^2 \frac{\partial R}{\partial r}\right) + \frac{1}{\Phi} \cdot \frac{\partial^2 \Phi}{\partial \phi^2} + \frac{\sin \theta}{\Theta} \cdot \frac{\partial}{\partial \theta}\left(\sin \theta \cdot \frac{\partial \Theta}{\partial \theta}\right)$$
$$+ r^2 \sin^2 \theta \cdot \frac{2m_0}{\hbar^2}(E - V) = 0 \tag{2-74}$$

We may note that the second term in Equation (2–74) is a function of $\phi$ only, while all the other terms are functions of either $r$ or $\theta$. We may then write that

$$\frac{1}{\Phi} \cdot \frac{\partial^2 \Phi}{\partial \phi^2} = -m^2 \tag{2-75}$$

where $m$ is a separation of variables constant.[4] The solution to Equation (2–75) is of the form

$$\Phi = e^{jm\phi} \tag{2-76}$$

Since the wave function must be single-valued, we impose the condition that $m$ is an integer, or

$$m = 0, \pm 1, \pm 2, \pm 3, \ldots \tag{2-77}$$

Incorporating the separation of variables constant $m$ into Equation (2–74), we can further separate the variables $\theta$ and $r$ and generate two

---

[3] The mass should be the rest mass of the two-particle system, but since the proton mass is much greater than the electron mass, the equivalent mass reduces to that of the electron.

[4] Using $m$ to mean the separation of variables constant developed historically and will be retained here even though there may be some confusion with the electron mass. In general, the mass parameter will be used in conjunction with a subscript.

additional separation of variables constants $l$ and $n$. The separation of variables constants $n$, $l$, and $m$ are known as *quantum numbers* and are related by

$$n = 1, 2, 3, \ldots$$
$$l = n - 1, n - 2, n - 3, \ldots, 0 \qquad (2\text{--}78)$$
$$|m| = l, l - 1, \ldots, 0$$

Each set of quantum numbers corresponds to a quantum state which the electron may occupy.

The electron energy may be written in the form

$$E_n = \frac{-m_0 e^4}{(4\pi\varepsilon_0)^2 2\hbar^2 n^2} \qquad (2\text{--}79)$$

where $n$ is the principal quantum number. The negative energy indicates that the electron is bound to the nucleus and we again see that the energy of the bound electron is quantized. If the energy were to become positive, then the electron would no longer be a bound particle and the total energy would no longer be quantized. Since the parameter $n$ in Equation (2–79) is an integer, the total energy of the electron can only take on discrete values. The quantized energy is again a result of the particle being bound in a finite region of space.

The solution of the wave equation may be designated by $\psi_{nlm}$, where $n$, $l$, and $m$ are again the various quantum numbers. For the lowest energy state, $n = 1$, $l = 0$, and $m = 0$; and the wave function is given by

$$\psi_{100} = \frac{1}{\sqrt{\pi}} \cdot \left(\frac{1}{a_0}\right)^{3/2} e^{-r/a_0} \qquad (2\text{--}80)$$

This function is spherically symmetric and the parameter $a_0$ is given by

$$a_0 = \frac{4\pi\varepsilon_0\hbar^2}{m_0 e^2} = 0.529 \text{ Å} \qquad (2\text{--}81)$$

and is equal to the Bohr radius.

The radial probability density function, or the probability of finding the electron at a particular distance from the nucleus, is proportional to the product $\psi_{100} \cdot \psi_{100}^*$ and also to the differential volume of the shell around the nucleus. The probability density function for the lowest energy state is plotted in Figure 2–11. The most probable distance from the nucleus is at $r = a_0$ which is the same as the Bohr theory. Considering this spherically symmetric probability function, we may now begin to conceive the concept of an electron cloud, or energy shell, surrounding the nucleus rather than a discrete particle orbiting around the nucleus.

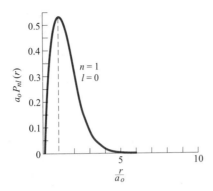

**Figure 2–11** The radial probability density function for the one-electron atom in the lowest energy state. (From Eisberg and Resnick [4]).

The radial probability density function for the next higher, spherically symmetric wave function, corresponding to $n = 2$, $l = 0$, and $m = 0$, is shown in Figure 2–12. This figure shows the idea of the next higher energy shell of the electron. The second energy shell is at a greater radius from the nucleus than the first energy shell. As indicated in the figure, though, there is still a small probability that the electron will exist at the smaller radius. For the case of $n = 2$ and $l = 1$, there are three possible states corresponding to the three allowed values of the quantum number $m$. These wave functions are no longer spherically symmetric.

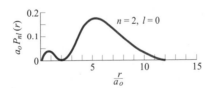

**Figure 2–12** The radial probability density function for the one-electron atom in the next higher energy state. (From Eisberg and Resnick [4]).

Although we have not gone into a great deal of mathematical detail for the one-electron atom, three results are important for the further analysis of semiconductor materials. The first is the solution of Schrodinger's wave equation, which again yields electron probability functions, as it did for the simpler potential functions. In developing the physics of semiconductor materials in later chapters, we will also be considering electron probability functions. The second result is the quantization of allowed energy levels for the bound electron. The third is the concept of quantum numbers and quantum states, which evolved from the separation of variables technique. We will consider this concept again in the next section and in later chapters when we deal with the semiconductor material physics.

### 2.4.2 The Periodic Table

The initial portion of the periodic table of elements may be determined using the results of the one-electron atom plus two additional concepts. The first concept needed is that of electron spin. The electron has an intrinsic angular momentum, or spin, which is quantized and may take on one of two possible values. The spin is designated by a quantum number $s$ which has a value of $s = +\frac{1}{2}$ or $s = -\frac{1}{2}$. We now have four basic quantum numbers which are $n$, $l$, $m$, and $s$.

The second concept needed is the *Pauli exclusion principle*. The Pauli exclusion principle states that, in any given system (an atom, molecule, or crystal), no two electrons may occupy the same quantum state. In an atom, the exclusion principle means that no two electrons may have the same set of quantum numbers. In the next chapter we will see that the exclusion principle is also an important factor in determining the distribution of electrons among available energy states in a crystal.

The first few elements of the periodic table are shown in Table 2–1. For the first element, hydrogen, we have one electron in the lowest energy state

Table 2–1

| Element | Notation | $n$ | $l$ | $m$ | $s$ |
|---|---|---|---|---|---|
| Hydrogen | $1s^1$ | 1 | 0 | 0 | $+\frac{1}{2}$ or $-\frac{1}{2}$ |
| Helium | $1s^2$ | 1 | 0 | 0 | $+\frac{1}{2}$ and $-\frac{1}{2}$ |
| Lithium | $1s^2 2s^1$ | 2 | 0 | 0 | $+\frac{1}{2}$ or $-\frac{1}{2}$ |
| Beryllium | $1s^2 2s^2$ | 2 | 0 | 0 | $+\frac{1}{2}$ and $-\frac{1}{2}$ |
| Boron | $1s^2 2s^2 2p^1$ | 2 | 1 | | |
| Carbon | $1s^2 2s^2 2p^2$ | 2 | 1 | | |
| Nitrogen | $1s^2 2s^2 2p^3$ | 2 | 1 | $m = 0, -1, +1$ | |
| Oxygen | $1s^2 2s^2 2p^4$ | 2 | 1 | $s = +\frac{1}{2}, -\frac{1}{2}$ | |
| Fluorine | $1s^2 2s^2 2p^5$ | 2 | 1 | | |
| Neon | $1s^2 2s^2 2p^6$ | 2 | 1 | | |

corresponding to $n = 1$. From Equation (2–78) both quantum numbers $l$ and $m$ must be zero. However, the electron can take on either spin factor $+\frac{1}{2}$ or $-\frac{1}{2}$. For helium, two electrons may exist in the lowest energy state. For this case, $l = m = 0$, so now both electron spin states are occupied and the lowest energy shell is full. The chemical activity of an element is determined primarily by the valence, or outermost, electrons. Since the valence energy shell of helium is full, helium does not react with other elements and is an inert element.

The third element, lithium, has three electrons. The third electron must go into the second energy shell corresponding to $n = 2$. When $n = 2$, the quantum number $l$ may be 0 or 1, and when $l = 1$, the quantum number $m$ may be $-1$, 0, or $+1$. In each case, the electron spin factor may be $+\frac{1}{2}$ or $-\frac{1}{2}$. For $n = 2$, then, there are eight possible quantum states. Neon has ten electrons. Two electrons are in the $n = 1$ energy shell and eight electrons are in the $n = 2$ energy shell. The second energy shell is now full, which means that neon is also an inert element.

From the solution of Schrodinger's wave equation for the one electron atom, plus the concepts of electron spin and the Pauli exclusion principle, we can begin to build up the periodic table of elements. As the atomic numbers of the elements increase, electrons will begin to interact with each other, so that the buildup of the periodic table will deviate somewhat from this simple method.

## 2.5 SUMMARY AND REVIEW

In this chapter we have considered some of the basic concepts of quantum mechanics, which can be used to describe certain experimental results that cannot be predicted using classical mechanics. An important element of quantum mechanics is the wave–particle duality principle. In some cases, electromagnetic waves exhibit particle-like behavior, and in other situations, particles show wave-like characteristics. The particle-like packet of electromagnetic energy is called the photon and particles are described by a de Broglie wavelength. Based on the wave–particle duality principle, the motion of electrons in a crystal can be described using wave theory.

The formulation for developing the properties of electrons in a crystal is provided by Schrodinger's wave quation. Max Born postulated that the square of the absolute value of the wave function is a probability density function. We cannot, from the uncertainly principle, describe with absolute accuracy the relationship between sets of conjugate variables, such as momentum and position. Instead, we consider probability functions for locating particles rather than knowing exact positions.

Schrodinger's wave equation was applied to several potential functions in order to gain an understanding of the behavior of electrons. Two basic results evolved that are important in the development of semiconductor physics. The first result is that, if a particle is bound within a finite volume of space, the allowed energies of the particle are quantized; a bound particle will possess only certain discrete energies. The second result is that there may be a finite probability of a particle existing in a region that is classically not allowed. This leads, for example, to the concept of tunneling through a potential barrier.

Schrodinger's wave equation was then applied to the one-electron atom. The concept of quantized energy levels was also developed for this potential function as was the concept of quantum numbers. Using these results and the Pauli exclusion principle, which says that no two electrons can have the same set of quantum numbers, the electron arrangement in the first few elements of the periodic table can be determined.

The concepts of quantum mechanics presented in this chapter will be applied in the next chapter to the electron in a single crystal. We eventually wish to describe the current-voltage characteristics of a semiconductor device. In order to do this, we must be able to describe the behavior and characteristics of the electron in a crystal. As we get into the chapters dealing with semiconductor devices, we will not, in most cases, consider quantum mechanics directly. However, it is important to keep in mind that quantum mechanics is the basis for describing the physics of the semiconductor material, as we will see in the next chapter.

## GLOSSARY OF IMPORTANT TERMS

**de Broglie wavelength:** The wavelength of a particle given as the ratio of Planck's constant to momentum.

**Heisenberg uncertainty principle:** The principle which states that we cannot describe with absolute accuracy the relationship between sets of conjugate variables that describe the behavior of particles, such as momentum and position.

**Pauli exclusion principle:** The principle which states that no two electrons can occupy the same quantum state.

**Photon:** The particle-like packet of electromagnetic energy.

**Quanta:** The particle-like packet of thermal radiation.

**Quantized energies:** The allowed discrete energy levels which bound particles may occupy.

**Quantum numbers:** A set of numbers which describes the quantum state of a particle, such as an electron in an atom.

**Quantum state:** A particular state of an electron which may be described, for example, by a set of quantum numbers.

**Tunneling:** The quantum mechanical phenomenon by which a particle may penetrate through a thin potential barrier.

**Wave–particle duality:** The characteristic by which electromagnetic waves sometimes exhibit particle-like behavior and particles sometimes exhibit wave-like behavior.

## PROBLEMS

1. The classical wave equation for a two-wire transmission line is given by $\partial^2 V(x,t)/\partial x^2 = LC \cdot \partial^2 V(x,t)/\partial t^2$. One possible solution is given by $V(x,t) = (\sin Kx) \cdot (\sin \omega t)$ where $K = n\pi/a$ and $\omega =$

$K/\sqrt{LC}$. Sketch, on the same graph, the function $V(x,t)$ as a function of $x$ for $0 \leq x \leq a$ and $n = 1$ when (*i*) $\omega t = 0$, (*ii*) $\omega t = \pi/2$, (*iii*) $\omega t = \pi$, (*iv*) $\omega t = 3\pi/2$, and (*v*) $\omega t = 2\pi$.

2. The function $V(x,t) = \cos(2\pi x/\lambda - \omega t)$ is also a solution to the classical wave equation. Sketch on the same graph the function $V(x,t)$ as a function of $x$ for $0 \leq x \leq 3\lambda$ when: (*i*) $\omega t = 0$, (*ii*) $\omega t = 0.25\pi$, (*iii*) $\omega t = 0.5\pi$, (*iv*) $\omega t = 0.75\pi$, and (*v*) $\omega t = \pi$.

## Section 2.1

3. The work function of a material refers to the minimum energy required to remove an electron from the material. Assume that the work function of gold is 4.90 eV and that of cesium is 1.90 eV. Calculate the maximum wavelength of light for the photoelectric emission of electrons for gold and cesium.

4. Calculate the de Broglie wavelength, $\lambda = h/p$, for: (*a*) An electron with kinetic energy of (*i*) 1.0 eV, and (*ii*) 100 eV. (*b*) A proton with kinetic energy of 1.0 eV. (*c*) A singly ionized tungsten atom with kinetic energy of 1.0 eV. (*d*) A 2000 kg truck traveling at 20 m/sec.

*5. An electron and a photon have the same energy. At what value of energy (in eV) will the wavelength of the photon be ten times that of the electron?

6. It is desired to produce X-ray radiation with a wavelength of 1 Å. (*a*) Through what potential voltage difference must the electron be accelerated in vacuum so that it can, upon colliding with a target, generate such a photon? (Assume that all of the electron's energy is transferred to the photon.) (*b*) What is the de Broglie wavelength of the electron in part (*a*) just before it hits the target?

7. When the uncertainty principle is considered, it is not possible to locate a photon in space more precisely than about one wavelength. Consider a photon with wavelength $\lambda = 1\mu m$. What is the uncertainty in the photon's (*a*) momentum and (*b*) energy?

8. An automobile has a mass of 1500 kg. What is the uncertainty in the velocity (in mph) when its center of mass is located with an uncertainty no greater than 1 cm?

9. (*a*) The uncertainty in the position of an electron is no greater than 1 Å. Determine the minimum uncertainty in its momentum. (*b*) The electron's energy is measured with an uncertainty no greater than 1 eV. Determine the minimum uncertainty in the time over which the measurement is made.

## Section 2.2

10. Assume that $\Psi_1(x,t)$ and $\Psi_2(x,t)$ are solutions of the one-dimensional time-dependent Schrodinger's wave equation. (*a*) Show that

$\Psi_1 + \Psi_2$ is also a solution. (b) Is $\Psi_1 \cdot \Psi_2$ a solution of the Schrodinger's equation in general? Why or why not?

11. Consider the wave function $\Psi(x,t) = A(\sin \pi x)e^{-j\omega t}$ for $-1 \le x \le +1$. Determine A so that $\int_{-1}^{1}|\Psi(x,t)|^2 \, dx = 1$.

12. Consider the wave function $\Psi(x,t) = A(\sin n\pi x)e^{-j\omega t}$ for $0 \le x \le 1$. Determine A so that $\int_{0}^{1}|\Psi(x,t)|^2 \, dx = 1$.

## Section 2.3

13. Consider a particle with mass of 10 mg in an infinite potential well 1.0 cm wide. (a) If the energy of the particle is 10 mJ, calculate the value of $n$ for that state. (b) What is the kinetic energy of the $(n + 1)$ state? (c) Would quantum effects be observable for this particle?

14. Calculate the lowest energy level for a neutron in a nucleus, by treating it as if it were in an infinite potential well of width equal to $10^{-14}$ m. Compare this with the lowest energy level for an electron in the same infinite potential well.

15. Consider the particle in the infinite potential well as shown in Figure 2–13. Derive and sketch the wave functions corresponding to the four lowest energy levels. (Do not normalize the wave functions.)

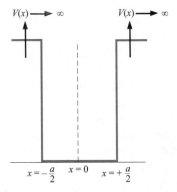

**Figure 2–13**   Potential function for problem 15.

16. Consider a three-dimensional infinite potential well. The potential function is given by $V(x) = 0$ for $0 < x < a$, $0 < y < a$, $0 < z < a$, and $V(x) = \infty$ elsewhere. Start with Schrodinger's wave equation, use the separation of variables technique, and show that the energy is quantized and is given by

$$E_{n_x n_y n_z} = \frac{\hbar^2 \pi^2}{2ma^2}\left(n_x^2 + n_y^2 + n_z^2\right)$$

where $n_x = 1,2,3, \ldots$, $n_y = 1,2,3, \ldots$, $n_z = 1,2,3, \ldots$.

17. Consider a proton in a one-dimensional infinite potential well shown in Figure 2–6. (*a*) Derive the expression for the allowed energy states of the proton. (*b*) Calculate the energy difference (in units of eV) between the lowest possible energy and the next higher energy state for (*i*) $a = 4$ Å, and (*ii*) $a = 0.5$ cm.

18. For the step potential function shown in Figure 2–14, assume that $E > V_0$ and that particles are incident from the $+x$ direction traveling in the $-x$ direction. (*a*) Write the wave solutions for each region. (*b*) Derive expressions for the transmission and reflection coefficients.

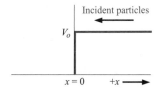

Figure 2–14   Potential function for problem 18.

19. Consider the penetration of a step potential function of height 2.4 eV by an electron whose energy is 2.1 eV. Determine the relative probability of finding the electron at the distance (*a*) 12 Å beyond the barrier, and (*b*) 48 Å beyond the barrier, compared to the probability of finding the incident particle at the barrier edge.

20. Evaluate the transmission coefficient for an electron of energy 2.2 eV impinging on a potential barrier of height 6.0 eV and thickness $10^{-10}$ meters. Repeat the calculation for a barrier thickness of $10^{-9}$ meters. Assume that Equation (2–69) is valid.

21. A proton attempts to penetrate a rectangular potential barrier of height 10 MeV and thickness $10^{-14}$ m. The particle has a total energy of 3 MeV. Calculate the probability that the particle will penetrate the potential barrier. Assume that Equation (2–69) is valid.

*22. An electron with energy $E$ is incident on a rectangular potential barrier as shown in Figure 2–9. The potential barrier is of width $a$ and height $V_0 \gg E$. (*a*) Write the form of the wave function in each of the three regions. (*b*) For this geometry, determine what coefficient in the wave function solutions is zero. (*c*) Derive the expression for the transmission coefficient for the electron (tunneling probability). (*d*) Sketch the wave function for the electron in each region.

*23. A potential function is shown in Figure 2–15 with incident particles coming from $-\infty$ with a total energy $E > V_2$. The constants $k$ are defined as

$$k_1 = \sqrt{\frac{2mE}{\hbar^2}} \qquad k_2 = \sqrt{\frac{2m}{\hbar^2}(E - V_1)} \qquad k_3 = \sqrt{\frac{2m}{\hbar^2}(E - V_2)}$$

Assume a special case for which $k_2a = 2n\pi$, $n = 1,2,3, \ldots$. Derive the expression, in terms of the constants, $k_1$, $k_2$, and $k_3$, for the transmission coefficient. The transmission coefficient is defined as the ratio of the flux of particles in region III to the incident flux in region I.

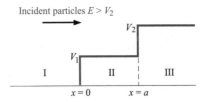

**Figure 2–15**   Potential function for problem 23.

*24.   Consider the one-dimensional potential function shown in Figure 2–16. Assume the total energy of an electron is $E < V_0$. (a) Write the wave solutions which apply in each region. (b) Write the set of equations which result from applying the boundary conditions. (c) Show explicitly why, or why not, the energy levels of the electron are quantized.

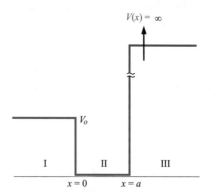

**Figure 2–16**   Potential function for problem 24.

### Section 2.4

25.   Calculate the energy of the electron in the hydrogen atom (in units of eV) for the first four allowed energy levels.

26.   Show that the most probable value of the radius, $r$, for the 1s electron in a hydrogen atom is equal to the Bohr radius, $a_0$.

27.   Show that the wave function for $\psi_{100}$ given by Equation (2–80) is a solution to the differential equation given by Equation (2–72).

# READING LIST

*1.  Datta, S. *Quantum Phenomena*. Vol. 8 of *Modular Series on Solid State Devices*. Reading, Mass.: Addison-Wesley, 1989.

*2.  deCogan, D. *Solid State Devices: A Quantum Physics Approach*. New York: Springer-Verlag, 1987.

3.  Eisberg, R. M. *Fundamentals of Modern Physics*. New York: Wiley, 1961.

4.  Eisberg, R., and R. Resnick. *Quantum Physics of Atoms, Molecules, Solids, Nuclei, and Particles*. New York: Wiley, 1974.

5.  McKelvey, J. P. *Solid State and Semiconductor Physics*. New York: Harper and Row, 1966.

6.  Pauling, L., and E. B. Wilson. *Introduction to Quantum Mechanics*. New York: McGraw-Hill, 1935.

7.  Pierret, R. F. *Advanced Semiconductor Fundamentals*. Vol. 6 of *Modular Series on Solid State Devices*. Reading, Mass.: Addison-Wesley, 1987.

8.  Pohl, H. A. *Quantum Mechanics for Science and Engineering*. Englewood Cliffs, N.J.: Prentice Hall, 1967.

9.  Schiff, L. I. *Quantum Mechanics*. New York: McGraw-Hill, 1955.

# INTRODUCTION TO THE QUANTUM THEORY OF SOLIDS

In the last chapter, we applied quantum mechanics and the Schrodinger's wave equation to determine the behavior of electrons in the presence of various potential functions. We found that one important characteristic of an electron bound to an atom or bound within a finite space is that the electron can take on only discrete values of energy; that is, the energies are quantized. We also discussed the Pauli exclusion principle, which stated that only one electron is allowed to occupy any given quantum state. In this chapter, we will generalize these concepts to the electron in a crystal lattice.

One of our goals is to determine the electrical properties of semiconductor material, which we will then use to develop the current-voltage characteristics of semiconductor devices. Toward this end, we have two tasks in this chapter: to determine the properties of electrons in a crystal lattice, and to determine the statistical characteristics of the very large number of electrons in a crystal.

To start, we will expand the concept of discrete allowed electron energies that occur in a single atom to a band of allowed electron energies in a single-crystal solid. First we will qualitatively discuss the feasibility of the allowed energy bands in a crystal and then we will develop a more rigorous mathematical derivation of this theory using Schrodinger's wave equation. This energy band theory is a basic principle of semiconductor material physics and can also be used to explain differences in electrical characteristics between metals, insulators, and semiconductors.

Since current in a solid is due to the net flow of charge, it is important to determine the response of an electron in the crystal to an applied external force, such as an electric field. The movement of an electron in a lattice is different than that of an electron in free space. We will develop a concept allowing us to relate the quantum mechanical behavior of electrons in a crystal to classical Newtonian mechanics. This analysis leads to a parameter called the electron effective mass. As part of this development, we will find that we can define a new particle in a semiconductor called a *hole*. The motion of both electrons and holes gives rise to currents in a semiconductor.

Because the number of electrons in a semiconductor is very large, it is impossible to follow the motion of each individual particle. We will develop the statistical behavior of electrons in a crystal, noting that the Pauli exclusion principle is an important factor in determining the statistical law the electrons must follow. The resulting probability function will determine the distribution of electrons among the available energy states. The energy band theory and the probability function will be used extensively in the next chapter, when we develop the theory of the semiconductor in equilibrium.

## 3.1. ALLOWED AND FORBIDDEN ENERGY BANDS

In Chapter 2 we treated the one-electron, or hydrogen, atom. That analysis showed that the energy of the bound electron is quantized: only discrete values of electron energy are allowed. The radial probability density for the electron was also determined. This function gives the probability of finding the electron at a particular distance from the nucleus and shows that the electron is not localized at a given radius. We can extrapolate these single-atom results to a crystal and qualitatively derive the concepts of allowed and forbidden energy bands. We will then apply quantum mechanics and Schrodinger's wave equation to the problem of an electron in a single crystal, which will be modeled as a one-dimensional perfect periodic potential function. We will find from this analysis that the electronic energy states occur in bands of allowed states that are separated by forbidden energy bands.

### 3.1.1 Formation of Energy Bands

Figure 3–1a shows the radial probability density function for the lowest electron energy state of the single, noninteracting hydrogen atom, and Figure 3–1b shows the same probability curves for two atoms that are in close proximity to each other. The wave functions of the two atom electrons overlap, which means that the two electrons will interact. This interaction or perturbation results in the discrete quantized energy level splitting into two discrete energy levels, schematically shown in Figure 3–1c. The splitting of the discrete state into two states is consistent with the Pauli exclusion principle.

Now, if we somehow start with a regular periodic arrangement of hydrogen-type atoms that are initially very far apart, and begin pushing the atoms together, the initial quantized energy level will split into a band of discrete energy levels. This effect is shown schematically in Figure 3–2 where the parameter $r_0$ represents the equilibrium interatomic distance in the crystal. At the equilibrium interatomic distance, there is a band of allowed energies, but within the allowed band, the energies are at discrete levels. The Pauli exclusion principle states that the joining of atoms to form a system (crystal)

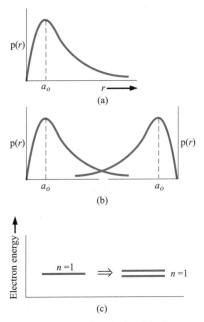

Figure 3–1 (a) Probability density function of an isolated hydrogen atom. (b) Overlapping probability density functions of two adjacent hydrogen atoms. (c) The splitting of the $n = 1$ state.

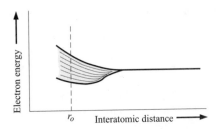

Figure 3–2 The splitting of an energy state into a band of allowed energies.

does not alter the total number of quantum states regardless of size. However, since no two electrons can have the same quantum number, the discrete energy must split into a band of energies in order that each electron can occupy a distinct quantum state.

We have seen previously that, at any energy level, the number of allowed quantum states is relatively small. In order to accommodate all of the electrons in a crystal, then, we must have many energy levels within the allowed band. As an example, suppose that we have a system with $10^{19}$ one-electron atoms and also suppose that, at the equilibrium interatomic distance, the width of the allowed energy band is 1 eV. For simplicity, we will assume that each electron in the system occupies a different energy level and, if the discrete energy states are equidistance apart, then the energy levels are separated by $10^{-19}$ eV. This energy difference is extremely small, so that for all practical purposes, we have a quasi-continuous energy distribution through the allowed energy band. The fact that $10^{-19}$ eV is a very small difference between two energy states can be seen from the following example.

---

**Example 3–1**

**Objective:** To calculate the change in kinetic energy of an electron when the velocity changes by a small value.

Consider an electron traveling at a velocity of $10^7$ cm/sec. Assume the velocity increases by a value of 1 cm/sec. The increase in kinetic energy is given by

$$\Delta E = \tfrac{1}{2} m v_2^2 - \tfrac{1}{2} m v_1^2 = \tfrac{1}{2} m(v_2^2 - v_1^2)$$

Let $v_2 = v_1 + \Delta v$. Then

$$v_2^2 = (v_1 + \Delta v)^2 = v_1^2 + 2v_1\Delta v + (\Delta v)^2$$

But $\Delta v \ll v_1$, so we have that

$$\Delta E \cong \tfrac{1}{2} m(2v_1\Delta v) = m v_1 \Delta v$$

**Solution:** Substituting the numbers into this equation, we obtain

$$\Delta E = (9.11 \times 10^{-31})(10^5)(0.01) = 9.11 \times 10^{-28} \text{ J}$$

which may be converted to units of electron volts as

$$\Delta E = \frac{9.11 \times 10^{-28}}{1.6 \times 10^{-19}} = 5.7 \times 10^{-9} \text{ eV}$$

**Comment:** A change in velocity of 1 cm/sec compared to $10^7$ cm/sec results in a change in energy of $5.7 \times 10^{-9}$ eV, which is orders of magnitude larger than the change in energy of $10^{-19}$ eV between energy states in the allowed energy band. This example serves to demonstrate that a difference in adjacent energy states of $10^{-19}$ eV is indeed very small, so that the discrete energies within an allowed band may be treated as a quasi-continuous distribution.

---

Consider again a regular periodic arrangement of atoms, in which each atom now contains more than one electron. Suppose the atom in this imaginary crystal contains electrons up through the $n = 3$ energy level. If the atoms are initially very far apart, the electrons in adjacent atoms will not interact and will occupy the discrete energy levels. If these atoms are brought closer together, the outermost electrons in the $n = 3$ energy shell will begin to interact initially, so that this discrete energy level will split into a band of allowed energies. If the atoms continue to move closer together, the electrons in the $n = 2$ shell may begin to interact and will also split into a band of allowed energies. Finally, if the atoms become sufficiently close together, the innermost electrons in the $n = 1$ level may interact, so that this energy level may also split into a band of allowed energies. The splitting of these discrete energy levels is qualitatively shown in Figure 3–3. If the

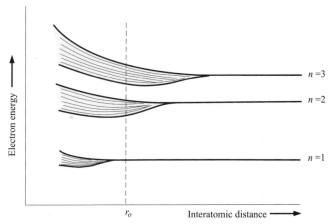

**Figure 3–3**  Schematic showing the splitting of three energy states into allowed bands of energies.

equilibrium interatomic distance is $r_0$, then we have bands of allowed energies that the electrons may occupy separated by bands of forbidden energies. This energy-band splitting and the formation of allowed and forbidden bands is the energy-band theory of single-crystal materials.

The actual band splitting in a crystal is much more complicated than indicated in Figure 3–3. The band splitting of silicon, for example, is shown in Figure 3–4. We need only consider the $n = 3$ level for the valence elec-

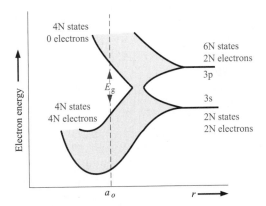

**Figure 3–4**  The splitting of the 3s and 3p states of silicon into the allowed and forbidden energy bands. (From Shockley [6]).

trons, since the first two energy shells are completely full and are tightly bound to the nucleus. The 3s state corresponds to $n = 3$ and $l = 0$ and contains two quantum states per atom. This state will contain two electrons at $T = 0°K$. The 3p state corresponds to $n = 3$ and $l = 1$ and contains six quantum states per atom. This state will contain the remaining two electrons in the individual silicon atom.

As the interatomic distance decreases, the 3s and 3p states interact and overlap. At the equilibrium interatomic distance, the bands have again split, but now four quantum states per atom are in the lower band and four quantum states per atom are in the upper band. At absolute zero degrees, electrons are in the lowest energy state so that all states in the lower band (the valence band) will be full and all states in the upper band (the conduction band) will be empty. The bandgap energy $E_g$ between the top of the valence band and the bottom of the conduction band is the width of the forbidden energy band.

So far we have discussed qualitatively how and why bands of allowed and forbidden energies are formed in a crystal. The formation of these energy bands is directly related to the electrical characteristics of the crystal, as we will see later in our discussion.

### *3.1.2  The Kronig-Penney Model

The concept of allowed and forbidden energy levels can be developed more rigorously by considering quantum mechanics and Schrodinger's wave equation.

The potential function of a single, noninteracting, one-electron atom is shown in Figure 3–5a. Also indicated on the figure are the discrete energy levels allowed for the electron. Figure 3–5b shows the same type of potential function for the case when several atoms are in close proximity arranged in a one-dimensional array. The potential functions of adjacent atoms overlap so that the net potential function for this case is shown in Figure 3–5c. It is this potential function we would need to use in Schrodinger's wave equation to model a one-dimensional single-crystal material.

The solution to Schrodinger's wave equation, for this one-dimensional single-crystal lattice, is made more tractable by considering a simpler potential function. Figure 3–6 is the one-dimensional Kronig-Penney model of the periodic potential function, which is used to represent a one-dimensional single-crystal lattice. We need to solve Schrodinger's wave equation in each region. As with previous quantum mechanical problems, the more interesting solution occurs for the case when $E < V_o$, which corresponds to a particle being bound within the crystal. The electrons are contained in the potential wells, but we have the possibility of tunneling between wells. The Kronig-Penney model is an idealized periodic potential representing a one-dimensional single crystal, but the results will illustrate many of the important features of the quantum behavior of electrons in a periodic lattice.

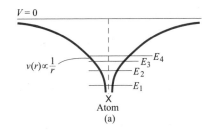

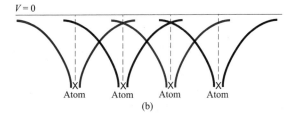

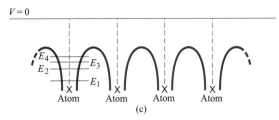

**Figure 3–5** (a) Potential function of a single isolated atom. (b) Overlapping potential functions of adjacent atoms. (c) Net potential function of a one-dimensional single crystal.

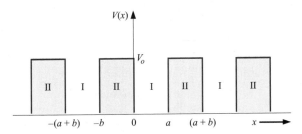

**Figure 3–6** The one-dimensional periodic potential function of the Kronig-Penney model.

To obtain the solution to Schrodinger's wave equation, we make use of a mathematical theorem by Bloch. The theorem states that all one-electron wave functions, for problems involving periodically varying potential energy functions, must be of the form

$$\psi(x) = u(x)e^{jkx} \qquad (3\text{–}1)$$

The parameter $k$ is called a constant of the motion and will be considered in more detail as we develop the theory. The function $u(x)$ is a periodic function with period $(a + b)$.

We stated in Chapter 2 that the total solution to the wave equation is the product of the time-independent solution and the time-dependent solution, or

$$\Psi(x, t) = \psi(x)\phi(t) = u(x)e^{jkx} \cdot e^{-j(E/\hbar)t} \qquad (3\text{--}2)$$

which may be written as

$$\Psi(x, t) = u(x)e^{j(kx-(E/\hbar)t)} \qquad (3\text{--}3)$$

This traveling-wave solution represents the motion of an electron in a single-crystal material. The amplitude of the traveling wave is a periodic function and the parameter $k$ is also referred to as a wave number.

We can now begin to determine a relation between the parameter $k$, the total energy $E$, and the potential $V_o$. If we consider region I in Figure 3–6 $(0 < x < a)$ in which $V(x) = 0$, take the second derivative of Equation (3–1), and substitute this result into the time-independent Schrodinger's wave equation given by Equation (2–13), we obtain the relation

$$\frac{d^2u_1(x)}{dx^2} + 2jk\frac{du_1(x)}{dx} - (k^2 - \alpha^2)u_1(x) = 0 \qquad (3\text{--}4)$$

The function $u_1(x)$ is the amplitude of the wave function in region I and the parameter $\alpha$ is defined as

$$\alpha^2 = \frac{2mE}{\hbar^2} \qquad (3\text{--}5)$$

Consider now a specific region II, $-b < x < 0$, in which $V(x) = V_o$, and apply Schrodinger's wave equation. We obtain the relation

$$\frac{d^2u_2(x)}{dx^2} + 2jk\frac{du_2(x)}{dx} - \left(k^2 - \alpha^2 + \frac{2mV_o}{\hbar^2}\right)u_2(x) = 0 \qquad (3\text{--}6)$$

where $u_2(x)$ is the amplitude of the wave function in region II. We may define

$$\frac{2m}{\hbar^2}(E - V_o) = \alpha^2 - \frac{2mV_o}{\hbar^2} = \beta^2 \qquad (3\text{--}7)$$

so that Equation (3–6) may be written as

$$\frac{d^2u_2(x)}{dx^2} + 2jk\frac{du_2(x)}{dx} - (k^2 - \beta^2)u_2(x) = 0 \qquad (3\text{--}8)$$

Note that from Equation (3–7), if $E > V_o$, the parameter $\beta$ is real, whereas if $E < V_o$, then $\beta$ is imaginary.

The solution to Equation (3–4), for region I, is of the form

$$u_1(x) = Ae^{j(\alpha-k)x} + Be^{-j(\alpha+k)x} \text{ for } (0 < x < a) \qquad (3\text{--}9)$$

and the solution to Equation (3–8), for region II, is of the form

$$u_2(x) = Ce^{j(\beta-k)x} + De^{-j(\beta+k)x} \text{ for } (-b < x < 0) \qquad (3-10)$$

Since the potential function $V(x)$ is everywhere finite, both the wave function $\psi(x)$ and its first derivative $\partial\psi(x)/\partial x$ must be continuous. This continuity condition implies that the wave amplitude function $u(x)$ and its first derivative $\partial u(x)/\partial x$ must also be continuous.

If we consider the boundary at $x = 0$ and apply the continuity condition to the wave amplitude, we have

$$u_1(0) = u_2(0) \qquad (3-11)$$

Substituting Equations (3–9) and (3–10) into Equation (3–11), we obtain

$$A + B - C - D = 0 \qquad (3-12)$$

Now applying the condition that

$$\frac{du_1}{dx}\bigg|_{x=0} = \frac{du_2}{dx}\bigg|_{x=0} \qquad (3-13)$$

we obtain

$$(\alpha - k)A - (\alpha + k)B - (\beta - k)C + (\beta + k)D = 0 \qquad (3-14)$$

We have considered region I as $0 < x < a$ and region II as $-b < x < 0$. The periodicity and the continuity condition mean that the function $u_1$, as $x \to a$, is equal to the function $u_2$, as $x \to -b$. This condition may be written as

$$u_1(a) = u_2(-b) \qquad (3-15)$$

Applying the solutions for $u_1(x)$ and $u_2(x)$ to the boundary condition in Equation (3–15) yields

$$Ae^{j(\alpha-k)a} + Be^{-j(\alpha+k)a} - Ce^{-j(\beta-k)b} - De^{j(\beta+k)b} = 0 \qquad (3-16)$$

The last boundary condition is

$$\frac{du_1}{dx}\bigg|_{x=a} = \frac{du_2}{dx}\bigg|_{x=-b} \qquad (3-17)$$

which gives

$$(\alpha - k)Ae^{j(\alpha-k)a} - (\alpha + k)Be^{-j(\alpha+k)a} - (\beta - k)Ce^{-j(\beta-k)b}$$
$$+ (\beta + k)De^{j(\beta+k)b} = 0 \qquad (3-18)$$

We now have four homogeneous equations, Equations (3–12), (3–14), (3–16), and (3–18), with four unknowns as a result of applying the four boundary conditions. In a set of simultaneous, linear, homogeneous equations, there is a nontrivial solution if, and only if, the determinant of the coefficients is zero. In our case, the coefficients in question are the coefficients of the parameters $A$, $B$, $C$, and $D$.

The evaluation of this determinant is extremely laborious and will not be considered in detail. The result is

$$\frac{-(\alpha^2 + \beta^2)}{2\alpha\beta} (\sin \alpha a)(\sin \beta b) + (\cos \alpha a)(\cos \beta b) = \cos k(a + b) \qquad (3\text{--}19)$$

Equation (3–19) relates the parameter $k$ to the total energy $E$ (through the parameter $\alpha$) and the potential function $V_o$ (through the parameter $\beta$).

As we mentioned, the more interesting solutions occur for $E < V_o$, which applies to the electron bound within the crystal. From Equation (3–7), the parameter $\beta$ is then an imaginary quantity. We may define

$$\beta = j\gamma \qquad (3\text{--}20)$$

where $\gamma$ is a real quantity. Equation (3–19) can be written in terms of $\gamma$ as

$$\frac{\gamma^2 - \alpha^2}{2\alpha\gamma} (\sin \alpha a)(\sinh \gamma b) + (\cos \alpha a)(\cosh \gamma b) = \cos k(a + b) \qquad (3\text{--}21)$$

Equation (3–21) does not lend itself to an analytical solution, but must be solved using numerical or graphical techniques to obtain the relation between $k$, $E$, and $V_o$. The solution of Schrodinger's wave equation for a single bound particle resulted in discrete allowed energies. The solution of Equation (3–21) will result in a band of allowed energies.

To obtain an equation that is more susceptible to a graphical solution and thus will illustrate the nature of the results, let the potential barrier width $b \to 0$ and the barrier height $V_o \to \infty$, but such that the product $bV_o$ remains finite. Equation (3–21) then reduces to

$$\left(\frac{mV_oba}{\hbar^2}\right) \frac{\sin \alpha a}{\alpha a} + \cos \alpha a = \cos ka \qquad (3\text{--}22)$$

We may define a parameter $P'$ as

$$P' = \frac{mV_oba}{\hbar^2} \qquad (3\text{--}23)$$

Then, finally, we have the relation

$$P' \frac{\sin \alpha a}{\alpha a} + \cos \alpha a = \cos ka \qquad (3\text{--}24)$$

Equation (3–24) again gives the relation between the parameter $k$, total energy $E$ (through the parameter $\alpha$), and the potential barrier $bV_o$. We may note that Equation (3–24) is not a solution of Schrodinger's wave equation but gives the conditions for which Schrodinger's wave equation will have a solution. If we assume the crystal is infinitely large, then $k$ in Equation (3–24) can assume a continuum of values and must be real.

### 3.1.3 The k-Space Diagram

To begin to understand the nature of the solution, initially consider the special case for which $V_o = 0$. In this case $P' = 0$, which corresponds to a free particle since there are no potential barriers. From Equation (3–24), we have that

$$\cos \alpha a = \cos ka \qquad (3\text{–}25)$$

or

$$\alpha = k \qquad (3\text{–}26)$$

Since the potential is equal to zero, the total energy $E$ is equal to the kinetic energy, so that, using Equation (3–5), Equation (3–26) may be written as

$$\alpha = \sqrt{\frac{2mE}{\hbar^2}} = \sqrt{\frac{2m(\frac{1}{2}mv^2)}{\hbar^2}} = \frac{p}{\hbar} = k \qquad (3\text{–}27)$$

where $p$ is the particle momentum. The constant of the motion parameter $k$ is related to the particle momentum for the free electron. The parameter $k$ is also referred to as a wave number.

We can also relate the energy and momentum as

$$E = \frac{p^2}{2m} = \frac{k^2\hbar^2}{2m} \qquad (3\text{–}28)$$

Figure 3–7 shows the parabolic relation of Equation (3–28) between the energy $E$ and momentum $p$ for the free particle. Since the momentum and wave number are linearly related, Figure 3–7 is also the $E$ versus $k$ curve for the free particle.

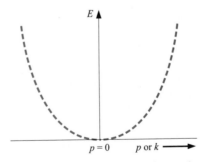

$p = 0$  $p$ or $k$ →

Figure 3–7 The parabolic $E$ versus $k$ curve for the free electron.

We now want to consider the relation between $E$ and $k$ from Equation (3–24) for the particle in the single-crystal lattice. As the parameter $P'$ increases, the particle becomes more tightly bound to the potential well or atom. We may define the left side of Equation (3–24) to be a function $f(\alpha a)$, so that

$$f(\alpha a) = P' \frac{\sin \alpha a}{\alpha a} + \cos \alpha a \tag{3-29}$$

Figure 3–8a is a plot of the first term of Equation (3–29) versus $\alpha a$. Figure 3–8b shows a plot of the $\cos \alpha a$ term and Figure 3–8c is the sum of the two terms, or $f(\alpha a)$.

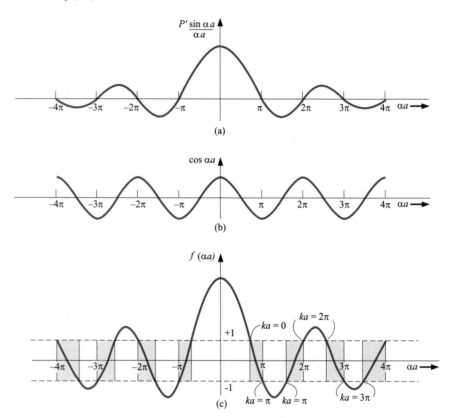

**Figure 3–8**   A plot of (a) the first term in Equation (3–29), (b) the second term in Equation (3–29), and (c) the entire $f(\alpha a)$ function. The shaded areas show the allowed values of $(\alpha a)$ corresponding to real values of $k$.

Now from Equation (3–24), we also have that

$$f(\alpha a) = \cos ka \tag{3-30}$$

For Equation (3–30) to be valid, the allowed values of the $f(\alpha a)$ function must be bounded between $+1$ and $-1$. Figure 3–8c shows the allowed values of $f(\alpha a)$ and the allowed values of $\alpha a$ in the shaded areas. Also shown on the figure are the values of $ka$ from the right side of Equation (3–30) which correspond to the allowed values of $f(\alpha a)$.

The parameter $\alpha$ is related to the total energy $E$ of the particle through

Equation (3–5), which is $\alpha^2 = 2mE/\hbar^2$. A plot of the energy $E$ of the particle as a function of the wave number $k$ can be generated from Figure 3–8c. Figure 3–9 shows this plot and shows the concept of allowed energy bands for the particle propagating in the crystal lattice. Since the energy $E$ has discontinuities, we also have the concept of forbidden energies for the particles in the crystal.

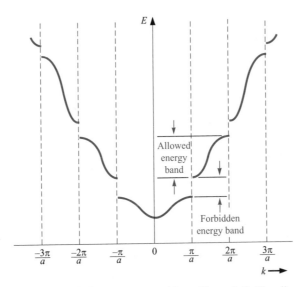

**Figure 3–9** The $E$ versus $k$ diagram generated from Figure 3–8. The allowed energy bands and forbidden energy bandgaps are indicated.

Consider again the right side of Equation (3–24), which is the function $\cos ka$. The cosine function is periodic so that

$$\cos ka = \cos (ka + 2n\pi) = \cos (ka - 2n\pi) \qquad (3\text{–}31)$$

where $n$ is a positive integer. We may consider Figure 3–9 and displace portions of the curve by $2\pi$. Mathematically, Equation (3–24) is still satisfied. Figure 3–10 shows how various segments of the curve can be displaced by the $2\pi$ factor. Figure 3–11 shows the case in which the entire $E$ versus $k$ plot is contained within $-\pi/a < k < \pi/a$. This plot is referred to as a reduced $k$-space diagram, or a reduced-zero representation.

We noted that for a free electron, the particle momentum and the wave number $k$ were related by $p = \hbar k$. Given the similarity between the free electron solution and the results of the single crystal shown in Figure 3–9, the parameter $\hbar k$ in a single crystal is referred to as the *crystal momentum*. This parameter is not the actual momentum of the electron in the crystal, but is a constant of the motion that includes the crystal interaction.

We have been considering the Kronig-Penney model, which is a one-dimensional periodic potential function used to model a single-crystal lat-

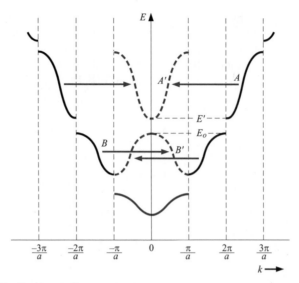

**Figure 3–10** The $E$ versus $k$ diagram showing $2\pi$ displacements of several sections of allowed energy bands.

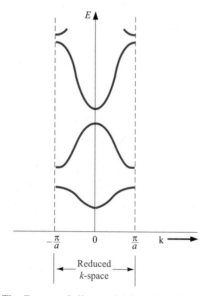

**Figure 3–11** The $E$ versus $k$ diagram in the reduced-zone representation.

tice. The principle result of this analysis, so far, is that electrons in the crystal occupy certain allowed energy bands and are excluded from the forbidden energy bands. For real three-dimensional single-crystal materials, a similar energy-band theory exists. We will obtain additional electron properties from the Kronig-Penney model in the next sections.

## 3.2 ELECTRICAL CONDUCTION IN SOLIDS

Again, we are eventually interested in determining the current-voltage characteristics of semiconductor devices. We will need to consider electrical conduction in solids as it relates to the band theory we have just developed. Let us begin by considering the motion of electrons in the various allowed energy bands.

### 3.2.1 The Energy Band and the Bond Model

In Chapter 1, we discussed the covalent bonding of silicon. Figure 3–12 shows a two-dimensional representation of the covalent bonding in a single-crystal silicon lattice. This figure represents silicon at $T = 0°K$ in which each silicon atom is surrounded by eight valence electrons that are in their lowest

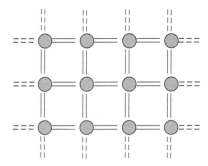

**Figure 3–12**    Two-dimensional representation of the covalent bonding in a semiconductor at $T = 0°K$.

energy state and are directly involved in the covalent bonding. Figure 3–4 represented the splitting of the discrete silicon energy states into bands of allowed energies as the silicon crystal is formed. At $T = 0°K$, the 4N states in the lower band, the valence band, are filled with the valence electrons. All of the valence electrons schematically shown in Figure 3–12 are in the valence band. The upper energy band, the conduction band, is completely empty at $T = 0°K$.

As the temperature increases above $0°K$, a few valence band electrons may gain enough thermal energy to break the covalent bond and jump into the conduction band. Figure 3–13a shows a two-dimensional representation of this bond-breaking effect and Figure 3–13b, a simple line representation of the energy-band model, shows the same effect.

The semiconductor is neutrally charged. This means that as the negatively charged electron breaks away from its covalent bonding position, a positively charged "empty state" is created in the original covalent bonding position in the valence band. As the temperature further increases, more

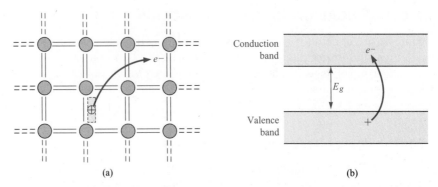

(a)                                                    (b)

**Figure 3–13**   (a) Two-dimensional representation of the breaking of a covalent bond. (b) The corresponding line representation of the energy band and the generation of a negative and positive charge with the breaking of a covalent bond.

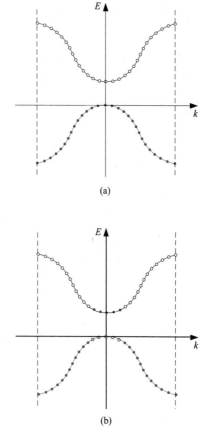

**Figure 3–14**   The $E$ versus $k$ diagram of the conduction and valence bands of a semiconductor at (a) $T = 0°K$ and (b) $T > 0°K$.

covalent bonds are broken, more electrons jump to the conduction band, and more positive "empty states" are created in the valence band.

We can also relate this bond breaking to the $E$ versus $k$ energy bands developed through the Kronig-Penney model. Figure 3–14a shows the $E$ versus $k$ diagram of the conduction and valence bands at $T = 0°K$. The energy states in the valence band are completely full and the states in the conduction band are empty. Figure 3–14b shows these same bands for $T > 0°K$, in which some electrons have gained enough energy to jump to the conduction band and have left empty states in the valence band. We are assuming at this point that no external forces are applied so the electron and "empty state" distributions are symmetrical with $k$.

### 3.2.2  Drift Current

Current is due to the net flow of charge. If we had a collection of positively charged ions with a volume density $N(cm^{-3})$ and an average drift velocity $v_d(cm/sec)$ then the drift current density would be

$$J = qNv_d \ (A/cm^2) \qquad (3-32)$$

If, instead of considering the average drift velocity, we considered the individual ion velocities, then we could write the drift current density as

$$J = q \sum_{i=1}^{N} v_i \qquad (3-33)$$

where $v_i$ is the velocity of the $i^{th}$ ion. The summation in Equation (3–33) is taken over a unit volume so that the current density $J$ is still in units of $A/cm^2$.

Since electrons are charged particles, a net drift of electrons in the conduction band will give rise to a current. The electron distribution in the conduction band, as shown in Figure 3–14b, is an even function of $k$ when no external force is applied. Recall that $k$ for a free electron is related to momentum so that, since there are as many electrons with a $+|k|$ value as there are with a $-|k|$ value, the net drift current density due to these electrons is zero. This result is certainly expected since there is no externally applied force.

If a force is applied to a particle and the particle moves, it must gain energy. This effect is expressed as

$$dE = Fdx = Fvdt \qquad (3-34)$$

where $F$ is the applied force, $dx$ is the differential distance the particle moves, $v$ is the velocity, and $dE$ is the increase in energy. If an external force is applied to the electrons in the conduction band, there are empty energy states into which the electrons can move; therefore, due to the external force, electrons can gain energy and a net momentum. The electron distribu-

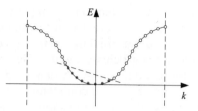

**Figure 3–15**   The asymmetric distribution of electrons in the $E$ versus $k$ diagram when an external force is applied.

tion in the conduction band may look like that shown in Figure 3–15, which implies that the electrons have gained a net momentum.

We may write the drift current density due to the motion of electrons as

$$J = -e \sum_{i=1}^{n} v_i \tag{3-35}$$

where $e$ is the magnitude of the electronic charge and $n$ is the number of electrons per unit volume in the conduction band. Again, the summation is taken over a unit volume so the current density is A/cm$^2$. We may note from Equation (3–35) that the current is directly related to the electron velocity; that is, the current is related to how well the electron can move in the crystal.

### 3.2.3 Electron Effective Mass

The movement of an electron in a lattice will, in general, be different than that of an electron in free space. In addition to an externally applied force, there are internal forces in the crystal due to positively charged ions or protons and negatively charged electrons, which will influence the motion of electrons in the lattice. We can write

$$F_{total} = F_{ext} + F_{int} = ma \tag{3-36}$$

where $F_{total}$, $F_{ext}$, and $F_{int}$ are the total force, the externally applied force, and the internal forces, respectively, acting on a particle in a crystal. The parameter $a$ is the acceleration and $m$ is the rest mass of the particle.

Since it is difficult to take into account all of the internal forces, we will write the equation

$$F_{ext} = m^*a \tag{3-37}$$

where the acceleration $a$ is now directly related to the external force. The parameter $m^*$, called the effective mass, takes into account the particle mass and also takes into account the effect of the internal forces.

To use an analogy for the effective mass concept, consider the difference in motion between a glass marble in a container filled with water and in a container filled with oil. In general, the marble will drop through the water at a faster rate than through the oil. The external force in this example is the gravitational force and the internal forces are related to the viscosity of the liquids. Because of the difference in motion of the marble in these two cases, the mass of the marble would appear to be different in water than in oil. (As with any analogy, we must be careful not to be too literal.)

We can also relate the effective mass of an electron in a crystal to the $E$ versus $k$ curves, such as was shown in Figure 3–10. In a semiconductor material, we will be dealing with allowed energy bands that are almost empty of electrons and other energy bands that are almost full of electrons.

To begin, consider the case of a free electron whose $E$ versus $k$ curve was shown in Figure 3–7. Recalling Equation (3–28), the energy and momentum are related by $E = p^2/2m = \hbar^2 k^2/2m$, where $m$ is the mass of the electron. The momentum and wave number $k$ are related by $p = \hbar k$. If we take the derivative of Equation (3–28) with respect to $k$, we obtain

$$\frac{dE}{dk} = \frac{\hbar^2 k}{m} = \frac{\hbar p}{m} \tag{3–38}$$

Relating momentum to velocity, Equation (3–38) can be written as

$$\frac{1}{\hbar}\frac{dE}{dk} = \frac{p}{m} = v \tag{3–39}$$

where $v$ is the velocity of the particle. The first derivative of $E$ with respect to $k$ is related to the velocity of the particle.

If we now take the second derivative of $E$ with respect to $k$, we have

$$\frac{d^2E}{dk^2} = \frac{\hbar^2}{m} \tag{3–40}$$

We may rewrite Equation (3–40) as

$$\frac{1}{\hbar^2}\frac{d^2E}{dk^2} = \frac{1}{m} \tag{3–41}$$

The second derivative of $E$ with respect to $k$ is inversely proportional to the mass of the particle. For the case of a free electron, the mass is a constant (nonrelativistic effect), so the second derivative function is a constant. We may also note from Figure 3–7 that $d^2E/dk^2$ is a positive quantity, which implies that the mass of the electron is also a positive quantity.

If we apply an electric field to the free electron and use Newton's classical equation of motion, we can write

$$F = ma = -e\mathrm{E} \tag{3–42}$$

where $a$ is the acceleration, E is the applied electric field, and $e$ is the magnitude of the electronic charge. Solving for the acceleration, we have

$$a = \frac{-e\mathrm{E}}{m} \tag{3–43}$$

The motion of the free electron is in the opposite direction to the applied electric field because of the negative charge.

Now apply the results from the Kronig-Penney model to the electron in the bottom of an energy band. Consider the allowed energy band in Figure 3–10 denoted by A which is equivalent to A'. The energy near the bottom of this energy band may be approximated by a parabola, just as a free particle. We may write

$$(E - E') = C_1 (k - k_o)^2 \tag{3–44}$$

The energy $E'$ is the energy at the bottom of the band. Since $E > E'$, the parameter $C_1$ is a positive quantity.

Taking the second derivative of $E$ with respect to $k$ from Equation (3–44), we obtain

$$\frac{d^2E}{dk^2} = 2C_1 \tag{3–45}$$

We may put Equation (3–45) in the form

$$\frac{1}{\hbar^2} \frac{d^2E}{dk^2} = \frac{2C_1}{\hbar^2} \tag{3–46}$$

Comparing Equation (3–46) to Equation (3–41), we may equate $\hbar^2/2C_1$ to the mass of the particle. However, the curvature of curve A in Figure 3–10 will not, in general, be the same as the curvature of the free-particle curve. We may write

$$\frac{1}{\hbar^2} \frac{d^2E}{dk^2} = \frac{2C_1}{\hbar^2} = \frac{1}{m^*} \tag{3–47}$$

where $m^*$ is called the effective mass. Since $C_1 > 0$, we have that $m^* > 0$ also.

The effective mass is a parameter that relates the quantum mechanical results derived from the Kronig-Penney model to the classical force equations. In most instances, the electron in the bottom of the conduction band can be thought of as a classical particle whose motion can be modeled using Newtonian mechanics, provided that the internal forces and quantum mechanical properties are taken into account through the effective mass. If we apply an electric field to the electron in the bottom of the allowed energy band, we may write the acceleration as

$$a = \frac{-e\mathrm{E}}{m_n^*} \tag{3–48}$$

where $m_n^*$ is the effective mass of the electron. The effective mass $m_n^*$ of the electron near the bottom of the conduction band is a constant.

### 3.2.4 Concept of the Hole

In considering the two-dimensional representation of the covalent bonding shown in Figure 3–13a, a positively charged "empty state" was created when a valence electron was elevated into the conduction band. For $T > 0°K$, all valence electrons may gain thermal energy; if a valence electron gains a small amount of thermal energy, it may hop into the "empty state." The movement of a valence electron into the "empty state" is equivalent to the movement of the positively charged "empty state" itself. Figure 3–16

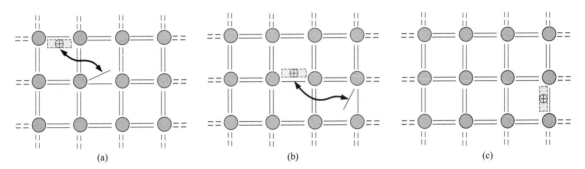

(a)  (b)  (c)

**Figure 3–16**  Visualization of the movement of a hole in a semiconductor.

shows the movement of valence electrons in the crystal alternately filling one "empty state" and creating a new "empty state," a motion equivalent to a positive charge moving in the valence band. The crystal now has a second equally important charge carrier that can give rise to a current. This charge carrier is called a *hole* and, as we will see, can also be thought of as a classical particle whose motion can be modeled using Newtonian mechanics.

The drift current density due to electrons in the valence band, such as shown in Figure 3–14b, can be written as

$$J = -e \sum_{i(\text{filled})} v_i \tag{3–49}$$

where the summation extends over all filled states. This summation is inconvenient since it extends over a nearly full valence band and takes into account a very large number of states. We may rewrite Equation (3–49) in the form

$$J = -e \sum_{i(\text{total})} v_i + e \sum_{i(\text{empty})} v_i \tag{3–50}$$

If we consider a band that is totally full, all available states are occupied by electrons. The individual electrons can be thought of as moving with a velocity as given by Equation (3–39)

$$v(E) = \left(\frac{1}{\hbar}\right)\left(\frac{dE}{dk}\right) \tag{3-39}$$

The band is symmetric in $k$ and each state is occupied so that, for every electron with a velocity $|v|$, there is a corresponding electron with a velocity $-|v|$. Since the band is full, the distribution of electrons with respect to $k$ cannot be changed with an externally applied force. The net drift current density generated from a completely full band, then, is zero, or

$$-e \sum_{i(\text{total})} v_i \equiv 0 \tag{3-51}$$

We can now write the drift current density from Equation (3–50) for an almost full band as

$$J = +e \sum_{i(\text{empty})} v_i \tag{3-52}$$

where the $v_i$ in the summation is the

$$v(E) = \left(\frac{1}{\hbar}\right)\left(\frac{dE}{dk}\right)$$

associated with the empty state. Equation (3–52) is entirely equivalent to placing a positively charged particle in the empty states and assuming all other states in the band are empty, or neutrally charged. This concept is shown in Figure 3–17. Figure 3–17a shows the valence band with the conventional electron-filled states and empty states while Figure 3–17b shows the new concept of positive charges occupying the original empty states.

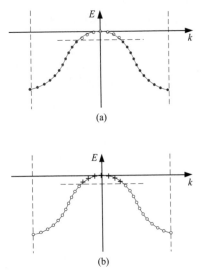

(a)

(b)

**Figure 3–17**   (a) Valence band with conventional electron-filled states and empty states. (b) Concept of positive charges occupying the original empty states.

This concept is consistent with the discussion of the positively charged "empty state" in the valence band as shown in Figure 3–16.

The $v_i$ in the summation of Equation (3–52) is related to how well this positively charged particle moves in the semiconductor. Now consider an electron near the top of an allowed energy band denoted as B in Figure 3–10. The curve B' is entirely equivalent to B. The energy near the top of the allowed energy band may again be approximated by a parabola so that we may write

$$(E - E_o) = -C_2(k - k_o)^2 \qquad (3-53)$$

The energy $E_o$ is the energy at the top of the energy band. Since $E < E_o$ for electrons in this band, then the parameter $C_2$ must be a positive quantity.

Taking the second derivative of $E$ with respect to $k$ from Equation (3–53), we obtain

$$\frac{d^2E}{dk^2} = -2C_2 \qquad (3-54)$$

We may rearrange this equation so that

$$\frac{1}{\hbar^2}\frac{d^2E}{dk^2} = \frac{-2C_2}{\hbar^2} \qquad (3-55)$$

Comparing Equation (3–55) with Equation (3–41), we may write

$$\frac{1}{\hbar^2}\frac{d^2E}{dk^2} = \frac{-2C_2}{\hbar^2} = \frac{1}{m^*} \qquad (3-56)$$

where $m^*$ is again an effective mass. We have argued that $C_2$ is a positive quantity which now implies that $m^*$ is a negative quantity. An electron moving near the top of an allowed energy band behaves as if it has a negative mass.

We must keep in mind that the effective mass parameter is used to relate quantum mechanics and classical mechanics. The attempt to relate these two theories leads to this strange result of a negative effective mass. However, we must recall that solutions to Schrodinger's wave equation from the last chapter also led to results which contradicted classical mechanics. The negative effective mass is another such example.

In discussing the concept of effective mass in the last section, we used an analogy of marbles moving through two liquids. Now consider placing an ice cube in the center of a container filled with water: the ice cube will move upward toward the surface in a direction opposite to the gravitational force. The ice cube appears to have a negative effective mass since its acceleration is opposite to the external force. The effective mass parameter takes into account all internal forces acting on the particle.

If we again consider an electron near the top of an allowed energy band and use Newton's force equation for an applied electric field, we will have

$$F = m^*a = -e\mathrm{E} \tag{3-57}$$

However, $m^*$ is now a negative quantity, so we may write

$$a = \frac{-e\mathrm{E}}{-|m^*|} = \frac{+e\mathrm{E}}{|m^*|} \tag{3-58}$$

An electron moving near the top of an allowed energy band moves in the same direction as the applied electric field.

The net motion of electrons in a nearly full band can be described by considering just the empty states, provided that a positive electronic charge is associated with each state and that the negative of $m^*$ from Equation (3–56) is associated with each state. We now can model this band as having particles with a positive electronic charge and a positive effective mass. The density of these particles in the valence band is the same as the density of empty electronic energy states. This new particle is the *hole*. The hole, then, has a positive effective mass denoted by $m_p^*$ and a positive electronic charge so that it will move in the same direction as an applied field.

### 3.2.5 Metals, Insulators, and Semiconductors

Each crystal has its own energy-band structure. We noted that the splitting of the energy states in silicon, for example, to form the valence and conduction bands, was complex. Complex band splitting occurs in other crystals, leading to large variations in band structures between various solids, and to a wide range of electrical characteristics observed in these various materials. We can qualitatively begin to understand some basic differences in electrical characteristics caused by variations in band structure by considering some simplified energy bands.

There are several possible energy-band conditions to consider. Figure 3–18a shows an allowed energy band that is completely empty of electrons. If an electric field is applied, there are no particles to move, so there will be no current. Figure 3–18b shows another allowed energy band whose energy states are completely full of electrons. We argued in the previous section that a completely full energy band will also not give rise to a current. A material that has energy bands either completely empty or completely full is an insulator. The resistivity of an insulator is very large or, conversely, the conductivity of an insulator is very small. There are essentially no charged particles that can contribute to a drift current. Figure 3–18c shows a simplified energy-band diagram of an insulator. The bandgap energy $E_g$ of an insulator is usually on the order of 3.5 to 6 eV or larger, so that at room temperature, there are essentially no electrons in the conduction band and the valence band remains completely full. There are very few thermally generated electrons and holes in an insulator.

Figure 3–19a shows an energy band with relatively few electrons near the bottom of the band. Now, if an electric field is applied, the electrons can gain

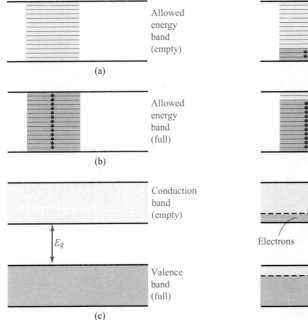

**Figure 3–18**   Allowed energy bands showing (a) an empty band, (b) a completely full band, and (c) the bandgap energy between the two allowed bands.

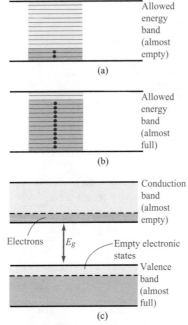

**Figure 3–19**   Allowed energy bands showing (a) an almost empty band, (b) an almost full band, and (c) the bandgap energy between the two allowed bands.

energy, move to higher energy states, and move through the crystal. The net flow of charge is a current. Figure 3–19b shows an allowed energy band that is almost full of electrons, which means that we can consider the holes in this band. If an electric field is applied, the holes can move and give rise to a current. Figure 3–19c shows the simplified energy-band diagram for this case. The bandgap energy may be on the order of 1 eV. This energy-band diagram represents a semiconductor for $T > 0K$. The resistivity of a semiconductor, as we will see in the next chapter, can be controlled and varied over many orders of magnitude.

The characteristics of a metal include a very low value of resistivity. The energy-band diagram for a metal may be in one of two forms. Figure 3–20a shows the case of a partially full band in which there are many electrons available for conduction, so that the material can exhibit a large electrical conductivity. Figure 3–20b shows another possible energy-band diagram of a metal. The band splitting into allowed and forbidden energy bands is a complex phenomenon and Figure 3–20b shows a case in which the conduction and valence bands overlap at the equilibrium interatomic distance. As in

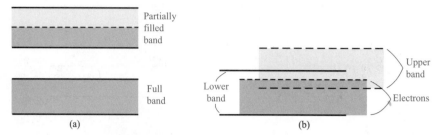

**Figure 3–20**    Two possible energy bands of a metal showing (a) a partially filled band and (b) overlapping allowed energy bands.

the case shown in Figure 3–20a, there are large numbers of electrons as well as large numbers of empty energy states into which the electrons can move, so that this material can also exhibit a very high electrical conductivity.

## 3.3 EXTENSION TO THREE DIMENSIONS

The basic concept of allowed and forbidden energy bands and the basic concept of effective mass have been developed in the last sections using the one-dimensional periodic potential function of the Kronig-Penney model. In this section, we will extend these concepts to three dimensions and to real crystals. There will be no attempt to extend the mathematics to the three-dimensional case. We will, instead, qualitatively consider particular characteristics of the three-dimensional crystal in terms of the $E$ versus $k$ plots, bandgap energy, and effective mass. We must emphasize that we will only briefly touch on the basic three-dimensional concepts; therefore, many details will not be considered.

One problem encountered in extending the potential function to a three-dimensional crystal is that the distance between atoms varies as the direction through the crystal changes. Figure 3–21 shows a face-centered cubic structure with the [100] and [110] directions indicated. Electrons traveling in different directions encounter different potential patterns and therefore different $k$-space boundaries. The $E$ versus $k$ diagrams are in general a function of the $k$-space direction in a crystal.

### 3.3.1 The $k$-Space Diagrams of Si and GaAs

Figure 3–22 shows an $E$ versus $k$ diagram of gallium arsenide and of silicon. These simplified diagrams show the basic properties considered in this text, but do not show many of the details more appropriate for advanced-level courses.

Note that in place of the usual positive and negative $k$-axes, we now show two different crystal directions. The $E$ versus $k$ diagram for the one-dimen-

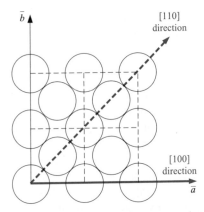

**Figure 3–21** The (100) plane of a face-centered cubic crystal showing the [100] and [110] directions.

sional model was symmetric in $k$ so that no new information is obtained by displaying the negative axis. It is normal practice to plot the [100] direction along the normal $+k$ axis and to plot the [111] portion of the diagram so the $+k$ points to the left. In the case of diamond or zincblende lattices, the maxima in the valence band energy and minima in the conduction band energy occur at $k = 0$ or along one of these two directions.

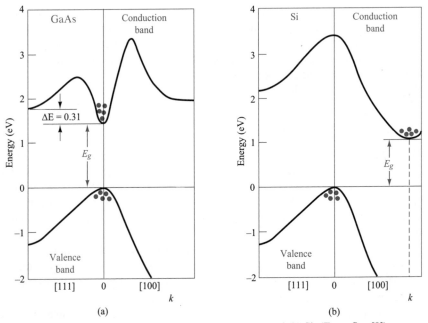

**Figure 3–22** Energy band structures of (a) GaAs and (b) Si. (From Sze [9]).

Figure 3–22a shows the $E$ versus $k$ diagram for GaAs. The valence band maximum and the conduction band minimum both occur at $k = 0$. The electrons in the conduction band tend to settle at the minimum conduction band energy which is at $k = 0$. Similarly, holes in the valence band tend to congregate at the uppermost valence band energy. In GaAs, the minimum conduction band energy and maximum valence band energy occur at the same $k$-value. A semiconductor with this property is said to be a *direct* bandgap semiconductor; transitions between the two allowed bands can take place with no change in crystal momentum. This direct nature has significant effect on the optical properties of the material. GaAs and other direct bandgap materials are ideally suited for use in semiconductor lasers and other optical devices, as we will see in later chapters.

The $E$ versus $k$ diagram for silicon is shown in Figure 3–22b. The maximum in the valence band energy occurs at $k = 0$ as before. The minimum in the conduction band energy occurs not at $k = 0$, but along the [100] direction. The difference between the minimum conduction band energy and the maximum valence band energy is still defined as the bandgap energy $E_g$. A semiconductor whose maximum valence band energy and minimum conduction band energy do not occur at the same $k$-value is called an *indirect* bandgap semiconductor. When electrons make a transition between the conduction and valence bands, we must invoke the law of conservation of momentum. A transition in an indirect bandgap material must necessarily include an interaction with the crystal so that crystal momentum is conserved.

Germanium is also an indirect bandgap material, whose valence band maximum occurs at $k = 0$ and whose conduction band minimum occurs along the [111] direction. GaAs is a direct bandgap semiconductor, but other compound semiconductors, such as GaP and AlAs, have indirect bandgaps. We will consider the compound semiconductor materials in more detail when we consider optical devices.

### 3.3.2 Additional Effective Mass Concepts

The curvature of the $E$ versus $k$ diagrams near the minimum of the conduction band energy is related to the effective mass of the electron. We may note from Figure 3–22a that the curvature of the conduction band at its minimum value for GaAs is larger than that of silicon so the effective mass of an electron in the conduction band of GaAs will be smaller than that in silicon.

For the one-dimensional $E$ versus $k$ diagram, the effective mass was defined by Equation (3–41) as $1/m^* = 1/\hbar^2 \cdot d^2E/dk^2$. A complication occurs in the effective mass concept in a real crystal. A three-dimensional crystal can be described by three $k$-vectors. The curvature of the $E$ versus $k$ diagram at the conduction band minimum may not be the same in the three $k$-directions. We will not consider the details of the various effective mass parameters here. In later sections and chapters, the effective mass parameters used in calculations will be a kind of statistical average that is adequate for most device calculations.

## 3.4 DENSITY OF STATES FUNCTION

As we have stated, we eventually wish to describe the current-voltage characteristics of semiconductor devices. Since current is due to the flow of charge, an important step in the process is to determine the number of electrons and holes in the semiconductor that will be available for conduction. The number of carriers that can contribute to the conduction process is a function of the number of available energy or quantum states since, by the Pauli exclusion principle, only one electron can occupy a given quantum state. When we discussed the splitting of energy levels into bands of allowed and forbidden energies, we indicated that the band of allowed energies was actually made up of discrete energy levels. We must determine the density of these allowed energy states as a function of energy in order to calculate the electron and hole concentrations.

### 3.4.1 Mathematical Derivation

To determine the density of allowed quantum states as a function of energy, we need to consider an appropriate mathematical model. Electrons are allowed to move relatively freely in the conduction band of a semiconductor, but are confined to the crystal. As a first step, we will consider a free electron confined to a three-dimensional infinite potential well, where the potential well represents the crystal. The potential of the infinite potential well is defined as

$$V(x,y,z) = 0 \quad \text{for} \quad \begin{array}{l} 0 < x < a \\ 0 < y < a \\ 0 < z < a \end{array} \tag{3-59}$$

$$V(x,y,z) = \infty \quad \textit{elsewhere}$$

where the crystal is assumed to be a cube with length $a$. Schrodinger's wave equation in three dimensions can be solved using the separation of variables technique. Extrapolating the results from the one-dimensional infinite potential well, we can show (see problem 9) that

$$\frac{2mE}{\hbar^2} = k^2 = k_x^2 + k_y^2 + k_z^2 = \left(n_x^2 + n_y^2 + n_z^2\right)\left(\frac{\pi^2}{a^2}\right) \tag{3-60}$$

where $n_x$, $n_y$, and $n_z$ are positive integers. (Negative values of $n_x$, $n_y$, and $n_z$ yield the same wave function, except for the sign, as the positive integer values, resulting in the same probability function and energy, so the negative integers do not represent a different quantum state.)

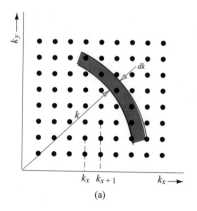

(a)

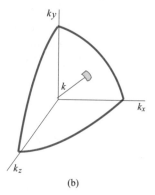

(b)

**Figure 3–23**   (a) A two-dimensional array of allowed quantum states in $k$-space. (b) The positive one eighth of the spherical $k$-space.

We can schematically plot the allowed quantum states in $k$-space. Figure 3–23a shows a two-dimensional plot as a function of $k_x$ and $k_y$. Each point represents an allowed quantum state corresponding to various integral values of $n_x$ and $n_y$. Positive and negative values of $k_x$, $k_y$, or $k_z$ have the same energy and represent the same energy state. Since negative values of $k_x$, $k_y$, or $k_z$ do not represent additional quantum states, the density of quantum states will be determined by considering only the positive one eighth of the spherical $k$-space as shown in Figure 3–23b.

The distance between two quantum states in the $k_x$ direction, for example, is given by

$$k_{x+1} - k_x = (n_x + 1)\left(\frac{\pi}{a}\right) - n_x\left(\frac{\pi}{a}\right) = \frac{\pi}{a} \tag{3-61}$$

Generalizing this result to three dimensions, the volume $V_k$ of a single quantum state is

$$V_k = \left(\frac{\pi}{a}\right)^3 \tag{3-62}$$

We can now determine the density of quantum states in $k$-space. A differential volume in $k$-space is shown in Figure 3–23b and is given by $4\pi k^2 dk$ so that the differential density of quantum states in $k$-space can be written as

$$g_T(k)dk = 2\left(\frac{1}{8}\right)\frac{4\pi k^2 dk}{\left(\frac{\pi}{a}\right)^3} \tag{3–63}$$

The first factor, 2, takes into account the two spin states allowed for each quantum state; the next factor, $\frac{1}{8}$, takes into account that we are considering only the quantum states for positive values of $k_x$, $k_y$, and $k_z$. The factor $4\pi k^2 dk$ is again the differential volume and the factor $(\pi/a)^3$ is the volume of one quantum state. Equation (3–63) may be simplified to

$$g_T(k)dk = \frac{\pi k^2 dk}{\pi^3} \cdot a^3 \tag{3–64}$$

Equation (3–64) gives the density of quantum states as a function of momentum, through the parameter $k$. We can now determine the density of quantum states as a function of energy $E$. For a free electron, the parameters $E$ and $k$ are related by

$$k^2 = \frac{2mE}{\hbar^2} \tag{3–65a}$$

or

$$k = \frac{1}{\hbar}\sqrt{2mE} \tag{3–65b}$$

The differential $dk$ is

$$dk = \frac{1}{\hbar}\sqrt{\frac{m}{2E}}\,dE \tag{3–66}$$

Then, substituting the expressions for $k^2$ and $dk$ into Equation (3–64), the number of energy states between $E$ and $E + dE$ is given by

$$g_T(E)dE = \frac{\pi a^3}{\pi^3}\left(\frac{2mE}{\hbar^2}\right)\cdot\frac{1}{\hbar}\sqrt{\frac{m}{2E}}\,dE \tag{3–67}$$

Recalling that $\hbar = h/2\pi$, Equation (3–67) becomes

$$g_T(E)dE = \frac{4\pi a^3}{h^3}\cdot(2m)^{3/2}\cdot\sqrt{E}\cdot dE \tag{3–68}$$

Equation (3–68) gives the total number of quantum states between the energy $E$ and $E + dE$ in the crystal space volume of $a^3$. If we divide by the volume $a^3$, then we will obtain the density of quantum states per unit volume of the crystal. Equation (3–68) then becomes

$$g(E) = \frac{4\pi(2m)^{3/2}}{h^3}\sqrt{E} \tag{3–69}$$

The density of quantum states is a function of energy $E$. As the energy of this free electron becomes small, the number of available quantum states decreases. This density function is really a double density, in that the units are given in terms of states per unit energy per unit volume.

---

**Example 3–2**

**Objective:** To calculate the density of states per unit volume over a particular energy range.

Consider the density of states for a free electron given by Equation (3–69). Calculate the density of states per unit volume with energies between zero and 1 eV.

**Solution:** The volume density of quantum states, using Equation (3–69), can be found from

$$N = \int_0^{1eV} g(E)dE = \frac{4\pi(2m)^{3/2}}{h^3} \cdot \int_0^{1eV} \sqrt{E} \, dE$$

or

$$N = \frac{4\pi(2m)^{3/2}}{h^3} \cdot \frac{2}{3} \cdot E^{3/2}$$

The density of states is now

$$N = \frac{4\pi[2(9.11 \times 10^{-31})]^{3/2}}{(6.625 \times 10^{-34})^3} \cdot \frac{2}{3} \cdot (1.6 \times 10^{-19})^{3/2} = 4.5 \times 10^{27} \, m^{-3}$$

or

$$N = 4.5 \times 10^{21} \text{ states/cm}^3$$

**Comment:** The density of quantum states is typically a large number. An effective density of states in a semiconductor, as we will see in the following sections and in the next chapter, is also a large number but is usually less than the density of atoms in the semiconductor crystal.

---

### 3.4.2 Extension to Semiconductors

In the last section, we derived a general expression for the density of allowed electron quantum states using the model of a free electron with mass $m$ bounded in a three-dimensional infinite potential well. We can extend this same general model to a semiconductor to determine the density of quantum states in the conduction band and the density of quantum states in the valence band. Electrons and holes are confined within the semiconductor crystal so we will again use the basic model of the infinite potential well.

The parabolic relationship between energy and momentum of a free electron was given in Equation (3–28) as $E = p^2/2m = \hbar^2k^2/2m$. Figure 3–24a shows the conduction energy band in the reduced $k$-space which was derived from the Kronig-Penney model. The $E$ versus $k$ curve near $k = 0$ at the

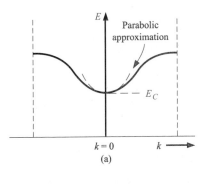

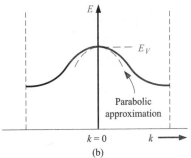

**Figure 3–24**  (a) The conduction band in reduced $k$-space and the parabolic approximation. (b) The valence band in reduced $k$-space and the parabolic approximation.

bottom of the conduction band can be approximated as a parabola, so we may write

$$E = E_c + \frac{\hbar^2 k^2}{2m_n^*} \qquad (3\text{--}70)$$

where $E_c$ is the bottom edge of the conduction band and $m_n^*$ is the electron effective mass. Equation (3–70) may be rewritten to give

$$E - E_c = \frac{\hbar^2 k^2}{2m_n^*} \qquad (3\text{--}71)$$

The general form of the $E$ versus $k$ relation for an electron in the bottom of a conduction band is the same as the free electron, except the mass is replaced by the effective mass. We can then think of the electron in the bottom of the conduction band as being a "free" electron with its own particular mass. The right side of Equation (3–71) is of the same form as the right side of Equation (3–28), which was used in the derivation of the density of states function. Because of this similarity which yields the "free" conduction electron model, we may generalize the free electron results of Equation

(3–69) and write the density of allowed electronic energy states in the conduction band as

$$g_c(E) = \frac{4\pi(2m_n^*)^{3/2}}{h^3} \sqrt{E - E_c} \tag{3–72}$$

Equation (3–72) is valid for $E \geq E_c$. As the energy of the electron in the conduction band decreases, the number of available quantum states also decreases.

The density of quantum states in the valence band can be obtained using the same infinite potential well model, since the hole is also confined in the semiconductor crystal and can be treated as a "free" particle. The effective mass of the hole is $m_p^*$. Figure 3–24b shows the valence energy band in the reduced $k$-space. We may also approximate the $E$ versus $k$ curve near $k = 0$ by a parabola for a "free" hole, so that

$$E = E_v - \frac{\hbar^2 k^2}{2m_p^*} \tag{3–73}$$

Equation (3–73) may be rewritten to give

$$E_v - E = \frac{\hbar^2 k^2}{2m_p^*} \tag{3–74}$$

Again, the right side of Equation (3–74) is of the same form used in the general derivation of the density of states function. We may then generalize

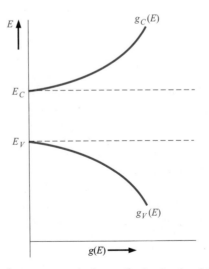

**Figure 3–25** The density of energy states in the conduction band and the density of energy states in the valence band as a function of energy.

the density of states function from Equation (3–69) to apply to the valence band, so that

$$g_v(E) = \frac{4\pi(2m_p^*)^{3/2}}{h^3} \sqrt{E_v - E} \qquad (3-75)$$

Equation (3–75) is valid for $E \leq E_v$.

We have argued that quantum states do not exist within the forbidden energy band, so $g(E) = 0$ for $E_v < E < E_c$. Figure 3–25 shows the plot of the density of quantum states as a function of energy. If the electron and hole effective masses were equal, then the functions $g_c(E)$ and $g_v(E)$ would be symmetrical about the energy midway between $E_c$ and $E_v$, or the midgap energy, $E_{midgap}$.

## 3.5 STATISTICAL MECHANICS

In dealing with large numbers of particles, we are interested only in the statistical behavior of the group as a whole rather than in the behavior of each individual particle. For example, gas within a container will exert an average pressure on the walls of the vessel. The pressure is actually due to the collisions of the individual gas molecules with the walls, but we do not follow each individual molecule as it collides with the wall. Likewise in a crystal, the electrical characteristics will be determined by the statistical behavior of a large number of electrons.

### 3.5.1 Statistical Laws

In determining the statistical behavior of particles, we must consider the laws that the particles obey. There are three distribution laws determining the distribution of particles among available energy states.

One distribution law is the Maxwell-Boltzmann probability function. In this case, the particles are considered to be distinguishable by being numbered, for example, for 1 to $N$, with no limit to the number of particles allowed in each energy state. The behavior of gas molecules in a container at fairly low pressure is an example of this distribution.

A second distribution law is the Bose-Einstein function. The particles in this case are indistinguishable and, again, there is no limit to the number of particles permitted in each quantum state. The behavior of photons, or black body radiation, is an example of this law.

The third distribution law is the Fermi-Dirac probability function. In this case, the particles are again indistinguishable, but now only one particle is permitted in each quantum state. Electrons in a crystal obey this law. In each case, the particles are assumed to be noninteracting.

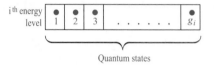

Figure 3–26    The $i^{th}$ energy level with $g_i$ quantum states.

### 3.5.2 The Fermi-Dirac Probability Function

Figure 3–26 shows the $i^{th}$ energy level with $g_i$ quantum states. A maximum of one particle is allowed in each quantum state by the Pauli exclusion principle. There are $g_i$ ways of choosing where to place the first particle, $(g_i - 1)$ ways of choosing where to place the second particle, $(g_i - 2)$ ways of choosing where to place the third particle, and so on. Then the total number of ways of arranging $N_i$ particles in the $i^{th}$ energy level (where $N_i \leq g_i$) is

$$(g_i)(g_i - 1) \cdots (g_i - (N_i - 1)) = \frac{g_i!}{(g_i - N_i)!} \qquad (3\text{–}76)$$

This expression includes all permutations of the $N_i$ particles among themselves.

However, since the particles are indistinguishable, the $N_i!$ number of permutations that the particles have among themselves in any given arrangement does not count as separate arrangements. The interchange of any two electrons, for example, does not produce a new arrangement. Therefore, the actual number of independent ways of realizing a distribution of $N_i$ particles in the $i^{th}$ level is

$$W_i = \frac{g_i!}{N_i!(g_i - N_i)!} \qquad (3\text{–}77)$$

---

### Example 3–3

**Objective:** To determine the possible number of ways of realizing a particular distribution.

Let $g_i = N_i = 10$. Then $(g_i - N_i)! = 1$.

**Solution:** Equation (3–77) becomes

$$\frac{g_i!}{N_i!(g_i - N_i)!} = \frac{10!}{10!} = 1$$

**Comment:** If we have 10 particles to be arranged in 10 quantum states, there is only one possible arrangement. Each quantum state contains one particle.

---

**Example 3–4**

**Objective:** To again determine the possible number of ways of realizing a particular distribution.

Let $g_i = 10$ and $N_i = 9$. In this case $g_i - N_i = 1$ so that $(g_i - N_i)! = 1$.

**Solution:** Equation (3–77) becomes

$$\frac{g_i!}{N_i!(g_i - N_i)!} = \frac{10!}{(9!)(1)} = \frac{(10)(9!)}{9!} = 10$$

**Comment:** In this case, if we have 10 quantum states and 9 particles, there is one empty quantum state. There are 10 possible arrangements, or positions, for the one empty state.

Equation (3–77) gives the number of independent ways of realizing a distribution of $N_i$ particles in the $i^{th}$ level. The total number of ways of arranging $(N_1, N_2, N_3, \ldots, N_n)$ indistinguishable particles among $n$ energy levels is the product of all distributions, or

$$W = \prod_{i=1}^{n} \frac{g_i!}{N_i!(g_i - N_i)!} \tag{3–78}$$

The parameter $W$ is the total number of ways in which $N$ electrons can be arranged in this system, where $N = \sum_{i=1}^{n} N_i$ is the total number of electrons in the system. We want to find the most probable distribution, which means that we want to find the maximum $W$. The maximum $W$ is found by varying $N_i$ among the $E_i$ levels which varies the distribution, but at the same time, we will keep the total number of particles and total energy constant.

We may write the most probable distribution function as

$$\frac{N(E)}{g(E)} = f_F(E) = \frac{1}{1 + \exp\left(\dfrac{E - E_F}{kT}\right)} \tag{3–79}$$

where $E_F$ is called the Fermi energy. The number density $N(E)$ is the number of particles per unit volume per unit energy and the function $g(E)$ is the number of quantum states per unit volume per unit energy. The function $f_F(E)$ is called the Fermi-Dirac distribution or probability function and gives the probability that a quantum state at the energy $E$ will be occupied by an electron. Another interpretation of the distribution function is that $f_F(E)$ is the ratio of filled to total quantum states at any energy $E$.

### 3.5.3 The Distribution Function and the Fermi Energy

To begin to understand the meaning of the distribution function and the Fermi energy, we can plot the distribution function versus energy. Initially, let $T = 0°K$ and consider the case when $E < E_F$. The exponential term in Equation (3–79) becomes $\exp[(E - E_F)/kT] \rightarrow \exp(-\infty) = 0$. The resulting distribution function is $f_F(E < E_F) = 1$. Again let $T = 0°K$ and consider the case when $E > E_F$. The exponential term in the distribution function becomes $\exp[(E - E_F)/kT] \rightarrow \exp(+\infty) \rightarrow +\infty$. The resulting Fermi-Dirac distribution function now becomes $f_F(E > E_F) = 0$.

The Fermi-Dirac distribution function for $T = 0°K$ is plotted in Figure 3–27. This result shows that, for $T = 0°K$, the electrons are in their lowest possible energy states. The probability of a quantum state being occupied is unity for $E < E_F$ and the probability of a state being occupied is zero for $E > E_F$. All electrons have energies below the Fermi energy at $T = 0°K$.

Figure 3–28 shows discrete energy levels of a particular system as well as the number of available quantum states at each energy. If we assume, for

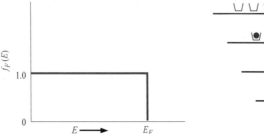

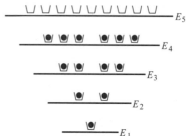

**Figure 3–27**   The Fermi probability function versus energy for $T = 0°K$.

**Figure 3–28**   Discrete energy states and quantum states for a particular system at $T = 0°K$.

this case, that the system contains 13 electrons, then Figure 3–28 shows how these electrons are distributed among the various quantum states at $T = 0°K$. The electrons will be in the lowest possible energy state so the probability of a quantum state being occupied in energy levels $E_1$–$E_4$ is unity and the probability of a quantum state being occupied in energy level $E_5$ is zero. The Fermi energy, for this case, must be above $E_4$ but less than $E_5$. The Fermi energy determines the statistical distribution of electrons and does not have to correspond to an allowed energy level.

Now consider a case in which the density of quantum states $g(E)$ is a continuous function of energy as shown in Figure 3–29. If we have $N_0$ electrons in this system, then the distribution of these electrons among the quantum states at $T = 0°K$ is shown by the dashed line. The electrons are in the lowest possible energy state so that all states below $E_F$ are filled and all

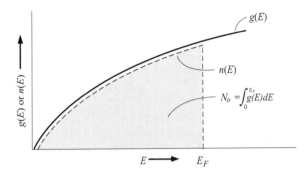

**Figure 3–29**  Density of quantum states and electrons in a continuous energy system at $T = 0°K$.

states above $E_F$ are empty. If $g(E)$ and $N_0$ are known for this particular system, then the Fermi energy $E_F$ can be determined.

Consider the situation when the temperature increases above $T = 0°K$. Electrons gain a certain amount of thermal energy so that some electrons can jump to higher energy levels, which means that the distribution of electrons among the available energy states will change. Figure 3–30 shows the

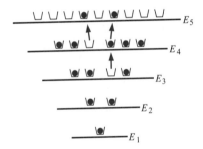

**Figure 3–30**  Discrete energy states and quantum states for the same system shown in Figure 3–28 for $T > 0°K$.

same discrete energy levels and quantum states as in Figure 3–28. The distribution of electrons among the quantum states has changed from the $T = 0°K$ case. Two electrons from the $E_4$ level have gained enough energy to jump to $E_5$ and one electron from $E_3$ has jumped to $E_4$. As the temperature changes, the distribution of electrons versus energy changes.

The change in the electron distribution among energy levels for $T > 0°K$ can be seen by plotting the Fermi-Dirac distribution function. If we let $E = E_F$ and $T > 0°K$, then Equation (3–79) becomes

$$f_F(E = E_F) = \frac{1}{1 + \exp(0)} = \frac{1}{1 + 1} = \frac{1}{2}$$

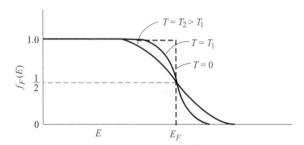

**Figure 3–31**   The Fermi probability function versus energy for different temperatures.

The probability of a state being occupied at $E = E_F$ is $\frac{1}{2}$. Figure 3–31 shows the Fermi-Dirac distribution function plotted for several temperatures, assuming the Fermi energy is independent of temperature.

We can see that for temperatures above absolute zero, there is a nonzero probability that some energy states above $E_F$ will be occupied by electrons and some energy states below $E_F$ will be empty. This result again means that some electrons have jumped to higher energy levels with increasing thermal energy.

---

**Example 3–5**

**Objective:** To calculate the probability that an energy state above $E_F$ is occupied by an electron.

Let $T = 300°K$. Determine the probability that an energy level $3kT$ above the Fermi energy is occupied by an electron.

**Solution:** From Equation (3–79), we can write

$$f_F(E) = \frac{1}{1 + \exp\left(\dfrac{E - E_F}{kT}\right)} = \frac{1}{1 + \exp\left(\dfrac{3kT}{kT}\right)}$$

which becomes

$$f_F(E) = \frac{1}{1 + 20.09} = 0.0474 = 4.74 \text{ percent}$$

**Comment:** At energies above $E_F$, the probability of a state being occupied by an electron can become significantly less than unity, or the ratio of electrons to available quantum states can be quite small.

---

We can see from Figure 3–31 that the probability of an energy above $E_F$ being occupied increases as the temperature increases and the probability of a state below $E_F$ being empty increases as the temperature increases.

**Example 3–6**

**Objective:** To determine the temperature at which there is a 1 percent probability that an energy state is empty.

Assume that the Fermi energy level for a particular material is 6.25 eV and that the electrons in this material follow the Fermi-Dirac distribution function. Calculate the temperature at which there is a 1 percent probability that a state 0.30 eV below the Fermi energy level will not contain an electron.

**Solution:** The probability that a state is empty is

$$1 - f_F(E) = 1 - \frac{1}{1 + \exp\left(\dfrac{E - E_F}{kT}\right)}$$

Then

$$0.01 = 1 - \frac{1}{1 + \exp\left(\dfrac{5.95 - 6.25}{kT}\right)}$$

Solving for $kT$, we find $kT = 0.06529$ eV so that the temperature is $T = 756°$K.

**Comment:** The Fermi probability function is a strong function of temperature.

---

We may note that the probability of a state a distance $dE$ above $E_F$ being occupied is the same as the probability of a state a distance $dE$ below $E_F$ being empty. The function $f_F(E)$ is symmetrical with the function $1 - f_F(E)$ about the Fermi energy, $E_F$. This symmetry effect is shown in Figure 3–32 and will be used in the next chapter.

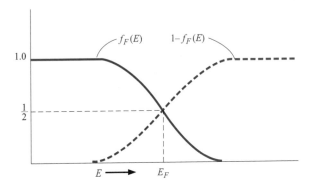

**Figure 3–32**   The probability of a state being occupied, $f_F(E)$, and the probability of a state being empty, $1 - f_F(E)$.

The Fermi-Dirac probability function determines the distribution of electrons among the available energy states. Again consider the example of the continuous density of states shown in Figure 3–29. If the temperature is not

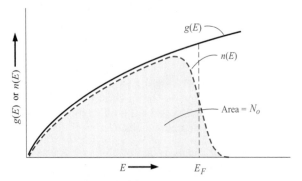

**Figure 3–33** Density of quantum states and electrons in a continuous energy system for $T > 0°K$.

too large, the Fermi energy is independent of temperature and the distribution of electrons among the quantum states for $T > 0°K$ will be as shown in Figure 3–33. Some states above $E_F$ are now occupied by electrons because electrons have gained thermal energy, while some states below $E_F$ are now empty. Note that the total number of electrons in the system remains constant, but the distribution of these electrons among the available states has changed.

Consider the case when $E - E_F >> kT$, where the exponential term in the denominator of Equation (3–79) is much greater than unity. We may neglect the 1 in the denominator, so the Fermi-Dirac distribution function becomes

$$f_F(E) \approx \exp\left[\frac{-(E - E_F)}{kT}\right] \tag{3–80}$$

Equation (3–80) is known as the Maxwell-Boltzmann approximation, or simply the Boltzmann approximation, to the Fermi-Dirac distribution function. Figure 3–34 shows the Fermi-Dirac probability function and the Boltz-

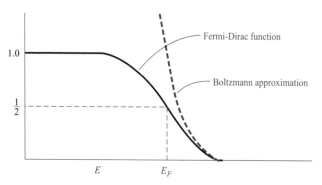

**Figure 3–34** The Fermi-Dirac probability function and the Maxwell-Boltzmann approximation.

mann approximation. This figure gives an indication of the range of energies over which the approximation is valid.

---

**Example 3–7**

**Objective:** To determine the energy at which the Boltzmann approximation may be considered valid.

Calculate the energy, in terms of $kT$ and $E_F$, at which the difference between the Boltzmann approximation and the Fermi-Dirac function is 5 percent of the Fermi function.

**Solution:** We can write

$$\frac{\exp\left[\dfrac{-(E - E_F)}{kT}\right] - \dfrac{1}{1 + \exp\left(\dfrac{E - E_F}{kT}\right)}}{\dfrac{1}{1 + \exp\left(\dfrac{E - E_F}{kT}\right)}} = 0.05$$

If we multiply both numerator and denominator by the $1 + \exp(\ )$ function, we have

$$\exp\left[\frac{-(E - E_F)}{kT}\right] \cdot \left\{1 + \exp\left[\frac{E - E_F}{kT}\right]\right\} - 1 = 0.05$$

which becomes

$$\exp\left[\frac{-(E - E_F)}{kT}\right] = 0.05$$

or

$$(E - E_F) = kT \exp\left(\frac{1}{0.05}\right) \cong 3kT$$

**Comment:** As seen in this example and in Figure 3–34, the $E - E_F \gg kT$ notation is somewhat misleading. The Maxwell-Boltzmann and Fermi-Dirac functions are within 5 percent of each other when $E - E_F \approx 3kT$.

---

The actual Boltzmann approximation is valid when $\exp[(E - E_F)/kT] \gg 1$. However, it is still common practice to use the $E - E_F \gg kT$ notation when applying the Boltzmann approximation. We will use this Boltzmann approximation in our discussion of semiconductors in the next chapter.

## 3.6 SUMMARY AND REVIEW

In this chapter, we have developed some concepts of the basic quantum theory of solids. These concepts are fundamental to the further development of our understanding of semiconductor materials and devices.

We first considered an intuitive approach to the concept of allowed and forbidden energy bands in a single-crystal material. This approach was developed using the results of the one-electron atom from the last chapter. The concept of allowed and forbidden energy bands was then developed with more mathematical rigor, using the Kronig-Penney model. This analysis gives the conditions, in terms of allowed energies, by which an electron can move through a one-dimensional periodic potential, a model of the single crystal. A concept derived from this analysis is that of effective mass. The effective mass is a parameter that allows us to relate the results of quantum mechanics to the classical force equation. Effective mass relates the motion of a particle in a crystal to an externally applied force and takes into account the effect of the crystal lattice on the motion of the particle. A portion of this analysis considered the motion of electrons near the top of the valence band. Electrons appear to move with a negative effective mass. It is convenient to define a new particle, called a hole, which corresponds to an empty electronic state near the top of the valence band. The hole has a positive electronic charge and a positive effective mass.

Energies within an allowed energy band are actually at discrete levels and each level contains a finite number of quantum states. Since only one electron is allowed per quantum state, the density of these quantum states determines the maximum number of electrons allowed in any particular energy band. An expression for the density of quantum states was derived using the three-dimensional infinite potential well as a model. This result was then generalized to the energy states in the conduction and valence bands of a semiconductor and the density of states functions, $g_c(E)$ and $g_v(E)$, were developed.

Finally, since we are dealing with large numbers of electrons and holes, we must consider the statistical behavior of these particles. The distribution of particles with respect to energy was derived and expressed as the Fermi-Dirac probability function. Associated with this function is the Fermi energy. At absolute zero degrees, all energy states below the Fermi energy are occupied by electrons and all energy states above the Fermi energy are empty of electrons. As temperature increases, the distribution of electrons changes and some electrons will occupy energy states above the Fermi level.

The concepts of quantum mechanics have been applied in this chapter to the electron in a single crystal. The concept of allowed and forbidden energy bands applies to all semiconductors and the electron effective mass will constantly appear in equations in the following chapters which describe semiconductor material and device characteristics. The concept of the hole as a positively charged particle in the valence band of a semiconductor is extremely important in that we now have two charge carriers in a semiconductor. The electron and hole will also be considered constantly throughout the remainder of the text.

The current in a semiconductor is a function of the density of electrons and holes in the crystal. These densities will be determined in the next chapter and are directly a function of the density of states and the Fermi-

Dirac probability function developed in this chapter. The quantum theory of solids is another essential step in the development of the semiconductor physics.

## GLOSSARY OF IMPORTANT TERMS

**Allowed energy band:** A band or range of energy levels that an electron in a crystal is allowed to occupy based on quantum mechanics.

**Density of states function:** The density of available quantum states as a function of energy, given in units of number per unit energy per unit volume.

**Electron effective mass:** The parameter that relates the acceleration of an electron in the conduction band of a crystal to an external force; a parameter that takes into account the effect of internal forces in the crystal.

**Fermi-Dirac probability function:** The function describing the statistical distribution of electrons among available energy states and the probability that an allowed energy state is occupied by an electron.

**Fermi energy:** In the simplest definition, the energy below which all states are filled with electrons and above which all states are empty at $T = 0°K$.

**Forbidden energy band:** A band or range of energy levels that an electron in a crystal is not allowed to occupy based on quantum mechanics.

**Hole:** The positively charged ''particle'' associated with an empty state in the top of the valence band.

**Hole effective mass:** The parameter that relates the acceleration of a hole in the valence band of a crystal to an applied external force (a positive quantity); a parameter that takes into account the effect of internal forces in a crystal.

**$k$-space diagram:** The plot of electron energy in a crystal versus $k$, where $k$ is the momentum-related constant of the motion that incorporates the crystal interaction.

**Kronig-Penney model:** The mathematical model representing a one-dimensional single-crystal lattice by a series of periodic step potential functions.

**Maxwell-Boltzmann approximation:** The condition in which the energy is several $kT$ above the Fermi energy or several $kT$ below the Fermi energy so that the Fermi-Dirac probability function can be approximated by a simple exponential function.

## PROBLEMS

### Section 3.1

1. Consider Figure 3–4, which shows the energy-band splitting of silicon. If the equilibrium lattice spacing were to change by a small

amount, discuss how you would expect the electrical properties of silicon to change. Determine at what point the material would behave like an insulator or like a metal.

2. Show that Equations (3–4) and (3–6) are derived from the Schrodinger's wave equation, using the form of solution given by Equation (3–3).

3. Show that Equations (3–9) and (3–10) are solutions of the differential equations given by Equations (3–4) and (3–8), respectively.

4. Show that Equations (3–12), (3–14), (3–16), and (3–18) result from the boundary conditions in the Kronig-Penney model.

5. Plot the function $f(\alpha a) = 9 \sin \alpha a / \alpha a + \cos \alpha a$ for $0 \leq \alpha a \leq 6\pi$. Also, given the function $f(\alpha a) = \cos ka$, indicate the allowed values of $\alpha a$ which will satisfy this equation.

6. Using the results of problem 5, determine approximately the difference in the values of $\alpha a$ between the allowed values at (a) $ka = \pi$, (b) $ka = 2\pi$, (c) $ka = 3\pi$, and (d) $ka = 4\pi$. Refer to Figure 3–8.

## Section 3.2

7. Two possible conduction bands are shown in the $E$ versus $k$ diagram given in Figure 3–35. State which band will result in the heavier electron effective mass. State why.

8. Two possible valence bands are shown in the $E$ versus $k$ diagram given in Figure 3–36. State which band will result in the heavier hole effective mass; state why.

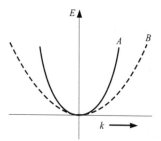

**Figure 3–35**  Conduction bands for problem 7.

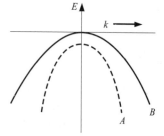

**Figure 3–36**  Valence bands for problem 8.

## Section 3.4

9. Starting with the three-dimensional infinite potential well function given by Equation (3–59) and using the separation of variables technique, derive Equation (3–60).

10. Show that Equation (3–69) can be derived from Equation (3–64).

## Section 3.5

11. Plot the Fermi-Dirac probability function, given by Equation (3–79), for $-6kT \leq (E - E_F) \leq + 6kT$.

12. Show that the probability of an energy state being occupied $\Delta E$ above the Fermi energy is the same as the probability of a state being empty $\Delta E$ below the Fermi level.

13. (a) Determine for what energy above $E_F$ (in terms of $kT$) the Fermi-Dirac probability function is within 1 percent of the Boltzmann approximation. (b) Give the value of the probability function at this energy.

14. The Fermi energy level for a particular material at $T = 300°K$ is 6.25 eV. The electrons in this material follow the Fermi-Dirac distribution function. (a) Find the probability of an energy level at 6.50 eV being occupied by an electron. (b) Repeat part (a) if the temperature is increased to $T = 950°K$. (Assume that $E_F$ is a constant.) (c) Calculate the temperature at which there is a 1 percent probability that a state 0.30 eV below the Fermi level will be empty of an electron.

15. The Fermi energy for copper at $T = 300°K$ is 7.0 eV. The electrons in copper follow the Fermi-Dirac distribution function. (a) Find the probability of an energy level at 7.15 eV being occupied by an electron. (b) Repeat part (a) for $T = 1000°K$. (Assume that $E_F$ is a constant.) (c) Repeat part (a) for $E = 6.85$ eV and $T = 300°K$. (d) Determine the probability of the energy state at $E = E_F$ being occupied at $T = 300°K$ and at $T = 1000°K$.

16. Consider the energy levels shown in Figure 3–37. Let $T = 300°K$. (a) If $E_1 - E_F = 0.30\ eV$, determine the probability that an energy state at $E = E_1$ is occupied by an electron and the probability that an energy state at $E = E_2$ is empty. (b) Repeat part (a) if $E_F - E_2 = 0.40\ eV$.

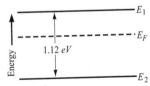

**Figure 3–37**    Energy levels for problem 16.

17. Repeat problem 16 for the case when $E_1 - E_2 = 1.42$ eV.

18. Assume the Fermi energy level is exactly in the center of the bandgap energy of a semiconductor at $T = 300°K$. (a) Calculate the probability that an energy state in the bottom of the conduction band is occupied by an electron for Si, Ge, and GaAs. (b) Calculate the probability that an energy state in the top of the valence band is empty for Si, Ge, and GaAs.

19. Calculate the temperature at which there is a $10^{-6}$ probability that an energy state 0.55 eV above the Fermi energy level is occupied by an electron.

20. Calculate the energy range (in eV) between $f_F(E) = 0.95$ and $f_F(E) = 0.05$ for $E_F = 7.0$ eV and for (a) $T = 300°$K and (b) $T = 500°$K.

## READING LIST

*1. Hess, K. *Advanced Theory of Semiconductor Devices*. Englewood Cliffs, N.J.: Prentice Hall, 1988.

2. Kittel, C. *Introduction to Solid State Physics*. 6th ed. New York: Wiley, 1986.

3. McKelvey, J. P. *Solid State and Semiconductor Physics*. New York: Harper and Row, 1966.

4. Pierret, R. F. *Semiconductor Fundamentals*. Vol. 1 of *Modular Series on Solid State Devices*. 2nd ed. Reading, Mass.: Addison-Wesley, 1988.

5. Pierret, R. F. *Advanced Semiconductor Fundamentals*. Vol. 6 of *Modular Series on Solid State Devices*. Reading, Mass.: Addison-Wesley, 1987.

*6. Shockley, W. *Electrons and Holes in Semiconductors*. New York: D. Van Nostrand, 1950.

*7. Shur, M. *Physics of Semiconductor Devices*. Englewood Cliffs, N.J.: Prentice Hall, 1990.

8. Streetman, B. G. *Solid State Electronic Devices*. 3rd ed. Englewood Cliffs, N.J.: Prentice Hall, 1990.

9. Sze, S. M. *Semiconductor Devices: Physics and Technology*. New York: Wiley, 1985.

*10. Wang, S. *Fundamentals of Semiconductor Theory and Device Physics*. Englewood Cliffs, N.J.: Prentice Hall, 1988.

## CHAPTER 4

# THE SEMICONDUCTOR IN EQUILIBRIUM

## PREVIEW

So far, we have been considering a general crystal and applying to it the concepts of quantum mechanics in order to determine a few of the characteristics of electrons in a single-crystal lattice. In this chapter, we will apply these concepts specifically to a semiconductor material. In particular, we will use the density of quantum states in the conduction band and the density of quantum states in the valence band along with the Fermi-Dirac probability function to determine the concentration of electrons and holes in the conduction and valence bands, respectively. We will also apply the concept of the Fermi energy to the semiconductor material.

We are eventually interested in determining the current-voltage characteristics, among others, of semiconductor devices. Since current is due to the net flow of charge, we need to determine the number of charge carriers available for generating the currents. We are familiar with the electron as the primary charge carrier in metallic wires. In a semiconductor, we have the electron as a charge carrier, but we also have a second, equally important type of charge carrier—the hole. This chapter will focus on electron and hole densities and will develop various terminologies associated with semiconductors.

This chapter deals with the semiconductor in equilibrium. Equilibrium, or thermal equilibrium, implies that no external forces such as voltages, electric fields, magnetic fields, or temperature gradients are acting on the semiconductor. All properties of the semiconductor will be independent of time in this case. Equilibrium is our starting point for developing the physics of the semiconductor. We will then be able to determine the characteristics that result when deviations from equilibrium occur, such as when a voltage is applied to a semiconductor device.

We will initially consider the properties of an intrinsic semiconductor; that is, a pure crystal with no impurity atoms or defects. We will see that the electrical properties of a semiconductor can be altered in desirable ways by adding controlled amounts of specific impurity atoms, called dopant atoms, to the crystal. Depending upon the type of dopant atom added, the dominant charge carrier in the semiconductor will be either electrons in the conduction band or holes in the valence band. Adding dopant atoms changes the distribution of electrons among the available energy states, so the Fermi energy becomes a function of the type and concentration of impurity atoms.

In thermal equilibrium, the semiconductor crystal is electrically neutral; that is, the net negative charge is equal to the net positive charge. This principle of charge neutrality is used to calculate the thermal-equilibrium electron and hole concentrations as a function of the impurity doping concentrations. Finally, as part of this discussion, we will attempt to add more insight into the significance of the Fermi energy.

## 4.1 CHARGE CARRIERS IN SEMICONDUCTORS

Current is the rate at which charge flows. In a semiconductor, two types of charge carrier, the electron and the hole, can contribute to a current. Since the current in a semiconductor is determined largely by the number of electrons in the conduction band and the number of holes in the valence band, an important characteristic of the semiconductor is the density of these charge carriers. The density of electrons and holes is related to the density of states function and the Fermi distribution function, both of which we have considered. A qualitative discussion of these relationships will be followed by a more rigorous mathematical derivation of the thermal-equilibrium concentration of electrons and holes.

### 4.1.1 Equilibrium Distribution of Electrons and Holes

The distribution (with respect to energy) of electrons in the conduction band is given by the density of allowed quantum states times the probability that a state will be occupied by an electron. This statement is written in equation form as

$$n(E) = g_c(E)f_F(E) \qquad (4-1)$$

where $f_F(E)$ is the Fermi-Dirac probability function and $g_c(E)$ is the density of quantum states in the conduction band. The total electron concentration per unit volume in the conduction band is then found by integrating Equation (4–1) over the entire conduction-band energy.

Similarly, the distribution (with respect to energy) of holes in the valence band is the density of allowed quantum states in the valence band multiplied by the probability that a state is not occupied by an electron. We may express this as

$$p(E) = g_v(E) \, [1 - f_F(E)] \qquad (4-2)$$

The total hole concentration per unit volume is found by integrating this function over the entire valence-band energy.

To find the thermal-equilibrium electron and hole concentrations, we need to determine the position of the Fermi energy with respect to the bottom of the conduction-band energy $E_c$ and the top of the valence-band energy $E_v$. To address this question, we will initially consider an intrinsic semiconductor. An ideal intrinsic semiconductor is a pure semiconductor with no impurity atoms and no lattice defects in the crystal (e.g., pure silicon). We have argued in the previous chapter that, for an intrinsic semiconductor at $T = 0°K$, all energy states in the valence band are filled with electrons and all energy states in the conduction band are empty of electrons. The Fermi energy must, therefore, be somewhere between $E_c$ and $E_v$. (The Fermi energy does not need to correspond to an allowed energy.)

As the temperature begins to increase above $0°K$, the valence electrons will gain thermal energy. A few electrons in the valence band may gain sufficient energy to jump to the conduction band. As an electron jumps from the valence band to the conduction band, an empty state, or hole, is created in the valence band. In an intrinsic semiconductor, then, electrons and holes are created in pairs by the thermal energy so that the number of electrons in the conduction band is equal to the number of holes in the valence band.

Figure 4–1 shows a plot of the density of states function in the conduction band $g_c(E)$, the density of states function in the valence band $g_v(E)$, and the Fermi-Dirac probability function for $T > 0°K$ when $E_F$ is approximately half way between $E_c$ and $E_v$. If we assume, for the moment, that the electron and hole effective masses are equal, then $g_c(E)$ and $g_v(E)$ are symmetrical functions about the midgap energy (the energy midway between $E_c$ and $E_v$). We noted previously that the function $f_F(E)$ for $E > E_F$ is symmetrical to the function $1 - f_F(E)$ for $E < E_F$ about the energy $E = E_F$. This also means that the function $f_F(E)$ for $E = E_F + dE$ is equal to the function $1 - f_F(E)$ for $E = E_F - dE$

Also seen in Figure 4–1 is a curve that is the product of $g_c(E)$ and $f_F(E)$. This curve shows the distribution of electrons in the conduction band, $n(E)$, which was given by Equation (4–1). The curve pictured as the product of $g_v(E)$ and $1 - f_F(E)$ shows the distribution of holes in the valence band, $p(E)$, which was given by Equation (4–2). The areas under these curves are then the total density of electrons in the conduction band and the total density of holes in the valence band. From this we see that if $g_c(E)$ and $g_v(E)$ are symmetrical, the Fermi energy must be at the midgap energy in order to obtain equal electron and hole concentrations. If the effective masses of the electron and hole are not exactly equal, then the effective density of states functions $g_c(E)$ and $g_v(E)$ will not be exactly symmetrical about the midgap energy. The Fermi level for the intrinsic semiconductor will then shift slightly from the midgap energy in order to obtain equal electron and hole concentrations.

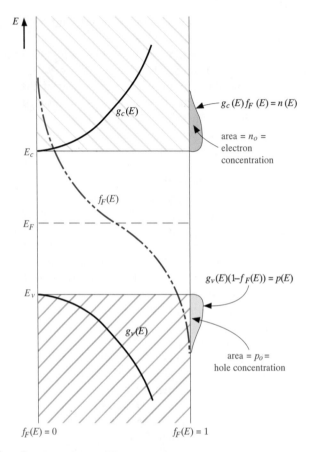

**Figure 4–1**   Density of states functions, Fermi-Dirac probability function, and areas representing electron and hole concentrations for the case when $E_F$ is near the midgap energy.

### 4.1.2  The $n_0$ and $p_0$ Equations

We have argued that the Fermi energy for an intrinsic semiconductor is near midgap. In deriving the equations for the thermal-equilibrium concentration of electrons $n_0$ and the thermal-equilibrium concentration of holes $p_0$, we will not be quite so restrictive. We will see later that, in particular situations, the Fermi energy can deviate from this midgap energy. We will assume initially, however, that the Fermi level remains within the bandgap energy.

The equation for the thermal-equilibrium concentration of electrons may be found by integrating Equation (4–1) over the conduction band energy, or

$$n_0 = \int g_c(E) f_F(E)\, dE \qquad (4-3)$$

The lower limit of integration is $E_c$ and the upper limit of integration should be the top of the allowed conduction band energy. However, since the Fermi probability function rapidly approaches zero with increasing energy as indicated in Figure 4–1, we can take the upper limit of integration to be infinity.

We are assuming that the Fermi energy is within the forbidden-energy bandgap. For electrons in the conduction band, we have $E > E_c$. If $(E_c - E_F) \gg kT$, then $(E - E_F) \gg kT$, so that the Fermi probability function reduces to the Boltzmann approximation[1], which is

$$f_F(E) = \cfrac{1}{1 + \exp \cfrac{(E - E_F)}{kT}} \cong \exp \frac{[-(E - E_F)]}{kT} \tag{4-4}$$

Applying the Boltzmann approximation to Equation (4–3), the thermal-equilibrium density of electrons in the conduction band is found from

$$n_0 = \int_{E_c}^{\infty} \frac{4\pi (2m_n^*)^{3/2}}{h^3} \sqrt{E - E_c} \exp \left[ \frac{-(E - E_F)}{kT} \right] dE \tag{4-5}$$

The integral of Equation (4–5) may be solved more easily by making a change of variable. If we let

$$\eta = \frac{E - E_c}{kT} \tag{4-6}$$

then Equation (4–5) becomes

$$n_0 = \frac{4\pi (2m_n^* kT)^{3/2}}{h^3} \exp \left[ \frac{-(E_c - E_F)}{kT} \right] \int_0^{\infty} \eta^{1/2} \exp(-\eta) d\eta \tag{4-7}$$

The integral is the gamma function, with a value of

$$\int_0^{\infty} \eta^{1/2} \exp(-\eta) d\eta = \frac{1}{2} \sqrt{\pi} \tag{4-8}$$

Then Equation (4–7) becomes

$$n_0 = 2 \left( \frac{2\pi m_n^* kT}{h^2} \right)^{3/2} \exp \left[ \frac{-(E_c - E_F)}{kT} \right] \tag{4-9}$$

We may define a parameter $N_c$ as

$$N_c = 2 \left( \frac{2\pi m_n^* kT}{h^2} \right)^{3/2} \tag{4-10}$$

so that the thermal-equilibrium electron concentration in the conduction band can be written as

---

[1] The Maxwell-Boltzmann and Fermi-Dirac distribution functions are within 5 percent of each other when $E - E_F \cong 3kT$ (see Figure 3–25). The $\gg$ notation is then somewhat misleading to indicate when the Boltzmann approximation is valid although it is commonly used.

$$n_0 = N_c \exp\left[\frac{-(E_c - E_F)}{kT}\right] \tag{4-11}$$

The parameter $N_c$ is called the effective density of states function in the conduction band. If we were to assume that $m_n^* = m_0$, then the value of the effective density of states function at $T = 300°K$ is $N_c = 2.5 \times 10^{19}$ cm$^{-3}$, which is the order of magnitude of $N_c$ for most semiconductors. If the effective mass of the electron is larger or smaller than $m_0$, then the value of the effective density of states function changes accordingly, but is still of the same order of magnitude.

The thermal-equilibrium concentration of holes in the valence band is found by integrating Equation (4–2) over the valence band energy, or

$$p_0 = \int g_v(E)[1 - f_F(E)]dE \tag{4-12}$$

We may note that

$$1 - f_F(E) = \frac{1}{1 + \exp\left(\dfrac{E_F - E}{kT}\right)} \tag{4-13a}$$

For energy states in the valence band, $E < E_v$. If $(E_F - E_v) \gg kT$ (the Fermi function is still assumed to be within the bandgap), then we have a slightly different form of the Boltzmann approximation. Equation (4–13a) may be written as

$$1 - f_F(E) = \frac{1}{1 + \exp\left(\dfrac{E_F - E}{kT}\right)} \approx \exp\left[\frac{-(E_F - E)}{kT}\right] \tag{4-13b}$$

Applying the Boltzmann approximation of Equation (4–13b) to Equation (4–12), the thermal-equilibrium concentration of holes in the valence band is

$$p_0 = \int_{-\infty}^{E_v} \frac{4\pi(2m_p^*)^{3/2}}{h^3} \sqrt{E_v - E} \exp\left[\frac{-(E_F - E)}{kT}\right] dE \tag{4-14}$$

where the lower limit of integration is taken as minus infinity instead of the bottom of the valence band. The exponential term decays fast enough so that this approximation is valid.

Equation (4–14) may be solved more easily by again making a change of variable. If we let

$$\eta' = \frac{E_v - E}{kT} \tag{4-15}$$

then Equation (4–14) becomes

$$p_0 = \frac{-4\pi(2m_p^*)^{3/2}}{h^3} \exp\left[\frac{-(E_F - E_v)}{kT}\right] \int_{+\infty}^{0} (\eta')^{1/2} \exp(-\eta')d\eta' \tag{4-16}$$

where the negative sign comes from the differential $dE = -kTd\eta'$. Note that the lower limit of $\eta'$ becomes $+\infty$ when $E = -\infty$. If we change the order of integration, we introduce another minus sign. Using Equation (4–8), Equation (4–16) becomes

$$p_0 = 2\left(\frac{2\pi m_p^* kT}{h^2}\right)^{3/2} \exp\left[\frac{-(E_F - E_v)}{kT}\right] \quad (4\text{–}17)$$

We may define a parameter $N_v$ as

$$N_v = 2\left(\frac{2\pi m_p^* kT}{h^2}\right)^{3/2} \quad (4\text{–}18)$$

which is called the effective density of states function in the valence band. The thermal-equilibrium concentration of holes in the valence band may now be written as

$$p_0 = N_v \exp\left[\frac{-(E_F - E_v)}{kT}\right] \quad (4\text{–}19)$$

The magnitude of $N_v$ is also on the order of $10^{19}$ cm$^{-3}$ at $T = 300°$K for most semiconductors.

The effective density of states functions, $N_c$ and $N_v$, are constant for a given semiconductor material at a fixed temperature. The values of the density of states function and of the effective masses for silicon, gallium arsenide, and germanium are given in Table 4–1. We may note that the value of $N_c$ for gallium arsenide is smaller than the typical $10^{19}$ cm$^{-3}$ value. This difference is due to the small electron effective mass in gallium arsenide.

**Table 4–1**  Effective density of states function and effective mass values

|  | $N_c$(cm$^{-3}$) | $N_v$(cm$^{-3}$) | $m_n^*/m_0$ | $m_p^*/m_0$ |
|---|---|---|---|---|
| Silicon | $2.8 \times 10^{19}$ | $1.04 \times 10^{19}$ | 1.08 | 0.56 |
| Gallium arsenide | $4.7 \times 10^{17}$ | $7.0 \times 10^{18}$ | 0.067 | 0.48 |
| Germanium | $1.04 \times 10^{19}$ | $6.0 \times 10^{18}$ | 0.55 | 0.37 |

The thermal equilibrium concentration both of electrons in the conduction band and of holes in the valence band is directly related to the effective density of states constants and to the Fermi energy level.

### 4.1.3 The Intrinsic Carrier Concentration

For an intrinsic semiconductor, the concentration of electrons in the conduction band is equal to the concentration of holes in the valence band. We may denote $n_i$ and $p_i$ as the electron and hole concentrations, respectively, in the intrinsic semiconductor. These parameters are usually referred to as the intrinsic electron concentration and intrinsic hole concentration. However, $n_i = p_i$, so normally we simply use the parameter $n_i$ as the intrinsic carrier concentration, which refers to either the intrinsic electron or hole concentration.

The Fermi energy level for the intrinsic semiconductor is called the intrinsic Fermi energy, or $E_F = E_{Fi}$. If we apply Equations (4–11) and (4–19) to the intrinsic semiconductor, then we can write

$$n_0 = n_i = N_c \exp\left[\frac{-(E_c - E_{Fi})}{kT}\right] \qquad (4\text{--}20)$$

and

$$p_0 = p_i = n_i = N_v \exp\left[\frac{-(E_{Fi} - E_v)}{kT}\right] \qquad (4\text{--}21)$$

If we take the product of Equations (4–20) and (4–21), we obtain

$$n_i^2 = N_c N_v \exp\left[\frac{-(E_c - E_{Fi})}{kT}\right] \cdot \exp\left[\frac{-(E_{Fi} - E_v)}{kT}\right] \qquad (4\text{--}22)$$

or

$$n_i^2 = N_c N_v \exp\left[\frac{-(E_c - E_v)}{kT}\right] = N_c N_v \exp\left[\frac{-E_g}{kT}\right] \qquad (4\text{--}23)$$

where $E_g$ is the bandgap energy. For a given semiconductor material at a constant temperature, the value of $n_i$ is a constant, and independent of the Fermi energy.

The intrinsic carrier concentration for silicon at $T = 300°K$ may be calculated using the effective density of states function values from Table 4–1. The value of $n_i$ calculated from Equation (4–23) for $E_g = 1.12$ eV is $n_i = 6.95 \times 10^9$ cm$^{-3}$. The commonly accepted value of $n_i$ for silicon at $T = 300°K$ is approximately $1.5 \times 10^{10}$ cm$^{-3}$. This discrepancy may arise from several sources. First, the values of the effective masses are determined at a low temperature where the cyclotron resonance experiments are performed. Since the effective mass is an experimentally determined parameter, and since the effective mass is a measure of how well a particle moves in a crystal, this parameter may be a slight function of temperature. Next, the density of states function for a semiconductor was obtained by generalizing the model of an electron in a three-dimensional infinite potential well. This theoretical function may also not agree exactly with experiment. However, the difference between the theoretical value and the experimental value of $n_i$ is approximately a factor of two, which, in many cases, is not significant. Table 4–2 lists the commonly accepted values of $n_i$ for silicon, gallium arsenide, and germanium at $T = 300°K$.

Table 4–2   Commonly accepted values of $n_i$ at $T = 300°K$

| | |
|---|---|
| Silicon | $n_i = 1.5 \times 10^{10}$ cm$^{-3}$ |
| Gallium arsenide | $n_i = 1.8 \times 10^6$ cm$^{-3}$ |
| Germanium | $n_i = 2.4 \times 10^{13}$ cm$^{-3}$ |

The intrinsic carrier concentration is a very strong function of temperature.

---

**Example 4–1**

**Objective:** To calculate the intrinsic carrier concentration in gallium arsenide at $T = 300°K$ and at $T = 450°K$.

The values of $N_c$ and $N_v$ at 300°K for gallium arsenide are $4.7 \times 10^{17}$ cm$^{-3}$ and $7.0 \times 10^{18}$ cm$^{-3}$, respectively. Both $N_c$ and $N_v$ vary as $T^{3/2}$. Assume the bandgap energy of gallium arsenide is 1.42 eV and does not vary with temperature over this range. The value of $kT$ at 450°K is

$$kT = (0.0259)\left(\frac{450}{300}\right) = 0.03885 \text{ eV}$$

**Solution:** Using Equation (4–23), we have for $T = 300°K$

$$n_i^2 = (4.7 \times 10^{17})(7.0 \times 10^{18}) \exp\left(\frac{-1.42}{0.0259}\right) = 5.09 \times 10^{12}$$

so that

$$n_i = 2.26 \times 10^6 \text{ cm}^{-3}$$

At $T = 450°K$, we have

$$n_i^2 = (4.7 \times 10^{17})(7.0 \times 10^{18})\left(\frac{450}{300}\right)^3 \exp\left(\frac{-1.42}{0.03885}\right) = 1.48 \times 10^{21}$$

so that

$$n_i = 3.85 \times 10^{10} \text{ cm}^{-3}$$

**Comment:** We may note from this example that the intrinsic carrier concentration increased by over four orders of magnitude as the temperature increased by 150°C.

---

Figure 4–2 is a plot of $n_i$ from Equation (4–23) for silicon, gallium arsenide, and germanium as a function of temperature. As seen in the figure, the value of $n_i$ for these semiconductors may easily vary over several orders of magnitude as the temperature changes over a reasonable range.

### 4.1.4 The Intrinsic Fermi-Level Position

We have qualitatively argued that the Fermi energy level is located near the center of the forbidden bandgap for the intrinsic semiconductor. We can specifically calculate the intrinsic Fermi-level position. Since the electron and hole concentrations are equal, setting Equations (4–20) and (4–21) equal to each other, we have

$$N_c \exp\left[\frac{-(E_c - E_{Fi})}{kT}\right] = N_v \exp\left[\frac{-(E_{Fi} - E_v)}{kT}\right] \tag{4–24}$$

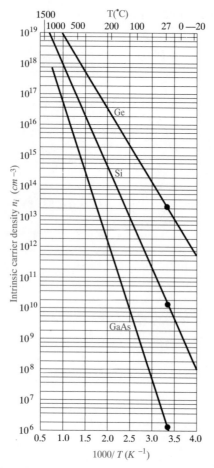

**Figure 4–2**   The intrinsic carrier concentration of Ge, Si, and GaAs as a function of temperature. (From Sze [9]).

If we take the natural log of both sides of this equation and solve for $E_{Fi}$, we obtain

$$E_{Fi} = \frac{1}{2}(E_c + E_v) + \frac{1}{2}kT\ln\left(\frac{N_v}{N_c}\right) \qquad (4\text{–}25)$$

From the definitions for $N_c$ and $N_v$ given by Equations (4–10) and (4–18), respectively, Equation (4–25) may be written as

$$E_{Fi} = \frac{1}{2}(E_c + E_v) + \frac{3}{4}kT\ln\left(\frac{m_p^*}{m_n^*}\right) \qquad (4\text{–}26a)$$

The first term, $\frac{1}{2}(E_c + E_v)$, is the energy exactly midway between $E_c$ and $E_v$, or the midgap energy. We can define

$$\frac{1}{2}(E_c + E_v) = E_{\text{midgap}}$$

so that

$$E_{Fi} - E_{\text{midgap}} = \frac{3}{4} kT \ln \left( \frac{m_p^*}{m_n^*} \right) \qquad (4\text{–}26b)$$

If we have a semiconductor in which $m_p^* = m_n^*$, then the intrinsic Fermi level would be exactly in the center of the bandgap. If $m_p^* > m_n^*$, the intrinsic Fermi level would be slightly above the center, and if $m_p^* < m_n^*$, it would be slightly below the center of the bandgap. The density of states function is directly related to the carrier effective mass; thus a larger effective mass would mean a larger density of states function. The intrinsic Fermi level must shift away from the band with the larger density of states in order to maintain equal numbers of electrons and holes.

---

**Example 4–2**

**Objective:** To calculate the position of the intrinsic Fermi level with respect to the center of the bandgap in silicon at $T = 300°K$.

   The density of states effective carrier masses in silicon are $m_n^* = 1.08m_0$ and $m_p^* = 0.56m_0$.

**Solution:** The intrinsic Fermi level with respect to the center of the bandgap is

$$E_{Fi} - E_{\text{midgap}} = \frac{3}{4} kT \ln \left( \frac{m_p^*}{m_n^*} \right) = \frac{3}{4} (0.0259) \ln \left( \frac{0.56}{1.08} \right)$$

or

$$E_{Fi} - E_{\text{midgap}} = -0.0128 \text{ eV} = -12.8 \text{ meV}$$

**Comment:** The intrinsic Fermi level in silicon is 12.8 meV below the midgap energy. If we compare 12.8 meV to 560 meV, which is one half of the bandgap energy of silicon, we can, in many applications, simply approximate the intrinsic Fermi level to be in the center of the bandgap.

---

## 4.2 DOPANT ATOMS AND ENERGY LEVELS

The intrinsic semiconductor may be an interesting material, but the real power of semiconductors is realized by adding small, controlled amounts of specific dopant, or impurity, atoms. This doping process, described briefly in Chapter 1, can greatly alter the electrical characteristics of the semiconductor. The doped semiconductor is called an extrinsic material, and is the primary reason we can fabricate the various semiconductor devices that we will consider in later chapters.

### 4.2.1 Qualitative Description

In Chapter 3, we discussed the covalent bonding of silicon and considered the simple two-dimensional representation of the single-crystal silicon lattice as shown in Figure 4–3. Now consider adding a group V element, such as

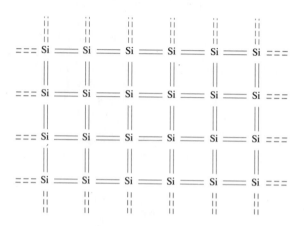

**Figure 4–3**   Two-dimensional representation of the intrinsic silicon lattice.

phosphorus, as a substitutional impurity. The group V element has five valence electrons. Four of these will contribute to the covalent bonding with the silicon atoms, leaving the fifth more loosely bound to the phosphorus atom. This effect is schematically shown in Figure 4–4. We will refer to the fifth valence electron as a donor electron.

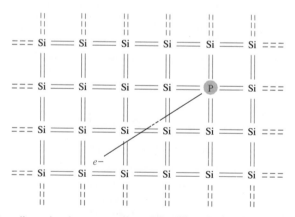

**Figure 4–4**   Two-dimensional representation of the silicon lattice doped with a phosphorus atom.

The phosphorus atom without the donor electron would be positively charged. At very low temperatures, the donor electron is bound to the phosphorus atom. However, by intuition, it should seem clear that the energy required to elevate the donor electron into the conduction band is considerably less than that for the electrons involved in the covalent bonding. Figure 4–5 shows the energy-band diagram that we would expect. The energy level, $E_d$, is the energy state of the donor electron.

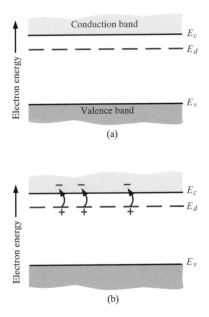

**Figure 4–5** The energy-band diagram showing (a) the discrete donor energy state and (b) the effect of a donor state being ionized.

If a small amount of energy, such as thermal energy, is added to the donor electron, it can be elevated into the conduction band leaving behind a positively charged phosphorus ion. The electron in the conduction band can now move through the crystal generating a current, while the positively charged ion is fixed in the crystal. This type of impurity atom donates an electron to the conduction band and so is called a donor impurity atom. The donor impurity atoms add electrons to the conduction band without creating holes in the valence band. The resulting material is referred to as an n-type semiconductor (n for the negatively charged electron).

Now consider adding a group III element, such as boron, as a substitutional impurity to silicon. The group III element has three valence electrons, which are all taken up in the covalent bonding. As shown in Figure 4–6a, one covalent bonding position appears to be empty. If an electron were to occupy this "empty" position, its energy would have to be greater than that of

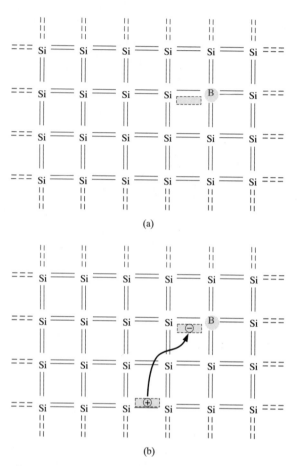

**Figure 4–6**   Two-dimensional representation of a silicon lattice (a) doped with a boron atom and (b) showing the ionization of the boron atom resulting in a hole.

the valence electrons, since the net charge state of the boron atom would now be negative. However, the electron occupying this "empty" position does not have sufficient energy to be in the conduction band, so its energy is far smaller than the conduction-band energy. Figure 4–6b shows how valence electrons may gain a small amount of thermal energy and move about in the crystal. The "empty" position associated with the boron atom becomes occupied, and other valence electron positions become vacated. These other vacated electron positions can be thought of as holes in the semiconductor material.

Figure 4–7 shows the expected energy state of the "empty" position and also the formation of a hole in the valence band. The hole can move through the crystal generating a current, while the negatively charged boron atom is fixed in the crystal. The group III atom accepts an electron from the valence band and so is referred to as an acceptor impurity atom. The acceptor atom can generate holes in the valence band without generating electrons in the

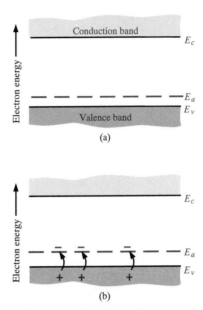

**Figure 4–7** The energy-band diagram showing (a) the discrete acceptor energy state and (b) the effect of an acceptor state being ionized.

conduction band. This type of semiconductor material is referred to as a p-type material (p for the positively charged hole).

The pure single-crystal semiconductor material is called an intrinsic material. Adding controlled amounts of dopant atoms, either donors or acceptors, creates a material called an extrinsic semiconductor. An extrinsic semiconductor will have either a preponderance of electrons (n-type) or a preponderance of holes (p-type).

### 4.2.2 Ionization Energy

We can calculate the approximate distance of the donor electron from the donor impurity ion, and also the approximate energy required to elevate the donor electron into the conduction band. This energy is referred to as the ionization energy. We will use the Bohr model of the atom for these calculations. The justification for using this model is that the most probable distance of an electron from the nucleus in a hydrogen atom, determined from quantum mechanics, is the same as the Bohr radius. The energy levels in the hydrogen atom determined from quantum mechanics are also the same as obtained from the Bohr theory.

In the case of the donor impurity atom, we may visualize the donor electron orbiting the donor ion which is embedded in the semiconductor material. We will need to use the permittivity of the semiconductor material in the calculations rather than the permittivity of free space as was used in the case of the hydrogen atom. We will also use the effective mass of the electron in the calculations.

The analysis begins by setting the coulomb force of attraction between the electron and ion equal to the centripetal force of the orbiting electron. This condition will give a steady orbit. We have

$$\frac{e^2}{4\pi\varepsilon r_n^2} = \frac{m^* v^2}{r_n} \tag{4-27}$$

where $v$ is the magnitude of the velocity and $r_n$ is the radius of the orbit. If we assume the angular momentum is also quantized, then we can write

$$m^* r_n v = n\hbar \tag{4-28}$$

where $n$ is a positive integer. Solving for $v$ from Equation (4-28), substituting into Equation (4-27), and solving for the radius, we obtain

$$r_n = \frac{n^2 \hbar^2 4\pi\varepsilon}{m^* e^2} \tag{4-29}$$

The assumption of the angular momentum being quantized leads to the radius also being quantized.

The Bohr radius is defined as

$$a_0 = \frac{4\pi\varepsilon_0 \hbar^2}{m_0 e^2} = 0.53\text{Å} \tag{4-30}$$

We can normalize the radius of the donor orbital to that of the Bohr radius, which gives

$$\frac{r_n}{a_0} = n^2 \varepsilon_r \left(\frac{m_0}{m^*}\right) \tag{4-31}$$

where $\varepsilon_r$ is the relative dielectric constant of the semiconductor material, $m_0$ is the rest mass of an electron, and $m^*$ is the conductivity effective mass of the electron in the semiconductor.

If we consider the lowest energy state in which n = 1, and if we consider silicon in which $\varepsilon_r = 11.7$ and the conductivity effective mass is $m^*/m_0 = 0.26$, then we have that

$$\frac{r_1}{a_0} = 45 \tag{4-32}$$

or $r_1 = 23.9\text{Å}$. This radius corresponds to approximately four lattice constants of silicon. Recall that one unit cell in silicon effectively contains eight atoms, so the radius of the orbiting donor electron encompasses many silicon atoms. The donor electron is not tightly bound to the donor atom.

The total energy of the orbiting electron is given by

$$E = T + V \tag{4-33}$$

where $T$ is the kinetic energy and $V$ is the potential energy of the electron. The kinetic energy is

$$T = \tfrac{1}{2} m^* v^2 \tag{4-34}$$

Using the velocity $v$ from Equation (4–28) and the radius $r_n$ from Equation (4–29), the kinetic energy becomes

$$T = \frac{m^*e^4}{2(n\hbar)^2(4\pi\varepsilon)^2} \tag{4-35}$$

The potential energy is

$$V = \frac{-e^2}{4\pi\varepsilon r_n} = \frac{-m^*e^4}{(n\hbar)^2(4\pi\varepsilon)^2} \tag{4-36}$$

The total energy is the sum of the kinetic and potential energies, so that

$$E = T + V = \frac{-m^*e^4}{2(n\hbar)^2(4\pi\varepsilon)^2} \tag{4-37}$$

For the hydrogen atom, $m^* = m_0$ and $\varepsilon = \varepsilon_0$. The ionization energy of the hydrogen atom in the lowest energy state is then $E = -13.6$ eV. If we consider silicon, the ionization energy is $E = -25.8$ meV, much less than the bandgap energy of silicon. This energy is the approximate ionization energy of the donor atom, or the energy required to elevate the donor electron into the conduction band.

For ordinary donor impurities such as phosphorus or arsenic in silicon or germanium, this hydrogenic model works quite well and gives some indication of the magnitudes of the ionization energies involved. Table 4–3 gives

Table 4–3  Impurity ionization energies in silicon and germanium

| Impurity | Semiconductor | |
|---|---|---|
| | Si | Ge |
| **Donors** | | |
| Phosphorus | 0.045 eV | 0.012 eV |
| Arsenic | 0.05 | 0.0127 |
| **Acceptors** | | |
| Boron | 0.045 | 0.0104 |
| Aluminum | 0.06 | 0.0102 |

the actual experimentally measured ionization energies for a few impurities in silicon and germanium. Germanium and silicon have different relative dielectric constants and effective masses; thus we would expect the ionization energies to differ.

### 4.2.3 Group III-V Semiconductors

In the previous sections, we have been discussing the donor and acceptor impurities in a group IV semiconductor, such as silicon. The situation in the group III-V compound semiconductors, such as gallium arsenide, is more complicated. Group II elements, such as beryllium, zinc, and cadmium, can enter the lattice as substitutional impurities, replacing the group III gallium element to become acceptor impurities. Similarly, group VI elements, such as selenium and tellurium, can enter the lattice substitutionally replacing the group V arsenic element to become donor impurities. The corresponding ionization energies for these impurities are smaller than for the impurities in silicon. The ionization energies for the donors in gallium arsenide are also smaller than the ionization energies for the acceptors, because of the smaller effective mass of the electron compared to that of the hole.

Group IV elements, such as silicon and germanium, can also be impurity atoms in gallium arsenide. If a silicon atom replaces a gallium atom, the silicon impurity will act as a donor, but if the silicon atom replaces an arsenic atom, then the silicon impurity will act as an acceptor. The same is true for germanium as an impurity atom. Such impurities are called amphoteric. Experimentally in gallium arsenide, it is found that germanium is predominantly an acceptor and silicon is predominantly a donor. Table 4–4 lists the ionization energies for the various impurity atoms in gallium arsenide.

**Table 4–4**  Impurity ionization energies in gallium arsenide

| Donor impurity | Ionization energy (eV) |
|---|---|
| Selenium | 0.0059 |
| Tellurium | 0.0058 |
| Silicon | 0.0058 |
| Germanium | 0.0061 |
| **Acceptor impurity** | |
| Beryllium | 0.028 |
| Zinc | 0.0307 |
| Cadmium | 0.0347 |
| Silicon | 0.0345 |
| Germanium | 0.0404 |

## 4.3  THE EXTRINSIC SEMICONDUCTOR

We defined an intrinsic semiconductor as a material with no impurity atoms present in the crystal. An extrinsic semiconductor is defined as a semiconductor in which controlled amounts of specific dopant or impurity atoms have been added so that the thermal-equilibrium electron and hole concentrations are different from the intrinsic carrier concentration. One type of carrier will predominate in an extrinsic semiconductor.

### 4.3.1  Equilibrium Distribution of Electrons and Holes

Adding donor or acceptor impurity atoms to a semiconductor will change the distribution of electrons and holes in the material. Since the Fermi energy is related to the distribution function, the Fermi energy will change as dopant atoms are added. If the Fermi energy changes from near the midgap value, the density of electrons in the conduction band and the density of holes in the valence band will change. These effects are shown in Figures 4–8 and 4–9. Figure 4–8 shows the case for $E_F > E_{midgap}$ and Figure 4–9 shows the case for $E < E_{midgap}$. When $E_F > E_{midgap}$, the electron concentration is larger than the hole concentration, and when $E < E_{midgap}$, the hole concentration is

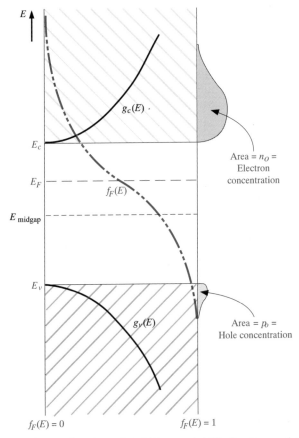

**Figure 4–8** Density of states functions, Fermi-Dirac probability function, and areas representing electron and hole concentrations for the case when $E_F$ is above the midgap energy.

larger than the electron concentration. When the density of electrons is greater than the density of holes, the semiconductor is n-type; donor impurity atoms have been added. When the density of holes is greater than the density of electrons, the semiconductor is p-type; acceptor impurity atoms have been added. The Fermi energy level in a semiconductor changes as the electron and hole concentrations change and, again, the Fermi energy changes as donor or acceptor impurities are added. The change in the Fermi level as a function of impurity concentrations will be considered in Section 4.6.

The expressions previously derived for the thermal-equilibrium concentration of electrons and holes, given by Equations (4–11) and (4–19), are general equations for $n_0$ and $p_0$ in terms of the Fermi energy. These equations are again given as

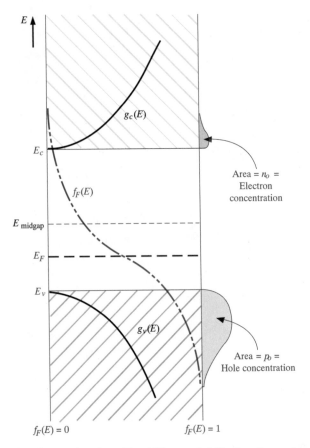

**Figure 4–9**  Density of states functions, Fermi-Dirac probability function, and areas representing electron and hole concentrations for the case when $E_F$ is below the midgap energy.

$$n_0 = N_c \exp\left[\frac{-(E_c - E_F)}{kT}\right]$$

and

$$p_0 = N_v \exp\left[\frac{-(E_F - E_v)}{kT}\right]$$

As we just discussed, the Fermi energy may vary through the bandgap energy, which will then change the values of $n_0$ and $p_0$.

---

### Example 4–3

**Objective:** To calculate the thermal equilibrium concentrations of electrons and holes for a given Fermi energy.

Consider silicon at $T = 300°K$ so that $N_c = 2.8 \times 10^{19}$ cm$^{-3}$ and $N_v = 1.04 \times 10^{19}$ cm$^{-3}$. Assume that the Fermi energy is 0.25 eV below the conduction band. If we assume that the bandgap energy of silicon is 1.12 eV, then the Fermi energy will be 0.87 eV above the valence band.

**Solution:** Using Equation (4–11), we have

$$n_0 = (2.8 \times 10^{19}) \exp \left( \frac{-0.25}{0.0259} \right) = 1.8 \times 10^{15} \text{ cm}^{-3}$$

From Equation (4–19), we can write

$$p_0 = (1.04 \times 10^{19}) \exp \left( \frac{-0.87}{0.0259} \right) = 2.7 \times 10^{4} \text{ cm}^{-3}$$

**Comment:** The change in the Fermi level is actually a function of the donor or acceptor impurity concentrations which are added to the semiconductor. However, this example shows that electron and hole concentrations change by orders of magnitude from the intrinsic carrier concentration as the Fermi energy changes by a few tenths of an electron-volt.

---

In this example, since $n_0 > p_0$, the semiconductor is n-type. In an n-type semiconductor, electrons are referred to as the majority carrier and holes as the minority carrier. By comparing the relative values of $n_0$ and $p_0$ in the example, it is easy to see how this designation came about. Similarly, in a p-type semiconductor where $p_0 > n_0$, holes are the majority carrier and electrons are the minority carrier.

We may derive another form of the equations for the thermal-equilibrium concentrations of electrons and holes. If we add and subtract an intrinsic Fermi energy in the exponent of Equation (4–11), we can write

$$n_0 = N_c \exp \left[ \frac{-(E_c - E_{Fi}) + (E_F - E_{Fi})}{kT} \right] \tag{4–38a}$$

or

$$n_0 = N_c \exp \left[ \frac{-(E_c - E_{Fi})}{kT} \right] \exp \left[ \frac{(E_F - E_{Fi})}{kT} \right] \tag{4–38b}$$

The intrinsic carrier concentration is given by Equation (4–20) as

$$n_i = N_c \exp \left[ \frac{-(E_c - E_{Fi})}{kT} \right]$$

so that the thermal-equilibrium electron concentration can be written as

$$n_0 = n_i \exp \left[ \frac{E_F - E_{Fi}}{kT} \right] \tag{4–39}$$

Similarly, if we add and subtract an intrinsic Fermi energy in the exponent of Equation (4–19), we will obtain

$$p_0 = n_i \exp \left[ \frac{-(E_F - E_{Fi})}{kT} \right] \tag{4-40}$$

As we will see, the Fermi level changes when donors and acceptors are added, but equations (4–39) and (4–40) show that, as the Fermi level changes from the intrinsic Fermi level, $n_0$ and $p_0$ change from the $n_i$ value. If $E_F > E_{Fi}$, then we will have $n_0 > n_i$ and $p_0 < n_i$. One characteristic of an n-type semiconductor is that $E_F > E_{Fi}$ so that $n_0 > p_0$. Similarly, in a p-type semiconductor, $E_F < E_{Fi}$ so that $p_0 > n_i$ and $n_0 < n_i$; thus $p_0 > n_0$.

The functional dependence of $n_0$ and $p_0$ with $E_F$ can be seen in Figures 4–8 and 4–9. As $E_F$ moves above or below $E_{Fi}$, the overlapping probability function with the density of states functions in the conduction band and valence band changes. As $E_F$ moves above $E_{Fi}$, the probability function in the conduction band increases, while the probability, $1 - f_F(E)$, of an empty state (hole) in the valence band decreases. As $E_F$ moves below $E_{Fi}$, the opposite occurs.

### 4.3.2 The $n_0 p_0$ Product

We may take the product of the general expressions for $n_0$ and $p_0$ as given in Equations (4–11) and (4–19), respectively. The result is

$$n_0 p_0 = N_c N_v \exp \left[ \frac{-(E_c - E_F)}{kT} \right] \exp \left[ \frac{-(E_F - E_v)}{kT} \right] \tag{4-41}$$

which may be written as

$$n_0 p_0 = N_c N_v \exp \left[ \frac{-E_g}{kT} \right] \tag{4-42}$$

As Equation (4–42) was derived for a general value of Fermi energy, the values of $n_0$ and $p_0$ are not necessarily equal. However, Equation (4–42) is exactly the same as Equation (4–23), which we derived for the case of an intrinsic semiconductor. We then have that, for the semiconductor in thermal equilibrium,

$$n_0 p_0 = n_i^2 \tag{4-43}$$

Equation (4–43) states that the product of $n_0$ and $p_0$ is always a constant for a given semiconductor material at a given temperature. Although this equation seems very simple, it is one of the fundamental principles of semiconductors in thermal equilibrium. The significance of this relation will become more apparent in the chapters that follow. It is important to keep in mind that Equation (4–43) was derived using the Boltzmann approximation. If the Boltzmann approximation is not valid, then likewise, Equation (4–43) is not valid.

An extrinsic semiconductor in thermal equilibrium does not, strictly speaking, contain an intrinsic carrier concentration, although some thermally generated carriers are present. The intrinsic electron and hole carrier concentrations are modified by the donor or acceptor impurities. However, we may think of the intrinsic concentration $n_i$ in Equation (4–43) simply as a parameter of the semiconductor material.

### *4.3.3 The Fermi-Dirac Integral

In the derivation of the Equations (4–11) and (4–19) for the thermal equilibrium electron and hole concentrations, we assumed that the Boltzmann approximation was valid. If the Boltzmann approximation does not hold, the thermal equilibrium electron concentration is written from Equation (4–3) as

$$n_0 = \frac{4\pi}{h^3}(2m_n^*)^{3/2}\int_{E_c}^{\infty}\frac{(E - E_c)^{1/2}dE}{1 + \exp\left(\dfrac{E - E_F}{kT}\right)} \tag{4–44}$$

If we again make a change of variable and let

$$\eta = \frac{E - E_c}{kT} \tag{4–45a}$$

and also define

$$\eta_F = \frac{E_F - E_c}{kT} \tag{4–45b}$$

then Equation (4–44) can be written as

$$n_0 = 4\pi\left(\frac{2m_n^* kT}{h^2}\right)^{3/2}\int_0^{\infty}\frac{\eta^{1/2}d\eta}{1 + \exp(\eta - \eta_F)} \tag{4–46}$$

The integral is defined as

$$F_{1/2}(\eta_F) = \int_0^{\infty}\frac{\eta^{1/2}d\eta}{1 + \exp(\eta - \eta_F)} \tag{4–47}$$

This function, called the Fermi-Dirac integral, is a tabulated function of the variable $\eta_F$. Figure 4–10 is a plot of the Fermi-Dirac integral. Note that if $\eta_F > 0$, then $E_F > E_c$; thus the Fermi energy is actually in the conduction band.

---

### Example 4–4

**Objective:** To calculate the electron concentration using the Fermi-Dirac integral.

Let $\eta_F = 2$ so that the Fermi energy is above the conduction band by approximately 52 meV at $T = 300°K$.

**Solution:** Equation (4–46) can be written as

$$n_0 = \frac{2}{\sqrt{\pi}}N_c F_{1/2}(\eta_F)$$

For silicon at 300°K, $N_c = 2.8 \times 10^{19}$ cm$^{-3}$ and from Figure 4–10, the Fermi-Dirac integral has a value of $F_{1/2}(2) = 2.3$. Then

$$n_0 = \frac{2}{\sqrt{\pi}}\,(2.8 \times 10^{19})(2.3) = 7.27 \times 10^{19}\ \text{cm}^{-3}$$

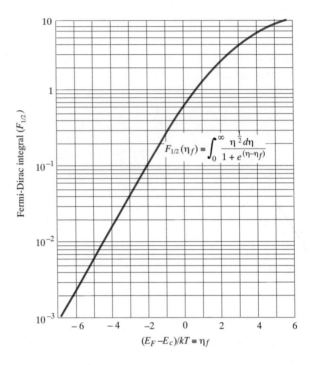

Figure 4–10   The Fermi-Dirac integral $F_{1/2}$ as a function of the Fermi energy. (From Sze [9]).

**Comment:** Note that if we had used Equation (4–11), the thermal equilibrium value of $n_0$ would be $n_0 = 2.08 \times 10^{20}$ cm$^{-3}$, which is incorrect since the Boltzmann approximation is not valid for this case.

We may use the same general method to calculate the thermal equilibrium concentration of holes. We obtain

$$p_0 = 4\pi \left(\frac{2m_p^* kT}{h^2}\right)^{3/2} \int_0^\infty \frac{(\eta')^{1/2}d\eta'}{1 + \exp(\eta' - \eta'_F)} \tag{4–48}$$

where

$$\eta' = \frac{E_v - E}{kT} \tag{4-49a}$$

and

$$\eta'_F = \frac{E_v - E_F}{kT} \tag{4-49b}$$

The integral in Equation (4–48) is the same Fermi-Dirac integral defined by Equation (4–47), although the variables have slightly different definitions. We may note that if $\eta'_F > 0$, then the Fermi level is in the valence band.

## 4.3.4 Degenerate and Nondegenerate Semiconductors

In our discussion of adding dopant atoms to a semiconductor, we have implicitly assumed that the concentration of dopant atoms added is small when compared to the density of host or semiconductor atoms. The small number of impurity atoms are spread far enough apart so that there is no interaction between donor electrons, for example, in an n-type material. We have assumed that the impurities introduce discrete, noninteracting donor energy states in the n-type semiconductor and discrete, noninteracting acceptor states in the p-type semiconductor. These types of semiconductors are referred to as nondegenerate semiconductors.

If the impurity concentration increases, the distance between the impurity atoms decreases and a point will be reached when donor electrons, for example, will begin to interact with each other. When this occurs, the single discrete donor energy will split into a band of energies. As the donor concentration further increases, the band of donor states widens and may overlap the bottom of the conduction band. This overlap occurs when the donor concentration becomes comparable with the effective density of states. When the concentration of electrons in the conduction band exceeds the density of states $N_c$, the Fermi energy lies within the conduction band. This type of semiconductor is called a degenerate n-type semiconductor.

In a similar way, as the acceptor doping concentration increases in a p-type semiconductor, the discrete acceptor energy states will split into a band of energies and may overlap the top of the valence band. The Fermi energy will lie in the valence band when the concentration of holes exceeds the density of states $N_v$. This type of semiconductor is called a degenerate p-type semiconductor.

A schematic model of the energy-band diagrams for a degenerate n-type and degenerate p-type semiconductor are shown in Figure 4–11. The energy states below $E_F$ are mostly filled with electrons and the energy states above $E_F$ are mostly empty. In the degenerate n-type semiconductor, the states between $E_F$ and $E_c$ are mostly filled with electrons, thus the electron concentration in the conduction band is very large. Similarly, in the degenerate p-type semiconductor, the energy states between $E_v$ and $E_F$ are mostly empty; thus the hole concentration in the valence band is very large.

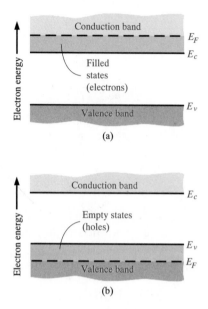

**Figure 4-11**   Simplified energy-band diagrams for degenerately doped (a) n-type and (b) p-type semiconductors.

## 4.4 STATISTICS OF DONORS AND ACCEPTORS

In the last chapter, we discussed the Fermi-Dirac distribution function which gives the probability that a particular energy state will be occupied by an electron. We need to reconsider this function and apply the probability statistics to the donor and acceptor energy states.

### 4.4.1 Probability Function

One postulate used in the derivation of the Fermi-Dirac probability function was the Pauli exclusion principle, which states that only one particle is permitted in each quantum state. The Pauli exclusion principle also applies to the donor and acceptor states.

Suppose we have $N_i$ electrons and $g_i$ quantum states, where the subscript $i$ indicates the $i$th energy level. There are $g_i$ ways of choosing where to put the first particle. Each donor level has two possible spin orientations for the donor electron; thus each donor level has two quantum states. The insertion of an electron into one quantum state, however, precludes putting an electron into the second quantum state. By adding one electron, the vacancy requirement of the atom is satisfied, and the addition of a second electron in the donor level is not possible. For this reason, there are only $(g_i - 2)$ ways of choosing where to put the second particle, $(g_i - 4)$ ways of choosing where to put the third particle, and so on. The total number of ways of arranging the $N_i$ electrons in the $g_i$ quantum states, under the conditions given above, is then

$$(g_i)(g_i - 2)(g_i - 4) \cdots (g_i - (2N_i - 2)) = \frac{2^{g_i/2}\left(\frac{g_i}{2}\right)!}{2^{((g_i/2)-N_i)} \cdot \left(\frac{g_i}{2} - N_i\right)!} \qquad (4\text{--}50)$$

As before, this expression includes all the permutations of the $N_i$ particles among themselves. The actual number of independent ways of realizing a particular distribution is given by Equation (4–50) divided by $N_i!$.

The most probable distribution is obtained by maximizing the particular distributions as we did in Chapter 3. We will obtain

$$N_i = \frac{\frac{1}{2} g_i}{1 + \frac{1}{2} \exp\left(\frac{E_i - E_F}{kT}\right)} \qquad (4\text{--}51)$$

The factor $\frac{1}{2} g_i$ is equal to the density of donor atoms in the semiconductor material. We may write the probability function of electrons occupying the donor state as

$$n_d = \frac{N_d}{1 + \frac{1}{2} \exp\left(\frac{E_d - E_F}{kT}\right)} \qquad (4\text{--}52)$$

where $n_d$ is the density of electrons occupying the donor level and $E_d$ is the energy of the donor level. The factor $\frac{1}{2}$ in this equation is a direct result of the spin factor which we discussed above. The $\frac{1}{2}$ factor is sometimes written as $1/g$, where $g$ is called a degeneracy factor.

Equation (4–52) can also be written in the form

$$n_d = N_d - N_d^+ \qquad (4\text{--}53)$$

where $N_d^+$ is the concentration of ionized donors. In many applications, we will be interested more in the concentration of ionized donors than in the concentration of electrons remaining in the donor states.

If we do the same type of analysis for acceptor atoms, we obtain the expression

$$p_a = \frac{N_a}{1 + \frac{1}{g} \exp\left(\frac{E_F - E_a}{kT}\right)} = N_a - N_a^- \qquad (4\text{--}54)$$

where $N_a$ is the concentration of acceptor atoms, $E_a$ is the acceptor energy level, $p_a$ is the concentration of holes in the acceptor states, and $N_a^-$ is the concentration of ionized acceptors. A hole in an acceptor state corresponds to an acceptor atom that is neutrally charged and still has an "empty" bonding position as we discussed in Section 4.2.1. The parameter g is, again, a degeneracy factor. The ground state degeneracy factor g is normally taken as four for the acceptor level in silicon and gallium arsenide because of the detailed band structure.

## 4.4.2 Complete Ionization and Freeze-out

The probability function for electrons in the donor energy state was just derived and given by Equation (4–52). If we assume that $(E_d - E_F) \gg kT$, then

$$n_d \approx \frac{N_d}{\frac{1}{2} \exp \left( \frac{E_d - E_F}{kT} \right)} = 2N_d \exp \left[ \frac{-(E_d - E_F)}{kT} \right] \qquad (4\text{–}55)$$

If $(E_d - E_F) \gg kT$, then the Boltzmann approximation is also valid for the electrons in the conduction band so that, from Equation (4–11),

$$n_0 = N_c \exp \left[ \frac{-(E_c - E_F)}{kT} \right]$$

We would like to determine the relative number of electrons in the donor state compared with the total number of electrons; therefore we need to consider the ratio of electrons in the donor state to the total number of electrons in the conduction band plus donor state. Using the expressions of Equations (4–55) and (4–11), we write

$$\frac{n_d}{n_d + n_0} = \frac{2N_d \exp \left[ \frac{-(E_d - E_F)}{kT} \right]}{2N_d \exp \left[ \frac{-(E_d - E_F)}{kT} \right] + N_c \exp \left[ \frac{-(E_c - E_F)}{kT} \right]} \qquad (4\text{–}56)$$

The Fermi energy cancels out of this expression. Dividing by the numerator term, we obtain

$$\frac{n_d}{n_d + n_0} = \frac{1}{1 + \frac{N_c}{2N_d} \exp \left[ \frac{-(E_c - E_d)}{kT} \right]} \qquad (4\text{–}57)$$

The factor $(E_c - E_d)$ is just the ionization energy of the donor electrons.

---

### Example 4–5

**Objective:** To determine the fraction of total electrons still in the donor states at $T = 300°K$.

Consider phosphorus doping in silicon, for $T = 300°K$, at a concentration of $N_d = 10^{16}$ cm$^{-3}$.

**Solution:** From Equation (4–57), we have

$$\frac{n_d}{n_0 + n_d} = \frac{1}{1 + \dfrac{2.8 \times 10^{19}}{2(10^{16})} \exp\left(\dfrac{-0.045}{0.0259}\right)} = 0.0041 = 0.41 \text{ percent.}$$

**Comment:** This example shows that there are very few electrons in the donor state compared with the conduction band. Essentially all of the electrons from the donor states are in the conduction band and, since only about 0.4 percent of the donor states contain electrons, the donor states are said to be completely ionized.

At room temperature, then, the donor states are essentially completely ionized and, for a typical doping of $10^{16}$ cm$^{-3}$, almost all donor impurity atoms have donated an electron to the conduction band.

At room temperature, there is also essentially complete ionization of the acceptor atoms. This means that each acceptor atom has accepted an electron from the valence band so that $p_a$ is zero. At typical acceptor doping concentrations, a hole is created in the valence band for each acceptor atom. This ionization effect and the creation of electrons and holes in the conduction band and valence band, respectively, are shown in Figure 4–12.

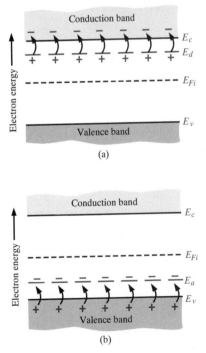

**Figure 4–12** Energy-band diagrams showing complete ionization of (a) donor states and (b) acceptor states.

The opposite of complete ionization occurs at $T = 0°K$. At absolute zero degrees, all electrons are in their lowest possible energy state; that is, for an n-type semiconductor, each donor state must contain an electron, therefore $n_d = N_d$ or $N_d^+ = 0$. We must have, then, from Equation (4–52) that $\exp[(E_d - E_F)/kT] = 0$. Since $T = 0°K$, this will occur for $\exp(-\infty) = 0$, which means that $E_F > E_d$. The Fermi energy level must be above the donor energy level at absolute zero. In the case of a p-type semiconductor at absolute zero temperature, the impurity atoms will not contain any electrons, so that the Fermi energy level must be below the acceptor energy state. The distribution of electrons among the various energy states, and hence the Fermi energy, is a function of temperature.

A detailed analysis, not given in this text, shows that at $T = 0°K$, the Fermi energy is halfway between $E_c$ and $E_d$ for the n-type material and halfway between $E_a$ and $E_v$ for the p-type material. These effects are shown in Figure 4–13. No electrons from the donor state are thermally elevated into the conduction band; this effect is called freeze-out. Similarly, when no electrons from the valence band are elevated into the acceptor states, the effect is also called freeze-out.

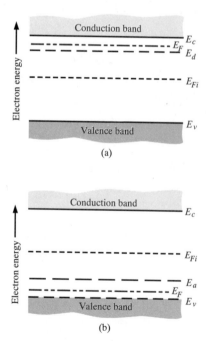

(a)

(b)

**Figure 4–13**   Energy-band diagram at $T = 0°K$ for (a) n-type and (b) p-type semiconductors.

Between $T = 0°K$, freeze-out, and $T = 300°K$, complete ionization, we have partial ionization of donor or acceptor atoms.

**Example 4-6**

**Objective:** To determine the temperature at which 90 percent of acceptor atoms are ionized.

Consider p-type silicon doped with boron at a concentration of $N_a = 10^{16}$ cm$^{-3}$.

**Solution:** Find the ratio of holes in the acceptor state to the total number of holes in the valence band plus acceptor state. Taking into account the Boltzmann approximation and assuming the degeneracy factor is $g = 4$, we have

$$\frac{p_a}{p_0 + p_a} = \frac{1}{1 + \dfrac{N_v}{4N_a} \cdot \exp\left[\dfrac{-(E_a - E_v)}{kT}\right]}$$

For 90 percent ionization,

$$\frac{p_a}{p_0 + p_a} = 0.10 = \frac{1}{1 + \dfrac{(1.04 \times 10^{19}) \left(\dfrac{T}{300}\right)^{3/2}}{4(10^{16})} \cdot \exp\left[\dfrac{-0.045}{0.0259 \left(\dfrac{T}{300}\right)}\right]}$$

Using trial and error, we find that $T = 193°$K.

**Comment:** This example shows that at approximately 100°C below room temperature, we still have 90 percent of the acceptor atoms ionized; in other words, 90 percent of the acceptor atoms have "donated" a hole to the valence band.

## 4.5 CHARGE NEUTRALITY

In thermal equilibrium, the semiconductor crystal is electrically neutral. The electrons are distributed among the various energy states creating negative and positive charges, but the net charge density is zero. This charge-neutrality condition is used to determine the thermal-equilibrium electron and hole concentrations as a function of the impurity doping concentration. We will define a compensated semiconductor and then determine the electron and hole concentrations.

### 4.5.1 Compensated Semiconductors

A compensated semiconductor is one that contains both donor and acceptor impurity atoms in the same region. A compensated semiconductor can be formed, for example, by diffusing acceptor impurities into an n-type material, or by diffusing donor impurities into a p-type material. An n-type compensated semiconductor occurs when $N_d > N_a$ and a p-type compensated semiconductor occurs when $N_a > N_d$. If $N_a = N_d$, we have a completely compensated semiconductor that has, as we will show, the characteristics of an intrinsic material. Compensated semiconductors are created quite naturally during device fabrication as we will see later.

### 4.5.2 Equilibrium Electron and Hole Concentrations

Figure 4–14 shows the energy-band diagram of a semiconductor when both donor and acceptor impurity atoms are added to the same region to form a compensated semiconductor. The figure shows how the electrons and holes can be distributed among the various states.

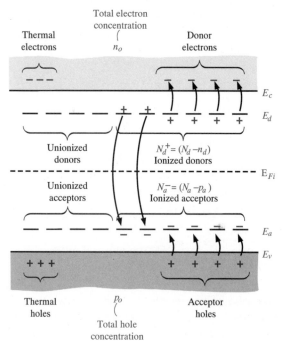

**Figure 4–14**   Energy-band diagram of a compensated semiconductor showing ionized and un-ionized donors and acceptors.

The charge neutrality condition is expressed by equating the density of negative charges to the density of positive charges. We then have

$$n_0 + N_a^- = p_0 + N_d^+ \qquad (4\text{–}58)$$

or

$$n_0 + (N_a - p_a) = p_0 + (N_d - n_d) \qquad (4\text{–}59)$$

where $n_0$ and $p_0$ are the thermal-equilibrium concentrations of electrons and holes in the conduction band and valence band, respectively. The parameter $n_d$ is the concentration of electrons in the donor energy states, so $N_d^+ = N_d - n_d$ is the concentration of positively charged donor states. Similarly, $p_a$ is the concentration of holes in the acceptor states, so $N_a^- = N_a - p_a$ is the concentration of negatively charged acceptor states. We have expressions for $n_0$, $p_0$, $n_d$, and $p_a$ in terms of the Fermi energy and temperature.

If we assume complete ionization, $n_d$ and $p_a$ are both zero, and Equation (4–59) becomes

$$n_0 + N_a = p_0 + N_d \qquad (4\text{–}60)$$

If we express $p_0$ as $n_i^2/n_0$, then Equation (4–60) can be written as

$$n_0 + N_a = \frac{n_i^2}{n_0} + N_d \qquad (4\text{–}61a)$$

which in turn can be written as

$$n_0^2 - (N_d - N_a)n_0 - n_i^2 = 0 \qquad (4\text{–}61b)$$

The electron concentration $n_0$ can be determined using the quadratic formula, or

$$n_0 = \frac{(N_d - N_a)}{2} + \sqrt{\left(\frac{N_d - N_a}{2}\right)^2 + n_i^2} \qquad (4\text{–}62)$$

The positive sign in the quadratic formula must be used, since, in the limit of an intrinsic semiconductor when $N_a = N_d = 0$, the electron concentration must be a positive quantity, or $n_0 = n_i$.

Equation (4–62) is used to calculate the electron concentration in an n-type semiconductor, or when $N_d > N_a$.

---

**Example 4–7**

**Objective:** To determine the thermal equilibrium electron and hole concentrations for a given doping concentration.

Consider an n-type silicon semiconductor at $T = 300°K$ in which $N_d = 10^{16}$ cm$^{-3}$ and $N_a = 0$. The intrinsic carrier concentration is assumed to be $n_i = 1.5 \times 10^{10}$ cm$^{-3}$.

**Solution:** From Equation (4–62), the majority carrier electron concentration is

$$n_0 = \frac{10^{16}}{2} + \sqrt{\left(\frac{10^{16}}{2}\right)^2 + (1.5 \times 10^{10})^2} \cong 10^{16} \text{ cm}^{-3}$$

The minority carrier hole concentration is found from

$$p_0 = \frac{n_i^2}{n_0} = \frac{(1.5 \times 10^{10})^2}{1 \times 10^{16}} = 2.25 \times 10^4 \text{ cm}^{-3}$$

**Comment:** In this example, $N_d \gg n_i$, so that the thermal-equilibrium majority carrier electron concentration is essentially equal to the donor impurity concentration. The thermal-equilibrium majority and minority carrier concentrations can differ by many orders of magnitude.

We have argued in our discussion and we may note from the results of Example 4–7 that the concentration of electrons in the conduction band increases above the intrinsic carrier concentration as we add donor impurity atoms. At the same time, the minority carrier hole concentration decreases below the intrinsic carrier concentration as we add donor atoms. We must keep in mind that as we add donor impurity atoms and the corresponding donor electrons, there is a redistribution of electrons among available energy states. Figure 4–15 shows a schematic of this physical redistribution. A few

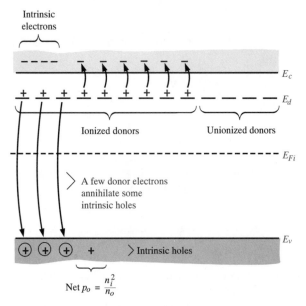

**Figure 4–15**   Energy-band diagram showing the redistribution of electrons when donors are added.

of the donor electrons will fall into the empty states in the valence band and, in doing so, will annihilate some of the intrinsic holes. The minority carrier hole concentration will therefore decrease as we have seen in Example 4–7. At the same time, because of this redistribution, the net electron concentration in the conduction band is *not* simply equal to the donor concentration plus the intrinsic electron concentration.

**Example 4–8**

**Objective:** To calculate the thermal-equilibrium electron and hole concentrations in germanium.

Consider a germanium sample at $T = 300°K$ in which $N_d = 5 \times 10^{13}$ cm$^{-3}$ and $N_a = 0$. Assume that $n_i = 2.4 \times 10^{13}$ cm$^{-3}$.

**Solution:** Again, from Equation (4–62), the majority carrier electron concentration is

$$n_0 = \frac{5 \times 10^{13}}{2} + \sqrt{\left(\frac{5 \times 10^{13}}{2}\right)^2 + (2.4 \times 10^{13})^2} = 5.97 \times 10^{13} \text{ cm}^{-3}$$

The minority carrier hole concentration is

$$p_0 = \frac{n_i^2}{n_0} = \frac{(2.4 \times 10^{13})^2}{5.97 \times 10^{13}} = 9.65 \times 10^{12} \text{ cm}^{-3}$$

**Comment:** If the donor impurity concentration is not too different in magnitude from the intrinsic carrier concentration, then the thermal-equilibrium majority carrier electron concentration is influenced by the intrinsic concentration.

---

We have seen that the intrinsic carrier concentration $n_i$ is a very strong function of temperature. As the temperature increases, additional electron-hole pairs are thermally generated so that the $n_i^2$ term in Equation (4–62) may begin to dominate. The semiconductor will eventually lose its extrinsic characteristics. Figure 4–16 shows the electron concentration versus temperature in silicon doped with $5 \times 10^{14}$ donors per cm$^3$. As the temperature

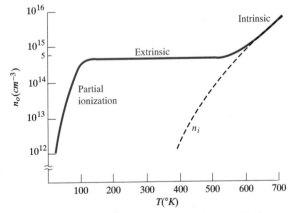

**Figure 4–16** Electron concentration versus temperature showing the three regions: partial ionization, extrinsic, and intrinsic.

increases, we can see where the intrinsic concentration begins to dominate. Also shown is the partial ionization, or the onset of freeze-out, at the low temperature.

If we reconsider Equation (4–60) and express $n_0$ as $n_i^2/p_0$, then we have

$$\frac{n_i^2}{p_0} + N_a = p_0 + N_d \qquad (4\text{–}63a)$$

which can be written as

$$p_0^2 - (N_a - N_d)p_0 - n_i^2 = 0 \qquad (4\text{–}63b)$$

Using the quadratic formula, the hole concentration is given by

$$p_0 = \frac{N_a - N_d}{2} + \sqrt{\left(\frac{N_a - N_d}{2}\right)^2 + n_i^2} \qquad (4\text{–}64)$$

where the positive sign, again, must be used. Equation (4–64) is used to calculate the thermal-equilibrium majority carrier hole concentration in a p-type semiconductor, or when $N_a > N_d$.

---

### Example 4–9

**Objective:** To determine the thermal-equilibrium hole and electron concentrations in silicon for a given doping density.

Consider a silicon semiconductor at $T = 300°K$ with $N_a = 10^{17}$ cm$^{-3}$ and $N_d = 0$. The semiconductor is p-type. Assume $n_i = 1.5 \times 10^{10}$ cm$^{-3}$.

**Solution:** Using Equation (4–64), the majority carrier hole concentration is

$$p_0 = \frac{10^{17}}{2} + \sqrt{\left(\frac{10^{17}}{2}\right)^2 + (1.5 \times 10^{10})^2} \cong 10^{17} \text{ cm}^{-3}$$

The minority carrier electron concentration is

$$n_0 = \frac{n_i^2}{p_0} = \frac{(1.5 \times 10^{10})^2}{10^{17}} = 2.25 \times 10^3 \text{ cm}^{-3}$$

**Comment:** The thermal-equilibrium majority carrier hole concentration is essentially equal to the impurity acceptor concentration if $N_a \gg n_i$. Note again that there are orders of magnitude difference between the minority carrier electron and majority carrier hole concentrations.

---

### Example 4–10

**Objective:** To calculate the thermal-equilibrium electron and hole concentrations in a compensated n-type semiconductor.

Consider a silicon semiconductor at $T = 300°K$ in which $N_d = 10^{16}$ cm$^{-3}$ and $N_a = 3 \times 10^{15}$ cm$^{-3}$. Assume $n_i = 1.5 \times 10^{10}$ cm$^{-3}$.

**Solution:** Since $N_d > N_a$, the compensated semiconductor is n-type and the thermal-equilibrium majority carrier electron concentration is given by Equation (4–62) as

$$n_0 = \frac{10^{16} - 3 \times 10^{15}}{2} + \sqrt{\left(\frac{10^{16} - 3 \times 10^{15}}{2}\right)^2 + (1.5 \times 10^{10})^2}$$

so that

$$n_0 \cong 7 \times 10^{15} \text{ cm}^{-3}$$

The minority carrier hole concentration is

$$p_0 = \frac{n_i^2}{n_0} = \frac{(1.5 \times 10^{10})^2}{7 \times 10^{15}} = 3.21 \times 10^4 \text{ cm}^{-3}$$

**Comment:** If we assume complete ionization and if $(N_d - N_a) \gg n_i$, then the majority carrier electron concentration is, to a very good approximation, just the difference between the donor and acceptor concentrations.

---

We may note that, for a compensated n-type semiconductor, the minority carrier hole concentration is determined from

$$p_0 = \frac{n_i^2}{n_0} = \frac{n_i^2}{(N_d - N_a)}$$

Equations (4–62) and (4–64) are used to calculate the majority carrier electron concentration in an n-type semiconductor and majority carrier hole concentration in a p-type semiconductor, respectively. The minority carrier hole concentration in an n-type semiconductor could, theoretically, be calculated from Equation (4–64). However, we would be subtracting two numbers on the order of $10^{16}$ cm$^{-3}$, for example, to obtain a number on the order of $10^4$ cm$^{-3}$, which from a practical point of view is not possible. The minority carrier concentrations are calculated from $n_0 p_0 = n_i^2$ once the majority carrier concentration has been determined.

## 4.6  POSITION OF FERMI ENERGY LEVEL

We discussed qualitatively in Section 4.3.1 how the electron and hole concentrations change as the Fermi energy level moves through the bandgap energy. Then, in Section 4.5, we calculated the electron and hole concentrations as a function of donor and acceptor impurity concentrations. We can now determine the position of the Fermi energy level as a function of the doping concentrations and as a function of temperature. The relevance of the Fermi energy level will be further discussed after the mathematical derivations.

### 4.6.1 Mathematical Derivation

The position of the Fermi energy level within the bandgap can be determined using the equations already developed for the thermal-equilibrium electron and hole concentrations. If we assume the Boltzmann approximation to be valid, then from Equation (4–11) we have $n_0 = N_c \exp\left[-(E_c - E_F)/kT\right]$. We can solve for $E_c - E_F$ from this equation and obtain

$$E_c - E_F = kT \ln\left(\frac{N_c}{n_0}\right) \tag{4–65}$$

where $n_0$ is given by Equation (4–62). If we consider an n-type semiconductor in which $N_d \gg n_i$, then $n_0 \cong N_d$, so that

$$E_c - E_F = kT \ln\left(\frac{N_c}{N_d}\right) \tag{4–66}$$

The distance between the bottom of the conduction band and the Fermi energy is a logarithmic function of the donor concentration. As the donor concentration increases, the Fermi level moves closer to the conduction band. Conversely, if the Fermi level moves closer to the conduction band, then the electron concentration in the conduction band is increasing. We may note that if we have a compensated semiconductor, then the $N_d$ term in Equation (4–66) is simply replaced by $N_d - N_a$, or the net effective donor concentration.

---

**Example 4–11**

**Objective:** To calculate the position of the Fermi energy level with respect to the bottom of the conduction band in an n-type semiconductor.

Consider silicon at $T = 300°K$. Let $N_d = 10^{16}$ cm$^{-3}$, $N_a = 0$, and $n_i = 1.5 \times 10^{10}$ cm$^{-3}$.

**Solution:** From Equation (4–66), we have

$$E_c - E_F = kT \ln\left(\frac{N_c}{N_d}\right) = (0.0259)\ln\left(\frac{2.8 \times 10^{19}}{10^{16}}\right)$$

or

$$E_c - E_F = 0.206 \text{ eV}$$

**Comment:** For this n-type material, the Fermi level is 0.206 eV below the conduction band edge, and approximately 0.161 eV below the donor energy level; therefore, the Boltzmann approximation is valid for this case.

---

We may develop a slightly different expression for the position of the Fermi level. We had from Equation (4–39) that $n_0 = n_i \exp\left[(E_F - E_{Fi})/kT\right]$. We can solve for $E_F - E_{Fi}$ as

$$E_F - E_{Fi} = kT\ln\left(\frac{n_0}{n_i}\right) \tag{4-67}$$

Equation (4–67) can be used specifically for an n-type semiconductor, where $n_0$ is given by Equation (4–62), to find the difference between the Fermi level and the intrinsic Fermi level as a function of the donor concentration. We may note that, if the net effective donor concentration is zero, that is $N_d - N_a = 0$, then $n_0 = n_i$ and $E_F = E_{Fi}$. A completely compensated semiconductor has the characteristics of an intrinsic material in terms of carrier concentration and Fermi level position.

We can derive the same types of equations for a p-type semiconductor. From Equation (4–19), we have $p_0 = N_v \exp\left[-(E_F - E_v)/kT\right]$, so that

$$E_F - E_v = kT\ln\left(\frac{N_v}{p_0}\right) \tag{4-68}$$

If we assume that $N_a \gg n_i$, then Equation (4–68) can be written as

$$E_F - E_v = kT\ln\left(\frac{N_v}{N_a}\right) \tag{4-69}$$

The distance between the Fermi level and the top of the valence-band energy for a p-type semiconductor is a logarithmic function of the acceptor concentration: as the acceptor concentration increases, the Fermi level moves closer to the valence band. Equation (4–69) still assumes that the Boltzmann approximation is valid. Again, if we have a compensated p-type semiconductor, then the $N_a$ term in Equation (4–69) is replaced by $N_a - N_d$, or the net effective acceptor concentration.

We can also derive an expression for the relationship between the Fermi level and the intrinsic Fermi level in terms of the hole concentration. We have from Equation (4–40) that $p_0 = n_i \exp\left[-(E_F - E_{Fi})/kT\right]$, which yields

$$E_{Fi} - E_F = kT\ln\left(\frac{p_0}{n_i}\right) \tag{4-70}$$

Equation (4–70) can be used to find the difference between the intrinsic Fermi level and the Fermi energy in terms of the acceptor concentration. The hole concentration $p_0$ in Equation (4–70) is given by Equation (4–64).

We may again note from Equation (4–67) that, for an n-type semiconductor, $n_0 > n_i$ and $E_F > E_{Fi}$. The Fermi level for an n-type semiconductor is above $E_{Fi}$. For a p-type semiconductor, $p_0 > n_i$, and from Equation (4–70) we see that $E_{Fi} > E_F$. The Fermi level for a p-type semiconductor is below $E_{Fi}$. These results are shown in Figure 4–17.

### 4.6.2 Variation of $E_F$ with Doping Concentration and Temperature

We may plot the position of the Fermi energy level as a function of the doping concentration. Figure 4–18 shows the Fermi energy level as a func-

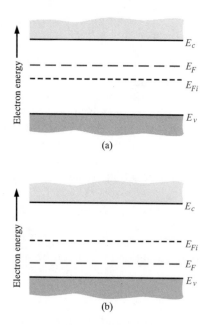

(a)

(b)

**Figure 4–17**   Position of Fermi level for an (a) n-type ($N_d > N_a$) and (b) p-type ($N_a > N_d$) semiconductor.

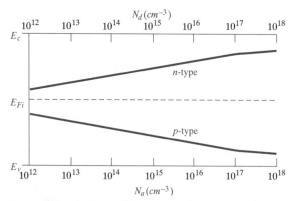

**Figure 4–18**   Position of Fermi level as a function of donor concentration (n-type) and acceptor concentration (p-type).

tion of donor concentration (n-type) and as a function of acceptor concentration (p-type) for silicon at $T = 300°K$. As the doping levels increase, the Fermi energy level moves closer to the conduction band for the n-type material and closer to the valence band for the p-type material. Keep in mind that the equations for the Fermi energy level that we have derived assume that the Boltzmann approximation is valid.

**Example 4–12**

**Objective:** To determine the Fermi-level position and the maximum doping at which the Boltzmann approximation is still valid.

Consider p-type silicon, at $T = 300°K$, doped with boron. We may assume that the limit of the Boltzmann approximation occurs when $E_F - E_a = 3kT$. (See section 4.1.2.)

**Solution:** From Table 4–3, we find the ionization energy is $E_a - E_v = 0.045$ eV for boron in silicon. If we assume that $E_{Fi} \cong E_{midgap}$, then from Equation (4–70), the position of the Fermi level at the maximum doping is given by

$$E_{Fi} - E_F = \frac{E_g}{2} - (E_a - E_v) - (E_F - E_a) = kT\ln\left(\frac{N_a}{n_i}\right)$$

or

$$0.56 - 0.045 - 3(0.0259) = 0.437 = (0.0259)\ln\left(\frac{N_a}{n_i}\right)$$

We can then solve for the doping as

$$N_a = n_i \exp\left(\frac{0.437}{0.0259}\right) = 3.2 \times 10^{17} \text{ cm}^{-3}$$

**Comment:** If the acceptor (or donor) concentration in silicon is greater than approximately $3 \times 10^{17}$ cm$^{-3}$, then the Boltzmann approximation of the distribution function becomes less valid and the equations for the Fermi-level position are no longer quite as accurate.

The intrinsic carrier concentration $n_i$, in Equations (4–67) and (4–70), is a strong function of temperature, so that $E_F$ is a function of temperature also. Figure 4–19 shows the variation of the Fermi energy level in silicon with temperature for several donor and acceptor concentrations. As the temperature increases, $n_i$ increases, and $E_F$ moves closer to the intrinsic Fermi level. At high temperature, the semiconductor material begins to lose its extrinsic characteristics and begins to behave more like an intrinsic semiconductor. At the very low temperature, freeze-out occurs; the Boltzmann approximation is no longer valid and the equations we derived for the Fermi-level position no longer apply. At the low temperature where freeze-out occurs, the Fermi level goes above $E_d$ for the n-type material and below $E_a$ for the p-type material. At absolute zero degrees, all energy states below $E_F$ are full and all energy states above $E_F$ are empty.

### 4.6.3 Relevance of the Fermi Energy

We have been calculating the position of the Fermi energy level as a function of doping concentrations and temperature. This analysis may seem some-

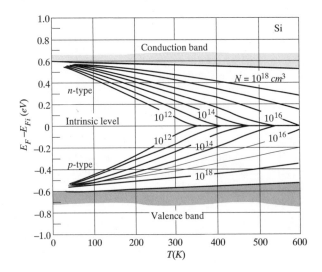

**Figure 4–19** Position of Fermi level as a function of temperature for various doping concentrations. (From Sze [9]).

what arbitrary and fictitious. However, these relations do become significant later in our discussion of pn junctions and the other semiconductor devices we will consider. An important point is that, in thermal equilibrium, the Fermi energy level is a constant throughout a system. We will not prove this statement, but we can intuitively see its validity by considering the following example.

Suppose we have a particular material, A, whose electrons are distributed in the energy states of an allowed band as shown in Figure 4–20a. Most of the energy states below $E_{FA}$ contain electrons and most of the energy states above $E_{FA}$ are empty of electrons. Consider another material, B, whose electrons are distributed in the energy states of an allowed band as shown in Figure 4–20b. The energy states below $E_{FB}$ are mostly full and the energy states above $E_{FB}$ are mostly empty. If these two materials are brought into intimate contact, the electrons in the entire system will tend to seek the lowest possible energy. Electrons from material A will flow into the lower energy states of material B, as indicated in Figure 4–20c, until thermal equilibrium is reached. Thermal equilibrium will occur when the distribution of electrons, as a function of energy, is the same in the two materials. This equilibrium state occurs when the Fermi energy is the same in the two materials as shown in Figure 4–20d. The Fermi energy, important in the physics of the semiconductor, also provides a good pictorial representation of the characteristics of the semiconductor materials and devices.

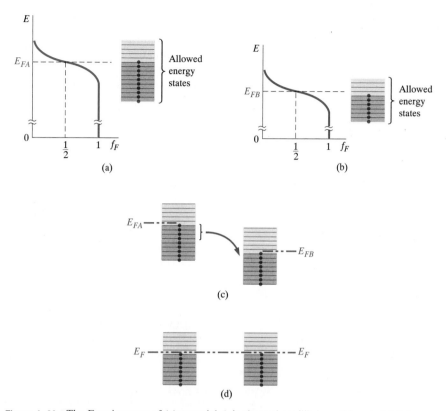

**Figure 4–20** The Fermi energy of (a) material A in thermal equilibrium, (b) material B in thermal equilibrium, (c) materials A and B at the instant they are placed in contact, and (d) materials A and B in contact at thermal equilibrium.

## 4.7 SUMMARY AND REVIEW

In this chapter we have determined the thermal-equilibrium concentrations of electrons in the conduction band and of holes in the valence band. These concentrations indicate the number of charge particles available to generate currents in a semiconductor.

The electron concentration in the conduction band is the integral over the conduction-band energy of the density of states function times the probability that a state is occupied by an electron, which is the Fermi-Dirac probability function. Using the Boltzmann approximation, we obtained an expression for the electron concentration in terms of the Fermi energy and an effective density of states function. We derived a similar expression for the density of holes in the valence band. These expressions were then applied to the intrinsic semiconductor and the intrinsic carrier concentration was determined as a function of the bandgap energy and temperature. The position of the intrinsic Fermi level within the bandgap was also determined.

We then considered adding controlled amounts of specific types of impurity atoms to create extrinsic semiconductors. The addition of donor impurity atoms creates the n-type semiconductor and the addition of acceptor impurity atoms creates the p-type semiconductor. Adding donors or acceptors allows us to vary the electron and hole concentrations by orders of magnitude. The statistical distribution of electrons and holes among the various energy states of the conduction and valence bands and donor and acceptor levels was determined. We found that, at room temperature, we have essentially complete ionization: all donor and acceptor atoms are ionized. Using the concepts of charge neutrality and complete ionization, equations for electron and hole concentrations as functions of the donor, acceptor, and intrinsic carrier concentrations were developed. The $n_0 p_0$ product was derived in which $n_0 p_0 = n_i^2$. This relationship is fundamental for semiconductors in thermal equilibrium. The position of the Fermi energy level was then derived also as a function of the donor, acceptor, and intrinsic carrier concentrations. Finally, we discussed the relevance of the Fermi energy, and concluded that, in any system that is in thermal equilibrium, the Fermi energy must be constant throughout that system.

We have considered the semiconductor in thermal equilibrium and determined the electron and hole concentrations as a function of donor and acceptor concentrations. We will use these expressions in the next chapter as well as throughout the text in determining the current-voltage relations of semiconductor materials and devices. We will see that, by controlling the donor or acceptor impurity concentrations in different regions of the semiconductor, we can choose its properties, thus giving us great flexibility in the design of semiconductor material characteristics.

## GLOSSARY OF IMPORTANT TERMS

**Acceptor atoms:** Impurity atoms added to a semiconductor to create a p-type material.

**Charge carrier:** The electron and/or hole that moves inside the semiconductor and gives rise to electrical currents.

**Compensated semiconductor:** A semiconductor that contains both donors and acceptors in the same semiconductor region.

**Complete ionization:** The condition when all donor atoms are positively charged by giving up their donor electrons and all acceptor atoms are negatively charged by accepting electrons.

**Degenerate semiconductor:** A semiconductor whose electron concentration or hole concentration is greater than the effective density of states, so that the Fermi level is in the conduction band (n-type) or in the valence band (p-type).

**Donor atoms:** Impurity atoms added to a semiconductor to create an n-type material.

**Effective density of states:** The parameter $N_c$ which results from integrating the density of quantum states $g_c(E)$ times the Fermi function $f_F(E)$ over the conduction-band energy, and the parameter $N_v$ which results from integrating the density of quantum states $g_v(E)$ times $[1 - f_F(E)]$ over the valence-band energy.

**Extrinsic semiconductor:** A semiconductor in which controlled amounts of donors and/or acceptors have been added so that the electron and hole concentrations change from the intrinsic carrier concentration, and a preponderance of either electrons (n-type) or holes (p-type) is created.

**Freeze-out:** The condition that occurs in a semiconductor when the temperature is lowered and the donors and acceptors become neutrally charged. The electron and hole concentrations become very small.

**Intrinsic carrier concentration $n_i$:** The electron concentration in the conduction band and the hole concentration in the valence band (equal values) in an intrinsic semiconductor.

**Intrinsic Fermi level $E_{Fi}$:** The position of the Fermi level in an intrinsic semiconductor.

**Intrinsic semiconductor:** A pure semiconductor material with no impurity atoms and no lattice defects in the crystal.

**Nondegenerate semiconductor:** A semiconductor in which a relatively small number of donors and/or acceptors have been added so that discrete, noninteracting donor states and/or discrete, noninteracting acceptor states are introduced.


## PROBLEMS

### Section 4.1

1. Calculate and plot the intrinsic carrier concentration, $n_i$, at $T = 200, 300, 400, 500,$ and $600°K$ for: (a) silicon, (b) germanium, and (c) gallium arsenide.

2. The intrinsic carrier concentration in silicon is to be no greater than $n_i = 1 \times 10^{12}$ cm$^{-3}$. Assume $E_g = 1.12$ eV. Determine the maximum temperature allowed for the silicon.

3. Repeat problem 2 for gallium arsenide. Assume $E_g = 1.4$ eV.

4. Find the ratio of the intrinsic carrier concentrations at two temperatures, $n_i(T_2)/n_i(T_1)$ where $T_2 = 300°K$ and $T_1 = 290°K$, when the semiconductor is (a) germanium and (b) gallium arsenide. Assume that the effective masses and bandgap energies are independent of temperature.

5.  Given the effective masses of electrons and holes in silicon, germanium, and gallium arsenide, calculate the position of the intrinsic Fermi energy level with respect to the center of the bandgap for each semiconductor at $T = 300°K$.

6.  Calculate the position of the intrinsic Fermi level with respect to the center of the bandgap at $T = 300°K$ if (a) $m_n^* = 1.10\ m_0$, $m_p^* = 0.25\ m_0$, and (b) $m_n^* = 0.45\ m_0$, $m_p^* = 1.21\ m_0$.

7.  Plot $E_{Fi}$ with respect to the center of the bandgap in silicon for $200°K \le T \le 600°K$.

8.  For a particular semiconductor material, $T = 300°K$, $N_c = 1 \times 10^{18}$ cm$^{-3}$, and $N_v = 1 \times 10^{19}$ cm$^{-3}$. Let $E_g = 1.45$ eV. Determine the position of the intrinsic Fermi level with respect to the center of the bandgap.

9.  If the density of states function in the conduction band of a particular semiconductor is a constant equal to K, derive the expression for the thermal-equilibrium concentration of electrons in the conduction band, assuming Fermi-Dirac statistics and assuming the Boltzmann approximation is valid.

10. Repeat problem 9 if the density of states function is given by $g_c(E) = C_1(E - E_c)$ for $E \ge E_c$ where $C_1$ is a constant.

## Section 4.2

11. Calculate the ionization energy and radius of the donor electron in germanium using the Bohr theory. (Use the density of states effective mass as a first approximation.)

12. Repeat problem 11 for gallium arsenide.

## Section 4.3

13. Calculate the values of $n_0$ and $p_0$ for germanium at $T = 300°K$ if $E_c - E_F = 0.195$ eV.

14. Determine the values of $n_0$ and $p_0$ for silicon at $T = 300°K$ if the Fermi energy is 0.22 eV above the valence band energy.

15. If $E_c - E_F = 0.25$ eV in gallium arsenide at $T = 400°K$, calculate the values of $n_0$ and $p_0$.

16. The value of $p_0$ in silicon at $T = 300°K$ is $10^{15}$ cm$^{-3}$. Determine (a) $E_c - E_F$ and (b) $n_0$.

*17. Assume that $E_F = E_v$ at $T = 300°K$ in silicon. Determine $p_0$.

*18. Consider silicon at $T = 300°K$ which has $n_0 = 5 \times 10^{19}$ cm$^{-3}$. Determine $E_c - E_F$.

## Section 4.4

*19. Consider n-type silicon with $E_c - E_d = 0.045$ eV. (a) Calculate the temperature at which 60 percent of the donor atoms are ionized. (b) Calculate the temperature at which 90 percent of the donor atoms are ionized. Let $N_d = 1 \times 10^{17}$ cm$^{-3}$ for both parts (a) and (b). (Note: You may have to use trial-and-error methods to solve the problem.)

*20. Assume that the donor ionization energy in gallium arsenide is 0.050 eV. Also assume that $E_c - E_F = 0.25$ eV. (a) Calculate the ratio of $n_d/N_d$ at $T = 300°$K. (b) Assuming that $E_c - E_F$ remains constant, determine the temperature at which the ratio in part (a) is 0.01.

## Section 4.5

21. Consider a germanium semiconductor at $T = 300°$K. Calculate the thermal equilibrium concentrations of $n_0$ and $p_0$ for: (a) $N_a = 10^{13}$ cm$^{-3}$, $N_d = 0$, and (b) $N_d = 5 \times 10^{15}$ cm$^{-3}$, $N_a = 0$.

22. Calculate and plot $n_0$ versus $T$ for $200°$K $\leq T \leq 600°$K when $N_d = 10^{14}$ cm$^{-3}$, $N_a = 0$ for (a) silicon and (b) germanium. Assume complete ionization. (Plot $n_0$ on a log scale.)

23. A silicon semiconductor at $T = 300°$K is doped with $N_d = 1 \times 10^{11}$ cm$^{-3}$, $N_a = 1 \times 10^{10}$ cm$^{-3}$. Assume complete ionization. (a) Is this p-type or n-type? (b) Calculate $n_0$ and $p_0$.

24. Assume that silicon, germanium, and gallium arsenide each have dopant concentrations of $N_d = 1 \times 10^{13}$ cm$^{-3}$ and $N_a = 2.5 \times 10^{13}$ cm$^{-3}$ at $T = 300°$K. For each of the three materials: (a) Is this material n-type or p-type? (b) Calculate $n_0$ and $p_0$.

25. Consider gallium arsenide at $T = 300°$K with a donor impurity concentration of $1 \times 10^{15}$ cm$^{-3}$. Acceptors are to be added to make the material a compensated p-type semiconductor. If the thermal-equilibrium hole concentration is to be $5 \times 10^{15}$ cm$^{-3}$, what concentration of acceptor atoms must be added? What is the thermal-equilibrium concentration of electrons in the compound material?

## Section 4.6

26. Consider germanium with an acceptor concentration of $N_a = 10^{15}$ cm$^{-3}$ and a donor concentration of $N_d = 0$. Consider temperatures of $T = 200, 300, 400, 500,$ and $600°$K. Calculate and plot the position of the Fermi energy with respect to the intrinsic Fermi level as a function of temperature.

27. Consider germanium at $T = 300°K$ with donor concentrations of $N_d = 10^{14}$, $10^{15}$, $10^{16}$, $10^{17}$, and $10^{18}$ cm$^{-3}$. Let $N_a = 0$. Calculate and plot the position of the Fermi energy level with respect to the intrinsic Fermi level as a function of the doping concentration.

28. Sketch the Fermi energy level for an n-type semiconductor as a function of temperature beginning at $T = 0°K$ and going to a "very high" temperature. Explain the characteristics.

29. For a particular semiconductor, $E_g = 1.50$ eV, $m_p^* = 10m_n^*$, $T = 300°K$, and $n_i = 1 \times 10^5$ cm$^{-3}$. (a) Determine the position of the intrinsic Fermi energy level with respect to the center of the bandgap. (b) Impurity atoms are added so that the Fermi energy level is 0.45 eV below the center of the bandgap. (i) Are acceptor or donor atoms added? (ii) What is the concentration of impurity atoms added?

30. Silicon at $T = 300°K$ contains acceptor atoms at a concentration of $N_a = 5 \times 10^{15}$ cm$^{-3}$. Donor atoms are added forming an n-type compensated semiconductor such that the Fermi level is 0.215 eV below the conduction band edge. What concentration of donor atoms are added?

31. Silicon at $T = 300°K$ is doped with acceptor atoms at a concentration of $N_a = 7 \times 10^{15}$ cm$^{-3}$. (a) Determine $E_F - E_v$. (b) Calculate the concentration of additional acceptor atoms which must be added to move the Fermi level a distance $kT$ closer to the valence-band edge.

32. Consider the parameters of a semiconductor given in problem 8. One type of impurity atom is to be added so that the Fermi level is 0.50 eV above the intrinsic level. (a) Are donors or acceptors to be added? (b) What must be the concentration of impurity atoms added?

33. Gallium arsenide at $T = 300°K$ contains acceptor impurity atoms at a density of $10^{15}$ cm$^{-3}$. Additional impurity atoms are to be added so that the Fermi level is 0.45 eV below the intrinsic level. Determine the concentration and type (donor or acceptor) of impurity atoms to be added.

## Summary and Review

34. A special semiconductor material is to be "designed." The semiconductor is to be n-type and doped with $1 \times 10^{15}$ cm$^{-3}$ donor atoms. Assume complete ionization and assume $N_a = 0$. The effective density of states functions are given by $N_c = N_v = 1.5 \times 10^{19}$ cm$^{-3}$ and are independent of temperature. A particular semiconductor device fabricated with this material requires the electron concentration to be no greater than $1.01 \times 10^{15}$ cm$^{-3}$ at $T = 400°K$. What is the minimum value of the bandgap energy?

35.  Silicon atoms, at a concentration of $10^{10}$ cm$^{-3}$, are added to gallium arsenide. Assume that the silicon atoms act as fully ionized dopant atoms and that 5 percent of the concentration added replace gallium atoms and 95 percent replace arsenic atoms. Let $T = 300°$K. (a) Determine the donor and acceptor concentrations. (b) Calculate the electron and hole concentrations and the position of the Fermi level with respect to $E_{Fi}$.

*36.  Defects in a semiconductor material introduce allowed energy states within the forbidden bandgap. Assume that a particular defect in silicon introduces two discrete levels: a donor level 0.25 eV above the top of the valence band, and an acceptor level 0.65 eV above the top of the valence band. The charge state of each defect is a function of the position of the Fermi level. (a) Sketch the charge density of each defect as the Fermi level moves from $E_v$ to $E_c$. Which defect level dominates in heavily doped n-type material? In heavily doped p-type material? (b) Determine the electron and hole concentrations and the location of the Fermi level in an n-type sample doped at $N_d = 10^{17}$ cm$^{-3}$. In a p-type sample doped at $N_a = 10^{17}$ cm$^{-3}$. (c) Determine the Fermi level position if no dopant atoms are added. Is the material n-type, p-type, or intrinsic?

## READING LIST

*1.  Hess, K. *Advanced Theory of Semiconductor Devices*. Englewood Cliffs, N.J.: Prentice Hall, 1988.

2.  McKelvey, J. P. *Solid State and Semiconductor Physics*. New York: Harper & Row, 1966.

3.  Navon, D. H. *Semiconductor Microdevices and Materials*. New York: Holt, Rinehart & Winston, 1986.

4.  Pierret, R. F. *Semiconductor Fundamentals*. Vol. 1 of *Modular Series on Solid State Devices*. 2nd ed. Reading, Mass.: Addison-Wesley, 1988.

5.  _____. *Advanced Semiconductor Fundamentals*. Vol. 6 of *Modular Series on Solid State Devices*. Reading, Mass.: Addison-Wesley, 1987.

*6.  Shur, M. *Physics of Semiconductor Devices*. Englewood Cliffs, N.J.: Prentice Hall, 1990.

7.  Smith, R. A. *Semiconductors*. 2nd ed. New York: Cambridge University Press, 1978.

8.  Streetman, B. G. *Solid State Electronic Devices*. 3rd ed. Englewood Cliffs, N.J.: Prentice Hall, 1990.

9.  Sze, S. M. *Physics of Semiconductor Devices*. 2nd ed. New York: Wiley, 1981.

*10.  Wang, S. *Fundamentals of Semiconductor Theory and Device Physics*. Englewood Cliffs, N.J.: Prentice Hall, 1989.

*11.  Wolfe, C. M.; N. Holonyak, Jr.; and G. E. Stillman. *Physical Properties of Semiconductors*. Englewood Cliffs, N.J.: Prentice Hall, 1989.

12.  Yang, E. S. *Microelectronic Devices*. New York: McGraw-Hill, 1988.

## CHAPTER 5

# CARRIER TRANSPORT PHENOMENA

In the previous chapter, we considered the semiconductor in equilibrium and determined electron and hole concentrations in the conduction and valence bands, respectively. A knowledge of the densities of these charged particles is important toward an understanding of the electrical properties of a semiconductor material. The net flow of the electrons and holes in a semiconductor will generate currents. The process by which these charged particles move is called transport. In this chapter we will consider the two basic transport mechanisms in a semiconductor crystal: drift—the movement of charge due to electric fields, and diffusion—the flow of charge due to density gradients. We should mention, in passing, that temperature gradients in a semiconductor can also lead to carrier movement. However, as the semiconductor device size becomes smaller, this effect can usually be ignored. The carrier transport phenomena are the foundation for finally determining the current-voltage characteristics of semiconductor devices. We will implicitly assume in this chapter that, though there will be a net flow of electrons and holes due to the transport processes, thermal equilibrium will not be substantially disturbed. Nonequilibrium processes will be considered in the next chapter.

The drift of electrons and holes caused by an electric field produces a drift current, which is a function of the concentrations of the mobile electrons and holes, and also a function of the net drift velocity of these charge carriers. The average drift velocity of a carrier is related to the electric field by a parameter called mobility. Electrons and holes do not move totally unimpeded through a semiconductor, but are involved in collisions with semiconductor atoms and with ionized impurity atoms. The mobility gives an indication of how well a carrier moves in a semiconductor.

Diffusion is the process whereby particles flow from a region of high concentration to a region of low concentration. If these particles are charged, as are electrons or holes, then diffusion of these particles leads to a diffusion current. The diffusion current density is related to the gradient of the carrier concentration by a parameter called the diffusion coefficient. The diffusion coefficient gives an indication of how well a particle will diffuse through a semiconductor due to a density gradient.

Finally in this chapter, we will discuss the Hall effect. The Hall effect involves the forces on a moving charged particle due to electric and magnetic fields that are perpendicular to each other. The Hall effect is used extensively to experimentally measure the majority carrier concentration and carrier mobility. The Hall effect is also used in computer keyboards as a Hall-effect sensor.

## 5.1 CARRIER DRIFT

An electric field applied to a semiconductor will produce a force on electrons and holes so that they will experience a net acceleration and net movement, provided there are available energy states in the conduction and valence bands. This net movement of charge due to an electric field is called *drift*. The net drift of charge gives rise to a *drift current*.

### 5.1.1 Drift Current Density

If we have a positive volume charge density $\rho$ moving at an average drift velocity $v_d$, the drift current density is given by

$$J_{drf} = \rho v_d \tag{5-1}$$

where $J$ is in units of coul/cm$^2$-sec or amps/cm$^2$. If the volume charge density is due to positively charged holes, then

$$J_{p|drf} = (ep)v_{dp} \tag{5-2}$$

where $J_{p|drf}$ is the drift current density due to holes and $v_{dp}$ is the average drift velocity of the holes.

The equation of motion of a positively charged hole in the presence of an electric field is

$$F = m_p^* a = eE \tag{5-3}$$

where $e$ is the magnitude of the electronic charge, $a$ is the acceleration, E is the electric field, and $m_p^*$ is the effective mass of the hole. If the electric field is constant, then we expect the velocity to increase linearly with time. However, charged particles in a semiconductor are involved in collisions with ionized impurity atoms and with thermally vibrating lattice atoms. These collisions, or scattering events, alter the velocity characteristics of the particle.

As the hole accelerates in a crystal due to the electric field, the velocity increases. When the charged particle collides with an atom in the crystal, for example, the particle loses most, or all, of its energy. The particle will again begin to accelerate and gain energy until it is again involved in a scattering process. This continues over and over again. Throughout this process, the particle will gain an average drift velocity which, for low electric fields, is directly proportional to the electric field. We may then write

$$v_{dp} = \mu_p E \qquad (5\text{--}4)$$

where $\mu_p$ is the proportionality factor and is called the hole mobility. The mobility is an important parameter of the semiconductor since it describes how well a particle will move due to an electric field. The unit of mobility is usually expressed in terms of $cm^2/V\text{-sec}$.

By combining Equations (5–2) and (5–4), the drift current density due to holes may be written as

$$J_{p|drf} = (ep)v_{dp} = e\mu_p p E \qquad (5\text{--}5)$$

The drift current due to holes is in the same direction as the applied electric field.

The same discussion of drift applies to electrons. We may write

$$J_{n|drf} = \rho v_{dn} = (-en)v_{dn} \qquad (5\text{--}6)$$

where $J_{n|drf}$ is the drift current density due to electrons and $v_{dn}$ is the average drift velocity of electrons. The net charge density of electrons is negative.

The average drift velocity of an electron is also proportional to the electric field for small fields. However, since the electron is negatively charged, the net motion of the electron is opposite to the electric field direction. We can then write

$$v_{dn} = -\mu_n E \qquad (5\text{--}7)$$

where $\mu_n$ is the electron mobility and is a positive quantity. Equation (5–6) may now be written as

$$J_{n|drf} = (-en)(-\mu_n E) = e\mu_n n E \qquad (5\text{--}8)$$

The conventional drift current due to electrons is also in the same direction as the applied electric field even though the electron movement is in the opposite direction.

Electron and hole mobilities are functions of temperature and doping concentrations, as we will see in the next section. Table 5–1 shows some typical mobility values at $T = 300°K$ for low doping concentrations.

Table 5–1 Typical mobility values at $T = 300°K$ and low doping concentrations

| | $\mu_n(cm^2/V\text{-sec})$ | $\mu_p(cm^2/V\text{-sec})$ |
|---|---|---|
| Silicon | 1350 | 480 |
| Gallium arsenide | 8500 | 400 |
| Germanium | 3900 | 1900 |

Since both electrons and holes contribute to the drift current, the total drift current density is the sum of the individual electron and hole drift current densities, so we may write

$$J_{drf} = e(\mu_n n + \mu_p p)E \qquad\qquad (5-9)$$

**Example 5–1**

**Objective:** To calculate the drift current density in a semiconductor for a given electric field.

Consider a germanium sample at $T = 300°K$ with doping concentrations of $N_d = 0$ and $N_a = 10^{16}$ cm$^{-3}$. Assume complete ionization and assume electron and hole mobilities given in Table 5–1. Calculate the drift current density if the applied electric field is $E = 50$ V/cm.

**Solution:** Since $N_a > N_d$, the semiconductor is p-type and the majority carrier hole concentration, from Chapter 4, is given by

$$p = \frac{N_a - N_d}{2} + \sqrt{\left(\frac{N_a - N_d}{2}\right)^2 + n_i^2} \cong 10^{16} \text{ cm}^{-3}$$

The minority carrier electron concentration is

$$n = \frac{n_i^2}{p} = \frac{(2.4 \times 10^{13})^2}{10^{16}} = 5.76 \times 10^{10} \text{ cm}^{-3}$$

For this extrinsic p-type semiconductor, the drift current density is

$$J_{drf} = e(\mu_n n + \mu_p p)E \cong e\mu_p N_a E$$

Then $\qquad J_{drf} = (1.6 \times 10^{-19})(1900)(10^{16})(50) = 152$ A/cm$^2$

**Comment:** Significant drift current densities can be obtained in a semiconductor applying relatively small electric fields. We may note from this example that the drift current will usually be due primarily to the majority carrier in an extrinsic semiconductor.

### 5.1.2 Mobility Effects

In the last section, we defined mobility, which relates the average drift velocity of a carrier to the electric field. Electron and hole mobilities are important semiconductor parameters in the characterization of carrier drift, as seen in Equation (5–9).

Equation (5–3) related the acceleration of a hole to a force such as an electric field. We may write this equation as

$$F = m_p^* \frac{dv}{dt} = eE \qquad\qquad (5-10)$$

where $v$ is the velocity of the particle due to the electric field and does not include the random thermal velocity. If we assume that the effective mass and electric field are constants, then we may integrate Equation (5–10) and obtain

$$v = \frac{eEt}{m_p^*} \qquad (5–11)$$

where we have assumed the initial drift velocity to be zero.

Figure 5–1a shows a schematic model of the random thermal velocity and motion of a hole in a semiconductor with zero electric field. There is a mean time between collisions which may be denoted by $\tau_{cp}$. If a small electric field (E-field) is applied as indicated in Figure 5–1b, there will be a net drift of the hole in the direction of the E-field, and the net drift velocity will be a small perturbation on the random thermal velocity, so the time between collisions will not be altered appreciably. If we use the mean time between collisions $\tau_{cp}$ in place of the time $t$ in Equation (5–11), then the mean peak velocity just prior to a collision or scattering event is

$$v_{d|peak} = \left(\frac{e\tau_{cp}}{m_p^*}\right)E \qquad (5–12a)$$

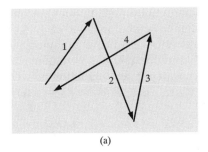

(a)

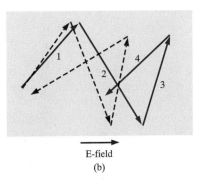

E-field

(b)

**Figure 5–1** Typical random behavior of a hole in a semiconductor (a) without an electric field and (b) with an electric field.

The average drift velocity is one half the peak value so that we can write

$$\langle v_d \rangle = \frac{1}{2} \left( \frac{e\tau_{cp}}{m_p^*} \right) E \qquad (5\text{--}12b)$$

However, if the proper averaging of the mean time between collisions over the random thermal velocities is performed, the $\frac{1}{2}$ in Equation (5–12b) is eliminated. The hole mobility is then given by

$$\mu_p = \frac{v_{dp}}{E} = \frac{e\tau_{cp}}{m_p^*} \qquad (5\text{--}13)$$

The same analysis applies to electrons; thus we can write the electron mobility as

$$\mu_n = \frac{e\tau_{cn}}{m_n^*} \qquad (5\text{--}14)$$

where $\tau_{cn}$ is the mean time between collisions for an electron.

There are two collision or scattering mechanisms that dominate in a semiconductor and affect the carrier mobility: phonon or lattice scattering, and ionized impurity scattering.

The atoms in a semiconductor crystal have a certain amount of thermal energy at temperatures above absolute zero that causes the atoms to randomly vibrate about their lattice position within the crystal. The lattice vibrations cause a disruption in the perfect periodic potential function we used in the Kronig-Penney model in Chapter 3. A perfect periodic potential in a solid allowed electrons to move unimpeded, or with no scattering, through the crystal. But the thermal vibrations cause a disruption of the potential function, resulting in an interaction between the electrons or holes and the vibrating lattice atoms. This lattice scattering is also referred to as phonon scattering.

Since lattice scattering is related to the thermal motion of atoms, the rate at which the scattering occurs will be a function of temperature. If we denote $\mu_L$ as the mobility that would be observed if only lattice scattering existed, then the scattering theory states that to first order

$$\mu_L \propto T^{-3/2} \qquad (5\text{--}15)$$

Mobility that is due to lattice scattering increases as the temperature decreases. Intuitively, we expect the lattice vibrations to decrease as the temperature decreases, which implies that the probability of a scattering event will also decrease, thus increasing mobility.

The temperature dependence of electron and hole mobilities in silicon is shown in Figure 5–2. In lightly doped semiconductors, lattice scattering dominates and the carrier mobility decreases with temperature as we have discussed. The temperature dependence of mobility is proportional to $T^{-n}$. The inserts in the figure show that the parameter $n$ is not equal to $\frac{3}{2}$ as the first order scattering theory predicted. However, mobility does increase as the temperature decreases.

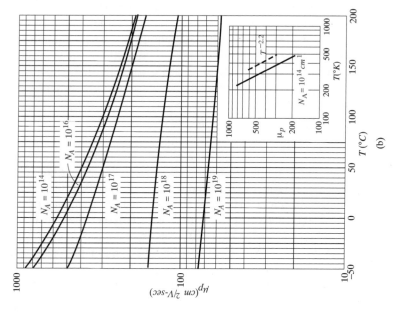

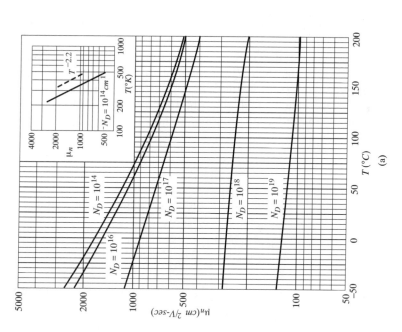

**Figure 5–2** (a) Electron and (b) hole mobilities in silicon versus temperature for various doping concentrations. Inserts show temperature dependence for "almost" intrinsic silicon. (From Pierret [4]).

The second interaction mechanism affecting carrier mobility is called ionized impurity scattering. We have seen that impurity atoms are added to the semiconductor to control or alter its characteristics. These impurities are ionized at room temperature so that a coulomb interaction exists between the electrons or holes and the ionized impurities. This coulomb interaction produces scattering or collisions and also alters the velocity characteristics of the charge carrier. If we denote $\mu_I$ as the mobility that would be observed if only ionized impurity scattering existed, then to first order we have

$$\mu_I \propto \frac{T^{+3/2}}{N_I} \tag{5-16}$$

where $N_I = N_d^+ + N_a^-$ is the total ionized impurity concentration in the semiconductor. If temperature increases, the random thermal velocity of a carrier increases, reducing the time the carrier spends in the vicinity of the ionized impurity center. The less time spent in the vicinity of a coulomb force, the smaller the scattering effect and the larger the expected value of $\mu_I$. If the number of ionized impurity centers increases, then the probability of a carrier encountering an ionized impurity center increases, implying a smaller value of $\mu_I$.

Figure 5–3 is a plot of electron and hole mobilities in germanium, silicon, and gallium arsenide at $T = 300°K$ as a function of impurity concentration. More accurately, these curves are of mobility versus ionized impurity concentration $N_I$. As the impurity concentration increases, the number of impurity scattering centers increases, thus reducing mobility.

If $\tau_L$ is the mean time between collisions due to lattice scattering, then $dt/\tau_L$ is the probability of a lattice scattering event occurring in a differential time $dt$. Likewise, if $\tau_I$ is the mean time between collisions due to ionized impurity scattering, then $dt/\tau_I$ is the probability of an ionized impurity scattering event occurring in the differential time $dt$. If these two scattering processes are independent, then the total probability of a scattering event occurring in the differential time $dt$ is the sum of the individual events, or

$$\frac{dt}{\tau} = \frac{dt}{\tau_I} + \frac{dt}{\tau_L} \tag{5-17}$$

where $\tau$ is the mean time between any scattering event.

Comparing Equation (5–17) with the definitions of mobility given by Equation (5–13) or (5–14), we can write

$$\frac{1}{\mu} = \frac{1}{\mu_I} + \frac{1}{\mu_L} \tag{5-18}$$

where $\mu_I$ is the mobility due to the ionized impurity scattering process and $\mu_L$ is the mobility due to the lattice scattering process. The parameter $\mu$ is the net mobility. With two or more independent scattering mechanisms, the inverse mobilities add, which means that the net mobility decreases.

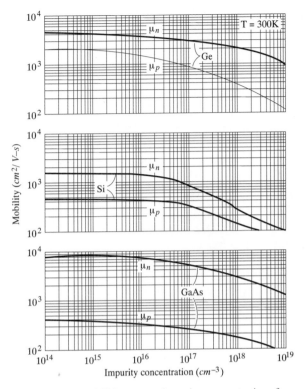

**Figure 5–3**   Electron and hole mobilities versus impurity concentrations for germanium, silicon, and gallium arsenide at $T = 300°$K. (From Sze [7]).

### 5.1.3  Conductivity

The drift current density, given by Equation (5–9), may be written as

$$J_{drf} = e(\mu_n n + \mu_p p)E = \sigma E \tag{5–19}$$

where $\sigma$ is the conductivity of the semiconductor material. The conductivity is given in units of (ohm-cm)$^{-1}$ and is a function of the electron and hole concentrations and mobilities. We have just seen that the mobilities are functions of impurity concentrations; conductivity, then, is a somewhat complicated function of impurity concentration.

The reciprocal of conductivity is resistivity, which is denoted by $\rho$ and is given in units of ohm-cm. We can write

$$\rho = \frac{1}{\sigma} = \frac{1}{e(\mu_n n + \mu_p p)} \tag{5–20}$$

Figure 5–4 is a plot of resistivity as a function of impurity concentration in silicon, germanium, gallium arsenide, and gallium phosphide at $T = 300°$K.

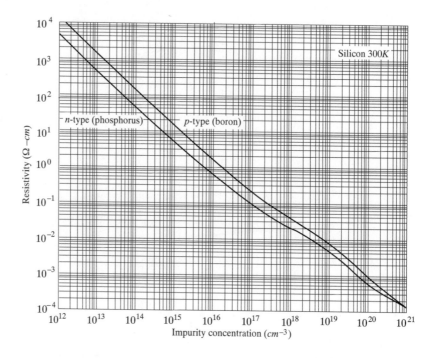

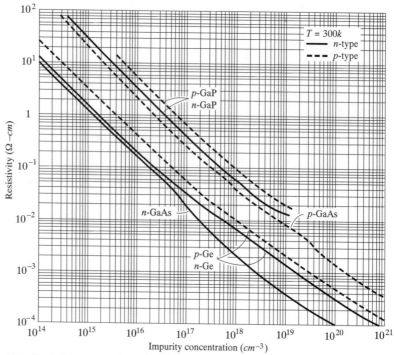

**Figure 5–4**    Resistivity versus impurity concentration at $T = 300°K$ in (a) silicon and (b) germanium, gallium arsenide, and gallium phosphide. (From Sze [7]).

Obviously, the curves are not linear functions of $N_d$ or $N_a$ due to mobility effects.

If we have a bar of semiconductor material as shown in Figure 5–5 with a voltage applied which produces a current $I$, then we can write

$$J = \frac{I}{A} \tag{5-21a}$$

and
$$E = \frac{V}{L} \tag{5-21b}$$

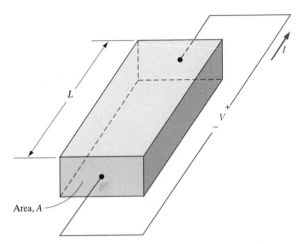

**Figure 5–5** Bar of semiconductor material as a resistor.

We can now rewrite Equation (5–19) as

$$\frac{I}{A} = \sigma \left(\frac{V}{L}\right) \tag{5-22a}$$

or
$$V = \left(\frac{L}{\sigma A}\right) I = \left(\frac{\rho L}{A}\right) I = IR \tag{5-22b}$$

Equation (5–22b) is Ohm's law for a semiconductor. The resistance is a function of resistivity, or conductivity, as well as the geometry of the semiconductor.

If we consider, for example, a p-type semiconductor with an acceptor doping $N_a(N_d = 0)$ in which $N_a \gg n_i$ and if we assume that the electron and hole mobilities are of the same order of magnitude, then the conductivity becomes

$$\sigma = e(\mu_n n + \mu_p p) \approx e\mu_p p \tag{5-23}$$

If we also assume complete ionization, then Equation (5–23) becomes

$$\sigma \cong e\mu_p N_a \cong \frac{1}{\rho} \tag{5–24}$$

The conductivity and resistivity of an extrinsic semiconductor are a function primarily of the majority carrier parameters.

We may plot the carrier concentration and conductivity of a semiconductor as a function of temperature for a particular doping concentration. Figure 5–6 shows the electron concentration and conductivity of silicon as a function of inverse temperature for the case when $N_d = 10^{15}$ cm$^{-3}$. In the mid-temperature range, or extrinsic range, as shown, we have complete ionization—the electron concentration remains essentially constant. However, the mobility is a function of temperature so the conductivity varies with temperature in this range. At higher temperatures, the intrinsic carrier concentration increases and begins to dominate the electron concentration as well as the conductivity. In the lower temperature range, freeze-out begins to occur; the electron concentration and conductivity decrease with decreasing temperature.

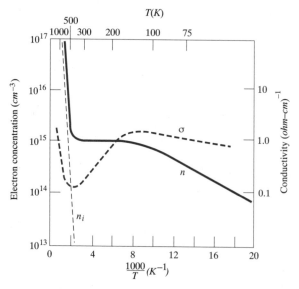

**Figure 5–6**  Electron concentration and conductivity versus inverse temperature for silicon. (After Sze [7]).

---

## Example 5–2

**Objective:** To determine the majority carrier concentration and mobility given the type and resistivity of a semiconductor.

Consider p-type silicon at $T = 300°K$ with a resistivity of $\rho = 100$ ohm-cm. Determine the p-type doping concentration and hole mobility assuming $N_d = 0$.

**Solution:** For a p-type semiconductor, the resistivity is

$$\rho \cong \frac{1}{e\mu_p p} = \frac{1}{e\mu_p N_a}$$

From Figure 5–4, we may note that a resistivity of 100 ohm-cm corresponds to a low doping concentration. As a first approximation, let $\mu_p = 480$ cm$^2$/V-sec. Then

$$N_a = \frac{1}{e\mu_p p} = \frac{1}{(1.6 \times 10^{-19})(480)(100)} = 1.3 \times 10^{14} \text{ cm}^{-3}$$

**Comment:** We may note from Figure 5–3 that the hole mobility for this doping concentration is essentially equal to our assumed value. For high-resistivity semi-conductors, the mobility corresponding to a low impurity concentration is a good first approximation.

---

## Example 5–3

**Objective:** To determine the doping concentration and majority carrier mobility given the type and conductivity of a compensated semiconductor.

Consider compensated n-type silicon at $T = 300°K$, with a conductivity of $\sigma = 16$ (ohm-cm)$^{-1}$ and an acceptor doping concentration of $10^{17}$ cm$^{-3}$. Determine the donor concentration and the electron mobility.

**Solution:** For n-type silicon at $T = 300°K$, we can assume complete ionization; therefore the conductivity, assuming $N_d - N_a \gg n_i$, is given by

$$\sigma \cong e\mu_n n = e\mu_n(N_d - N_a)$$

We have that

$$16 = (1.6 \times 10^{-19}) \, \mu_n(N_d - 10^{17})$$

Since mobility is a function of the ionized impurity concentration, we can use Figure 5–3 along with trial and error to determine $\mu_n$ and $N_d$. For example, if we choose $N_d = 2 \times 10^{17}$, then $N_I = N_d^+ + N_a^- = 3 \times 10^{17}$ so that $\mu_n \approx 510$ cm$^2$/V-sec which gives $\sigma = 8.16$ (ohm-cm)$^{-1}$. If we choose $N_d = 5 \times 10^{17}$, then $N_I = 6 \times 10^{17}$ so that $\mu_n \approx 325$ cm$^2$/V-sec which gives $\sigma = 20.8$ (ohm-cm)$^{-1}$. The doping is bounded between these two values. Further trial and error yields

$$N_d \cong 3.5 \times 10^{17} \text{ cm}^{-3}$$

and

$$\mu_n \approx 400 \text{ cm}^2/\text{V-sec}$$

which gives

$$\sigma \cong 16 \text{ (ohm-cm)}^{-1}$$

**Comment:** We can see from this example that, in high-conductivity semiconductor material, mobility is a strong function of carrier concentration.

For an intrinsic material, the conductivity can be written as

$$\sigma_i = e(\mu_n + \mu_p)n_i \tag{5-25}$$

The concentrations of electrons and holes are equal in an intrinsic semiconductor, so the intrinsic conductivity includes both the electron and hole mobility. Since, in general, the electron and hole mobilities are not equal, the intrinsic conductivity is not the minimum value possible at a given temperature.

### 5.1.4 Velocity Saturation

So far in our discussion of drift velocity, we have assumed that mobility is not a function of electric field, meaning that the drift velocity will increase linearly with applied electric field. The total velocity of a particle is the sum of the random thermal velocity and drift velocity. At $T = 300°K$, the average random thermal energy is given by

$$\tfrac{1}{2} m v_{th}^2 = \tfrac{3}{2} kT = \tfrac{3}{2} (0.0259) = 0.03885 \text{ eV} \tag{5-26}$$

This energy translates into a mean thermal velocity of approximately $10^7$ cm/sec for an electron in silicon. If we assume an electron mobility of $\mu_n = 1350$ cm$^2$/V-sec in low-doped silicon, a drift velocity of $10^5$ cm/sec, or 1 percent of the thermal velocity, will be achieved if the applied electric field is approximately 75 V/cm. This applied electric field does not appreciably alter the energy of the electron.

Figure 5–7 is a plot of average drift velocity as a function of applied electic field for electrons and holes in silicon, gallium arsenide, and germanium. At low electric fields where there is a linear variation of velocity with

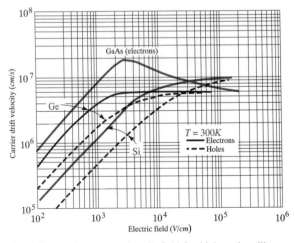

**Figure 5–7**  Carrier drift velocity versus electric field for high purity silicon, germanium, and gallium arsenide. (From Sze [7]).

electric field, the slope of the drift velocity versus electric field curve is the mobility. The behavior of the drift velocity of carriers at high electric fields deviates substantially from the linear relationship observed at low fields. The drift velocity of electrons in silicon, for example, saturates at approximately $10^7$ cm/sec at an electric field of approximately 30 kV/cm. If the drift velocity of a charge carrier saturates, then the drift current density will also saturate and become independent of the applied electric field.

The drift velocity versus electric field characteristic of gallium arsenide is more complicated than for silicon or germanium. At low fields, the slope of the drift velocity versus E-field is constant and is the low-field electron mobility, which is approximately 8500 cm²/V-sec for gallium arsenide. The low-field electron mobility in gallium arsenide is much larger than in silicon. As the field increases, the electron drift velocity in gallium arsenide reaches a peak and then decreases. A differential mobility is the slope of the $v_d$ versus E curve at a particular point on the curve and the negative slope of the drift velocity versus electric field represents a negative differential mobility. The negative differential mobility produces a negative differential resistance; this characteristic is used in the design of oscillators.

The negative differential mobility can be understood by considering the E versus k diagram for gallium arsenide which is shown again in Figure 5–8. The density of states effective mass of the electron in the lower valley is $m_n^* = 0.067\, m_0$. The small effective mass leads to a large mobility. As the E-field increases, the energy of the electron increases and the electron can be

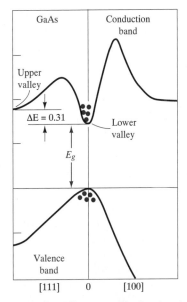

**Figure 5–8**   Energy-band structure for gallium arsenide showing the upper valley and lower valley in the conduction band. (From Sze [8]).

scattered into the upper valley, where the density of states effective mass is $0.55\ m_0$. The larger effective mass in the upper valley yields a smaller mobility. This intervalley transfer mechanism results in a decreasing average drift velocity of electrons with electric field, or the negative differential mobility characteristic. We will see in a later chapter how this negative differential mobility is used to make a high-frequency GUNN effect oscillator.

## 5.2 CARRIER DIFFUSION

There is a second mechanism, in addition to drift, that can induce a current in a semiconductor. We may consider a classic physics example in which a container, as shown in Figure 5–9, is divided into two parts by a membrane.

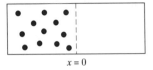

$x = 0$

**Figure 5–9**   Container divided by a membrane with gas molecules on one side.

The left side contains gas molecules at a particular temperature and the right side is initially empty. The gas molecules are in continual random thermal motion so that, when the membrane is broken, the gas molecules flow into the right side of the container. *Diffusion* is the process whereby particles flow from a region of high concentration toward a region of low concentration. If the gas molecules were electrically charged, the net flow of charge would result in a *diffusion current*.

### 5.2.1 Diffusion Current Density

To begin to understand the diffusion process in a semiconductor, we will consider a simplified analysis. Assume that an electron concentration varies in one dimension as shown in Figure 5–10. The temperature is assumed to be uniform so that the average thermal velocity of electrons is independent of $x$. To calculate the current, we will determine the net flow of electrons per unit time per unit area crossing the plane at $x = 0$. If the distance $l$ shown in Figure 5–10 is the mean-free path of an electron, that is, the average distance an electron travels between collisions ($l = v_{th}\tau_{cn}$), then on the average, electrons moving to the right at $x = -l$ and electrons moving to the left at $x = +l$ will cross the $x = 0$ plane. One half of the electrons at $x = -l$ will be traveling to the right at any instant of time and one half of the electrons at $x = +l$ will be traveling to the left at any given time. The net rate of electron flow, $F_n$, in the $+x$ direction at $x = 0$ is given by

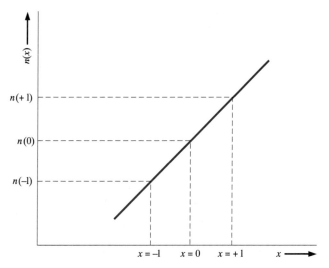

**Figure 5–10** Electron concentration versus distance.

$$F_n = \tfrac{1}{2}\, n(-l)v_{th} - \tfrac{1}{2}\, n(+l)v_{th} = \tfrac{1}{2}\, v_{th}[n(-l) - n(+l)] \qquad (5\text{–}27)$$

If we expand the electron concentration in a Taylor series about $x = 0$ keeping only the first two terms, then Equation (5–27) can be written as

$$F_n = \frac{1}{2}\, v_{th} \left\{ \left[ n(0) - l\,\frac{dn}{dx} \right] - \left[ n(0) + l\,\frac{dn}{dx} \right] \right\} \qquad (5\text{–}28)$$

which becomes

$$F_n = -v_{th}l\,\frac{dn}{dx} \qquad (5\text{–}29)$$

Each electron has a charge $(-e)$ so the current is

$$J = -eF_n = +ev_{th}l\,\frac{dn}{dx} \qquad (5\text{–}30)$$

The current described by Equation (5–30) is the electron diffusion current and is proportional to the spatial derivative, or density gradient, of the electron concentration.

The diffusion of electrons from a region of high concentration to a region of low concentration produces a flux of electrons flowing in the negative x direction for this example. Since electrons have a negative charge, the conventional current direction is in the positive $x$ direction. These one-dimensional flux and current directions are shown in Figure 5–11a. We may write the electron diffusion current density for this one-dimensional case, in the form

$$J_{nx|dif} = eD_n\,\frac{dn}{dx} \qquad (5\text{–}31)$$

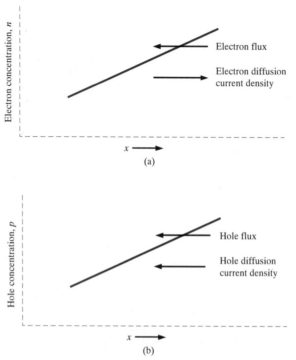

**Figure 5–11**   (a) Diffusion of electrons due to a density gradient. (b) Diffusion of holes due to a density gradient.

where $D_n$ is called the electron diffusion coefficient, has units of cm²/sec, and is a positive quantity. If the electron density gradient becomes negative, the electron diffusion current density will be in the negative $x$ direction.

Figure 5–11b shows an example of a hole concentration as a function of distance in a semiconductor. The diffusion of holes, from a region of high concentration to a region of low concentration, produces a flux of holes in the negative $x$ direction. Since holes are positively charged particles, the conventional diffusion current density is also in the negative $x$ direction. The hole diffusion current density is proportional to the hole density gradient and to the electronic charge, so we may write

$$J_{px|dif} = -eD_p \frac{dp}{dx} \tag{5–32}$$

for the one-dimensional case. The parameter $D_p$ is called the hole diffusion coefficient, has units of cm²/sec, and is a positive quantity. If the hole density gradient becomes negative, the hole diffusion current density will be in the positive x direction.

**Example 5–4**

**Objective:** To calculate the diffusion current density given a density gradient.

Assume that, in an n-type gallium arsenide semiconductor at $T = 300°K$, the electron concentration varies linearly from $1 \times 10^{18}$ to $7 \times 10^{17}$ cm$^{-3}$ over a distance of 0.10 cm. Calculate the diffusion current density if the electron diffusion coefficient is $D_n = 225$ cm$^2$/sec.

**Solution:** The diffusion current density is given by

$$J_{n|dif} = eD_n \frac{dn}{dx} \approx eD_n \frac{\Delta n}{\Delta x}$$

$$= (1.6 \times 10^{-19})(225) \left( \frac{1 \times 10^{18} - 7 \times 10^{17}}{0.10} \right) = 108 \text{ A/cm}^2$$

**Comment:** A significant diffusion current density can be generated in a semiconductor material with only a modest density gradient.

## 5.2.2 Total Current Density

We now have four possible independent current mechanisms in a semiconductor. These components are electron drift and diffusion currents and hole drift and diffusion currents. The total current density is the sum of these four components, or for the one-dimensional case, we have

$$J = en\mu_n E_x + ep\mu_p E_x + eD_n \frac{dn}{dx} - eD_p \frac{dp}{dx} \qquad (5\text{–}33)$$

This equation may be generalized to three dimensions as

$$J = en\mu_n E + ep\mu_p E + eD_n \nabla n - eD_p \nabla p \qquad (5\text{–}34)$$

The electron mobility gives an indication of how well an electron moves in a semiconductor due to the force of an electric field. The electron diffusion coefficient gives an indication of how well an electron moves in a semiconductor as a result of a density gradient. The electron mobility and diffusion coefficient are not independent parameters. Similarly, the hole mobility and diffusion coefficient are not independent parameters. The relationship between mobility and the diffusion coefficient will be developed in the next section.

The expression for the total current in a semiconductor contains four terms. Fortunately in most situations, we will only need to consider one term at any one time at a particular point in a semiconductor.

## 5.3 GRADED IMPURITY DISTRIBUTION

In most cases so far, we have assumed that the semiconductor is uniformly doped. In many semiconductor devices, however, there may be regions that are nonuniformly doped. We will investigate how a nonuniformly doped semiconductor reaches thermal equilibrium and, from this analysis, we will derive the Einstein relation which relates mobility and the diffusion coefficient.

### 5.3.1 Induced Electric Field

Consider a semiconductor that is nonuniformly doped with donor impurity atoms. If the semiconductor is in thermal equilibrium, the Fermi energy level is constant through the crystal so the energy-band diagram may qualitatively look like that shown in Figure 5–12. The doping concentration decreases as $x$ increases in this case. There will be a diffusion of majority carrier electrons from the region of high concentration to the region of low concentration which is in the $+x$ direction. The flow of negative electrons leaves behind positively charged donor ions. The separation of positive and negative charge induces an electric field that is in a direction to oppose the diffusion process. When equilibrium is reached, the mobile carrier concentration is not exactly equal to the fixed impurity concentration and the induced electric field prevents any further separation of charge. In most cases of interest, the space charge induced by this diffusion process is a small fraction of the impurity concentration, thus the mobile carrier concentration is not too different from the impurity dopant density.

The electric potential $\phi$ is related to electron potential energy by the charge $(-e)$, so we can write

$$\phi = + \frac{1}{e} (E_F - E_{Fi}) \tag{5–35}$$

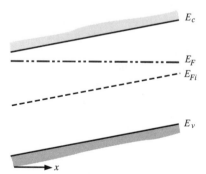

**Figure 5–12**   Energy-band diagram for a semiconductor in thermal equilibrium with a nonuniform donor impurity concentration.

The electric field for the one-dimensional situation is defined as

$$E_x = -\frac{d\phi}{dx} = \frac{1}{e}\frac{dE_{Fi}}{dx} \tag{5-36}$$

If the intrinsic Fermi level changes as a function of distance through a semiconductor in thermal equilibrium, an electric field exists in the semiconductor.

If we assume a quasi-neutrality condition in which the electron concentration is almost equal to the donor impurity concentration, then we can still write

$$n_0 = n_i \exp\left[\frac{E_F - E_{Fi}}{kT}\right] \approx N_d(x) \tag{5-37}$$

Solving for $E_F - E_{Fi}$, we obtain

$$E_F - E_{Fi} = kT\ln\left(\frac{N_d(x)}{n_i}\right) \tag{5-38}$$

The Fermi level is constant for thermal equilibrium so when we take the derivative with respect to $x$ we obtain

$$-\frac{dE_{Fi}}{dx} = \frac{kT}{N_d(x)}\frac{dN_d(x)}{dx} \tag{5-39}$$

The electric field can then be written, combining Equations (5–39) and (5–36), as

$$E_x = -\left(\frac{kT}{e}\right)\frac{1}{N_d(x)}\frac{dN_d(x)}{dx} \tag{5-40}$$

Since we have an electric field, there will be a potential difference through the semiconductor due to the nonuniform doping.

---

**Example 5–5**

**Objective:** To determine the induced electric field in a semiconductor in thermal equilibrium, given a linear variation in doping concentration.

Assume that the donor concentration in an n-type semiconductor at $T = 300°K$ is given by

$$N_d(x) = 10^{16} - 10^{19}x \quad (cm^{-3})$$

where $x$ is given in cm and ranges between $0 \le x \le 1 \ \mu m$.

**Solution:** Taking the derivative of the donor concentration, we have

$$\frac{dN_d(x)}{dx} = -10^{19} \quad (cm^{-4})$$

The electric field is given by Equation (5–40) so that we have

$$E_x = \frac{-(0.0259)(-10^{19})}{(10^{16} - 10^{19}x)}$$

At $x = 0$, for example, we find

$$E_x = 25.9 \text{ V/cm}$$

**Comment:** We may recall from our previous discussion of drift current that fairly small electric fields can produce significant drift current densities, so that an induced electric field from nonuniform doping can significantly influence semiconductor device characteristics.

### 5.3.2 The Einstein Relation

If we consider the nonuniformly doped semiconductor represented by the energy-band diagram shown in Figure 5–12 and assume there are no electrical connections so that the semiconductor is in thermal equilibrium, then the individual electron and hole currents must be zero. We can write

$$J_n = 0 = en\mu_n E_x + eD_n \frac{dn}{dx} \tag{5–41}$$

If we assume quasi-neutrality so that $n \cong N_d(x)$, then Equation (5–41) can be rewritten as

$$J_n = 0 = e\mu_n N_d(x)E_x + eD_n \frac{dN_d(x)}{dx} \tag{5–42}$$

Substituting the expression for the electric field from Equation (5–40) into Equation (5–42), we obtain

$$0 = -e\mu_n N_d(x) \left(\frac{kT}{e}\right) \frac{1}{N_d(x)} \frac{dN_d(x)}{dx} + eD_n \frac{dN_d(x)}{dx} \tag{5–43}$$

Equation (5–43) is valid for the condition

$$\frac{D_n}{\mu_n} = \frac{kT}{e} \tag{5–44a}$$

The hole current must also be zero in the semiconductor. From this condition, we can show that

$$\frac{D_p}{\mu_p} = \frac{kT}{e} \tag{5–44b}$$

Combining Equations (5-44a) and (5-44b), we have

$$\frac{D_n}{\mu_n} = \frac{D_p}{\mu_p} = \frac{kT}{e} \tag{5–45}$$

The diffusion coefficient and mobility are not independent parameters. This relation between the mobility and diffusion coefficient, given by Equation (5–45), is known as the *Einstein relation*.

---

**Example 5–6**

**Objective:** To determine the diffusion coefficient given the carrier mobility.
Assume that the mobility of a particular carrier is 1000 cm$^2$/V-sec at $T = 300°$K.

**Solution:** Using the Einstein relation, we have that

$$D = \left(\frac{kT}{e}\right) \mu = (0.0259)(1000) = 25.9 \text{ cm}^2/\text{sec}$$

**Comment:** Although this example is fairly simple and straightforward, it is important to keep in mind the relative order of magnitudes of the mobility and diffusion coefficient. The diffusion coefficient is approximately 40 times smaller than the mobility at room temperature.

---

Table 5–2 shows the diffusion coefficient values at $T = 300°$K corresponding to the mobilities listed in Table 5–1 for silicon, gallium arsenide, and germanium.

Table 5–2  Typical mobility and diffusion coefficient values at $T = 300°$K ($\mu$ = cm$^2$/V-sec and $D$ = cm$^2$/sec)

|  | $\mu_n$ | $D_n$ | $\mu_p$ | $D_p$ |
|---|---|---|---|---|
| Silicon | 1350 | 35 | 480 | 12.4 |
| Gallium arsenide | 8500 | 220 | 400 | 10.4 |
| Germanium | 3900 | 101 | 1900 | 49.2 |

The relation between the mobility and diffusion coefficient given by Equation (5–45) contains temperature. It is important to keep in mind that the major temperature effects are a result of lattice scattering and ionized impurity scattering processes, as discussed in Section 5.1.2. As the mobilities are strong functions of temperature due to the scattering processes, the diffusion coefficients are also strong functions of temperature. The specific temperature dependence given in Equation (5–45) is a small fraction of the real temperature characteristic.

## *5.4 THE HALL EFFECT

The Hall effect is a consequence of the forces that are exerted on moving charges by electric and magnetic fields. The Hall effect is used to distinguish whether a semiconductor is n-type or p-type[1] and is used to measure the

---

[1] We will assume an extrinsic semiconductor material in which the majority carrier concentration is much larger than the minority carrier concentration.

majority carrier concentration and majority carrier mobility. The Hall effect device, as discussed in this section, is used to experimentally measure semi-conductor parameters. However, the Hall effect device is used extensively in engineering applications as a magnetic probe and in other circuit applications.

The force on a particle having a charge $q$ and moving in a magnetic field is given by

$$F = qv \times B \tag{5-46}$$

where the cross product is taken between velocity and magnetic field so that the force vector is perpendicular to both the velocity and magnetic field.

Figure 5–13 illustrates the Hall effect. A semiconductor with a current $I_x$ is placed in a magnetic field perpendicular to the current. In this case, the

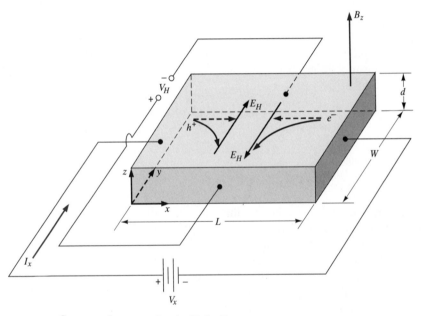

**Figure 5-13**    Geometry for measuring the Hall effect.

magnetic field is in the $z$-direction. Electrons and holes flowing in the semi-conductor will experience a force as indicated in the figure. The force on both electrons and holes is in the $(-y)$ direction. In a p-type semiconductor $(p_0 > n_0)$, there will be a buildup of positive charge on the $y = 0$ surface of the semiconductor and, in an n-type semiconductor $(n_0 > p_0)$, there will be a buildup of negative charge on the $y = 0$ surface. This net charge induces an electric field in the $y$-direction as shown in the figure. In steady state, the magnetic field force will be exactly balanced by the induced electric field force. This balance may be written as

$$F = q[E + v \times B] = 0 \qquad (5.47a)$$

which becomes

$$qE_y = qv_x B_z \qquad (5.47b)$$

The induced electric field in the $y$-direction is called the Hall field. The Hall field produces a voltage across the semiconductor which is called the Hall voltage. We can write

$$V_H = +E_H W \qquad (5-48)$$

where $E_H$ is assumed positive in the $+y$ direction and $V_H$ is positive with the polarity shown.

In a p-type semiconductor in which holes are the majority carrier, the Hall voltage will be positive as defined in Figure 5–13. In an n-type semiconductor in which electrons are the majority carrier, the Hall voltage will have the opposite polarity. The polarity of the Hall voltage is used to determine whether an extrinsic semiconductor is n-type or p-type.

Substituting Equation (5–48) into Equation (5–47), we have

$$V_H = v_x W B_z \qquad (5-49)$$

For a p-type semiconductor, the drift velocity of holes can be written as

$$v_{dx} = \frac{J_x}{ep} = \frac{I_x}{(ep)(Wd)} \qquad (5-50)$$

where $e$ is the magnitude of the electronic charge. Combining Equations (5–50) and (5–49), we have

$$V_H = \frac{I_x B_z}{epd} \qquad (5-51)$$

or, solving for the hole concentration, we obtain

$$p = \frac{I_x B_z}{edV_H} \qquad (5-52)$$

The majority carrier hole concentration is determined from the current, magnetic field, and Hall voltage.

For an n-type semiconductor, the Hall voltage is given by

$$V_H = -\frac{I_x B_z}{ned} \qquad (5-53)$$

so that the electron concentration is

$$n = -\frac{I_x B_z}{edV_H} \qquad (5-54)$$

We may note that the Hall voltage is negative for the n-type semiconductor;

therefore, the electron concentration determined from Equation (5–54) is actually a positive quantity.

Once the majority carrier concentration has been determined, we can calculate the low-field majority carrier mobility. For a p-type semiconductor, we can write

$$J_x = ep\mu_p E_x \qquad (5\text{–}55)$$

The current density and electric field can be converted to current and voltage so that Equation (5–55) becomes

$$\frac{I_x}{Wd} = \frac{ep\mu_p V_x}{L} \qquad (5\text{–}56)$$

The hole mobility is then given by

$$\mu_p = \frac{I_x L}{epV_x Wd} \qquad (5\text{–}57)$$

Similarly for an n-type semiconductor, the low-field electron mobility is determined from

$$\mu_n = \frac{I_x L}{enV_x Wd} \qquad (5\text{–}58)$$

---

**Example 5–7**

**Objective:** To determine the majority carrier concentration and mobility given Hall effect parameters.

Consider the geometry shown in Figure 5–13. Let $L = 10^{-1}$ cm, $W = 10^{-2}$ cm, and $d = 10^{-3}$ cm. Also assume that $I_x = 1.0$ mA, $V_x = 12.5$ volts, $B_z = 500$ gauss $= 5 \times 10^{-2}$ tesla, and $V_H = -6.25$ mV.

**Solution:** A negative Hall voltage for this geometry implies that we have an n-type semiconductor. Using Equation (5–54), we can calculate the electron concentration as

$$n = \frac{-(10^{-3})(5 \times 10^{-2})}{(1.6 \times 10^{-19})(10^{-5})(-6.25 \times 10^{-3})} = 5 \times 10^{21} \text{ m}^{-3} = 5 \times 10^{15} \text{ cm}^{-3}$$

The electron mobility is then determined from Equation (5–58) as

$$\mu_n = \frac{(10^{-3})(10^{-3})}{(1.6 \times 10^{-19})(5 \times 10^{21})(12.5)(10^{-4})(10^{-5})} = 0.10 \text{ m}^2/\text{V-sec}$$

or

$$\mu_n = 1000 \text{ cm}^2/\text{V-sec}$$

**Comment:** It is important to note that the MKS units must be used consistently in the Hall effect equations to yield correct results.

## 5.5 SUMMARY AND REVIEW

In this chapter we have considered carrier transport, the process by which charge moves in a semiconductor to produce current. We explored the theory of the two transport mechanisms, drift and diffusion.

We first considered drift. The motion or drift of electrons and holes due to electric fields leads to drift currents. The drift current density in a semiconductor is a function of the concentration of electrons and holes and is also a function of the average drift velocity of these charge carriers. The net flow of electrons and holes, due to an applied electric field, is in opposite directions because of the difference in charge, but the drift currents generated by the electrons and holes are in the same direction as the electric field.

The average drift velocity of a carrier is a function of the electric field and carrier mobility. Carrier mobility is an indicator of how well an electron or hole will drift through the semiconductor, and a function of the scattering or collision processes within the semiconductor. Two scattering mechanisms are lattice scattering and impurity scattering. Lattice scattering occurs because atoms in a semiconductor contain a certain amount of thermal energy in the form of atomic vibrations. The interatomic distance between atoms is then fluctuating, the perfect periodic potential function is altered, and the motion of the carrier through the semiconductor is disrupted. Since this scattering mechanism is a function of the thermal motion of atoms, the associated component of mobility is a strong function of temperature. Mobility decreases as temperature increases. Impurity scattering occurs because donor and acceptor impurities are ionized: a coulomb interaction occurs between the charged impurities and electrons and holes. When carriers drift into the vicinity of a charged impurity atom, they will be deflected or scattered. Mobility decreases as the impurity concentration increases.

Conductivity and resistivity are parameters used extensively in semiconductor material specifications, resistivity being the reciprocal of conductivity. These parameters are functions of electron and hole concentrations and mobilities.

We then considered diffusion. Diffusion of electrons and holes in a semiconductor leads to diffusion currents proportional to the gradients in electron and hole concentrations. The constant of proportionality is the electron diffusion coefficient or hole diffusion coefficient. The diffusion coefficients indicate how well the charge carriers diffuse in a semiconductor. The carrier diffusion coefficient and mobility are not independent parameters, but are related by the Einstein relation.

The Hall effect is used to experimentally determine conductivity type, majority carrier concentration, and majority carrier mobility. A magnetic field is applied perpendicular to a current in the semiconductor. A moving charge particle in a magnetic field experiences a force perpendicular to both the magnetic field and direction of motion. An electric field, the Hall field, is induced in the semiconductor due to these forces. The induced field produces a voltage, the Hall voltage, across the semiconductor. The polarity of the Hall voltage is a function of the semiconductor conductivity type and the magnitude of the Hall voltage is inversely proportional to the majority carrier concentration.

The two current processes resulting from drift and diffusion in a semiconductor are the basis for the current-voltage characteristics of semiconductor devices. We have implicitly assumed in these discussions that, though there is a net flow of electrons and holes, thermal equilibrium has not been substantially disturbed. In the next chapter, we will consider nonequilibrium processes in a semiconductor and then apply these two transport processes to both the equilibrium and nonequilibrium carriers in semiconductor devices.

## GLOSSARY OF IMPORTANT TERMS

**Conductivity:** A material parameter related to carrier drift; quantitatively, the ratio of drift current density to electric field.

**Diffusion:** The process whereby particles flow from a region of high concentration to a region of low concentration.

**Diffusion coefficient:** The parameter relating particle flux to the particle density gradient.

**Diffusion current:** The current that results from the diffusion of charged particles.

**Drift:** The process whereby charged particles move while under the influence of an electric field.

**Drift current:** The current that results from the drift of charged particles.

**Drift velocity:** The average velocity of charged particles in the presence of an electric field.

**Einstein relation:** The relation between the mobility and the diffusion coefficient.

**Hall voltage:** The voltage induced across a semiconductor in a Hall-effect measurement.

**Ionized impurity scattering:** The interaction between a charged carrier and an ionized impurity center.

**Lattice scattering:** The interaction between a charged carrier and a thermally vibrating lattice atom.

**Mobility:** The parameter relating carrier drift velocity and electric field.

**Resistivity:** The reciprocal of conductivity; a material parameter that is a measure of the resistance to current.

**Velocity saturation:** The saturation of carrier drift velocity with increasing electric field.

## PROBLEMS

(Note: Use the semiconductor parameters given in Appendix B if the parameters are not specifically given in a problem.)

### Section 5.1

1.  Consider a homogeneous gallium arsenide semiconductor at $T = 300°K$ with $N_d = 10^{16}$ cm$^{-3}$ and $N_a = 0$. (a) Calculate the thermal-equilibrium values of electron and hole concentrations. (b) For an applied E-field of 10 V/cm, calculate the drift current density. (c) Repeat parts (a) and (b) if $N_d = 0$ and $N_a = 10^{16}$ cm$^{-3}$.

2.  A silicon crystal having a cross-sectional area of 0.001 cm$^2$ and a length of $10^{-3}$ cm is connected at its ends to a 10-volt battery. At $T = 300°K$, we want a current of 100 mA in the silicon. Calculate: (a) the required resistance R, (b) the required conductivity, (c) the density of donor atoms to be added to achieve this conductivity, and (d) the concentration of acceptor atoms to be added to form a compensated p-type material with the conductivity given from part (b) if the initial concentration of donor atoms is $N_d = 10^{15}$ cm$^{-3}$.

3.  Calculate the average drift velocity (magnitude) of holes in a bar of silicon with a cross-sectional area of $10^{-4}$ cm$^2$, containing a hole concentration of $4.5 \times 10^{15}$ cm$^{-3}$ and carrying a current of 45 mA.

4.  A silicon semiconductor material at $T = 300°K$ is doped with donor atoms at $N_d = 8 \times 10^{10}$ cm$^{-3}$. Acceptor atoms are added so that the semiconductor becomes a compensated p-type material. (Assume complete ionization). If the conductivity of the final material is $7 \times 10^{-6}$ (ohm-cm)$^{-1}$, calculate: (a) the thermal-equilibrium majority carrier concentration, (b) the thermal-equilibrium minority carrier concentration, and (c) the concentration of acceptor atoms which need to be added.

5.  (a) In a p-type gallium arsenide semiconductor, the conductivity is $\sigma = 5$ (ohm-cm)$^{-1}$ at $T = 300°K$. Calculate the thermal-equilibrium values of the electron and hole concentrations. (b) Repeat part (a) for n-type silicon if the resistivity is $\rho = 8$ ohm-cm.

6.  A particular semiconductor has the following parameters:

$$N_c = 1.25 \times 10^{14}(T)^{3/2} \text{ cm}^{-3} \qquad \mu_n = 4200 \text{ cm}^2/\text{V-sec}$$
$$N_v = 8.08 \times 10^{13}(T)^{3/2} \text{ cm}^{-3} \qquad \mu_p = 1900 \text{ cm}^2/\text{V-sec}$$

where $\mu_n$ and $\mu_p$ are assumed to be independent of temperature. The minimum resistivity of the intrinsic material at $T = 450°K$ is to be $6.1 \times 10^8$ ohm-cm. (a) What is the maximum value of the intrinsic carrier concentration at this temperature? (b) What is the minimum value of the bandgap energy, $E_g$?

7.  Consider a semiconductor that is uniformly doped with $N_d = 10^{14}$ cm$^{-3}$ and $N_a = 0$, with an applied electric field of E = 100 V/cm. Assume that $\mu_n = 1000$ cm$^2$/V-sec and $\mu_p = 0$. Also assume the following parameters:

$$N_c = 2 \times 10^{19}(T/300)^{3/2} \text{ cm}^{-3}$$
$$N_v = 1 \times 10^{19}(T/300)^{3/2} \text{ cm}^{-3}$$
$$E_g = 1.10 \text{ eV}$$

(a) Calculate the electric-current density at $T = 300°$K. (b) At what temperature will this current increase by 5 percent? (Assume the mobilities are independent of temperature.)

8.  A semiconductor material has electron and hole mobilities $\mu_n$ and $\mu_p$, respectively. When the conductivity is considered as a function of the hole concentration, $p_0$, (a) show that the minimum value of conductivity, $\sigma_{min}$, can be written as

$$\sigma_{min} = \frac{2\sigma_i(\mu_n\mu_p)^{1/2}}{(\mu_n + \mu_p)}$$

where $\sigma_i$ is the intrinsic conductivity, and (b) show that the corresponding hole concentration is $p_0 = n_i(\mu_n/\mu_p)^{1/2}$.

9.  Assume that the mobility of electrons in silicon at $T = 300°$K is $\mu_n = 1300$ cm$^2$/V-sec. Also assume that the mobility is limited by lattice scattering and varies as $T^{-3/2}$. Determine the electron mobility at (a) $T = 200°$K and (b) $T = 400°$K.

10. Consider the drift of electrons at $T = 300°$K in silicon. Using Figure 5–7 and Equation (5–7), determine the effective electron mobility at (a) E = $10^3$ V/cm and (b) E = $10^5$ V/cm.

## Section 5.2

11. Consider a sample of silicon at $T = 300°$K. Assume that the electron concentration varies linearly with distance, as shown in Figure 5–14. The diffusion current density is found to be $J_n = 0.19$ A/cm$^2$.

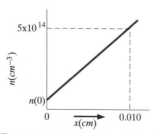

**Figure 5–14**   Figure for problem 11.

If the electron diffusion coefficient is $D_n = 25$ cm$^2$/sec, determine the electron concentration at $x = 0$.

12. The hole concentration in germanium at $T = 300°K$ varies as

$$p(x) = 10^{15} \exp\left(\frac{-x}{22.5}\right) \text{ cm}^{-3}$$

where $x$ is measured in $\mu m$. If the hole diffusion coefficient is $D_p = 48$ cm$^2$/sec, determine the hole diffusion current density as a function of x.

13. The electron concentration in silicon at $T = 300°K$ is given by

$$n(x) = 10^{16} \exp\left(\frac{-x}{18}\right) \text{ cm}^{-3}$$

where $x$ is measured in $\mu m$ and is limited to $0 \le x \le 25$ $\mu m$. The electron diffusion coefficient is $D_n = 25$ cm$^2$/sec and the electron mobility is $\mu_n = 960$ cm$^2$/V-sec. The total electron current density through the semiconductor is constant and equal to $J_n = -40$ A/cm$^2$. The electron current has both diffusion and drift current components. Determine the electric field as a function of $x$ which must exist in the semiconductor.

14. The total current in a semiconductor is constant and is composed of electron drift current and hole diffusion current. The electron concentration is constant and is equal to $10^{16}$ cm$^{-3}$. The hole concentration is given by

$$p(x) = 10^{15} \exp\left(\frac{-x}{L}\right) \text{ cm}^{-3} \qquad (x \ge 0)$$

where $L = 12$ $\mu m$. The hole diffusion coefficient is $D_p = 12$ cm$^2$/sec and the electron mobility is $\mu_n = 1000$ cm$^2$/V-sec. The total current density is $J = 4.8$ A/cm$^2$. Calculate: (a) the hole diffusion current density versus $x$, (b) the electron current density versus $x$, and (c) the electric field versus $x$.

*15. A constant electric field, E $= 12$ V/cm, exists in the $+x$ direction of an n-type gallium arsenide semiconductor for $0 \le x \le 50$ $\mu m$. The total current density is a constant and is $J = 100$ A/cm$^2$. At $x = 0$, the drift and diffusion currents are equal. Let $T = 300°K$ and $\mu_n = 8000$ cm$^2$/V-sec. (a) Determine the expression for the electron concentration $n(x)$. (b) Calculate the electron concentration at $x = 0$ and at $x = 50$ $\mu m$. (c) Calculate the drift and diffusion current densities at $x = 50$ $\mu m$.

## Section 5.3

16. Consider a semiconductor in thermal equilibrium (no current). Assume that the donor concentration varies exponentially as

$$N_d(x) = N_{d0} \exp(-ax)$$

over the range $0 \le x \le 1/\alpha$ where $N_{d0}$ is a constant. (a) Calculate the electric field as a function of $x$ for $0 \le x \le 1/\alpha$. (b) Calculate the potential difference between $x = 0$ and $x = 1/\alpha$.

17. Using the data in Example 5–5, calculate the potential difference between $x = 0$ and $x = 1$ $\mu m$.

18. Assume that the diffusion coefficient of a carrier at $T = 300°K$ is $D = 12.5$ cm$^2$/sec. Calculate the carrier mobility.

## Section 5.4

(Note: Refer to Figure 5–13 for the geometry of the Hall effect).

19. A sample of silicon is doped with $10^{16}$ boron atoms per cm$^3$. The Hall sample has the same geometrical dimensions given in Example 5–7. The current is $I_x = 1$ mA with $B_z = 350$ gauss $= 3.5 \times 10^{-2}$ tesla. Determine (a) the Hall voltage and (b) the Hall field.

20. Germanium is doped with $5 \times 10^{15}$ donor atoms per cm$^3$ at $T = 300°K$. The dimensions of the Hall device are $d = 5 \times 10^{-3}$ cm, $W = 2 \times 10^{-2}$ cm, and $L = 10^{-1}$ cm. The current is $I_x = 250$ $\mu A$, the applied voltage is $V_x = 100$ mV, and the magnetic flux density is $B_z = 500$ gauss $= 5 \times 10^{-2}$ tesla. Calculate: (a) the Hall voltage, (b) the Hall field, and (c) the carrier mobility.

21. A silicon Hall device at $T = 300°K$ has the following geometry: $d = 10^{-3}$ cm, $W = 10^{-2}$ cm, and $L = 10^{-1}$ cm. The following parameters are measured: $I_x = 0.75$ mA, $V_x = 15$ volts, $V_H = +5.8$ mV, and $B_z = 1000$ gauss $= 10^{-1}$ tesla. Determine: (a) the conductivity type, (b) the majority carrier concentration, and (c) the majority carrier mobility.

22. Consider silicon at $T = 300°K$. A Hall effect device is fabricated with the following geometry: $d = 5 \times 10^{-3}$ cm, $W = 5 \times 10^{-2}$ cm, and $L = 0.50$ cm. The electrical parameters measured are: $I_x = 0.50$ mA, $V_x = 1.25$ volts, and $B_z = 650$ gauss $= 6.5 \times 10^{-2}$ tesla. The Hall field is $E_H = -16.5$ mV/cm. Determine (a) the Hall voltage, (b) the conductivity type, (c) the majority carrier concentration, and (d) the majority carrier mobility.

23. Consider a gallium arsenide sample at $T = 300°K$. A Hall effect device has been fabricated with the following geometry: $d = 0.01$ cm, $W = 0.05$ cm, and $L = 0.5$ cm. The electrical parameters are: $I_x = 2.5$ mA, $V_x = 2.2$ volts, and $B_z = 2.5 \times 10^{-2}$ tesla. The Hall voltage is $V_H = -4.5$ mV. Find: (a) the conductivity type, (b)

the majority carrier concentration, (c) the mobility, and (d) the resistivity.

## Summary and Review

24. An n-type silicon semiconductor resistor is to be designed so that it carries a current of 5 mA with an applied voltage of 5V. (a) If $N_d = 3 \times 10^{14}$ cm$^{-3}$ and $N_a = 0$, design a resistor to meet the required specifications. (b) If $N_d = 3 \times 10^{16}$ cm$^{-3}$ and $N_a = 2.5 \times 10^{16}$ cm$^{-3}$, redesign the resistor. (c) Discuss the relative lengths of the two designs compared to the doping concentration. Is there a linear relationship?

25. When fabricating a Hall-effect device, the two points at which the Hall voltage is measured may not be lined up exactly perpendicular to the current $I_x$ (see Figure 5–13). Discuss the effect this misalignment will have on the Hall voltage. Show that a valid Hall voltage can be obtained from two measurements: first with the magnetic field in the $+z$ direction, and then in the $-z$ direction.

26. Another technique for determining the conductivity type of a semiconductor is called the hot probe method. It consists of two probes and an ammeter that indicates the direction of current. One probe is heated and the other is at room temperature. No voltage is applied, but a current will exist when the probes touch the semiconductor. Explain the operation of this hot probe technique and sketch a diagram indicating the direction of current for p- and n-type semiconductor samples.

## READING LIST

1. Muller, R. S., and T. I. Kamins. *Device Electronics for Integrated Circuits.* 2nd ed. New York: Wiley, 1986.

2. Navon, D. H. *Semiconductor Microdevices and Materials.* New York: Holt, Rinehart & Winston, 1986.

3. Pierret, R. F. *Semiconductor Fundamentals.* Vol. 1 of *Modular Series on Solid State Devices.* 2nd ed. Reading, Mass.: Addison-Wesley, 1988.

4. ———. *Advanced Semiconductor Fundamentals.* Vol. 6 of *Modular Series on Solid State Devices.* Reading, Mass.: Addison-Wesley, 1987.

*5. Shur, M. *Physics of Semiconductor Devices.* Englewood Cliffs, N.J.: Prentice Hall, 1990.

6. Streetman, B. G. *Solid State Electronic Devices.* 3rd ed. Englewood Cliffs, N.J.: Prentice Hall, 1990.

7. Sze, S. M. *Physics of Semiconductor Devices*. 2nd ed. New York: Wiley, 1981.

8. ————. *Semiconductor Devices: Physics and Technology*. New York: Wiley, 1985.

9. van der Ziel, A. *Solid State Physical Electronics*. 2nd ed. Englewood Cliffs, N.J.: Prentice Hall, 1968.

*10. Wang, S. *Fundamentals of Semiconductor Theory and Device Physics*. Englewood Cliffs, N.J.: Prentice Hall, 1989.

11. Yang, E. S. *Microelectronic Devices*. New York: McGraw-Hill 1988.

# NONEQUILIBRIUM EXCESS CARRIERS IN SEMICONDUCTORS

## CHAPTER 6

## PREVIEW

Our discussion of the physics of semiconductors in Chapter 4 was based on thermal equilibrium. When a voltage is applied or a current exists in a semiconductor device, the semiconductor is operating under nonequilibrium conditions. In our discussion of current transport in Chapter 5, we did not address nonequilibrium conditions but implicitly assumed that equilibrium was not significantly disturbed. In this chapter, we will discuss the behavior of nonequilibrium electron and hole concentrations as functions of time and space coordinates.

Excess electrons in the conduction band and excess holes in the valence band may exist in addition to the thermal equilibrium concentrations if an external excitation is applied to the semiconductor. We will begin with a general discussion of the generation and recombination of excess carriers and the rates at which they are created and recombine. We will derive the continuity equations and time-dependent diffusion equations for electrons and holes, which describe the behavior of these particles in spatial coordinates and in time. These equations relate diffusion, drift, generation, and recombination to the rate of change of the carrier concentrations.

Excess electrons and excess holes do not move independently of each other. They diffuse, drift, and recombine with the same effective diffusion coefficient, drift mobility, and lifetime. This phenomenon is called ambipolar transport. We will develop the ambipolar transport equation which describes the behavior of the excess electrons and holes. The behavior of excess carriers is fundamental to the operation of semiconductor devices. Several examples of the generation of excess carriers will be explored to illustrate the characteristics of the ambipolar transport phenomenon.

The Fermi energy was previously defined for a semiconductor in thermal equilibrium. The creation of excess electrons and holes means that the semiconductor is no longer in thermal equilibrium. We can define two new parameters that apply to the nonequilibrium semiconductor: the quasi-Fermi energy for electrons and the quasi-Fermi energy for holes.

Another parameter that characterizes the nonequilibrium semiconductor is the excess-carrier lifetime, describing the mean time that an excess electron or hole will exist before recombining. The basic theory relating the excess-carrier lifetime to other semiconductor parameters was developed by Shockley, Read, and Hall. We will develop this theory and apply the results to specific situations.

Semiconductor devices are generally fabricated at or near a surface. We will consider the effect of these surfaces on the characteristics of excess electrons and holes. These effects can significantly influence the semiconductor device properties.

## 6.1 CARRIER GENERATION AND RECOMBINATION

Any deviation from thermal equilibrium will tend to change the electron and hole concentrations in a semiconductor. A sudden increase in temperature, for example, will increase the rate at which electrons and holes are thermally generated so that their concentrations will change with time until new equilibrium values are reached. An external excitation, such as light (a flux of photons), can also generate electrons and holes, creating a nonequilibrium condition. To understand the generation and recombination processes, we will first consider direct band-to-band generation and recombination, and then later, the effect of allowed electronic energy states within the bandgap, referred to as traps or recombination centers.

### 6.1.1 The Semiconductor in Equilibrium

We have determined the thermal-equilibrium concentration of electrons and holes in the conduction and valence bands, respectively. In thermal equilibrium, these concentrations are independent of time. However, electrons are continually being thermally excited from the valence band into the conduction band by the random nature of the thermal process. At the same time, electrons moving randomly through the crystal in the conduction band may come in close proximity to holes and "fall" into the empty states in the valence band. This recombination process annihilates both the electron and hole. Since the net carrier concentrations are independent of time in thermal equilibrium, the rate at which electrons and holes are generated and the rate at which they recombine must be equal. The generation and recombination processes are schematically shown in Figure 6–1.

Let $G_{n0}$ and $G_{p0}$ be the thermal-generation rates of electrons and holes, respectively, given in units of #/cm$^3$-sec. For the direct band-to-band generation, the electrons and holes are created in pairs so we must have that

$$G_{n0} = G_{p0} \tag{6-1}$$

Let $R_{n0}$ and $R_{p0}$ be the recombination rates of electrons and holes, respectively, for a semiconductor in thermal equilibrium, again given in units of

**Figure 6–1**  Electron-hole generation and recombination.

$\#/cm^3$-sec. In direct band-to-band recombination, electrons and holes recombine in pairs so that

$$R_{n0} = R_{p0} \qquad (6\text{–}2)$$

In thermal equilibrium, the concentrations of electrons and holes are independent of time; therefore, the generation and recombination rates are equal, so we have

$$G_{n0} = G_{p0} = R_{n0} = R_{p0} \qquad (6\text{–}3)$$

## 6.1.2 Excess Carrier Generation and Recombination

Electrons in the valence band may be excited into the conduction band when, for example, high-energy photons are incident on a semiconductor. When this happens, not only is an electron created in the conduction band, but a hole is created in the valence band; thus an electron-hole pair is generated. The additional electrons and holes created are called *excess electrons* and *excess holes*.

The excess electrons and holes are generated by an external force at a particular rate. Let $g_n'$ be the generation rate of excess electrons and $g_p'$ be that of excess holes. These generation rates also have units of $\#/cm^3$-sec. For the direct band-to-band generation, the excess electrons and holes are also created in pairs, so we must have

$$g_n' = g_p' \qquad (6\text{–}4)$$

When excess electrons and holes are created, the concentration of electrons in the conduction band and of holes in the valence band increase above their thermal-equilibrium value. We may write

$$n = n_0 + \delta n \qquad (6\text{–}5a)$$

and

$$p = p_0 + \delta p \qquad (6\text{–}5b)$$

where $n_0$ and $p_0$ are the thermal-equilibrium concentrations, and $\delta n$ and $\delta p$ are the excess electron and hole concentrations. Figure 6–2 shows the ex-

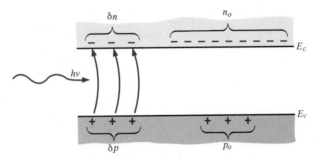

**Figure 6–2**   Creation of excess electron and hole densities by photons.

cess electron-hole generation process and the resulting carrier concentrations. The external force has perturbed the equilibrium condition so that the semiconductor is no longer in thermal equilibrium. We may note from Equations (6–5a) and (6–5b) that, in a nonequilibrium condition, $np \neq n_0 p_0 = n_i^2$.

A steady-state generation of excess electrons and holes will not cause a continual buildup of the carrier concentrations. As in the case of thermal equilibrium, an electron in the conduction band may "fall down" into the valence band, leading to the process of excess electron-hole recombination. This process is shown in Figure 6–3. The recombination rate for excess

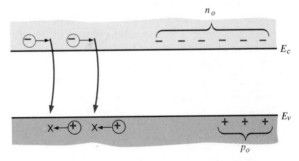

**Figure 6–3**   Recombination of excess carriers reestablishing thermal equilibrium.

electrons is denoted by $R_n'$ and for excess holes by $R_p'$. Both parameters have units of #/cm³-sec. The excess electrons and holes recombine in pairs so the recombination rates must be equal. We can then write

$$R_n' = R_p' \tag{6–6}$$

In the direct band-to-band recombination that we are considering, the recombination occurs spontaneously; thus the probability of an electron and hole recombining is constant with time. The rate at which electrons recombine must be proportional to the electron concentration and must also be proportional to the hole concentration. If there are no electrons or holes, there can be no recombination.

The net rate of change in the electron concentration can be written as

$$\frac{dn(t)}{dt} = \alpha_r[n_i^2 - n(t)p(t)] \tag{6-7}$$

where

$$n(t) = n_0 + \delta n(t) \tag{6-8a}$$

and

$$p(t) = p_0 + \delta p(t) \tag{6-8b}$$

The first term, $\alpha_r n_i^2$, in Equation (6–7) is the thermal-equilibrium generation rate. Since excess electrons and holes are created and recombine in pairs, we have that $\delta n(t) = \delta p(t)$. (Excess electron and hole concentrations are equal so we can simply use the phrase excess carriers to mean either.) The thermal-equilibrium parameters, $n_0$ and $p_0$, being independent of time, Equation (6–7) becomes

$$\frac{d(\delta n(t))}{dt} = \alpha_r[n_i^2 - (n_0 + \delta n(t))(p_0 + \delta p(t))] \tag{6-9}$$
$$= -\alpha_r \delta n(t)[(n_0 + p_0) + \delta n(t)]$$

Equation (6–9) can easily be solved if we impose the condition of *low-level injection*. Low-level injection puts limits on the magnitude of the excess carrier concentration compared to the thermal equilibrium carrier concentrations. In an extrinsic n-type material, we generally have $n_0 \gg p_0$ and, in an extrinsic p-type material, we generally have $p_0 \gg n_0$. Low-level injection means that the excess carrier concentration is much less than the thermal equilibrium majority carrier concentration. Conversely, high-level injection occurs when the excess carrier concentration becomes comparable to or greater than the thermal equilibrium majority carrier concentration.

If we consider a p-type material ($p_0 \gg n_0$) under low-level injection ($\delta n(t) \ll p_0$), then Equation (6–9) becomes

$$\frac{d(\delta n(t))}{dt} = -\alpha_r p_0 \delta n(t) \tag{6-10}$$

The solution to the equation is an exponential decay from the initial excess concentration, or

$$\delta n(t) = \delta n(0)e^{-\alpha_r p_0 t} = \delta n(0)e^{-t/\tau_{n0}} \tag{6-11}$$

where $\tau_{n0} = (\alpha_r p_0)^{-1}$ and is a constant for the low-level injection. Equation (6–11) describes the decay of excess minority carrier electrons so that $\tau_{n0}$ is often referred to as the *excess minority carrier lifetime*.[1]

---

[1] In chapter 5 we defined $\tau$ as a mean time between collisions. We define $\tau$ here as the mean time before a recombination event occurs. The two parameters are not related.

The recombination rate, which is defined as a positive quantity, of excess minority carrier electrons can be written using Equation (6–10) as

$$R_n' = \frac{-d(\delta n(t))}{dt} = +\alpha_r p_0 \delta n(t) = \frac{\delta n(t)}{\tau_{n0}} \tag{6–12}$$

For the direct band-to-band recombination, the excess majority carrier holes recombine at the same rate, so that

$$R_n' = R_p' = \frac{\delta n(t)}{\tau_{n0}} \tag{6–13}$$

In the case of an n-type material ($n_0 \gg p_0$) under low-level injection ($\delta n(t) \ll n_0$), the decay of minority carrier holes occurs with a time constant $\tau_{p0} = (\alpha_r n_0)^{-1}$, where $\tau_{p0}$ is also referred to as the excess minority carrier lifetime. The recombination rate of the majority carrier electrons will be the same as that of the minority carrier holes, so we have

$$R_n' = R_p' = \frac{\delta n(t)}{\tau_{p0}} \tag{6–14}$$

The generation rates of excess carriers are not functions of electron or hole concentrations. In general, the generation and recombination rates may be functions of the space coordinates and time.

## 6.2 MATHEMATICAL ANALYSIS OF EXCESS CARRIERS

We will develop the equations that describe the behavior of excess electrons and holes. Since generation rates may be a function of time and space coordinates, the excess carrier concentration may also be a function of the same variables.

### 6.2.1 Continuity Equations

Figure 6–4 shows a differential volume element in which a one-dimensional hole-particle flux is entering the differential element at $x$ and is leaving the

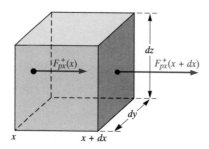

**Figure 6–4** Differential volume showing $x$-component of the hole-particle flux.

element at $x + dx$. The parameter $F_{px}^+$ is the hole-particle flux, or flow, and has units of # of holes/cm²-sec. For the $x$-component of the particle current density shown, we may write

$$F_{px}^+(x + dx) = F_{px}^+(x) + \frac{\partial F_{px}^+}{\partial x} \cdot dx \qquad (6-15)$$

This equation is a Taylor expansion of $F_{px}^+(x + dx)$, where the differential length $dx$ is small, so that only the first two terms in the expansion are significant. The net increase in the number of holes per unit time within the differential volume element due to the $x$-component of hole flux is given by

$$\frac{\partial p}{\partial t} dxdydz = [F_{px}^+(x) - F_{px}^+(x + dx)]dydz = - \frac{\partial F_{px}^+}{\partial x} dxdydz \qquad (6-16)$$

If $F_{px}^+(x) > F_{px}^+(x + dx)$, for example, there will be a net increase in the number of holes in the differential volume element with time. If we generalize to a three-dimensional hole flux, then the right side of Equation (6–16) may be written as $-\nabla \cdot F_p^+ dxdydz$, where $\nabla \cdot F_p^+$ is the divergence of the flux vector. We will limit ourselves to a one-dimensional analysis.

The generation rate and recombination rate of holes will also affect the hole concentration in the differential volume. The net increase in the number of holes per unit time in the differential volume element is then given by

$$\frac{\partial p}{\partial t} dxdydz = - \frac{\partial F_p^+}{\partial x} dxdydz + g_p dxdydz - \frac{p}{\tau_{pt}} dxdydz \qquad (6-17)$$

where $p$ is the density of holes. The first term on the right side of Equation (6–17) is the increase in the number of holes per unit time due to the hole flux, the second term is the increase in the number of holes per unit time due to the generation of holes, and the last term is the decrease in the number of holes per unit time due to the recombination of holes. The recombination rate for holes is given by $p/\tau_{pt}$ where $\tau_{pt}$ includes the thermal equilibrium carrier lifetime and the excess carrier lifetime.

If we divide both sides of Equation (6–17) by the differential volume $dxdydz$, the net increase in the hole concentration per unit time is

$$\frac{\partial p}{\partial t} = - \frac{\partial F_p^+}{\partial x} + g_p - \frac{p}{\tau_{pt}} \qquad (6-18)$$

Equation (6–18) is known as the continuity equation for holes.

Similarly, the one-dimensional continuity equation for electrons is given by

$$\frac{\partial n}{\partial t} = - \frac{\partial F_n^-}{\partial x} + g_n - \frac{n}{\tau_{nt}} \qquad (6-19)$$

where $F_n^-$ is the electron-particle flow, or flux, also given in units of # of electrons/cm²-sec.

## 6.2.2 Time-Dependent Diffusion Equations

In Chapter 5, we derived the hole and electron current densities, which are given, in one dimension, by

$$J_p = e\mu_p p \mathrm{E} - eD_p \frac{\partial p}{\partial x} \tag{6-20}$$

and

$$J_n = e\mu_n n \mathrm{E} + eD_n \frac{\partial n}{\partial x} \tag{6-21}$$

If we divide the hole current density by $(+e)$ and the electron current density by $(-e)$, we obtain each particle flux. These equations become

$$\frac{J_p}{+e} = F_p^+ = \mu_p p \mathrm{E} - D_p \frac{\partial p}{\partial x} \tag{6-22}$$

and

$$\frac{J_n}{(-e)} = F_n^- = -\mu_n n \mathrm{E} - D_n \frac{\partial n}{\partial x} \tag{6-23}$$

Taking the divergence of Equations (6–22) and (6–23), and substituting back into the continuity equations of (6–18) and (6–19), we obtain

$$\frac{\partial p}{\partial t} = -\mu_p \frac{\partial(p\mathrm{E})}{\partial x} + D_p \frac{\partial^2 p}{\partial x^2} + g_p - \frac{p}{\tau_{pt}} \tag{6-24}$$

and

$$\frac{\partial n}{\partial t} = +\mu_n \frac{\partial(n\mathrm{E})}{\partial x} + D_n \frac{\partial^2 n}{\partial x^2} + g_n - \frac{n}{\tau_{nt}} \tag{6-25}$$

Keeping in mind that we are limiting ourselves to a one-dimensional analysis, we can expand the derivative of the product as

$$\frac{\partial(p\mathrm{E})}{\partial x} = \mathrm{E} \frac{\partial p}{\partial x} + p \frac{\partial \mathrm{E}}{\partial x} \tag{6-26}$$

In a more generalized three dimensional analysis, Equation (6–26) would have to be replaced by a vector identity. Equations (6–24) and (6–25) can be written in the form

$$D_p \frac{\partial^2 p}{\partial x^2} - \mu_p \left( \mathrm{E} \frac{\partial p}{\partial x} + p \frac{\partial \mathrm{E}}{\partial x} \right) + g_p - \frac{p}{\tau_{pt}} = \frac{\partial p}{\partial t} \tag{6-27}$$

and

$$D_n \frac{\partial^2 n}{\partial x^2} + \mu_n \left( \mathrm{E} \frac{\partial n}{\partial x} + n \frac{\partial \mathrm{E}}{\partial x} \right) + g_n - \frac{n}{\tau_{nt}} = \frac{\partial n}{\partial t} \tag{6-28}$$

Equations (6–27) and (6–28) are the time-dependent diffusion equations for holes and electrons, respectively. Since both the hole concentration $p$ and the electron concentration $n$ contain the excess concentrations, Equations (6–27) and (6–28) describe the space and time behavior of the excess carriers.

The hole and electron concentrations are functions of both the thermal equilibrium and the excess values as given in Equations (6–5a) and (6–5b). The thermal-equilibrium concentrations, $n_0$ and $p_0$, are not functions of time. For the special case of a homogeneous semiconductor, $n_0$ and $p_0$ are also independent of the space coordinates. Equations (6–27) and (6–28) may then be written in the form

$$D_p \frac{\partial^2(\delta p)}{\partial x^2} - \mu_p \left( \mathrm{E} \frac{\partial(\delta p)}{\partial x} + p \frac{\partial \mathrm{E}}{\partial x} \right) + g_p - \frac{p}{\tau_{pt}} = \frac{\partial(\delta p)}{\partial t} \qquad (6\text{--}29)$$

and

$$D_n \frac{\partial^2(\delta n)}{\partial x^2} + \mu_n \left( \mathrm{E} \frac{\partial(\delta n)}{\partial x} + n \frac{\partial \mathrm{E}}{\partial x} \right) + g_n - \frac{n}{\tau_{nt}} = \frac{\partial(\delta n)}{\partial t} \qquad (6\text{--}30)$$

Note that Equations (6–29) and (6–30) contain terms involving the total concentrations, $p$ and $n$, and terms involving only the excess concentrations, $\delta p$ and $\delta n$.

## 6.3  AMBIPOLAR TRANSPORT

Originally, we assumed that the electric field in the current Equations (6–20) and (6–21) was an applied electric field. This electric field term appears in the time-dependent diffusion equations given by Equations (6–29) and (6–30). If a pulse of excess electrons and a pulse of excess holes are created at a particular point in a semiconductor with an applied electric field, the excess holes and electrons *will tend* to drift in opposite directions. However, because the electrons and holes are charged particles, any separation will induce an internal electric field between the two sets of particles. This internal electric field will create a force attracting the electrons and holes back toward each other. The electric field term in Equations (6–29) and (6–30) is then composed of the externally applied field plus the induced internal field. This E-field may be written as

$$\mathrm{E} = \mathrm{E}_{app} + \mathrm{E}_{int} \qquad (6\text{--}31)$$

where $\mathrm{E}_{app}$ is the applied electric field and $\mathrm{E}_{int}$ is the induced internal electric field.

Since the internal E-field creates a force attracting the electrons and holes, this E-field will hold the pulses of excess electrons and excess holes together. The negatively charged electrons and positively charged holes then will drift or diffuse together with a single effective mobility or diffusion coefficient. This phenomenon is called *ambipolar diffusion* or *ambipolar transport*.

### 6.3.1 Derivation of the Ambipolar Transport Equation

The time-dependent diffusion Equations (6–29) and (6–30) describe the behavior of the excess carriers. However, a third equation is required to relate the excess electron and hole concentrations to the internal electric field—Poisson's equation, which may be written as

$$\nabla \cdot E_{int} = \frac{e(\delta p - \delta n)}{\varepsilon_s} = \frac{\partial E_{int}}{\partial x} \tag{6–32}$$

where $\varepsilon_s$ is the permittivity of the semiconductor material.

To make the solution of Equations (6–29), (6–30), and (6–32) more tractable, we need to make some approximations. We can show that only a relatively small internal electric field is sufficient to keep the excess electrons and holes drifting and diffusing together. Hence, we can assume that

$$|E_{int}| \ll |E_{app}| \tag{6–33}$$

However, the $\nabla \cdot E_{int}$ term may not be negligible. We will impose the condition of charge neutrality: we will assume that the excess electron concentration is just balanced by an equal excess hole concentration at any point in space and time. If this assumption were exactly true, there would be no induced internal electric field to keep the two sets of particles together. However, only a very small difference in the excess electron concentration and excess hole concentration will set up an internal E-field sufficient to keep the particles diffusing and drifting together. We can show that a 1 percent difference in $\delta p$ and $\delta n$, for example, will result in non-negligible values of the $\nabla \cdot E = \nabla \cdot E_{int}$ term in Equations (6–29) and (6–30).

We can combine Equations (6–29) and (6–30) to eliminate the $\nabla \cdot E$ term. Considering Equations (6–1) and (6–4), we can define

$$g_n = g_p \equiv g \tag{6–34}$$

and considering Equations (6–2) and (6–6), we can define

$$R_n = \frac{n}{\tau_{nt}} = R_p = \frac{p}{\tau_{pt}} \equiv R \tag{6–35}$$

The lifetimes in Equation (6–35) include the thermal-equilibrium carrier lifetimes and the excess-carrier lifetimes. If we impose the charge neutrality condition, then $\delta n \cong \delta p$. We will denote both the excess electron and excess hole concentrations in Equations (6–29) and (6–30) by $\delta n$. We may then rewrite Equations (6–29) and (6–30) as

$$D_p \frac{\partial^2(\delta n)}{\partial x^2} - \mu_p \left( E \frac{\partial(\delta n)}{\partial x} + p \frac{\partial E}{\partial x} \right) + g - R = \frac{\partial(\delta n)}{\partial t} \qquad (6\text{-}36)$$

and

$$D_n \frac{\partial^2(\delta n)}{\partial x^2} + \mu_n \left( E \frac{\partial(\delta n)}{\partial x} + n \frac{\partial E}{\partial x} \right) + g - R = \frac{\partial(\delta n)}{\partial t} \qquad (6\text{-}37)$$

If we multiply Equation (6–36) by $\mu_n n$, multiply Equation (6–37) by $\mu_p p$, and add the two equations, the $\nabla \cdot E = \partial E / \partial x$ term will be eliminated. The result of this addition gives

$$(\mu_n n D_p + \mu_p p D_n) \frac{\partial^2(\delta n)}{\partial x^2} + (\mu_n \mu_p)(p - n) \, E \, \frac{\partial(\delta n)}{\partial x}$$
$$+ (\mu_n n + \mu_p p)(g - R) = (\mu_n n + \mu_p p) \frac{\partial(\delta n)}{\partial t} \qquad (6\text{-}38)$$

If we divide Equation (6–38) by the term $(\mu_n n + \mu_p p)$, this equation becomes

$$D' \frac{\partial^2(\delta n)}{\partial x^2} + \mu' E \frac{\partial(\delta n)}{\partial x} + g - R = \frac{\partial(\delta n)}{\partial t} \qquad (6\text{-}39)$$

where

$$D' = \frac{\mu_n n D_p + \mu_p p D_n}{\mu_n n + \mu_p p} \qquad (6\text{-}40)$$

and

$$\mu' = \frac{\mu_n \mu_p (p - n)}{\mu_n n + \mu_p p} \qquad (6\text{-}41)$$

Equation (6–39) is called the *ambipolar transport equation* and describes the behavior of the excess electrons and holes in time and space. The parameter $D'$ is called the *ambipolar diffusion coefficient* and $\mu'$ is called the *ambipolar mobility*.

The Einstein relation relates the mobility and diffusion coefficient by

$$\frac{\mu_n}{D_n} = \frac{\mu_p}{D_p} = \frac{e}{kT} \qquad (6\text{-}42)$$

Using these relations, the ambipolar diffusion coefficient may be written in the form

$$D' = \frac{D_n D_p (n + p)}{D_n n + D_p p} \qquad (6\text{-}43)$$

The ambipolar diffusion coefficient, $D'$, and the ambipolar mobility, $\mu'$, are functions of the electron and hole concentrations, $n$ and $p$, respectively. Since both $n$ and $p$ contain the excess-carrier concentration $\delta n$, the coefficients in the ambipolar transport equation are not constants. The ambipolar transport equation, given by Equation (6–39), then, is a nonlinear differential equation.

### 6.3.2 Limits of Extrinsic Doping and Low Injection

The ambipolar transport equation may be simplified and linearized by considering an extrinsic semiconductor and by considering low-level injection. The ambipolar diffusion coefficient, from Equation (6–43), may be written as

$$D' = \frac{D_n D_p[(n_0 + \delta n) + (p_0 + \delta n)]}{D_n(n_0 + \delta n) + D_p(p_0 + \delta n)} \tag{6–44}$$

where $n_0$ and $p_0$ are the thermal-equilibrium electron and hole concentrations, respectively, and $\delta n$ is the excess carrier concentration. If we consider a p-type semiconductor, we can assume that $p_0 \gg n_0$. The condition of low-level injection, or just low injection, means that the excess carrier concentration is much smaller than the thermal-equilibrium majority carrier concentration. For the p-type semiconductor, then, low injection implies that $\delta n \ll p_0$. Assuming that $n_0 \ll p_0$ and $\delta n \ll p_0$, and assuming that $D_n$ and $D_p$ are on the same order of magnitude, the ambipolar diffusion coefficient from Equation (6–44) reduces to

$$D' = D_n \tag{6–45}$$

If we apply the conditions of an extrinsic p-type semiconductor and low injection to the ambipolar mobility, Equation (6–41) reduces to

$$\mu' = \mu_n \tag{6–46}$$

It is important to note that for an extrinsic p-type semiconductor under low injection, the ambipolar diffusion coefficient and the ambipolar mobility coefficient reduce to the minority-carrier electron parameter values, which are constants. The ambipolar transport equation reduces to a linear differential equation with constant coefficients.

If we now consider an extrinsic n-type semiconductor under low injection, we may assume that $p_0 \ll n_0$ and $\delta n \ll n_0$. The ambipolar diffusion coefficient from Equation (6–43) reduces to

$$D' = D_p \tag{6–47}$$

and the ambipolar mobility from Equation (6–41) reduces to

$$\mu' = -\mu_p \tag{6–48}$$

The ambipolar parameters again reduce to the minority-carrier values, which are constants. Note that, for the n-type semiconductor, the ambipolar mobility is a negative value. The ambipolar mobility term is associated with carrier drift, therefore the sign of the drift term depends on the charge of the particle. The equivalent ambipolar particle is negatively charged as can be seen by comparing Equations (6–30) and (6–39). If the ambipolar mobility reduces to that of a positively charged hole, a negative sign is introduced as shown in Equation (6–48).

The remaining terms we need to consider in the ambipolar transport equation are the generation rate and the recombination rate. Recall that the electron and hole recombination rates are equal and were given by Equation (6–35) as $R_n = R_p = n/\tau_{nt} = p/\tau_{pt} \equiv R$, where $\tau_{nt}$ and $\tau_{pt}$ are the mean electron and hole lifetimes, respectively. If we consider the inverse lifetime functions, then $1/\tau_{nt}$ is the probability per unit time that an electron will encounter a hole and recombine. Likewise, $1/\tau_{pt}$ is the probability per unit time that a hole will encounter an electron and recombine. If we again consider an extrinsic p-type semiconductor under low injection, the concentration of majority carrier holes will be essentially constant, even when excess carriers are present. Then, the probability per unit time of a minority carrier electron encountering a majority carrier hole will be essentially constant. Hence $\tau_{nt} \equiv \tau_n$, the minority carrier electron lifetime, will remain a constant for the extrinsic p-type semiconductor under low injection.

Similarly, if we consider an extrinsic n-type semiconductor under low injection, the minority carrier hole lifetime, $\tau_{pt} \equiv \tau_p$, will remain constant. Even under the condition of low injection, the minority carrier hole concentration may increase by several orders of magnitude. The probability per unit time of a majority carrier electron encountering a hole may change drastically. The majority carrier lifetime, then, may change substantially when excess carriers are present.

Consider, again, the generation and recombination terms in the ambipolar transport equation. For electrons we may write

$$g - R = g_n - R_n = (G_{n0} + g'_n) - (R_{n0} + R'_n) \tag{6–49}$$

where $G_{n0}$ and $g'_n$ are the thermal-equilibrium electron and excess electron generation rates, respectively. The terms $R_{n0}$ and $R'_n$ are the thermal-equilibrium electron and excess electron recombination rates, respectively. For thermal equilibrium, we have that

$$G_{n0} = R_{n0} \tag{6–50}$$

so Equation (6–49) reduces to

$$g - R = g'_n - R'_n = g'_n - \frac{\delta n}{\tau_n} \tag{6–51}$$

where $\tau_n$ is the excess minority carrier electron lifetime.

For the case of holes, we may write

$$g - R = g_p - R_p = (G_{p0} + g'_p) - (R_{p0} + R'_p) \tag{6–52}$$

where $G_{p0}$ and $g'_p$ are the thermal-equilibrium hole and excess hole generation rates, respectively. The terms $R_{p0}$ and $R'_p$ are the thermal-equilibrium hole and excess hole recombination rates, respectively. Again, for thermal equilibrium, we have that

$$G_{p0} = R_{p0} \tag{6-53}$$

so that Equation (6–52) reduces to

$$g - R = g'_p - R'_p = g'_p - \frac{\delta p}{\tau_p} \tag{6-54}$$

where $\tau_p$ is the excess minority carrier hole lifetime.

The generation rate for excess electrons must equal the generation rate for excess holes. We may then define a generation rate for excess carriers as $g'$, so that $g'_n = g'_p \equiv g'$. We also determined that the minority carrier lifetime is essentially a constant for low injection. Then the term $g - R$ in the ambipolar transport equation may be written in terms of the minority-carrier parameters.

The ambipolar transport equation, given by Equation (6–39), for a p-type semiconductor under low injection then becomes

$$D_n \frac{\partial^2(\delta n)}{\partial x^2} + \mu_n E \frac{\partial(\delta n)}{\partial x} + g' - \frac{\delta n}{\tau_{n0}} = \frac{\partial(\delta n)}{\partial t} \tag{6-55}$$

The parameter $\delta n$ is the excess minority carrier electron concentration, the parameter $\tau_{n0}$ is the minority carrier lifetime under low injection, and the other parameters are the usual minority carrier electron parameters.

Similarly, for an extrinsic n-type semiconductor under low injection, the ambipolar transport equation becomes

$$D_p \frac{\partial^2(\delta p)}{\partial x^2} - \mu_p E \frac{\partial(\delta p)}{\partial x} + g' - \frac{\delta p}{\tau_{p0}} = \frac{\partial(\delta p)}{\partial t} \tag{6-56}$$

The parameter $\delta p$ is the excess minority carrier hole concentration, the parameter $\tau_{p0}$ is the minority carrier hole lifetime under low injection, and the other parameters are the usual minority carrier hole parameters.

It is extremely important to note that the transport and recombination parameters in Equations (6–55) and (6–56) are those of the minority carrier. Equations (6–55) and (6–56) describe the drift, diffusion, and recombination of excess minority carriers as a function of spatial coordinates and as a function of time. Recall that we had imposed the condition of charge neutrality; the excess minority carrier concentration is equal to the excess majority carrier concentration. The excess majority carriers, then, diffuse and drift with the excess minority carriers; thus the behavior of the excess majority carrier is determined by the minority carrier parameters. This ambipolar phenomenon is extremely important in semiconductor physics, and is the basis for describing the characteristics and behavior of semiconductor devices.

### 6.3.3 Applications of the Ambipolar Transport Equation

The ambipolar transport equation will be solved for several problems. These examples will help illustrate the behavior of excess carriers in a semiconductor material, and the results will be used later in the discussion of the pn junction and the other semiconductor devices.

---

**Example 6–1**

**Objective:** To determine the time behavior of excess carriers as a semiconductor returns to thermal equilibrium.

Consider an infinitely large, homogeneous n-type semiconductor with zero applied electric field. Assume that at time $t = 0$, a uniform concentration of excess carriers exists in the crystal, but assume that $g' = 0$ for $t > 0$. If we assume that the concentration of excess carriers is much smaller than the thermal-equilibrium electron concentration, then the low-injection condition applies. Calculate the excess carrier concentration as a function of time for $t \geq 0$.

**Solution:** For the n-type semiconductor, we need to consider the ambipolar transport equation for the minority carrier holes, which was given by Equation (6–56). The equation is

$$D_p \frac{\partial^2(\delta p)}{\partial x^2} - \mu_p E \frac{\partial(\delta p)}{\partial x} + g' - \frac{\delta p}{\tau_{p0}} = \frac{\partial(\delta p)}{\partial t}$$

We are assuming a uniform concentration of excess holes so that $\partial^2(\delta p)/\partial x^2 = \partial(\delta p)/\partial x = 0$. For $t > 0$, we are also assuming that $g' = 0$. Equation (6–56) reduces to

$$\frac{d(\delta p)}{dt} = -\frac{\delta p}{\tau_{p0}} \tag{6–57}$$

Since there is no spatial variation, the total time derivative may be used. At low injection, the minority carrier hole lifetime, $\tau_{p0}$, is a constant. The solution to Equation (6–57) is

$$\delta p(t) = \delta p(0)e^{-t/\tau_{p0}} \tag{6–58}$$

where $\delta p(0)$ is the uniform concentration of excess carriers which exists at time $t = 0$. The concentration of excess holes decays exponentially with time, with a time constant equal to the minority carrier hole lifetime.

From the charge-neutrality condition, we have that $\delta n = \delta p$, so the excess electron concentration is given by

$$\delta n(t) = \delta p(0)e^{-t/\tau_{p0}} \tag{6–59}$$

**Numerical Calculation:** Consider n-type gallium arsenide doped at $N_d = 10^{16}$ cm$^{-3}$. Assume that $10^{14}$ electron-hole pairs per cm$^3$ have been created at $t = 0$ and assume the minority carrier hole lifetime is $\tau_{p0} = 10$ ns.

We may note that $\delta p(0) \ll n_0$ so low injection applies. Then from Equation (6–58) we can write

$$\delta p(t) = 10^{14}e^{-t/10^{-8}} \text{ cm}^{-3}$$

The excess hole and excess electron concentrations will decay to $1/e$ of their initial value in 10 ns.

**Comment:** The excess electrons and holes recombine at the rate determined by the excess minority carrier hole lifetime in the n-type semiconductor.

---

**Example 6–2**

**Objective:** To determine the time dependence of excess carriers in reaching a steady-state condition.

   Again consider an infinitely large, homogeneous n-type semiconductor with a zero applied electric field. Assume that, for $t < 0$, the semiconductor is in thermal equilibrium and that, for $t \geq 0$, a uniform generation rate exists in the crystal. Calculate the excess carrier concentration as a function of time assuming the condition of low injection.

**Solution:** The condition of a uniform generation rate and a homogeneous semiconductor again implies that $\partial^2(\delta p)/\partial x^2 = \partial(\delta p)/\partial x = 0$ in Equation (6–56). The equation, for this case, reduces to

$$g' - \frac{\delta p}{\tau_{p0}} = \frac{d(\delta p)}{dt} \qquad (6\text{--}60)$$

The solution to this differential equation is

$$\delta p(t) = g'\tau_{p0} \left(1 - e^{-t/\tau_{p0}}\right) \qquad (6\text{--}61)$$

**Numerical Calculation:** Consider n-type silicon at $T = 300°K$ doped at $N_d = 2 \times 10^{16}$ cm$^{-3}$. Assume that $\tau_{p0} = 10^{-7}$ sec and $g' = 5 \times 10^{21}$ cm$^{-3}$ sec$^{-1}$. From Equation (6–61), we can write

$$\delta p(t) = (5 \times 10^{21})(10^{-7})[1 - e^{-t/10^{-7}}] = 5 \times 10^{14} [1 - e^{-t/10^{-7}}] \text{ cm}^{-3}$$

**Comment:** We may note that for $t \to \infty$, we will create a steady state excess hole and electron concentration of $5 \times 10^{14}$ cm$^{-3}$. We may note that $\delta p \ll n_0$ so low injection is valid.

---

The excess minority carrier hole concentration increases with time with the same time constant $\tau_{p0}$, which is the excess minority carrier lifetime. The excess carrier concentration reaches a steady-state value as time goes to infinity, even though a steady-state generation of excess electrons and holes exists. This steady-state effect can be seen from Equation (6–60) by setting $d(\delta p)/dt = 0$. The remaining terms simply state that, in steady state, the generation rate is equal to the recombination rate.

---

**Example 6–3**

**Objective:** To determine the steady-state spatial dependence of the excess carrier concentration.

Consider a p-type semiconductor that is homogeneous and infinite in extent. Assume a zero applied electric field. For a one-dimensional crystal, assume that excess carriers are being generated at $x = 0$ only, as indicated in Figure 6–5. The excess carriers being generated at $x = 0$ will begin diffusing in both the $+x$ and $-x$ directions. Calculate the steady-state excess carrier concentration as a function of $x$.

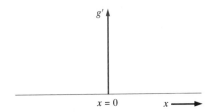

**Figure 6–5** Steady-state generation rate at $x = 0$.

**Solution:** The ambipolar transport equation for excess minority carrier electrons was given by Equation (6–55), and is written as

$$D_n \frac{\partial^2(\delta n)}{\partial x^2} + \mu_n E \frac{\partial(\delta n)}{\partial x} + g' - \frac{\delta n}{\tau_{n0}} = \frac{\partial(\delta n)}{\partial t}$$

From our assumptions, we have $E = 0$, $g' = 0$ for $x \neq 0$, and $\partial(\delta n)/\partial t = 0$ for steady state. Assuming a one-dimensional crystal, Equation (6–55) reduces to

$$D_n \frac{d^2(\delta n)}{dx^2} - \frac{\delta n}{\tau_{n0}} = 0 \tag{6–62}$$

Dividing by the diffusion coefficient, Equation (6–62) may be written as

$$\frac{d^2(\delta n)}{dx^2} - \frac{\delta n}{D_n \tau_{n0}} = \frac{d^2(\delta n)}{dx^2} - \frac{\delta n}{L_n^2} = 0 \tag{6–63}$$

where we have defined $L_n^2 = D_n \tau_{n0}$. The parameter $L_n$ has the unit of length and is called the minority carrier electron diffusion length. The general solution to Equation (6–63) is

$$\delta n(x) = A e^{-x/L_n} + B e^{x/L_n} \tag{6–64}$$

As the minority carrier electrons diffuse away from $x = 0$, they will recombine with the majority carrier holes. The minority carrier electron concentration will then decay toward zero at both $x = +\infty$ and $x = -\infty$. These boundary conditions mean that $B \equiv 0$ for $x > 0$ and $A \equiv 0$ for $x < 0$. The solution to Equation (6–63) may then be written as

$$\delta n(x) = \delta n(0) e^{-x/L_n} \quad x \geq 0 \tag{6–65a}$$

and

$$\delta n(x) = \delta n(0) e^{+x/L_n} \quad x \leq 0 \tag{6–65b}$$

where $\delta n(0)$ is the value of the excess electron concentration at $x = 0$. The steady-state excess electron concentration decays exponentially with distance away from the source at $x = 0$.

**Numerical Calculation:** Consider p-type silicon at $T = 300°K$ doped at $N_a = 5 \times 10^{16}$ cm$^{-3}$. Assume that $\tau_{n0} = 5 \times 10^{-7}$ sec, $D_n = 25$ cm$^2$/sec, and $\delta n(0) = 10^{15}$ cm$^{-3}$.

The minority carrier diffusion length is

$$L_n = \sqrt{D_n \tau_{n0}} = \sqrt{(25)(5 \times 10^{-7})} = 35.4 \ \mu m$$

Then for $x \geq 0$, we have

$$\delta n(x) = 10^{15} e^{-x/35.4 \times 10^{-4}} \ \text{cm}^{-3}$$

**Comment:** We may note that the steady-state excess concentration decays to $1/e$ of its value at $x = 35.4 \ \mu m$.

---

As before, we will assume charge neutrality; thus, the steady-state excess majority carrier hole concentration also decays exponentially with distance with the same characteristic minority carrier electron diffusion length $L_n$. Figure 6–6 is a plot of the total electron and hole concentrations as a function of distance. We are assuming low injection, that is, $\delta n(0) \ll p_0$ in the p-type semiconductor. The total concentration of majority carrier holes barely changes. However, we may have $\delta n(0) \gg n_0$ and still satisfy the low injection condition. The minority carrier concentration may change by many orders of magnitude.

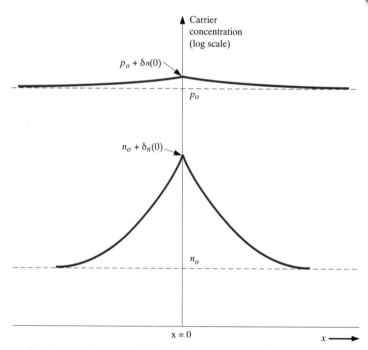

**Figure 6–6**   Steady-state electron and hole concentrations for the case when excess electrons and holes are generated at $x = 0$.

The three previous examples, which applied the ambipolar transport equation to specific situations, assumed either a homogeneous or a steady-state condition; only the time variation or the spatial variation was considered. Now consider an example in which both the time and spatial dependence are considered in the same problem.

**Example 6–4**

**Objective:** To determine both the time dependence and spatial dependence of the excess carrier concentration.

Assume that a finite number of electron-hole pairs is generated instantaneously at time $t = 0$ and at $x = 0$, but assume $g' = 0$ for $t > 0$. Assume we have an n-type semiconductor with a constant applied electric field equal to $E_0$, which is applied in the $+x$ direction. Calculate the excess carrier concentration as a function of $x$ and $t$.

**Solution:** The one-dimensional ambipolar transport equation for the minority carrier holes can be written from Equation (6–56) as

$$D_p \frac{\partial^2(\delta p)}{\partial x^2} - \mu_p E_0 \frac{\partial(\delta p)}{\partial x} - \frac{\delta p}{\tau_{p0}} = \frac{\partial(\delta p)}{\partial t} \tag{6–66}$$

The solution to this partial differential equation is of the form

$$\delta p(x,t) = p'(x,t)e^{-t/\tau_{p0}} \tag{6–67}$$

By substituting Equation (6–67) into Equation (6–66), we are left with the partial differential equation

$$D_p \frac{\partial^2 p'(x,t)}{\partial x^2} - \mu_p E_0 \frac{\partial p'(x,t)}{\partial x} = \frac{\partial p'(x,t)}{\partial t} \tag{6–68}$$

Equation (6–68) is normally solved using Laplace transform techniques. The solution, without going through the mathematical details, is

$$p'(x,t) = \frac{1}{(4\pi D_p t)^{1/2}} \exp\left[ \frac{-(x - \mu_p E_0 t)^2}{4 D_p t} \right] \tag{6–69}$$

The total solution, from Equations (6–67) and (6–69), for the excess minority carrier hole concentration is

$$\delta p(x,t) = \frac{e^{-t/\tau_{p0}}}{(4\pi D_p t)^{1/2}} \exp\left[ \frac{-(x - \mu_p E_0 t)^2}{4 D_p t} \right] \tag{6–70}$$

**Comment:** We could show that Equation (6–70) is a solution by direct substitution back into the partial differential equation, Equation (6–66).

Equation (6–70) can be plotted as a function of distance $x$, for various times. Figure 6–7 shows such a plot for the case when the applied electric field is zero. For $t > 0$, the excess minority carrier holes diffuse in both the $+x$ and $-x$ directions. During this time, the excess majority carrier elec-

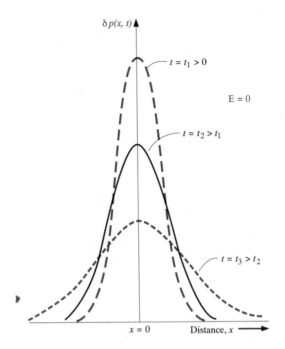

**Figure 6–7**  Excess hole concentration versus distance at various times for zero applied electric field.

trons, which were generated, diffuse at exactly the same rate as the holes. As time proceeds, the excess holes recombine with the excess electrons so that at $t = \infty$ the excess hole concentration is zero. In this particular example, both diffusion and recombination processes are occurring at the same time.

Figure 6–8 shows a plot of Equation (6–70) as a function of distance $x$ at various times for the case when the applied electric field is not zero. In this case, the pulse of excess minority carrier holes is drifting in the $+x$ direction, which is the direction of the electric field. We still have the same diffusion and recombination processes as we had before. An important point to consider is that, with charge neutrality, $\delta n = \delta p$ at any instant of time and at any point in space. The excess-electron concentration is equal to the excess-hole concentration. In this case, then, the excess-electron pulse is moving in the same direction as the applied electric field even though the electrons have a negative charge. In the ambipolar transport process, the excess carriers are characterized by the minority carrier parameters. In this example, the excess carriers behave according to the minority carrier hole parameters, which include $D_p$, $\mu_p$, and $\tau_{p0}$. The excess majority carrier electrons are being pulled along by the excess minority carrier holes.

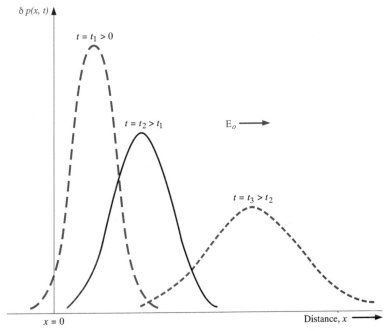

**Figure 6–8** Excess-hole concentration versus distance at various times for a constant applied electric field.

### *6.3.4 Haynes-Shockley Experiment

We have derived the mathematics describing the behavior of excess carriers in a semiconductor. The Haynes-Shockley experiment was one of the first experiments to actually measure excess-carrier behavior.

Figure 6–9 shows the basic experimental arrangement. The voltage source, $V_1$, establishes an applied electric field, $E_0$, in the $+x$ direction in the n-type semiconductor sample. Excess carriers are effectively injected into the semiconductor at contact A. Contact B is a rectifying contact that is under reverse bias by the voltage source $V_2$. The contact B will collect a fraction of the excess carriers as they drift through the semiconductor. The collected carriers will generate an output voltage, $V_0$.

This experiment corresponds to the problem we discussed in Example 6–4. Figure 6–10 shows the excess-carrier concentrations at contacts A and B for two conditions. Figure 6–10a shows the idealized excess-carrier pulse at contact A at time $t = 0$. For a given electric field, $E_{01}$, the excess carriers will drift along the semiconductor producing an output voltage as a function of time given in Figure 6–10b. The peak of the pulse will arrive at contact B at time $t_0$. If the applied electric field is reduced to a value $E_{02}$, $E_{02} < E_{01}$, the

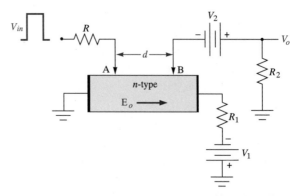

**Figure 6–9**   The basic Haynes-Shockley experimental arrangement.

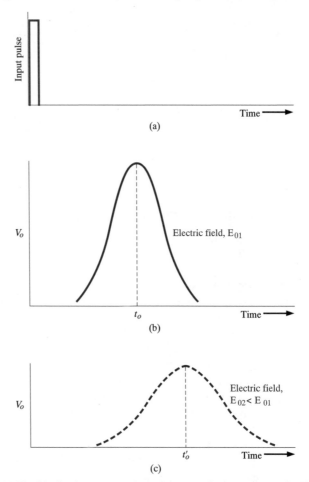

**Figure 6–10**   (a) The idealized excess carrier pulse at terminal A at $t = 0$. (b) The excess-carrier pulse versus time at terminal B for a given applied electric field. (c) The excess-carrier pulse versus time at terminal B for a smaller applied electric field.

output voltage response at contact B will look approximately as shown in Figure 6–10c. For the smaller electric field, the drift velocity of the pulse of excess carriers is smaller, and so it will take a longer time for the pulse to reach the contact B. During this longer time period, there is more diffusion and more recombination. The excess-carrier pulse shapes shown in Figures 6–10b and 6–10c are different for the two electric field conditions.

The minority carrier mobility, lifetime, and diffusion coefficient can be determined from this single experiment. As a good first approximation, the peak of the minority carrier pulse will arrive at contact B when the exponent involving distance and time in Equation (6–70) is zero, or

$$x - \mu_p E_0 t = 0 \qquad (6\text{–}71a)$$

In this case, $x = d$, where $d$ is the distance between contacts A and B, and $t = t_0$, where $t_0$ is the time at which the peak of the pulse reaches contact B. The mobility may be calculated as

$$\mu_p = \frac{d}{E_0 t_0} \qquad (6\text{–}71b)$$

Figure 6–11 again shows the output response as a function of time. At times $t_1$ and $t_2$, the magnitude of the excess concentration is $e^{-1}$ of its peak

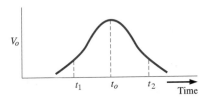

**Figure 6–11** The output excess-carrier pulse versus time to determine the diffusion coefficient.

value. If the time difference between $t_1$ and $t_2$ is not too large, $e^{-t/\tau_{p0}}$ and $(4\pi D_p t)^{1/2}$ do not change appreciably during this time; then the equation

$$(d - \mu_p E_0 t)^2 = 4 D_p t \qquad (6\text{–}72)$$

is satisfied at both $t = t_1$ and $t = t_2$. If we set $t = t_1$ and $t = t_2$ in Equation (6–72) and add the two resulting equations, we may show that the diffusion coefficient is given by

$$D_p = \frac{(\mu_p E_0)^2 (\Delta t)^2}{16 t_0} \qquad (6\text{–}73)$$

where

$$\Delta t = t_2 - t_1 \qquad (6\text{–}74)$$

The area $S$ under the curve shown in Figure 6–11 is proportional to the number of excess holes which have not recombined with majority carrier electrons. We may write

$$S = K \exp\left(\frac{-t_0}{\tau_{p0}}\right) = K \exp\left(\frac{-d}{\mu_p E_0 \tau_{p0}}\right) \qquad (6\text{–}75)$$

where $K$ is a constant. By varying the electric field, the area under the curve will change. A plot of $\ln(S)$ as a function of $(d/\mu_p E_0)$ will yield a straight line whose slope is $(1/\tau_{p0})$, so the minority carrier lifetime can also be determined from this experiment.

The Haynes-Shockley experiment is elegant in the sense that the three basic processes of drift, diffusion, and recombination are all observed in a single experiment. The determination of mobility is straightforward and can yield accurate values. The determination of the diffusion coefficient and lifetime is more complicated and may lead to some inaccuracies.

## 6.4  QUASI-FERMI ENERGY LEVELS

The thermal-equilibrium electron and hole concentrations are functions of the Fermi energy level. We can write

$$n_0 = n_i \exp\left[\frac{E_F - E_{Fi}}{kT}\right] \qquad (6\text{–}76a)$$

and

$$p_0 = n_i \exp\left[\frac{E_{Fi} - E_F}{kT}\right] \qquad (6\text{–}76b)$$

where $E_F$ and $E_{Fi}$ are the Fermi energy and intrinsic Fermi energy, respectively, and $n_i$ is the intrinsic carrier concentration. Figure 6–12a shows the energy-band diagram for an n-type semiconductor in which $E_F > E_{Fi}$. For this case, we may note from Equations (6–76a) and (6–76b) that $n_0 > n_i$ and $p_0 < n_i$, as we would expect. Similarly, Figure 6–12b shows the energy-band diagram for a p-type semiconductor in which $E_F < E_{Fi}$. Again we may note from Equations (6–76a) and (6–76b) that $n_0 < n_i$ and $p_0 > n_i$, as we would expect for the p-type material. These results are for thermal equilibrium.

If excess carriers are created in a semiconductor, we are no longer in thermal equilibrium and the Fermi energy is strictly no longer defined. However, we may define a quasi-Fermi level for electrons and a quasi-Fermi level for holes that apply for nonequilibrium. If $\delta n$ and $\delta p$ are the excess electron and hole concentrations, respectively, we may write

$$n_0 + \delta n = n_i \exp\left[\frac{E_{Fn} - E_{Fi}}{kT}\right] \qquad (6\text{–}77a)$$

and

$$p_0 + \delta p = n_i \exp\left[\frac{E_{Fi} - E_{Fp}}{kT}\right] \qquad (6\text{–}77b)$$

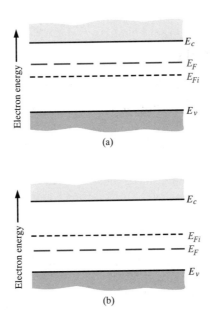

**Figure 6–12** Thermal-equilibrium energy-band diagrams for (a) n-type semiconductor and (b) p-type semiconductor.

where $E_{Fn}$ and $E_{Fp}$ are the quasi-Fermi energy levels for electrons and holes, respectively. The total electron concentration and the total hole concentration are functions of the quasi-Fermi levels.

---

### Example 6–5

**Objective:** To calculate the quasi-Fermi energy levels.

Consider an n-type semiconductor at $T = 300°K$ with carrier concentrations of $n_0 = 10^{15}$ cm$^{-3}$, $n_i = 10^{10}$ cm$^{-3}$, and $p_0 = 10^5$ cm$^{-3}$. In nonequilibrium, assume that the excess carrier concentrations are $\delta n = \delta p = 10^{13}$ cm$^{-3}$.

**Solution:** The Fermi level for thermal equilibrium can be determined from Equation (6–76a). We have

$$E_F - E_{Fi} = kT\ln\left(\frac{n_0}{n_i}\right) = 0.2982 \text{ eV}$$

We can use Equation (6–77a) to determine the quasi-Fermi level for electrons in nonequilibrium. We can write

$$E_{Fn} - E_{Fi} = kT\ln\left(\frac{n_0 + \delta n}{n_i}\right) = 0.2984 \text{ eV}$$

Equation (6–77b) can be used to calculate the quasi-Fermi level for holes in nonequilibrium. We can write

$$E_{Fi} - E_{Fp} = kT\ln\left(\frac{p_0 + \delta p}{n_i}\right] = 0.179 \text{ eV}$$

**Comment:** We may note that the quasi-Fermi level for electrons is above $E_{Fi}$ while the quasi-Fermi level for holes is below $E_{Fi}$.

Figure 6–13a shows the energy-band diagram with the Fermi energy level corresponding to thermal equilibrium. Figure 6–13b now shows the energy-band diagram under the nonequilibrium condition. Since the majority carrier electron concentration does not change significantly for this low-injection condition, the quasi-Fermi level for electrons is not much different from the thermal-equilibrium Fermi level. The quasi-Fermi energy level for the minority carrier holes is significantly different from the Fermi level and illustrates the fact that we have deviated from thermal equilibrium significantly. Since the electron concentration has increased, the quasi-Fermi level for electrons has moved slightly closer to the conduction band. The hole concentration has increased significantly so that the quasi-Fermi level for holes has moved much closer to the valence band. The quasi-Fermi energy levels will be considered again when we discuss forward-biased pn junctions.

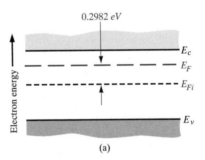

(a)

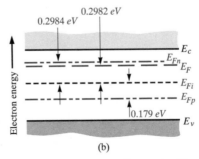

(b)

**Figure 6–13**   (a) Thermal-equilibrium energy-band diagram for $N_d = 10^{15}$ cm$^{-3}$ and $n_i = 10^{10}$ cm$^{-3}$. (b) Quasi-Fermi levels for electrons and holes if $10^{13}$ cm$^{-3}$ excess carriers are present.

## *6.5 EXCESS-CARRIER LIFETIME

The rate at which excess electrons and holes recombine is an important characteristic of the semiconductor and will influence many of the device characteristics as we will see in later chapters. We considered recombination briefly at the beginning of this chapter and argued that the recombination rate is inversely proportional to the mean carrier lifetime. We have assumed up to this point that the mean carrier lifetime is simply a parameter of the semiconductor material.

We have been considering an ideal semiconductor in which electronic energy states do not exist within the forbidden-energy bandgap. This ideal effect is present in a perfect single-crystal material with an ideal periodic-potential function. In a real semiconductor material, defects occur within the crystal and disrupt the perfect periodic-potential function. If the density of these defects is not too great, the defects will create discrete electronic energy states within the forbidden-energy band. These allowed energy states may be the dominant effect in determining the mean carrier lifetime. The mean carrier lifetime may be determined from the Shockley-Read-Hall theory of recombination.

### 6.5.1 Shockley-Read-Hall Theory of Recombination

An allowed energy state, also called a trap, within the forbidden bandgap may act as a recombination center, capturing both electrons and holes with almost equal probability. This equal probability of capture means that the capture cross sections for electrons and holes are approximately equal. The Shockley-Read-Hall theory of recombination assumes that a single recombination center, or trap, exists at an energy $E_t$ within the bandgap. There are four basic processes, shown in Figure 6–14, that may occur at this single trap. We will assume that the trap is an acceptor-type trap; that is, it is negatively charged when it contains an electron and is neutral when it does not contain an electron.

The four basic processes are as follows:

Process 1: The capture of an electron from the conduction band by an initially neutral empty trap.

Process 2: The inverse of process 1—the emission of an electron which is initially occupying a trap level back into the conduction band.

Process 3: The capture of a hole from the valence band by a trap containing an electron. (Or we may consider the process to be the emission of an electron from the trap into the valence band.)

Process 4: The inverse of process 3—the emission of a hole from a neutral trap into the valence band. (Or we may consider this process to be the capture of an electron from the valence band.)

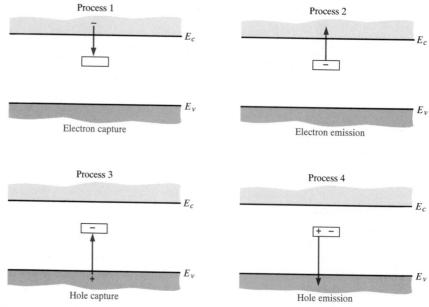

**Figure 6–14**    The four basic trapping and emission processes for the case of an acceptor-type trap.

In process 1, the rate at which electrons from the conduction band are captured by the traps is proportional to the density of electrons in the conduction band and proportional to the density of empty trap states. We can then write the electron capture rate as

$$R_{cn} = C_n N_t (1 - f_F(E_t)) n \qquad (6\text{–}78)$$

where

$R_{cn}$ = capture rate (#/cm³-sec)

$C_n$ = constant proportional to electron-capture cross section

$N_t$ = total concentration of trapping centers

$n$ = electron concentration in the conduction band

$f_F(E_t)$ = Fermi function at the trap energy

The Fermi function at the trap energy is given by

$$f_F(E_t) = \cfrac{1}{1 + \exp\left[\cfrac{E_t - E_F}{kT}\right]} \qquad (6\text{–}79)$$

which is the probability that a trap will contain an electron. The function $(1 - f_F(E_t))$ is then the probability that the trap is empty. In Equation (6–79), we have assumed that the degeneracy factor is one, which is the usual approximation made in this analysis. However, if a degeneracy factor is included, it will eventually be absorbed in other constants later in the analysis.

For process 2, the rate at which electrons are emitted from filled traps back into the conduction band is proportional to the number of filled traps, so that

$$R_{en} = E_n N_t f_F(E_t) \qquad (6\text{-}80)$$

where

$R_{en}$ = emission rate (#/cm$^3$-sec)

$E_n$ = constant

$f_F(E_t)$ = probability that the trap is occupied

In thermal equilibrium, the rate of electron capture from the conduction band and the rate of electron emission back into the conduction band must be equal. Then

$$R_{en} = R_{cn} \qquad (6\text{-}81)$$

so that

$$E_n N_t f_{F0}(E_t) = C_n N_t (1 - f_{F0}(E_t)) n_0 \qquad (6\text{-}82)$$

where $f_{F0}$ denotes the thermal-equilibrium Fermi function. Note that, in thermal equilibrium, the value of the electron concentration in the capture rate term is the equilibrium value $n_0$. Using the Boltzmann approximation for the Fermi function, we can find $E_n$ in terms of $C_n$ as

$$E_n = n' C_n \qquad (6\text{-}83)$$

where $n'$ is defined as

$$n' = N_c \exp\left[\frac{-(E_c - E_t)}{kT}\right] \qquad (6\text{-}84)$$

The parameter $n'$ is equivalent to an electron concentration that would exist in the conduction band if the trap energy $E_t$ coincided with the Fermi energy $E_F$.

In nonequilibrium, excess electrons exist, so that the net rate at which electrons are captured from the conduction band is given by

$$R_n = R_{cn} - R_{en} \qquad (6\text{-}85)$$

which is just the difference between the capture rate and the emission rate. Combining Equations (6–78) and (6–80) with (6–85) gives

$$R_n = [C_n N_t (1 - f_F(E_t)) n] - [E_n N_t f_F(E_t)] \qquad (6\text{-}86)$$

We may note that, in this equation, the electron concentration $n$ is the total concentration, which includes the excess electron concentration. The remaining constants and terms in Equation (6–86) are the same as defined previously and the Fermi energy in the Fermi probability function needs to be replaced by the quasi-Fermi energy for electrons. The constants $E_n$ and $C_n$ are related by Equation (6–83), so the net recombination rate can be written as

$$R_n = C_n N_t [n(1 - f_F(E_t)) - n'f_F(E_t)] \qquad (6\text{–}87)$$

If we consider processes 3 and 4 in the recombination theory, the net rate at which holes are captured from the valence band is given by

$$R_p = C_p N_t [p f_F(E_t) - p'(1 - f_F(E_t))] \qquad (6\text{–}88)$$

where $C_p$ is a constant proportional to the hole capture rate, and $p'$ is given by

$$p' = N_v \exp \left[ \frac{-(E_t - E_v)}{kT} \right] \qquad (6\text{–}89)$$

In a semiconductor in which the trap density is not too large, the excess electron and hole concentrations are equal and the recombination rates of electrons and holes are equal. If we set Equation (6–87) equal to Equation (6–88) and solve for the Fermi function, we obtain

$$f_F(E_t) = \frac{C_n n + C_p p'}{C_n(n + n') + C_p(p + p')} \qquad (6\text{–}90)$$

We may note that $n'p' = n_i^2$. Then, substituting Equation (6–90) back into either Equation (6–87) or (6–88) gives

$$R_n = R_p = \frac{C_n C_p N_t (np - n_i^2)}{C_n(n + n') + C_p(p + p')} \equiv R \qquad (6\text{–}91)$$

Equation (6–91) is the recombination rate of electrons and holes due to the recombination center at $E = E_t$. If we consider thermal equilibrium, then $np = n_0 p_0 = n_i^2$, so that $R_n = R_p = 0$. Equation (6–91), then, is the recombination rate of excess electrons and holes.

Since $R$ in Equation (6–91) is the recombination rate of the excess carriers, we may write

$$R = \frac{\delta n}{\tau} \qquad (6\text{–}92)$$

where $\delta n$ is the excess-carrier concentration and $\tau$ is the lifetime of the excess carriers.

## 6.5.2  Limits of Extrinsic Doping and Low Injection

The ambipolar transport equation, Equation (6–39), was simplified from a nonlinear differential equation to a linear differential equation by applying limits of extrinsic doping and low injection. We may apply these same limits to the recombination rate equation.

Consider an n-type semiconductor under low injection. Then

$$n_0 \gg p_0, \; n_0 \gg \delta p, \; n_0 \gg n', \; n_0 \gg p'$$

where $\delta p$ is the excess minority carrier hole concentration. The assumptions of $n_0 \gg n'$ and $n_0 \gg p'$ imply that the trap level energy is near midgap so that $n'$ and $p'$ are not too different from the intrinsic carrier concentration. With these assumptions, Equation (6–91) reduces to

$$R = C_p N_t \delta p \qquad (6\text{–}93)$$

The recombination rate of excess carriers in the n-type semiconductor is a function of the parameter $C_p$, which is related to the minority carrier hole capture cross section. The recombination rate, then, is a function of the minority carrier parameter in the same way that the ambipolar transport parameters reduced to their minority carrier values.

The recombination rate is related to the mean carrier lifetime. Comparing Equations (6–92) and (6–93), we may write

$$R = \frac{\delta n}{\tau} = C_p N_t \, \delta p \equiv \frac{\delta p}{\tau_{p0}} \qquad (6\text{–}94)$$

where

$$\tau_{p0} = \frac{1}{C_p N_t} \qquad (6\text{–}95)$$

and where $\tau_{p0}$ is defined as the excess minority carrier hole lifetime. If the trap concentration increases, the probability of excess carrier recombination increases; thus the excess minority carrier lifetime decreases.

Similarly, if we have a strongly extrinsic p-type material under low injection, we can assume that

$$p_0 \gg n_0, \; p_0 \gg \delta n, \; p_0 \gg n', \; p_0 \gg p'$$

The lifetime then becomes that of the excess minority carrier electron lifetime, or

$$\tau_{n0} = \frac{1}{C_n N_t} \qquad (6\text{–}96)$$

Again note that for the n-type material, the lifetime is a function of $C_p$, which is related to the capture rate of the minority carrier hole. And for the p-type material, the lifetime is a function of $C_n$, which is related to the capture rate of the minority carrier electron. The excess-carrier lifetime for an extrinsic material under low injection reduces to that of the minority carrier.

**Example 6–6**

**Objective:** To determine the excess-carrier lifetime in an intrinsic semiconductor.

If we substitute the definitions of excess-carrier lifetimes from Equations (6–95) and (6–96) into Equation (6–91), the recombination rate can be written as

$$R = \frac{(np - n_i^2)}{\tau_{p0}(n + n') + \tau_{n0}(p + p')} \tag{6–97}$$

Consider an intrinsic semiconductor containing excess carriers. Then $n = n_i + \delta n$ and $p = n_i + \delta n$. Also assume that $n' = p' = n_i$.

**Solution:** Equation (6–97) now becomes

$$R = \frac{2n_i\delta n + (\delta n)^2}{(2n_i + \delta n)(\tau_{p0} + \tau_{n0})}$$

If we also assume very low injection, so that $\delta n \ll 2n_i$, then we can write

$$R = \frac{\delta n}{\tau_{p0} + \tau_{n0}} = \frac{\delta n}{\tau}$$

where $\tau$ is the excess carrier lifetime. We see that $\tau = \tau_{p0} + \tau_{n0}$ in the intrinsic material.

**Comment:** The excess-carrier lifetime increases as we change from an extrinsic to an intrinsic semiconductor.

---

Intuitively, we can see that the number of majority carriers that are available for recombining with excess minority carriers decreases as the extrinsic semiconductor becomes intrinsic. Since there are fewer carriers available for recombining in the intrinsic material, the mean lifetime of an excess carrier increases.

## *6.6 SURFACE EFFECTS

In all previous discussions, we have implicitly assumed the semiconductors were infinite in extent; thus, we were not concerned with any boundary conditions at a semiconductor surface. In any real application of semiconductors, the material is not infinitely large and therefore surfaces do exist between the semiconductor and an adjacent medium.

### 6.6.1 Surface States

When a semiconductor is abruptly terminated, the perfect periodic nature of the idealized single-crystal lattice ends abruptly at the surface. The disruption of the periodic-potential function results in allowed electronic energy states within the energy bandgap. In the previous section, we argued that simple defects in the semiconductor would create discrete energy states within the bandgap. The abrupt termination of the periodic potential at the surface results in a distribution of allowed energy states within the bandgap, shown schematically in Figure 6–15 along with the discrete energy states in the bulk semiconductor.

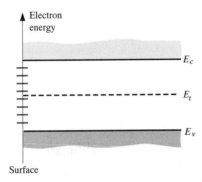

**Figure 6–15** Distribution of surface states within the forbidden bandgap.

The Shockley-Read-Hall recombination theory presented in the previous section showed that the excess minority carrier lifetime is inversely proportional to the density of trap states. We may argue that, since the density of traps at the surface is larger than in the bulk, the excess minority carrier lifetime at the surface will be smaller than the corresponding lifetime in the bulk material. If we consider an extrinsic n-type semiconductor, for example, the recombination rate of excess carriers in the bulk, given by Equation (6–94), is

$$R = \frac{\delta p}{\tau_{p0}} \equiv \frac{\delta p_B}{\tau_{p0}} \tag{6–98}$$

where $\delta p_B$ is the concentration of excess minority carrier holes in the bulk material. We may write a similar expression for the recombination rate of excess carriers at the surface as

$$R_s = \frac{\delta p_s}{\tau_{p0s}} \tag{6–99}$$

where $\delta p_s$ is the excess minority carrier hole concentration at the surface and $\tau_{p0s}$ is the excess minority carrier hole lifetime at the surface.

Assume that excess carriers are being generated at a constant rate throughout the entire semiconductor material. We showed that, in steady state, the generation rate is equal to the recombination rate for the case of a homogeneous, infinite semiconductor. Using this argument, the recombination rates at the surface and in the bulk material must be equal. Since $\tau_{p0s} < \tau_{p0}$, then the excess minority carrier concentration at the surface is smaller than the excess minority carrier concentration in the bulk region, or $\delta p_s <$ $\delta p_B$. Figure 6–16 shows an example of the excess-carrier concentration plotted as a function of distance from the semiconductor surface.

---

### Example 6–7

**Objective:** To determine the steady-state excess-carrier concentration as a function of distance from the surface of a semiconductor.

Consider Figure 6–16 in which the surface is at $x = 0$. Assume that in the n-type semiconductor $\delta p_B = 10^{14}$ cm$^{-3}$ and $\tau_{p0} = 10^{-6}$ sec in the bulk, and $\tau_{p0s} = 10^{-7}$ sec at the surface. Assume zero applied electric field and let $D_p = 10$ cm$^2$/sec

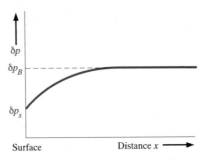

**Figure 6–16**   Steady-state excess hole concentration versus distance from a semiconductor surface.

**Solution:** From Equations (6–98) and (6–99), we have

$$\frac{\delta p_B}{\tau_{p0}} = \frac{\delta p_s}{\tau_{p0s}}$$

so that

$$\delta p_s = \delta p_B \left(\frac{\tau_{p0s}}{\tau_{p0}}\right) = (10^{14}) \left(\frac{10^{-7}}{10^{-6}}\right) = 10^{13} \text{ cm}^{-3}$$

From Equation (6–56), we can write

$$D_p \frac{d^2(\delta p)}{dx^2} + g' - \frac{\delta p}{\tau_{p0}} = 0 \qquad (6\text{–}100)$$

The generation rate can be determined from the steady-state conditions in the bulk, or

$$g' = \frac{\delta p_B}{\tau_{p0}} = \frac{10^{14}}{10^{-6}} = 10^{20} \text{ cm}^{-3} \text{ sec}^{-1}$$

The solution to Equation (6–100) is of the form

$$\delta p(x) = g'\tau_{p0} + Ae^{x/L_p} + Be^{-x/L_p} \qquad (6\text{–}101)$$

As $x \to +\infty$, $\delta p(x) = \delta p_B = g' \tau_{p0} = 10^{14} \, \text{cm}^{-3}$, which implies that $A = 0$. At $x = 0$, we have

$$\delta p(0) = \delta p_s = 10^{14} + B = 10^{13} \, \text{cm}^{-3}$$

so that $B = -9 \times 10^{13}$. The entire solution for the minority carrier hole concentration as a function of distance from the surface is

$$\delta p(x) = 10^{14} \, (1 - 0.9 e^{-x/L_p})$$

where

$$L_p = \sqrt{D_p \tau_{p0}} = \sqrt{(10)(10^{-6})} = 31.6 \, \mu\text{m}$$

**Comment:** The excess carrier concentration is smaller at the surface than in the bulk.

---

## 6.6.2 Surface Recombination Velocity

A gradient in the excess-carrier concentration exists near the surface as shown in Figure 6–16; excess carriers from the bulk region diffuse toward the surface where they recombine. This diffusion toward the surface can be described by the equation

$$-D_p \left[ \hat{n} \cdot \frac{d(\delta p)}{dx} \right] \Bigg|_{\text{surf}} = s \delta p |_{\text{surf}} \tag{6–102}$$

where each side of the equation is evaluated at the surface. The parameter $\hat{n}$ is the unit outward vector normal to the surface. Using the geometry of Figure 6–16, $d(\delta p)/dx$ is a positive quantity and $\hat{n}$ is negative so that the parameter $s$ is a positive quantity.

A dimensional analysis of Equation (6–102) shows that the parameter $s$ has units of cm/sec, or velocity. The parameter $s$ is called the *surface recombination velocity*. If the excess concentrations at the surface and in the bulk region were equal, then the gradient term would be zero and the surface recombination velocity would be zero. As the excess concentration at the surface becomes smaller, the gradient term becomes larger, and the surface recombination velocity increases. The surface recombination velocity gives some indication of the surface characteristics as compared with the bulk region.

Equation (6–102) may be used as a boundary condition to the general solution given by Equation (6–101) in Example 6–7. Using Figure 6–16, we have that $\hat{n} = -1$, and Equation (6–102) becomes

$$D_p \frac{d(\delta p)}{dx} \Bigg|_{\text{surf}} = s \delta p |_{\text{surf}} \tag{6–103}$$

We have argued that the coefficient $A$ is zero in Equation (6–101). Then, from Equation (6–101), we can write that

$$\delta p |_{\text{surf}} = \delta p(0) = g' \tau_{p0} + B \tag{6–104a}$$

and

$$\frac{d(\delta p)}{dx}\bigg|_{\text{surf}} = \frac{d(\delta p)}{dx}\bigg|_{x=0} = -\frac{B}{L_p} \qquad (6\text{--}104b)$$

Substituting Equations (6–104a) and (6–104b) into Equation (6–103) and solving for the coefficient $B$, we obtain

$$B = \frac{-sg'\tau_{p0}}{(D_p/L_p) + s} \qquad (6\text{--}105)$$

The excess minority carrier hole concentration can then be written as

$$\delta p(x) = g'\tau_{p0}\left[1 - \frac{sL_p e^{-x/L_p}}{D_p + sL_p}\right] \qquad (6\text{--}106)$$

---

### Example 6–8

**Objective:** To determine the steady-state excess concentration versus distance from the surface of a semiconductor as a function of surface recombination velocity.

Consider, initially, the case when the surface recombination velocity is zero, or $s = 0$.

**Solution:** Substituting $s = 0$ into Equation (6–106), we obtain

$$\delta p(x) = g'\tau_{p0}$$

Now consider the case when the surface recombination velocity is infinite, or $s = \infty$.

**Solution:** Substituting $s = \infty$ into Equation (6–106), we obtain

$$\delta p(x) = g'\tau_{p0}(1 - e^{-x/L_p})$$

**Comment:** For the case when $s = 0$, the surface has no effect and the excess minority carrier concentration at the surface is the same as in the bulk. In the other extreme when $s = \infty$, the excess minority carrier hole concentration at the surface is zero.

---

An infinite surface recombination velocity implies that the excess minority carrier concentration and lifetime at the surface are zero.

---

### Example 6–9

**Objective:** To determine the value of surface recombination velocity corresponding to the parameters given in Example 6–7.

From Example 6–7, we have that $g'\tau_{p0} = 10^{14}$ cm$^{-3}$, $D_p = 10$ cm$^2$/sec, $L_p = 31.6$ $\mu$m, and $\delta p(0) = 10^{13}$ cm$^{-3}$.

**Solution:** Writing Equation (6–106) at the surface, we have

$$\delta p(0) = g'\tau_{p0}\left[1 - \frac{s}{(D_p/L_p) + s}\right]$$

Solving for the surface recombination velocity, we find that

$$s = \frac{D_p}{L_p}\left[\frac{g'\tau_{p0}}{\delta p(0)} - 1\right]$$

which becomes

$$s = \frac{10}{31.6 \times 10^{-4}}\left[\frac{10^{14}}{10^{13}} - 1\right] = 2.85 \times 10^4 \text{ cm/sec}$$

**Comments:** This example shows that a surface recombination velocity of approximately $s = 3 \times 10^4$ cm/sec could seriously degrade the performance of semiconductor devices, such as solar cells, since these devices tend to be fabricated close to a surface.

---

In the above example, the surface influences the excess-carrier concentration to the extent that, even at a distance of $L_p = 31.6$ $\mu$m from the surface, the excess-carrier concentration is only two thirds of the value in the bulk. We will see in later chapters that device performance is dependent in large part on the properties of excess carriers.

## 6.7 SUMMARY AND REVIEW

We have considered the behavior and characteristics of excess carriers at nonequilibrium in a semiconductor material. Two of these characteristics are the generation and recombination rates. Excess electrons and holes are generated by some external force, such as incident photons or an applied voltage. Once they are formed, there is a finite probability that the carriers will recombine. The mean time that an excess electron-hole pair exists is the excess-carrier lifetime. The recombination rate of excess carriers is inversely proportional to the lifetime.

Excess electrons and holes do not move independently of each other. If a pulse of excess electrons and a pulse of excess holes were to drift in opposite directions, an internal electric field would develop due to the net space charge. This induced electric field holds the excess electrons and excess holes together, so that the two types of excess carriers move together. This common movement is called ambipolar transport and is described by the ambipolar transport equation. The ambipolar transport equation was derived by combining the continuity equations for electrons and holes with the current equations and imposing the charge neutrality condition.

Limits of extrinsic doping and low injection were applied to the ambipolar transport equation; its coefficients then reduce to the minority carrier parameter values. This result means that for an extrinsic semiconductor operating under low-injection conditions, excess electrons and holes diffuse and drift together with the characteristics of the minority carrier, a result that is fundamental to the behavior of semiconductor devices. Several examples of space- and time-dependent generation rates were considered in order to characterize the behavior of excess carriers under various conditions.

When excess electrons and holes are created, the semiconductor is no longer in thermal equilibrium. This nonequilibrium condition implies that the concept of the Fermi energy, which is defined only for thermal equilibrium, is no longer valid. However, in nonequilibrium, we can define two new parameters—the quasi-Fermi level for electrons and the quasi-Fermi level for holes—which characterize the total electron and hole concentrations in a semiconductor in nonequilibrium.

We considered the Shockley-Read-Hall theory of recombination. In silicon, for example, the lifetime of excess carriers is a function, primarily, of the number of defects, or recombination centers, in the semiconductor. This theory of recombination results in an expression for the recombination rate that is inversely proportional to the excess-carrier lifetime and directly proportional to the number of recombination centers in the semiconductor. Limits of extrinsic doping and low injection were applied to the recombination rate. We found, as a result, that the excess-carrier lifetime reduces to the minority carrier lifetime value—the same result as for the ambipolar transport equation.

The abrupt termination of a semiconductor results in a distribution of allowed electronic energy states in the forbidden energy bandgap. These interface states, or surface states, alter the behavior of excess carriers at the surface. In general, the excess-carrier concentration at the surface will be smaller than in the bulk due to a smaller excess-carrier lifetime. This produces a diffusion of excess carriers from the bulk toward the surface where they more easily recombine, an effect that is described by the surface recombination velocity. As semiconductor devices are usually fabricated at or near a surface, interface states and surface recombination velocity can affect device characteristics.

We have considered the behavior of excess carriers in a semiconductor, a characteristic that is important to the operation of semiconductor devices. Up to this point we have considered only the semiconductor material. From this point on, we will be considering semiconductor devices, which are formed by joining n- and p-regions. The ambipolar transport equation will be applied extensively in the following chapters to describe device properties.

## GLOSSARY OF IMPORTANT TERMS

**Ambipolar diffusion coefficient:** The effective diffusion coefficient of excess carriers.

**Ambipolar mobility:** The effective mobility of excess carriers.

**Ambipolar transport:** The process whereby excess electrons and holes diffuse, drift, and recombine with the same effective diffusion coefficient, mobility, and lifetime.

**Ambipolar transport equation:** The equation describing the behavior of excess carriers as a function of time and space coordinates.

**Carrier generation:** The process of elevating electrons from the valence band into the conduction band, creating an electron-hole pair.

**Carrier recombination:** The process whereby an electron "falls" into an empty state in the valence band (a hole) so that an electron-hole pair are annihilated.

**Excess carriers:** The term describing both excess electrons and excess holes.

**Excess electrons:** The concentration of electrons in the conduction band over and above the thermal-equilibrium concentration.

**Excess holes:** The concentration of holes in the valence band over and above the thermal-equilibrium concentration.

**Excess minority carrier lifetime:** The average time that an excess minority carrier exists before it recombines.

**Generation rate:** The rate ($\#/cm^3$-sec) at which electron-hole pairs are created.

**Low-level injection:** The condition in which the excess-carrier concentration is much smaller than the thermal-equilibrium majority carrier concentration.

**Minority carrier diffusion length:** The average distance a minority carrier diffuses before recombining: a parameter equal to $\sqrt{D\tau}$ where $D$ and $\tau$ are the minority carrier diffusion coefficient and lifetime, respectively.

**Quasi-Fermi level:** The quasi-Fermi level for electrons and the quasi-Fermi level for holes relate the nonequilibrium electron and hole concentrations, respectively, to the intrinsic carrier concentration and the intrinsic Fermi level.

**Recombination rate:** The rate ($\#/cm^3$-sec) at which electron-hole pairs recombine.

**Surface recombination velocity:** A parameter that relates the gradient of the excess carrier concentration at a surface to the surface concentration of excess carriers.

**Surface states:** The electronic energy states that exist within the bandgap at a semiconductor surface.

## PROBLEMS

(Note: Use the semiconductor parameters listed in Appendix B if they are not specifically given in a problem. Assume $T = 300°K$.)

### Section 6.1

1.   Consider a semiconductor in which $n_0 = 10^{15}$ cm$^{-3}$ and $n_i = 10^{10}$ cm$^{-3}$. Assume that the excess-carrier lifetime is $10^{-6}$ sec. Determine the electron-hole recombination rate if the excess-hole concentration is $\delta p = 5 \times 10^{13}$ cm$^{-3}$.

2.   A semiconductor, in thermal equilibrium, has a hole concentration of $p_0 = 10^{16}$ cm$^{-3}$ and an intrinsic concentration of $n_i = 10^{10}$ cm$^{-3}$. The minority carrier lifetime is $2 \times 10^{-7}$ sec. Determine the thermal-equilibrium generation rate of electrons.

### Section 6.2

3.   Derive Equation (6–27) from Equations (6–18) and (6–20).

4.   Consider a one-dimensional hole flux as shown in Figure 6–4. If the generation rate of holes in this differential volume is $g_p = 10^{20}$ cm$^{-3}$ sec$^{-1}$ and the recombination rate is zero, what must be the gradient in the particle current density to maintain a steady-state hole concentration?

### Section 6.3

5.   Starting with the continuity equations given by Equations (6–29) and (6–30), derive the ambipolar transport equation given by Equation (6–39).

6.   Assume that an n-type semiconductor is uniformly illuminated, producing a uniform excess generation rate $g'$. Show that in steady state the change in the semiconductor conductivity is given by

$$\Delta \sigma = e(\mu_n + \mu_p)\tau_{p0}g'$$

7.   Light is incident on a silicon sample starting at $t = 0$ and generating excess carriers uniformly throughout the silicon for $t > 0$. The generation rate is $g' = 5 \times 10^{21}$ cm$^{-3}$ sec$^{-1}$. The silicon ($T = 300°$K) is n-type with $N_d = 5 \times 10^{16}$ cm$^{-3}$ and $N_a = 0$. Let $n_i = 1.5 \times 10^{10}$ cm$^{-3}$, $\tau_{n0} = 10^{-6}$ sec, and $\tau_{p0} = 10^{-7}$ sec. Also let $\mu_n = 1000$ cm$^2$/V-sec and $\mu_p = 420$ cm$^2$/V-sec. Calculate the conductivity of the silicon as a function of time for $t \geq 0$.

8.   An n-type gallium arsenide semiconductor is doped with $N_d = 10^{16}$ cm$^{-3}$ and $N_a = 0$. The minority carrier lifetime is $\tau_{p0} = 2 \times 10^{-7}$ sec. Calculate the steady-state increase in conductivity if a uniform generation rate, $g' = 2 \times 10^{21}$ cm$^{-3}$ sec$^{-1}$, is incident on the semiconductor.

9.   A silicon sample at $T = 300°$K is n-type with $N_d = 5 \times 10^{16}$ cm$^{-3}$ and $N_a = 0$. The sample has a length of 0.1 cm and a cross-sec-

tional area of $10^{-4}$ cm$^2$. A voltage of 5 volts is applied between the ends of the sample. For $t < 0$, the sample has been illuminated with light producing an excess-carrier generation rate of $g' = 5 \times 10^{21}$ cm$^{-3}$ sec$^{-1}$ uniformly throughout the entire silicon. The minority carrier lifetime is $\tau_{p0} = 3 \times 10^{-7}$ sec. At $t = 0$, the light is turned off. Derive the expression for the current in the sample as a function of time for $t \geq 0$. (Neglect surface effects.)

10. Consider a homogeneous gallium arsenide semiconductor at $T = 300°K$ with $N_a = 10^{16}$ cm$^{-3}$ and $N_d = 0$. A light source is turned on at $t = 0$ producing a uniform generation rate of $g' = 10^{20}$ cm$^{-3}$ sec$^{-1}$. The electric field is zero. (a) Derive the expression for the excess-carrier concentration as a function of time. (b) If the maximum, steady-state, excess-carrier concentration is to be $1 \times 10^{14}$ cm$^{-3}$, determine the maximum value of the minority carrier lifetime. (c) Determine at what time the excess minority carrier concentration will be equal to one half of the steady-state value.

11. Consider a homogeneous gallium arsenide semiconductor at $T = 300°K$ doped at $N_a = 10^{14}$ cm$^{-3}$ and $N_d = 0$. A uniform excess carrier generation rate of $g' = 10^{20}$ cm$^{-3}$ sec$^{-1}$ exists for $t < 0$. The excess minority carrier lifetime is $\tau_{n0} = 0.1$ $\mu$sec. The generation rate is turned off at $t = 0$. Derive the expression for the excess electron concentration versus time, and calculate the electron concentration at (a) $t = 0.05$ $\mu$sec and (b) $t = 0.30$ $\mu$sec.

12. Consider a silicon material doped with $3 \times 10^{16}$ cm$^{-3}$ donor atoms. At $t = 0$, a light source is turned on producing a uniform generation rate of $g' = 2 \times 10^{20}$ cm$^{-3}$ sec$^{-1}$. At $t = 10^{-7}$ sec, the light source is turned off. Determine the excess minority carrier concentration as a function of $t$ for $0 \leq t \leq \infty$. Let $\tau_{p0} = 10^{-7}$ sec. Plot the excess minority carrier concentration as a function of time.

13. A semiconductor has the following properties:

$$D_n = 25 \text{ cm}^2/\text{sec} \qquad \tau_{n0} = 10^{-6} \text{ sec}$$
$$D_p = 10 \text{ cm}^2/\text{sec} \qquad \tau_{p0} = 10^{-7} \text{ sec}$$

The semiconductor is a homogeneous, p-type ($N_a = 10^{17}$ cm$^{-3}$) material in thermal equilibrium for $t \leq 0$. At $t = 0$, an external source is turned on which produces excess carriers uniformly at the rate of $g' = 10^{20}$ cm$^{-3}$ sec$^{-1}$. At $t = 2 \times 10^{-6}$ sec, the external source is turned off. (a) Derive the expression for the excess-electron concentration as a function of time for $0 \leq t \leq \infty$. (b) Determine the value of the excess-electron concentration at (i) $t = 0$, (ii) $t = 2 \times 10^{-6}$ sec, and (iii) $t = \infty$. (c) Plot the excess-electron concentration as a function of time.

14. Consider a bar of p-type silicon material that is homogeneously doped to a value of $3 \times 10^{15}$ cm$^{-3}$ at $T = 300°K$. The applied elec-

tric field is zero. A light source is incident on the end of the semiconductor as shown in Figure 6–17. The excess-carrier concentra-
tion generated at $x = 0$ is $\delta p(0) = \delta n(0) = 10^{13}$ cm$^{-3}$. Assume the
following parameters (neglect surface effects):

$$\mu_n = 1200 \text{ cm}^2/\text{V-sec} \qquad \tau_{n0} = 5 \times 10^{-7} \text{ sec}$$
$$\mu_p = 400 \text{ cm}^2/\text{V-sec} \qquad \tau_{p0} = 1 \times 10^{-7} \text{ sec}$$

(a) Calculate the steady-state excess electron and hole concentrations as a function of distance into the semiconductor. (b) Calculate
the electron diffusion current density as a function of $x$.

15.  The $x = 0$ end of an $N_a = 1 \times 10^{14}$ cm$^{-3}$ doped semi-infinite ($x \geq 0$)
bar of silicon maintained at $T = 300°$K is attached to a "minority
carrier digester" which makes $n_p = 0$ at $x = 0$ ($n_p$ is the minority
carrier electron concentration in a p-type semiconductor). The electric field is zero. (a) Determine the thermal-equilibrium values of $n_{p0}$
and $p_{p0}$. (b) What is the excess minority carrier concentration at
$x = 0$. (c) Derive the expression for the steady-state excess minority carrier concentration as a function of $x$.

16.  In a p-type silicon semiconductor, excess carriers are being generated at the end of the semiconductor bar at $x = 0$ as shown in Figure 6–17. The doping concentration is $N_a = 5 \times 10^{16}$ cm$^{-3}$ and
$N_d = 0$. The steady-state excess-carrier concentration at $x = 0$ is
$10^{15}$ cm$^{-3}$. (Neglect surface effects.) The applied electric field is
zero. Assume that $\tau_{n0} = \tau_{p0} = 8 \times 10^{-7}$ sec. (a) Calculate the
steady-state excess-electron concentration at (i) $x = 5 \times 10^{-4}$ cm,
(ii) $x = 20 \times 10^{-4}$ cm, and (iii) $x = 100 \times 10^{-4}$ cm. (b) Calculate
the hole current density as a function of $x$.

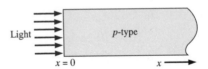

Light ⟶ p-type

$x = 0$                          $x \longrightarrow$

**Figure 6–17**   Figure for problems 14 and 16.

*17.  Consider the semiconductor described in problem 14. Assume a
constant electric field $E_0$ is applied in the $+x$ direction. (a) Derive
the expression for the steady-state excess-electron concentration.
(Assume the solution is of the form $e^{-\alpha x}$.) (b) Plot $\delta n$ versus $x$ for
(i) $E_0 = 0$ and (ii) $E_0 = 12$ V/cm. (c) Explain the general characteristics of the two curves plotted in part (b).

18.  Assume that a p-type semiconductor is in thermal equilibrium for
$t < 0$ and has an infinite minority carrier lifetime. Also assume that

the semiconductor is uniformly illuminated resulting in a uniform generation rate, $g'(t)$, which is given by

$$g'(t) = G_0' \quad \text{for } 0 < t < T$$
$$g'(t) = 0 \quad \text{for } t < 0 \text{ and } t > T$$

where $G_0'$ is a constant. Find the excess minority carrier concentration as a function of time.

*19. Consider the n-type semiconductor shown in Figure 6–18. Illumination produces a constant excess-carrier generation rate, $G_0'$, in the region $-L < x < +L$. Assume that the minority carrier lifetime is infinite and assume that the excess minority carrier hole concentration is zero at $x = -3L$ and at $x = +3L$. Find the steady-state excess minority carrier concentration versus $x$, for the case of low injection and for zero applied electric field.

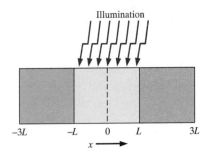

**Figure 6–18** Figure for problem 19.

20. An n-type germanium sample is used in the Haynes-Shockley experiment. The length of the sample is 1 cm and the applied voltage is $V_1 = 2.5$ V. The contacts A and B are separated by 0.75 cm. The peak of the pulse arrives at contact B 160 $\mu$sec after carrier injection at contact A. The width of the pulse is $\Delta t = 75.5$ $\mu$sec. Determine the hole mobility and diffusion coefficient. Compare the results with the Einstein relation.

## Section 6.4

21. An n-type silicon sample with $N_d = 10^{16}$ cm$^{-3}$ is steadily illuminated such that $g' = 10^{21}$ cm$^{-3}$ sec$^{-1}$. If $\tau_{n0} = \tau_{p0} = 10^{-6}$ sec, calculate the position of the quasi-Fermi levels for electrons and holes with respect to the intrinsic level (assume that $n_i = 1.5 \times 10^{10}$ cm$^{-3}$). Plot these levels on an energy-band diagram.

22. Consider a p-type silicon semiconductor at $T = 300°$K doped at $N_a = 5 \times 10^{15}$ cm$^{-3}$. (a) Determine the position of the Fermi level

with respect to the intrinsic Fermi level. (b) Excess carriers are generated such that the excess-carrier concentration is 10 percent of the thermal-equilibrium majority carrier concentration. Determine the quasi-Fermi levels with respect to the intrinsic Fermi level. (c) Plot the Fermi level and quasi-Fermi levels with respect to the intrinsic level.

23.  Consider an n-type gallium arsenide semiconductor at $T = 300°K$ doped at $N_d = 5 \times 10^{16}$ cm$^{-3}$. (a) Determine $E_{Fn} - E_F$ if the excess-carrier concentration is $0.1 N_d$. (b) Determine $E_{Fi} - E_{Fp}$.

24.  A p-type gallium arsenide semiconductor at $T = 300°K$ is doped at $N_a = 10^{16}$ cm$^{-3}$. The excess-carrier concentration is $10^{14}$ cm$^{-3}$. Determine the position of the quasi-Fermi levels with respect to the intrinsic Fermi level. Plot these levels on an energy-band diagram.

25.  Consider p-type silicon at $T = 300°K$ doped to $N_a = 5 \times 10^{14}$ cm$^{-3}$. Assume excess carriers are present and assume that $E_F - E_{Fp} = (0.01) kT$. (a) Does this condition correspond to low injection? Why or why not? (b) Determine $E_{Fn} - E_{Fi}$.

### Section 6.5

26.  Consider Equation (6–91) and the definitions of $\tau_{p0}$ and $\tau_{n0}$ given by Equations (6–95) and (6–96). Let $n' = p' = n_i$. Assume that in a particular region of a semiconductor, $n = p = 0$. (a) Determine the recombination rate $R$. (b) Explain what this result means physically.

27.  Again consider Equation (6–91) and the definitions of $\tau_{p0}$ and $\tau_{n0}$ given by Equations (6–95) and (6–96). Let $\tau_{p0} = 10^{-7}$ sec and $\tau_{n0} = 5 \times 10^{-7}$ sec. Also let $n' = p' = n_i = 10^{10}$ cm$^{-3}$. Assume very low injection so that $\delta n \ll n_i$. Calculate $R/\delta n$ for a semiconductor which is (a) n-type ($n_0 \gg p_0$), (b) intrinsic ($n_0 = p_0 = n_i$), and (c) p-type ($p_0 \gg n_0$).

### Section 6.6

*28.  Consider an n-type semiconductor as shown in Figure 6–19 doped at $N_d = 10^{16}$ cm$^{-3}$ and with a uniform excess-carrier generation rate equal to $g' = 10^{21}$ cm$^{-3}$ sec$^{-1}$. Assume that $D_p = 10$ cm$^2$/sec and $\tau_{p0} = 10^{-7}$ sec. The electric field is zero. (a) Determine the steady-

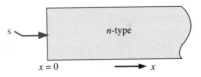

**Figure 6–19** Figure for problem 28.

state excess minority carrier concentration versus $x$ if the surface recombination velocity at $x = 0$ is (i) $s = 0$, (ii) $s = 2000$ cm/sec, and (iii) $s = \infty$. (b) Calculate the excess minority carrier concentration at $x = 0$ for (i) $s = 0$, (ii) $s = 2000$ cm/sec, and (iii) $s = \infty$.

*29. (a) Consider the p-type semiconductor shown in Figure 6–20 with the following parameters: $N_a = 5 \times 10^{16}$ cm$^{-3}$, $D_n = 25$ cm$^2$/sec, and $\tau_{n0} = 5 \times 10^{-7}$ sec. The surface recombination velocities at the

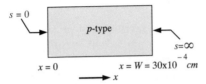

$s = 0$

p-type

$s = \infty$

$x = 0$     $x = W = 30 \times 10^{-4}$ cm

$x$

**Figure 6–20**    Figure for problem 29.

two surfaces are shown. The electric field is zero. The semiconductor is illuminated at $x = 0$ with an excess-carrier generation rate equal to $g' = 2 \times 10^{21}$ cm$^{-3}$ sec$^{-1}$. Determine the excess minority carrier electron concentration versus $x$ in steady state. (b) Repeat part (a) for $\tau_{n0} = \infty$.

*30. Consider the n-type semiconductor shown in Figure 6–21. Assume that $D_p = 10$ cm$^2$/sec and $\tau_{p0} = \infty$. The electric field is zero. Assume that a flux of excess electrons and holes is incident at $x = 0$. Let the flux of each carrier type be $10^{19}$ carriers/cm$^2$-sec. Determine the minority carrier hole current versus $x$ if the surface recombination velocity is (a) $s(W) = \infty$ and (b) $s(W) = 2000$ cm/sec.

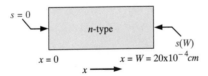

$s = 0$

n-type

$s(W)$

$x = 0$     $x = W = 20 \times 10^{-4}$ cm

$x$

**Figure 6–21**    Figure for problem 30.

*31. A p-type semiconductor is shown in Figure 6–22. The surface recombination velocities are shown. The semiconductor is uniformly illuminated for $-W < x < 0$ producing a constant excess-carrier generation rate $G_0'$. Determine the steady-state excess-carrier concentration versus $x$ if the minority carrier lifetime is infinite and if the electric field is zero.

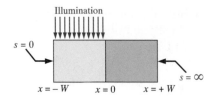

**Figure 6–22**   Figure for problem 31.

## Summary and Review

*32.   Consider an n-type semiconductor as shown in Figure 6–19. The material is doped at $N_d = 3 \times 10^{16}$ cm$^{-3}$ and $N_a = 0$. Assume that $D_p = 12$ cm$^2$/sec and $\tau_{p0} = 2 \times 10^{-7}$ sec. The electric field is zero. "Design" the surface recombination velocity so that the minority carrier diffusion current density at the surface is no greater than $J_p = -0.18$ A/cm$^2$ with a uniform excess-carrier generation rate equal to $g' = 3 \times 10^{21}$ cm$^{-3}$ sec$^{-1}$.

33.   Consider a semiconductor with excess carriers present. From the definition of carrier lifetimes and recombination rates, determine the average time that an electron stays in the conduction band and the average time that a hole stays in the valence band. Discuss these relations for (*a*) an intrinsic semiconductor and (*b*) an n-type semiconductor.

34.   Design a gallium arsenide photoconductor which is 5 $\mu$m thick. Assume that $\tau_{n0} = \tau_{p0} = 10^{-7}$ sec and $N_d = 5 \times 10^{15}$ cm$^{-3}$. With an excitation of $g' = 10^{21}$ cm$^{-3}$ sec$^{-1}$, a photocurrent of at least 1 $\mu$A is desired with an applied voltage of 1V.

## READING LIST

1.   Bube, R. H. *Photoconductivity in Solids*. New York: Wiley, 1957.

*2.   deCogan, D. *Solid State Devices: A Quantum Physics Approach*. New York: Springer-Verlag, 1987.

3.   Hall, R. H. "Electron-Hole Recombination." *Physical Review* 87, no. 2 (July 15, 1952), p. 387.

4.   Haynes, J. R., and W. Shockley. "The Mobility and Life of Injected Holes and Electrons in Germanium." *Physical Review* 81, no. 5 (March 1, 1951), pp. 835–43.

*5.   Hess, K. *Advanced Theory of Semiconductor Devices*. Englewood Cliffs, N. J.: Prentice Hall, 1988.

6.   Kingston, R. H. *Semiconductor Surface Physics*. Philadelphia: University of Pennsylvania Press, 1957.

7.   McKelvey, J. P. *Solid-State and Semiconductor Physics*. New York: Harper and Row, 1966.

8.  Pierret, R. F. *Semiconductor Fundamentals*. Vol. 1 of *Modular Series on Solid State Devices*. 2nd ed. Reading, Mass.: Addison-Wesley, 1988.

9.  ———. *Advanced Semiconductor Fundamentals*. Vol. 6 of *Modular Series on Solid State Devices*. Reading, Mass.: Addison-Wesley, 1987.

10. Shockley, W., and W. T. Read, Jr. "Statistics of the Recombinations of Holes and Electrons." *Physical Review* 87, no. 5 (September 1, 1952), pp. 835–42.

11. Streetman, B. G. *Solid State Electronic Devices*. 3rd ed. Englewood Cliffs, N. J.: Prentice Hall, 1990.

*12. Wang, S. *Fundamentals of Semiconductor Theory and Device Physics*. Englewood Cliffs, N. J.: Prentice Hall, 1989.

# THE PN JUNCTION

## CHAPTER 7

## PREVIEW

Up to this point in the text, we have been considering the properties of the semiconductor material. We have calculated electron and hole concentrations in thermal equilibrium and determined the position of the Fermi level. We then considered the nonequilibrium condition in which excess electrons and holes are present in the semiconductor. We now wish to consider the situation in which a p-type and an n-type semiconductor are brought into contact with one another to form a pn junction.

Most semiconductor devices contain at least one junction between p-type and n-type semiconductor regions. The pn junction provides the characteristics needed for rectifiers, amplifiers, switching circuits, and many other electronic circuit functions. In this chapter, we will consider the electrostatics of the pn junction; the current-voltage characteristics of the junction will be developed in the following chapter.

We will begin our discussion by considering how thermal equilibrium is reached when the p-type and n-type materials are brought into contact. Thermal equilibrium implies that the Fermi energy is constant through the entire semiconductor system. Although a discussion of joining the two materials to form a contact is useful in qualitatively understanding what happens in reaching thermal equilibrium, the actual pn junction structure is formed from a single-crystal semiconductor in which one region is doped p-type and the adjacent region is doped n-type.

When a voltage is applied between the p- and n-regions, the pn junction is no longer in thermal equilibrium; that is, the Fermi energy is no longer constant through the device. We will determine how the energy-band diagram and the electrostatic properties of the pn junction change when a voltage is applied. There is a capacitance called the junction capacitance, a parameter that affects the frequency characteristics of any semiconductor device containing a pn junction. We will determine the junction capacitance as a function of applied voltage.

The theory of the pn junction is quite easily developed for the case when both the p- and n-regions are uniformly doped. In real devices, however, the p- and n-regions are rarely both uniformly doped. We will consider a few doping profiles and determine how the doping profile affects the pn junction characteristics. Special doping profiles are used to produce specific junction capacitance characteristics, which makes these devices useful in electronic circuits; for example, the tuning stage of a radio receiver.

## 7.1 BASIC STRUCTURE OF THE PN JUNCTION

Figure 7–1a schematically shows the pn junction. It is important to realize that the entire semiconductor is a single-crystal material in which one region is doped with acceptor impurity atoms to form the p-region and the adjacent region is doped with donor atoms to form the n-region. The interface separating the n- and p-regions is referred to as the *metallurgical junction*.

The impurity doping concentrations in the p- and n-regions are shown in Figure 7–1b. For simplicity, we will consider a *step junction* in which the doping concentration is uniform in each region and there is an abrupt change in doping at the junction. Initially, at the metallurgical junction, there is a very large density gradient in both the electron and hole concentrations. Majority carrier electrons in the n-region will begin diffusing into the p-region and majority carrier holes in the p-region will begin diffusing into the n-region. If we assume there are no external connections to the semiconductor, then this diffusion process cannot continue indefinitely. As electrons diffuse from the n-region, positively charged donor atoms are left behind. Similarly, as holes diffuse from the p-region, they uncover negatively

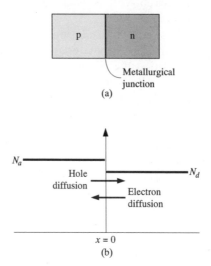

**Figure 7–1**   (a) Simplified geometry of a pn junction. (b) The doping profile of an ideal uniformly doped pn junction.

charged acceptor atoms. The net positive and negative charges in the n- and p-regions induce an electric field in the region near the metallurgical junction, in the direction from the positive to the negative charge, or from the n- to the p-region.

The net positively and negatively charged regions are shown in Figure 7–2. These two regions are referred to as the *space charge region*. Essentially all electrons and holes are swept out of the space charge region by the electric field. Since the space charge region is depleted of any mobile charge, this region is also referred to as the *depletion region;* these two terms will be used interchangeably. Density gradients still exist in the majority carrier concentrations at each edge of the space charge region. We can think of a density gradient as producing a "diffusion force" that acts on the majority carriers. These diffusion forces, acting on the electrons and holes at the edges of the space charge region, are shown in the figure. The electric field in the space charge region produces another force on the electrons and holes which is in the opposite direction to the "diffusion force" for each type of particle. In thermal equilibrium, the "diffusion force" and the E-field force exactly balance each other.

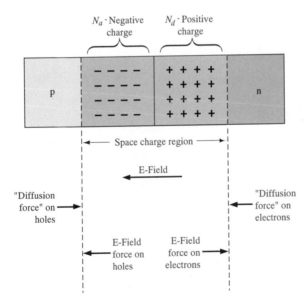

**Figure 7–2**   The space charge region, the electric field, and the forces acting on the charged carriers.

## 7.2 ZERO APPLIED BIAS

We have considered the basic pn junction structure and discussed briefly how the space charge region is formed. In this section we will examine the properties of the step junction in thermal equilibrium, where no currents exist and no external excitation is applied. We will determine the space charge region width, electric field, and potential through the depletion region.

### 7.2.1 Built-in Potential Barrier

If we assume that no voltage is applied across the pn junction, then the junction is in thermal equilibrium—the Fermi energy level is constant throughout the entire system. The energy-band diagram for the pn junction in thermal equilibrium is shown in Figure 7–3. The conduction and valance band energies must bend as we go through the space charge region, since the relative position of the conduction and valence bands with respect to the Fermi energy changes between the p- and n-regions.

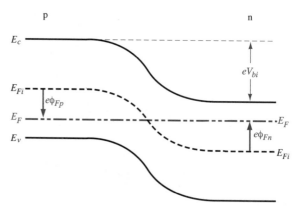

**Figure 7–3**   Energy-band diagram of a pn junction in thermal equilibrium.

Electrons in the conduction band of the n-region see a potential barrier in trying to move into the conduction band of the p-region. This potential barrier is referred to as the *built-in potential barrier* and is denoted by $V_{bi}$. The built-in potential barrier maintains equilibrium between majority carrier electrons in the n-region and minority carrier electrons in the p-region, and also between majority carrier holes in the p-region and minority carrier holes in the n-region. This potential difference across the junction cannot be measured with a voltmeter because new potential barriers will be formed between the probes and the semiconductor that will cancel $V_{bi}$. The potential $V_{bi}$ maintains equilibrium so no current is produced by this voltage.

The intrinsic Fermi level is equidistant from the conduction band edge through the junction, thus the built-in potential barrier can be determined as the difference between the intrinsic Fermi levels in the p- and n-regions. We can define the potentials $\phi_{Fn}$ and $\phi_{Fp}$ as shown in Figure 7–3, so we have

$$V_{bi} = |\phi_{Fn}| + |\phi_{Fp}| \tag{7-1}$$

In the n-region, the electron concentration in the conduction band is given by

$$n_0 = N_c \exp\left[\frac{-(E_c - E_F)}{kT}\right] \tag{7-2}$$

which can also be written in the form

$$n_0 = n_i \exp\left[\frac{E_F - E_{Fi}}{kT}\right] \tag{7-3}$$

where $n_i$ and $E_{Fi}$ are the intrinsic carrier concentration and the intrinsic Fermi energy, respectively. We may define the potential $\phi_{Fn}$ in the n-region as

$$e\phi_{Fn} = E_{Fi} - E_F \tag{7-4}$$

Equation (7–3) may then be written as

$$n_0 = n_i \exp\left[\frac{-(e\phi_{Fn})}{kT}\right] \tag{7-5}$$

Taking the natural log of both sides of Equation (7–5), setting $n_0 = N_d$, and solving for the potential, we obtain

$$\phi_{Fn} = -\frac{kT}{e} \ln\left(\frac{N_d}{n_i}\right) \tag{7-6}$$

Similarly, in the p-region, the hole concentration is given by

$$p_0 = N_a = n_i \exp\left[\frac{E_{Fi} - E_F}{kT}\right] \tag{7-7}$$

where $N_a$ is the acceptor concentration. We can define the potential $\phi_{Fp}$ in the p-region as

$$e\phi_{Fp} = E_{Fi} - E_F \tag{7-8}$$

Combining Equations (7–7) and (7–8), we find that

$$\phi_{Fp} = +\frac{kT}{e} \ln\left(\frac{N_a}{n_i}\right) \tag{7-9}$$

Finally, the built-in potential barrier for the step junction is found by substituting Equations (7–6) and (7–9) into Equation (7–1) which yields

$$V_{bi} = \frac{kT}{e} \ln\left(\frac{N_a N_d}{n_i^2}\right) = V_t \ln\left(\frac{N_a N_d}{n_i^2}\right) \tag{7-10}$$

where $V_t = \dfrac{kT}{e}$ and is defined as the thermal voltage.

At this time, we should note a subtle but important point concerning notation. Previously in the discussion of a semiconductor material, $N_d$ and $N_a$ denoted donor and acceptor impurity concentrations in the same region, thereby forming a compensated semiconductor. From this point on in the text, $N_d$ and $N_a$ will denote the net donor and acceptor concentrations in the individual n- and p-regions, respectively. If the p-region, for example, is a compensated material, then $N_a$ will represent the difference between the actual acceptor and donor impurity concentrations. The parameter $N_d$ is defined in a similar manner for the n-region.

---

**Example 7–1**

**Objective:** To calculate the built-in potential barrier in a pn junction.

Consider a silicon pn junction at $T = 300°K$ with doping densities $N_a = 1 \times 10^{18}$ cm$^{-3}$ and $N_d = 1 \times 10^{15}$ cm$^{-3}$. Assume that $n_i = 1.5 \times 10^{10}$ cm$^{-3}$,

**Solution:** The built-in potential barrier is determined from Equation (7–10) as

$$V_{bi} = (0.0259) \ln \left[ \frac{(10^{18})(10^{15})}{(1.5 \times 10^{10})^2} \right] = 0.754 \text{ volt}$$

If we change the acceptor doping from $N_a = 1 \times 10^{18}$ cm$^{-3}$ to $N_a = 1 \times 10^{16}$ cm$^{-3}$, but keep all other parameter values constant, then the built-in potential barrier becomes $V_{bi} = 0.635$ volt.

**Comment:** The built-in potential barrier changes only slightly as the doping concentrations change by orders of magnitude because of the logarithmic dependence.

---

### 7.2.2 Electric Field

An electric field is created in the depletion region due to the separation of positive and negative space charge densities. Figure 7–4 shows the volume charge density distribution in the pn junction assuming uniform doping and assuming an abrupt junction approximation. We will assume that the space

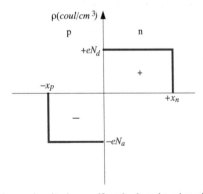

**Figure 7–4**  The space charge density in a uniformly doped pn junction assuming the abrupt junction approximation.

charge region abruptly ends in the n-region at $x = +x_n$ and abruptly ends in the p-region at $x = -x_p$ ($x_p$ is a positive quantity).

The electric field is determined from Poisson's equation which, for a one-dimensional analysis, is

$$\frac{d^2\phi(x)}{dx^2} = \frac{-\rho(x)}{\varepsilon_s} = -\frac{d\,E(x)}{dx} \qquad (7\text{--}11)$$

where $\phi(x)$ is the electric potential, $E(x)$ is the electric field, $\rho(x)$ is the volume charge density, and $\varepsilon_s$ is the permittivity of the semiconductor. From Figure (7–4), the charge densities are

$$\rho(x) = -eN_a \qquad -x_p < x < 0 \qquad (7\text{--}12a)$$

and

$$\rho(x) = eN_d \qquad 0 < x < x_n \qquad (7\text{--}12b)$$

The electric field in the p-region is found by integrating Equation (7–11). We have that

$$E = \int \frac{\rho(x)}{\varepsilon_s}\,dx = -\int \frac{eN_a}{\varepsilon_s}\,dx = \frac{-eN_a}{\varepsilon_s}x + C_1 \qquad (7\text{--}13)$$

where $C_1$ is a constant of integration. The electric field is assumed to be zero in the neutral p-region for $x < -x_p$ since the currents are zero in thermal equilibrium. As there are no surface charge densities within the pn junction structure, the electric field is a continuous function. The constant of integration is determined by setting $E = 0$ at $x = -x_p$. The electric field in the p-region is then given by

$$E = \frac{-eN_a}{\varepsilon_s}(x + x_p) \qquad -x_p \le x \le 0 \qquad (7\text{--}14)$$

In the n-region, the electric field is determined from

$$E = \int \frac{(eN_d)}{\varepsilon_s}\,dx = \frac{eN_d}{\varepsilon_s}x + C_2 \qquad (7\text{--}15)$$

where $C_2$ is again a constant of integration. The constant $C_2$ is determined by setting $E = 0$ at $x = x_n$, since the E-field is assumed to be zero in the n-region and is a continuous function. Then

$$E = \frac{-eN_d}{\varepsilon_s}(x_n - x) \qquad 0 \le x \le x_n \qquad (7\text{--}16)$$

The electric field is also continuous at the metallurgical junction, or at $x = 0$. Setting Equations (7–14) and (7–16) equal to each other at $x = 0$ gives

$$N_a x_p = N_d x_n \qquad (7\text{--}17)$$

Equation (7–17) states that the number of negative charges per unit area in the p-region is equal to the number of positive charges per unit area in the n-region.

Figure 7–5 is a plot of the electric field in the depletion region. The electric field direction is from the n- to the p-region, or in the negative x-direction for this geometry. For the uniformly doped pn junction, the E-field is a linear function of distance through the junction and the maximum (magnitude) electric field occurs at the metallurgical junction. An electric field exists in the depletion region even when no voltage is applied between the p- and n-regions.

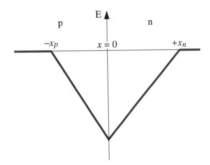

**Figure 7–5**   Electric field in the space charge region of a uniformly doped pn junction.

The potential in the junction is found by integrating the electric field. In the p-region then, we have

$$\phi(x) = -\int E(x)dx = \int \frac{eN_a}{\varepsilon_s}(x + x_p)dx \qquad (7\text{–}18)$$

or

$$\phi(x) = \frac{eN_a}{\varepsilon_s}\left(\frac{x^2}{2} + x_p \cdot x\right) + C_1' \qquad (7\text{–}19)$$

where $C_1'$ is again a constant of integration. The potential difference through the pn junction is the important parameter, rather than the absolute potential, so we may arbitrarily set the potential equal to zero at $x = -x_p$. The constant of integration is then found as

$$C_1' = \frac{eN_a}{2\varepsilon_s}x_p^2 \qquad (7\text{–}20)$$

so that the potential in the p-region can now be written as

$$\phi(x) = \frac{eN_a}{2\varepsilon_s}(x + x_p)^2 \qquad (-x_p \le x \le 0) \qquad (7\text{–}21)$$

The potential in the n-region is determined by integrating the electric field in the n-region, or

$$\phi(x) = \int \frac{eN_d}{\varepsilon_s}(x_n - x)dx \qquad (7\text{–}22)$$

Then
$$\phi(x) = \frac{eN_d}{\varepsilon_s}\left(x_n \cdot x - \frac{x^2}{2}\right) + C_2' \qquad (7\text{--}23)$$

where $C_2'$ is another constant of integration. The potential is a continuous function, so setting Equation (7–21) equal to Equation (7–23) at the metallurgical junction, or at $x = 0$, gives

$$C_2' = \frac{eN_a}{2\varepsilon_s} x_p^2 \qquad (7\text{--}24)$$

The potential in the n-region can be written as

$$\phi(x) = \frac{eN_d}{\varepsilon_s}\left(x_n \cdot x - \frac{x^2}{2}\right) + \frac{eN_a}{2\varepsilon_s} x_p^2 \qquad (0 \le x \le x_n) \qquad (7\text{--}25)$$

Figure 7–6 is a plot of the potential through the junction and shows the quadratic dependence on distance. The magnitude of the potential at $x = x_n$ is equal to the built-in potential barrier. Then from Equation (7–25), we have

$$V_{bi} = |\phi(x = x_n)| = \frac{e}{2\varepsilon_s}\left(N_d x_n^2 + N_a x_p^2\right) \qquad (7\text{--}26)$$

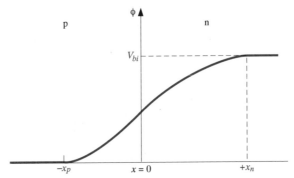

**Figure 7–6**  Electric potential through the space charge region of a uniformly doped pn junction.

The potential energy of an electron is given by $E = -e\phi$ which means that the electron potential energy also varies as a quadratic function of distance through the space charge region. The quadratic dependence on distance was shown in the energy-band diagram of Figure 7–3, although we did not explicitly know the shape of the curve at that time.

### 7.2.3 Space Charge Width

We can determine the distance that the space charge region extends into the p- and n-regions from the metallurgical junction. This distance is known as the space charge width. From Equation (7–17), we may write, for example,

$$x_p = \frac{N_d x_n}{N_a} \tag{7–27}$$

Then, substituting Equation (7–27) into Equation (7–26) and solving for $x_n$, we obtain

$$x_n = \left\{ \frac{2\varepsilon_s V_{bi}}{e} \left[ \frac{N_a}{N_d} \right] \left[ \frac{1}{N_a + N_d} \right] \right\}^{1/2} \tag{7–28}$$

Equation (7–28) gives the space charge width, or the width of the depletion region, $x_n$ extending into the n-type region for the case of zero applied voltage.

Similarly, if we solve for $x_n$ from Equation (7–17) and substitute into Equation (7–26), we find

$$x_p = \left\{ \frac{2\varepsilon_s V_{bi}}{e} \left[ \frac{N_d}{N_a} \right] \left[ \frac{1}{N_a + N_d} \right] \right\}^{1/2} \tag{7–29}$$

where $x_p$ is the width of the depletion region extending into the p-region for the case of zero applied voltage.

The total depletion or space charge width $W$ is the sum of the two components, or

$$W = x_n + x_p \tag{7–30}$$

Using Equations (7–28) and (7–29), we obtain

$$W = \left\{ \frac{2\varepsilon_s V_{bi}}{e} \left[ \frac{N_a + N_d}{N_a N_d} \right] \right\}^{1/2} \tag{7–31}$$

The built-in potential barrier can be determined from Equation (7–10) and then the total space charge region width is obtained using Equation (7–31).

---

**Example 7–2**

**Objective:** To calculate the space charge width and electric field in a pn junction.

Consider a silicon pn junction at $T = 300°K$ with doping concentrations of $N_a = 10^{16}$ cm$^{-3}$ and $N_d = 10^{15}$ cm$^{-3}$.

**Solution:** In Example 7–1, we determined the built-in potential barrier as $V_{bi} = 0.635$ volt. From Equation (7–31), the space charge width is

$$W = \left\{ \frac{2\varepsilon_s V_{bi}}{e} \left[ \frac{N_a + N_d}{N_a N_d} \right] \right\}^{1/2}$$

$$= \left\{ \frac{2(11.7)(8.85 \times 10^{-14})(0.635)}{1.6 \times 10^{-19}} \left[ \frac{10^{16} + 10^{15}}{(10^{16})(10^{15})} \right] \right\}^{1/2}$$

$$= 0.951 \times 10^{-4} \text{ cm} = 0.951 \ \mu\text{m}$$

Using Equations (7–28) and (7–29), we can find $x_n = 0.864 \ \mu$m, and $x_p = 0.086 \ \mu$m.

The peak electric field at the metallurgical junction, using Equation (7–16) for example, is

$$E_{max} = \frac{-eN_dx_n}{\varepsilon_s} = \frac{-(1.6 \times 10^{-19})(10^{15})(0.864 \times 10^{-4})}{(11.7)(8.85 \times 10^{-14})} = -1.34 \times 10^4 \ \text{V/cm}$$

**Comment:** The peak electric field in the space charge region of a pn junction is quite large. We must keep in mind, however, that there is no mobile charge in this region; hence there will be no drift current.

---

We may also note, from this example, that the width of each space charge region is a reciprocal function of the doping concentration: the depletion region will extend further into the lower-doped region.

## 7.3  REVERSE APPLIED BIAS

If we apply a potential between the p- and n-regions, we will no longer be in an equilibrium condition—the Fermi energy level will no longer be constant through the system. Figure 7–7 shows the energy-band diagram of the pn junction for the case when a positive voltage is applied to the n-region with respect to the p-region. As the positive potential is downward, the Fermi level on the n-side is below the Fermi level on the p-side. The difference between the two is equal to the applied voltage in units of energy.

The total potential barrier, indicated by $V_{\text{total}}$, has increased. This applied potential is the reverse-bias condition. The total potential barrier is now given by

$$V_{\text{total}} = |\phi_{Fn}| + |\phi_{Fp}| + V_R \qquad (7–32)$$

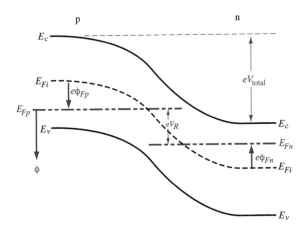

**Figure 7–7**   Energy-band diagram of a pn junction under reverse bias.

where $V_R$ is the magnitude of the applied reverse-bias voltage. Equation (7–32) can be rewritten as

$$V_{\text{total}} = V_{bi} + V_R \qquad (7\text{–}33)$$

where $V_{bi}$ is the same built-in potential barrier we had defined in thermal equilibrium.

### 7.3.1  Space Charge Width and Electric Field

Figure 7–8 shows a pn junction with an applied reverse-bias voltage $V_R$. Also indicated in the figure are the electric field in the space charge region and the electric field $E_{app}$, induced by the applied voltage. The electric fields in the neutral p- and n-regions are essentially zero, or at least very small, which means that the magnitude of the electric field in the space charge region must increase above the thermal-equilibrium value due to the applied voltage. The electric field originates on positive charge and terminates on negative charge; this means that the number of positive and negative charges must increase if the electric field increases. For given impurity doping concentrations, the number of positive and negative charges in the depletion region can be increased only if the space charge width $W$ increases. The space charge width $W$ increases, therefore, with an increasing reverse-bias voltage $V_R$. We are assuming that the electric field in the bulk n- and p-regions is zero. This assumption will become clearer in the next chapter when we discuss the current-voltage characteristics.

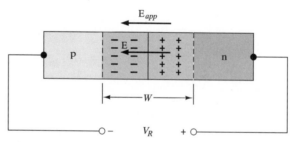

**Figure 7–8**   A pn junction with an applied reverse-bias voltage showing the directions of the electric field induced by $V_R$ and the space charge electric field.

In all of the previous equations, the built-in potential barrier can be replaced by the total potential barrier. The total space charge width can be written from Equation (7–31) as

$$W = \left\{ \frac{2\varepsilon_s(V_{bi} + V_R)}{e} \left[ \frac{N_a + N_d}{N_a N_d} \right] \right\}^{1/2} \qquad (7\text{–}34)$$

showing that the total space charge width increases as we apply a reverse-bias voltage. By substituting the total potential barrier $V_{total}$ into Equations (7–28) and (7–29), the space charge widths in the n- and p-regions, respectively, can be found as a function of applied reverse-bias voltage.

---

**Example 7–3**

**Objective:** To calculate the width of the space charge region in a pn junction when a reverse-bias voltage is applied.

Again consider a silicon pn junction at $T = 300°K$ with doping concentrations of $N_a = 10^{16}$ cm$^{-3}$ and $N_d = 10^{15}$ cm$^{-3}$. Assume that $n_i = 1.5 \times 10^{10}$ cm$^{-3}$ and let $V_R = 5$ volts.

**Solution:** The built-in potential barrier was calculated in Example 7–1 for this case and is $V_{bi} = 0.635$ volt. The space charge width is determined from Equation (7–34). We have

$$W = \left\{ \frac{2(11.7)(8.85 \times 10^{-14})(0.635 + 5)}{1.6 \times 10^{-19}} \left[ \frac{10^{16} + 10^{15}}{(10^{16})(10^{15})} \right] \right\}^{1/2}$$

so that

$$W = 2.83 \times 10^{-4} \text{ cm} = 2.83 \ \mu m$$

**Comment:** The space charge width has increased from 0.951 $\mu m$ at zero bias to 2.83 $\mu m$ at a reverse bias of 5 volts.

---

The magnitude of the electric field in the depletion region increases with an applied reverse-bias voltage. The electric field is still given by Equations (7–14) and (7–16) and is still a linear function of distance through the space charge region. Since $x_n$ and $x_p$ increase with reverse-bias voltage, the magnitude of the electric field also increases. The maximum electric field still occurs at the metallurgical junction.

The maximum electric field at the metallurgical junction, using Equations (7–14) and (7–16), is

$$\mathrm{E}_{max} = \frac{-eN_d x_n}{\varepsilon_s} = \frac{-eN_a x_p}{\varepsilon_s} \tag{7–35}$$

If we use either Equation (7–28) or (7–29) in conjunction with the total potential barrier, $V_{bi} + V_R$, then

$$\mathrm{E}_{max} = - \left\{ \frac{2e(V_{bi} + V_R)}{\varepsilon_s} \left( \frac{N_a N_d}{N_a + N_d} \right) \right\}^{1/2} \tag{7–36}$$

We can show that the maximum electric field in the pn junction can also be written as

$$E_{max} = \frac{-2(V_{bi} + V_R)}{W} \tag{7-37}$$

where $W$ is the total space charge width.

---

**Example 7-4**

**Objective:** To calculate the reverse-bias voltage required to obtain a particular maximum electric field.

Consider a silicon pn junction at $T = 300°K$ as shown in Figure 7-8 with doping concentrations $N_d = 5 \times 10^{15}$ cm$^{-3}$ and $N_a = 5 \times 10^{16}$ cm$^{-3}$. Assume $n_i = 1.5 \times 10^{10}$ cm$^{-3}$. Calculate the reverse-bias voltage that will produce a maximum electric field of $|E_{max}| = 1.25 \times 10^5$ V/cm.

**Solution:** The maximum electric field is given by Equation (7-36). If we solve for $V_{bi} + V_R$, we obtain

$$\begin{aligned}
V_{bi} + V_R &= \frac{\varepsilon_s E_{max}^2}{2e} \left(\frac{N_a + N_d}{N_a N_d}\right) \\
&= \frac{(11.7)(8.85 \times 10^{-14})(1.25 \times 10^5)^2}{2(1.6 \times 10^{-19})} \left[\frac{5 \times 10^{16} + 5 \times 10^{15}}{(5 \times 10^{16})(5 \times 10^{15})}\right] \\
&= 11.1 \text{ volts}
\end{aligned}$$

The built-in potential barrier, using Equation (7-10), is $V_{bi} = 0.718$ volt. Then we have that

$$V_R = 11.1 - 0.718 = 10.4 \text{ volts}$$

**Comment:** A moderate value of applied reverse-bias voltage can produce a relatively large electric field in the space charge region.

---

## 7.3.2 Junction Capacitance

Since we have a separation of positive and negative charges in the depletion region, a capacitance is associated with the pn junction. Figure 7-9 shows the charge densities in the depletion region for applied reverse-bias voltages of $V_R$ and $V_R + dV_R$. An increase in the reverse-bias voltage $dV_R$ will uncover additional positive charges in the n-region and additional negative charges in the p-region. The junction capacitance is defined as

$$C' = \frac{dQ'}{dV_R} \tag{7-38}$$

where
$$dQ' = eN_d dx_n = eN_a dx_p \tag{7-39}$$

The differential charge $dQ'$ is in units of coul/cm$^2$ so that the capacitance $C'$ is in units of farads/cm$^2$, or capacitance per unit area.

Using the total potential barrier, Equation (7-28) may be written as

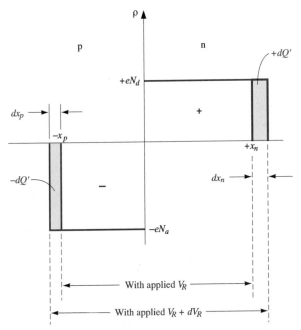

**Figure 7–9** Differential change in the space charge width with a differential change in reverse-bias voltage for a uniformly doped pn junction.

$$x_n = \left\{ \frac{2\varepsilon_s(V_{bi} + V_R)}{e} \left[ \frac{N_a}{N_d} \right] \left[ \frac{1}{N_a + N_d} \right] \right\}^{1/2} \tag{7–40}$$

The junction capacitance can be written as

$$C' = \frac{dQ'}{dV_R} = eN_d \frac{dx_n}{dV_R} \tag{7–41}$$

so that

$$C' = \left\{ \frac{e\varepsilon_s N_a N_d}{2(V_{bi} + V_R)(N_a + N_d)} \right\}^{1/2} \tag{7–42}$$

Exactly the same capacitance expression is obtained by considering the space charge region extending into the p-region $x_p$. The junction capacitance is also referred to as the depletion layer capacitance.

---

**Example 7–5**

**Objective:** To calculate the junction capacitance of a pn junction.

Consider the same pn junction which we considered in Example 7–3. Again assume that $V_R = 5$ volts.

**Solution:** The junction capacitance is found from Equation (7–42) as

$$C' = \left\{ \frac{(1.6 \times 10^{-19})(11.7)(8.85 \times 10^{-14})(10^{16})(10^{15})}{2(0.635 + 5)(10^{16} + 10^{15})} \right\}^{1/2}$$

or

$$C' = 3.66 \times 10^{-9} \text{ F/cm}^2$$

If the cross-sectional area of the pn junction is, for example, $A = 10^{-4} \text{ cm}^2$, then the total junction capacitance is

$$C = C' \cdot A = 0.366 \times 10^{-12} \text{ F} = 0.366 \text{ pF}$$

**Comment:** The value of junction capacitance is usually in the pF, or smaller, range.

---

Another expression for the junction capacitance that is sometimes convenient to use is

$$C' = \frac{\varepsilon_s}{W} \tag{7–43}$$

where, again, $W$ is the total space charge region width. Equation (7–43) is obtained by comparing Equations (7–34) and (7–42). The expression in Equation (7–43) is of the same form as the capacitance per unit area of a parallel plate capacitor where $W$ corresponds to the distance between capacitor plates. However, in the pn junction the space charge width $W$ is a function of applied bias; therefore, the capacitance is also a function of applied bias.

### 7.3.3 One-Sided Junctions

Consider a special pn junction called the one-sided junction. If, for example, $N_a \gg N_d$, this junction is referred to as a $p^+n$ junction. The total space charge width, from Equation (7–34), reduces to

$$W \approx \left\{ \frac{2\varepsilon_s(V_{bi} + V_R)}{eN_d} \right\}^{1/2} \tag{7–44}$$

Considering the expressions for $x_n$ and $x_p$, we have for the $p^+n$ junction

$$x_p \ll x_n \tag{7–45}$$

and

$$W \approx x_n \tag{7–46}$$

Almost the entire space charge layer extends into the low-doped region of the junction. This effect can be seen in Figure 7–10.

The junction capacitance of the $p^+n$ junction reduces to

$$C' \approx \left\{ \frac{e\varepsilon_s N_d}{2(V_{bi} + V_R)} \right\}^{1/2} \tag{7–47}$$

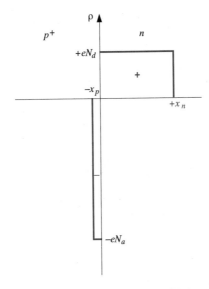

**Figure 7–10** Space charge density of a one-sided $p^+n$ junction.

The depletion layer capacitance of a one-sided junction is a function of the doping concentration in the low-doped region. Equation (7–47) may be manipulated to give

$$\left(\frac{1}{C'}\right)^2 = \frac{2(V_{bi} + V_R)}{e\varepsilon_s N_d} \qquad (7\text{–}48)$$

which shows that the inverse capacitance squared is a linear function of applied reverse-bias voltage.

Figure 7–11 shows a plot of Equation (7–48). The built-in potential of the junction can be determined by extrapolating the curve to the point where

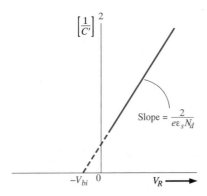

**Figure 7–11** $(1/C')^2$ versus $V_R$ of a uniformly doped pn junction.

$(1/C')^2 = 0$. The slope of the curve is inversely proportional to the doping concentration of the low-doped region in the junction; thus, this doping concentration can be experimentally determined. The assumptions used in the derivation of this capacitance include uniform doping in both semiconductor regions, the abrupt junction approximation, and a planar junction.

---

### Example 7–6

**Objective:** To determine the impurity doping concentrations in a $p^+n$ junction given the parameters from Figure 7–11.

Assume a silicon $p^+n$ junction at $T = 300°\text{K}$ with $n_i = 1.5 \times 10^{10}$ cm$^{-3}$. Assume that the intercept of the curve in Figure 7–11 gives $V_{bi} = 0.855$ volt and the slope is $1.32 \times 10^{15}[\text{F/cm}^2)^{-2}(\text{volt})^{-1}]$.

**Solution:** The slope of the curve in Figure 7–11 is given by $2/e\varepsilon_s N_d$, so we may write

$$N_d = \frac{2}{e\varepsilon_s(\text{slope})} = \frac{2}{(1.6 \times 10^{-19})(11.7)(8.85 \times 10^{-14})(1.32 \times 10^{15})}$$

or 
$$N_d = 9.15 \times 10^{15} \text{ cm}^{-3}$$

From the expression for $V_{bi}$, which is

$$V_{bi} = V_t \ln\left(\frac{N_a N_d}{n_i^2}\right) = \frac{kT}{e} \ln\left(\frac{N_a N_d}{n_i^2}\right)$$

we can solve for $N_a$ as

$$N_a = \frac{n_i^2}{N_d} \exp\left(\frac{eV_{bi}}{kT}\right) = \frac{(1.5 \times 10^{10})^2}{9.15 \times 10^{15}} \exp\left(\frac{0.855}{0.0259}\right)$$

which yields

$$N_a = 5.34 \times 10^{18} \text{ cm}^{-3}$$

**Comment:** The results of this example show that $N_a \gg N_d$; therefore the assumption of a one-sided junction was valid.

---

A one-sided pn junction is useful for experimentally determining the doping concentrations and built-in potential.

## *7.4 NONUNIFORMLY DOPED JUNCTIONS

In the pn junctions considered so far, we have assumed that each semiconductor region has been uniformly doped. In actual pn junction structures, this is not always true. In some electronic applications, specific nonuniform doping profiles are used to obtain special pn junction capacitance characteristics.

### 7.4.1 Linearly Graded Junction

If we start with a uniformly doped n-type semiconductor, for example, and diffuse acceptor atoms through the surface, the impurity concentrations will tend to be like those shown in Figure 7–12. The point $x = x'$ on the figure corresponds to the metallurgical junction. The depletion region extends into the p- and n-regions from the metallurgical junction as we have discussed previously. The net p-type doping concentration near the metallurgical junction may be approximated as a linear function of distance from the metallurgical junction. Likewise, as a first approximation, the net n-type doping concentration is also a linear function of distance extending into the n-region from the metallurgical junction. This effective doping profile is referred to as a linearly graded junction.

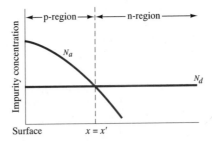

**Figure 7–12** Impurity concentrations of a pn junction with a nonuniformly doped p-region.

Figure 7–13 shows the space charge density in the depletion region of the linearly graded junction. For convenience, the metallurgical junction is placed at $x = 0$. The space charge density can be written as

$$\rho(x) = eax \qquad (7\text{–}49)$$

where $a$ is the gradient of the net impurity concentration.

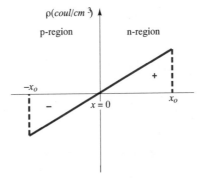

**Figure 7–13** Space charge density in a linearly graded pn junction.

The electric field and potential in the space charge region can be determined from Poisson's equation. We can write

$$\frac{dE}{dx} = \frac{\rho(x)}{\varepsilon_s} = \frac{eax}{\varepsilon_s} \tag{7-50}$$

so that the electric field can be found by integration as

$$E = \int \frac{eax}{\varepsilon_s} \, dx = \frac{ea}{2\varepsilon_s} (x^2 - x_0^2) \tag{7-51}$$

The electric field in the linearly graded junction is a quadratic function of distance rather than the linear function as found in the uniformly doped junction. The maximum electric field again occurs at the metallurgical junction. We may note that the electric field is zero at both $x = +x_0$ and at $x = -x_0$. The electric field in a nonuniformly doped semiconductor is not exactly zero, but the magnitude of this field is small, so setting $E = 0$ in the bulk regions is still a good approximation.

The potential is again found by integrating the electric field as

$$\phi(x) = -\int E dx \tag{7-52}$$

If we arbitrarily set $\phi = 0$ at $x = -x_0$, then the potential through the junction is

$$\phi(x) = \frac{-ea}{2\varepsilon_s} \left( \frac{x^3}{3} - x_0^2 x \right) + \frac{ea}{3\varepsilon_s} x_0^3 \tag{7-53}$$

The magnitude of the potential at $x = +x_0$ will equal the built-in potential barrier for this function. We then have that

$$\phi(x_0) = \frac{2}{3} \cdot \frac{eax_0^3}{\varepsilon_s} = V_{bi} \tag{7-54}$$

Another expression for the built-in potential barrier for a linearly graded junction can be approximated from the expression used for a uniformly doped junction. We can write

$$V_{bi} = V_t \ln \left[ \frac{N_d(x_0) N_a(-x_0)}{n_i^2} \right] \tag{7-55}$$

where $N_d(x_0)$ and $N_a(-x_0)$ are the doping concentrations at the edges of the space charge region. We can relate these doping concentrations to the gradient, so that

$$N_d(x_0) = ax_0 \tag{7-56a}$$

and

$$N_a(-x_0) = ax_0 \tag{7-56b}$$

Then the built-in potential barrier for the linearly graded junction becomes

$$V_{bi} = V_t \ln \left( \frac{ax_0}{n_i} \right)^2 \tag{7-57}$$

There may be situations in which the doping gradient is not the same on either side of the junction, but we will not consider that condition here.

If a reverse-bias voltage is applied to the junction, the potential barrier increases. The built-in potential barrier $V_{bi}$ in the above equations is then replaced by the total potential barrier $V_{bi} + V_R$. Solving for $x_0$ from Equation (7–54) and using the total potential barrier, we obtain

$$x_0 = \left\{ \frac{3}{2} \cdot \frac{\varepsilon_s}{ea} (V_{bi} + V_R) \right\}^{1/3} \qquad (7\text{--}58)$$

The junction capacitance per unit area can be determined using the same method as was used for the uniformly doped junction. Figure 7–14 shows the differential charge $dQ'$ which is uncovered as a differential voltage $dV_R$ is applied. The junction capacitance is then

$$C' = \frac{dQ'}{dV_R} = (eax_0) \frac{dx_0}{dV_R} \qquad (7\text{--}59)$$

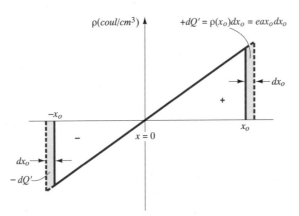

**Figure 7–14** Differential change in space charge width with a differential change in reverse-bias voltage for a linearly graded pn junction.

Using Equation (7–58), we obtain[1]

$$C' = \left\{ \frac{ea\varepsilon_s^2}{12(V_{bi} + V_R)} \right\}^{1/3} \qquad (7\text{--}60)$$

We may note that $C'$ is proportional to $(V_{bi} + V_R)^{-1/3}$ for the linearly graded junction as compared to $C' \alpha (V_{bi} + V_R)^{-1/2}$ for the uniformly doped junction. In the linearly graded junction, the capacitance is less dependent on reverse-bias voltage than in the uniformly doped junction.

---

[1] In a more exact analysis, $V_{bi}$ in Equation (7–60) is replaced by a gradient voltage. However, this analysis is beyond the scope of this text.

### 7.4.2 Hyperabrupt Junctions

The uniformly doped junction and linearly graded junction are not the only possible doping profiles. Figure 7–15 shows a generalized one-sided $p^+n$ junction where the generalized n-type doping concentration for $x > 0$ is given by

$$N = Bx^m \qquad (7\text{–}61)$$

The case of $m = 0$ corresponds to the uniformly doped junction and $m = +1$ corresponds to the linearly graded junction just discussed. The cases of $m = +2$ and $m = +3$ shown would approximate a fairly low-doped epitaxial n-type layer grown on a much more heavily doped $n^+$ substrate layer. When the value of $m$ is negative, we have what is referred to as a *hyperabrupt junction*. In this case, the n-type doping is larger near the metallurgical junction than in the bulk semiconductor. Equation (7–61) is used to approximate the n-type doping over a small region near $x = x_0$ and does not hold at $x = 0$ when $m$ is negative.

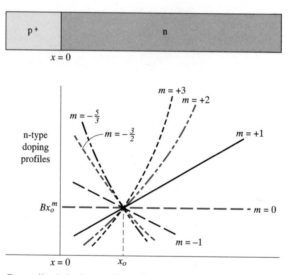

**Figure 7–15**   Generalized doping profiles of a one-sided $p^+n$ junction. (From Sze [6].)

The junction capacitance can be derived using the same analysis method as before and is given by

$$C' = \left\{ \frac{eB\varepsilon_s{}^{(m+1)}}{(m+2)(V_{bi} + V_R)} \right\}^{1/(m+2)} \qquad (7\text{–}62)$$

When $m$ is negative, the capacitance becomes a very strong function of reverse-bias voltage, a desired characteristic in *varactor diodes*. The term "varactor" comes from the words *variable reactor* and means a device whose reactance can be varied in a controlled manner with bias voltage.

If a varactor diode and an inductance are in parallel, the resonant frequency of the LC circuit is

$$f_r = \frac{1}{2\pi \sqrt{LC}} \qquad (7\text{–}63)$$

The capacitance of the diode, from Equation (7–62), can be written in the form

$$C = C_0(V_{bi} + V_R)^{-1/(m+2)} \qquad (7\text{–}64)$$

In a circuit application, we would, in general, like to have the resonant frequency be a linear function of reverse-bias voltage $V_R$, so we need

$$C \propto V^{-2} \qquad (7\text{–}65)$$

From Equation (7–64), the parameter $m$ required is found from

$$\frac{1}{m + 2} = 2 \qquad (7\text{–}66a)$$

or

$$m = -\frac{3}{2} \qquad (7\text{–}66b)$$

A specific doping profile will yield the desired capacitance characteristic.

## 7.5 SUMMARY AND REVIEW

In this chapter we have considered the electrostatics of the pn junction. We have assumed that the pn junction is formed from a single-crystal semiconductor in which one region is doped p-type and the adjacent region is doped n-type. This type of junction is called a homojunction.

A space charge region, or depletion region, exists on either side of the metallurgical junction separating the n- and p-regions. This region is essentially depleted on any mobile electrons or holes. A net positive charge density, due to the positively charged donor atoms, exists in the n-region and a net negative charge density, due to the negatively charged acceptor atoms, exists in the p-region. An electric field is induced in the depletion region due to the net space charge density, which means that there is a potential difference across this region. Under zero bias, this potential difference, known as the built-in potential barrier, establishes thermal equilibrium and holds back majority carrier electrons in the n-region and majority carrier holes in the p-region from crossing the depletion region. The built-in potential barrier is a function of the impurity doping concentrations, the intrinsic carrier concentration, and temperature.

When a reverse-bias voltage is applied across the pn junction (n-region positive with respect to the p-region), the potential barrier between the n-

and p-regions increases, the space charge width increases, and the magnitude of the electric field in the depletion region increases. For a uniformly doped pn junction, the electric field is a linear function of distance through the depletion region and the peak value of electric field occurs at the metallurgical junction. As the reverse-bias voltage changes, the amount of charge in the depletion region changes. This change in charge with voltage defines the junction capacitance. The junction capacitance is a function of impurity doping concentrations and applied reverse-bias voltage. We derived the capacitance expression, a parameter that will limit the switching speed of a pn junction.

The uniformly doped junction allows us to easily develop the electrostatics of the pn junction. However, in real pn junctions, the n- and p-regions are rarely both uniformly doped. The linearly graded junction represents a first approximation to a diffused junction. Expressions for the electric field and built-in potential barrier were derived. The expression for junction capacitance, derived for the linearly graded junction, varies as $(V_R)^{-1/3}$, whereas the junction capacitance varies as $(V_R)^{-1/2}$ for the uniformly doped junction. Other types of doping profiles can be used to obtain specific capacitance characteristics. A hyperabrupt junction is one in which the doping decreases away from the metallurgical junction. By using a specific doping profile, a junction capacitance that varies as $(V_R)^{-2}$ can be obtained. This characteristic is useful in varactor diodes which are used in resonant circuits.

The electrostatics of the pn junction provide a few of the properties of the junction under zero and reverse-bias conditions. In the next chapter, we will consider the pn junction under forward bias and develop the current-voltage characteristics of the device. The potential barrier across the junction decreases or increases with applied voltage and determines whether charge can easily flow across the junction or be essentially blocked. The resulting current-voltage relation will be highly nonlinear.

## GLOSSARY OF IMPORTANT TERMS

**Abrupt junction approximation:** The assumption that there is an abrupt discontinuity in space charge density between the space charge region and neutral semiconductor region.

**Built-in potential barrier:** The electrostatic potential difference between the p- and n-regions of a pn junction in thermal equilibrium.

**Depletion layer capacitance:** Another term for junction capacitance.

**Depletion region:** Another term for space charge region.

**Hyperabrupt junction:** A pn junction in which the doping concentration on one side decreases away from the metallurgical junction to achieve a specific capacitance-voltage characteristic.

**Junction capacitance:** The capacitance of the pn junction under reverse bias.

**Linearly graded junction:** A pn junction in which the doping concentrations on either side of the metallurgical junction are approximated by a linear distribution.

**Metallurgical junction:** The interface between the p- and n-doped regions of a pn junction.

**One-sided junction:** A pn junction in which one side of the junction is much more heavily doped than the adjacent side.

**Reverse bias:** The condition in which a positive voltage is applied to the n-region with respect to the p-region of a pn junction so that the potential barrier between the two regions increases above the thermal-equilibrium built-in potential barrier.

**Space charge region:** The region on either side of the metallurgical junction in which there is a net charge density due to ionized donors in the n-region and ionized acceptors in the p-region.

**Space charge width:** The width of the space charge region, a function of doping concentrations and applied voltage.

**Varactor diode:** A diode whose reactance can be varied in a controlled manner with bias voltage.

## PROBLEMS

### Section 7.2

1. Calculate $V_{bi}$ in a silicon pn junction at $T = 300°K$ for (A) $N_d = 10^{15}$ cm$^{-3}$ and $N_a = $ (a) $10^{15}$, (b)$10^{16}$, (c) $10^{17}$, (d) $10^{18}$ cm$^{-3}$. (B) Repeat part (A) for $N_d = 10^{18}$ cm$^{-3}$.

2. Calculate the built-in potential barrier, $V_{bi}$, for Si, Ge, and GaAs pn junctions if they each have the following dopant concentrations at $T = 300°K$:

   (a) $N_d = 10^{14}$ cm$^{-3}$     $N_a = 10^{17}$ cm$^{-3}$
   (b) $N_d = 5 \times 10^{16}$     $N_a = 5 \times 10^{16}$
   (c) $N_d = 10^{17}$     $N_a = 10^{17}$

3. An abrupt silicon pn junction at zero bias has dopant concentrations of $N_a = 10^{15}$ cm$^{-3}$ and $N_d = 2 \times 10^{17}$ cm$^{-3}$ at $T = 300°K$. Calculate (a) $V_{bi}$, (b) $x_n$, $x_p$, $W$, and (c) $E_{max}$.

4. Consider a uniformly doped GaAs junction at $T = 300°K$. At zero bias, only 20 percent of the total space charge region is to be in the p-region. The built-in potential barrier is $V_{bi} = 1.20$ volts. For zero bias, determine (a) $N_a$, (b) $N_d$, (c) $x_n$, (d) $x_p$, and (e) $E_{max}$.

*5. A uniformly doped silicon pn junction is doped to levels of $N_d = 5 \times 10^{15}$ cm$^{-3}$ and $N_a = 10^{16}$ cm$^{-3}$. The measured built-in potential barrier is $V_{bi} = 0.40$ volt. Determine the temperature at which this result occurs. (You may have to use trial and error to solve this problem.)

6. Consider a uniformly doped GaAs pn junction with doping concentrations of $N_a = 5 \times 10^{18}$ cm$^{-3}$ and $N_d = 5 \times 10^{16}$ cm$^{-3}$. Plot the built-in potential barrier, $V_{bi}$, versus temperature for $200°$K $\leq T \leq 500°$K.

## Section 7.3

7. An abrupt silicon pn junction has dopant concentrations of $N_a = 2 \times 10^{16}$ cm$^{-3}$ and $N_d = 2 \times 10^{15}$ cm$^{-3}$ at $T = 300°$K. Calculate (a) $V_{bi}$, (b) $W$ at $V_R = 0$ and $V_R = 8$ volts, and (c) the maximum electric field in the space charge region at $V_R = 0$ and $V_R = 8$ volts.

8. A silicon step junction at $T = 300°$K has doping levels of $N_a = 5 \times 10^{15}$ cm$^{-3}$ and $N_d = 1 \times 10^{15}$ cm$^{-3}$. The junction has a cross-sectional area of $10^{-4}$ cm$^2$ and has an applied reverse-bias voltage of $V_R = 5$ volts. Calculate (a) $V_{bi}$, (b) $x_n$, $x_p$, $W$, (c) $E_{max}$, and (d) the total junction capacitance.

9. An ideal one-sided silicon $n^+p$ junction has uniform doping on both sides of the abrupt junction. The doping relation is $N_d = 50 \ N_a$. The built-in potential barrier is $V_{bi} = 0.752$ volt. The maximum electric field in the junction is $E_{max} = 1.14 \times 10^5$ V/cm for a reverse-bias voltage of 10 volts. $T = 300°$K. Determine (a) $N_a$, $N_d$, (b) $x_p$ for $V_R = 10$, and (c) $C'_j$ for $V_R = 10$.

10. Consider a uniformly doped GaAs pn junction at $T = 300°$K. The junction capacitance at zero bias is $C_j(0)$ and the junction capacitance with a 10-volt reverse-bias voltage is $C_j(10)$. The ratio of the capacitances is

$$\frac{C_j(0)}{C_j(10)} = 3.13$$

Also under reverse bias, the space charge width into the p-region is 0.2 of the total space charge width. Determine (a) $V_{bi}$ and (b) $N_a$, $N_d$.

11. An abrupt silicon pn junction at $T = 300°$K is uniformly doped with $N_a = 10^{18}$ cm$^{-3}$ and $N_d = 10^{15}$ cm$^{-3}$. The pn junction area is $6 \times 10^{-4}$ cm$^2$. An inductance of 2.2 millihenry is placed in parallel with the pn junction. Calculate the resonant frequency of the circuit for reverse-bias voltages of (a) $V_R = 1$ volt and (b) $V_R = 10$ volts.

12. The junction capacitance of a one-sided $n^+p$ GaAs junction at $T = 300°$K was measured and had the following values: $C'_j = 1.12 \times 10^{-8}$ F/cm$^2$ at $V_R = 1$ volt, and $C'_j = 6.69 \times 10^{-9}$ F/cm$^2$ at $V_R = 5$ volts. Determine $N_d$, $N_a$, and $V_{bi}$.

13. A silicon pn junction is to be designed which meets the following specifications at $T = 300°K$. At a reverse-bias voltage of 1.2 volts, 10 percent of the total space charge region is to be in the n-region and the total junction capacitance is to be $3.5 \times 10^{-12}$ F with a cross-sectional area of $5.5 \times 10^{-4}$ cm$^2$. Determine (a) $N_a$, (b) $N_d$, and (c) $V_{bi}$.

14. A silicon pn junction at $T = 300°K$ has the doping profile shown in Figure 7–16. Calculate (a) $V_{bi}$, (b) $x_n$ and $x_p$ at zero bias, and (c) the applied bias required so that $x_n = 30$ $\mu$m.

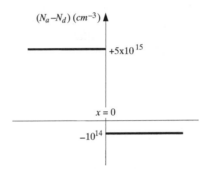

**Figure 7–16**   Figure for problem 14.

15. Consider a silicon pn junction with the doping profile shown in Figure 7–17. $T = 300°K$. (a) Calculate the applied reverse-bias voltage required so that the space charge region extends entirely through the p-region. (b) Determine the space charge width into the $n^+$-region with the reverse-bias voltage calculated in part (a). (c) Calculate the peak electric field for this applied voltage.

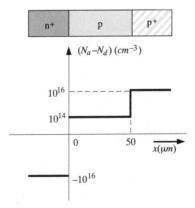

**Figure 7–17**   Figure for problem 15.

16. The total junction capacitance of a one-sided silicon pn junction at $T = 300°K$ is measured at $V_R = 50$ mV and found to be 1.3 pF. The junction area is $10^{-5}$ cm$^2$ and $V_{bi} = 0.95$ volt. (a) Find the impurity doping concentration of the low-doped side of the junction. (b) Find the impurity doping concentration of the higher-doped region.

17. The capacitance of a one-sided, uniformly doped $p^+n$ GaAs junction at $T = 300°K$ is $C' = 2.95 \times 10^{-8}$ F/cm$^2$ when a reverse-bias voltage of $V_R = 4$ volts is applied. The built-in potential barrier is $V_{bi} = 1.32$ volts. Determine (a) $N_d$, (b) $N_a$, and (c) $E_{max}$.

*18. A silicon PIN junction has the doping profile shown in Figure 7–18. The "I" corresponds to an ideal intrinsic region in which there is no impurity doping concentration. A reverse-bias voltage is applied to the PIN junction so that the total depletion width extends from $-2$ $\mu$m to $+2$ $\mu$m. (a) Using Poisson's equation, calculate the magnitude of the electric field at $x = 0$. (b) Sketch the electric field through the PIN junction. (c) Calculate the reverse-bias voltage that must be applied.

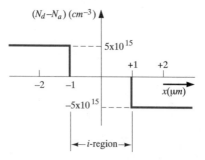

**Figure 7–18**    Figure for problem 18.

## Section 7.4

19. Consider a linearly graded junction. (a) Starting with Equation (7–49), derive the expression for the electric field given in Equation (7–51). (b) Derive the expression for the potential through the space charge region given by Equation (7–53).

20. The built-in potential barrier of a linearly graded silicon pn junction at $T = 300°K$ is $V_{bi} = 0.70$ volt. The junction capacitance measured at $V_R = 3.5$ volts is $C' = 7.2 \times 10^{-9}$ F/cm$^2$. Find the gradient, $a$, of the net impurity concentration.

## Summary and Review

21. A one-sided $p^+n$ silicon diode at $T = 300°K$ is doped at $N_a = 10^{18}$ cm$^{-3}$. Design the junction so that $C_j = 0.95$ pF at $V_R = 3.5$ volts. Calculate the junction capacitance when $V_R = 1.5$ volts.

22. A one-sided $p^+n$ junction with a cross-sectional area of $10^{-5}$ cm$^2$ has a measured built-in potential of $V_{bi} = 0.8$ volt at $T = 300°K$. A plot of $(1/C_j)^2$ versus $V_R$ is linear for $V_R < 1$ volt and is essentially constant for $V_R > 1$ volt. The capacitance is $C_j = 0.082$ pF at $V_R = 1$ volt. Determine the doping concentrations on either side of the metallurgical junction which will produce this capacitance characteristic.

*23. Silicon, at $T = 300°K$, is doped at $N_{d1} = 10^{15}$ cm$^{-3}$ for $x < 0$ and $N_{d2} = 5 \times 10^{16}$ cm$^{-3}$ for $x > 0$ to form an $n-n$ step junction. (a) Sketch the energy-band diagram. (b) Derive an expression for $V_{bi}$. (c) Sketch the charge density, electric field, and potential through the junction. (d) Explain where the charge density came from and is located.

*24. A diffused silicon pn junction has a linearly graded junction on the p-side with $a = 2 \times 10^{19}$ cm$^{-4}$, and a uniform doping of $10^{15}$ cm$^{-3}$ on the n-side. (a) If the depletion width on the p-side is 0.7 $\mu$m at zero bias, find the total depletion width, built-in potential, and maximum electric field at zero bias. (b) Plot the potential function through the junction.

## READING LIST

1. Muller, R. S., and T. I. Kamins. *Device Electronics for Integrated Circuits.* 2nd ed. New York: Wiley, 1986.

2. Navon, D. H. *Semiconductor Microdevices and Materials.* New York: Holt, Rinehart & Winston, 1986.

3. Neudeck, G. W. *The PN Junction Diode.* Vol. 2 of the *Modular Series on Solid State Devices.* 2nd ed. Reading, Mass.: Addison-Wesley, 1989.

*4. Shur, M. *Physics of Semiconductor Devices.* Englewood Cliffs, N.J.: Prentice Hall, 1990.

5. Streetman, B. G. *Solid State Electronic Devices.* 3rd ed. Englewood Cliffs, N.J.: Prentice Hall, 1990.

6. Sze, S. M. *Physics of Semiconductor Devices.* 2nd ed. New York: Wiley, 1981.

7. ———. *Semiconductor Devices: Physics and Technology.* New York: Wiley, 1985.

*8. Wang, S. *Fundamentals of Semiconductor Theory and Device Physics.* Englewood Cliffs, N.J.: Prentice Hall, 1989.

9. Yang, E. S. *Microelectronic Devices.* New York: McGraw-Hill, 1988.

# THE PN JUNCTION DIODE

**CHAPTER 8**

## PREVIEW

In the last chapter, we discussed the electrostatics of the pn junction in thermal equilibrium and under reverse bias. We determined the built-in potential barrier at thermal equilibrium and calculated the electric field in the space charge region. We also considered the junction capacitance. In this chapter, we will consider the pn junction with a forward-bias voltage applied and will determine the current-voltage characteristics. The potential barrier of the pn junction is lowered when a forward-bias voltage is applied, allowing electrons and holes to flow across the space charge region. When holes flow from the p-region across the space charge region into the n-region, they become excess minority carrier holes and are subject to the excess minority carrier diffusion, drift, and recombination processes discussed in Chapter 6. Likewise, when electrons from the n-region flow across the space charge region into the p-region, they become excess minority carrier electrons and are subject to these same processes.

When semiconductor devices with pn junctions are used in linear amplifiers, for example, time-varying signals are superimposed on the dc currents and voltages. A small sinusoidal voltage superimposed on a dc voltage applied across a pn junction will generate a small-signal sinusoidal current. The ratio of the sinusoidal current to voltage yields the small-signal admittance of the pn junction. The admittance of a forward-biased pn junction contains both conductance and capacitance terms. The capacitance, called a diffusion capacitance, differs from the junction capacitance discussed in the last chapter. Using the admittance function, the small-signal equivalent circuit of the pn junction will be developed.

Electrons and holes flow across the space charge region when the pn junction is forward-biased and there is a finite probability that a few of these electrons and holes will recombine within the space charge region. This recombination process generates an additional current component in the pn junction. Similarly, when a reverse-bias voltage is applied to the pn junction, electrons and holes are generated within the space charge region and are quickly swept out of this region by the electric field producing another current component in the diode. This generation process evolves from the Shockley-Read-Hall recombination theory discussed in Chapter 6.

The last three topics considered in this chapter are junction breakdown, switching transients, and the tunnel diode. When a sufficiently large reverse-bias voltage is applied across a pn junction, breakdown can occur, producing a large reverse-bias current in the junction, which can cause heating effects and catastrophic failure of the diode. Zener diodes, however, are designed to operate in the breakdown region. Breakdown puts limits on the amount of voltage that can be applied across a pn junction. When a pn junction is switched from one conducting state to the other, transients in the diode current and voltage occur. The switching time of the pn junction will be discussed here, and again in later chapters which deal with the switching of transistors. Finally, we will introduce a special pn junction diode: the tunnel diode.

## 8.1 IDEAL CURRENT-VOLTAGE RELATIONSHIP

The ideal current-voltage relationship of a pn junction is derived on the basis of four assumptions. (The last assumption has three parts, but each part deals with currents.) They are:

1. The abrupt depletion layer approximation applies. The space charge regions have abrupt boundaries and the semiconductor is neutral outside of the depletion region.
2. The Maxwell-Boltzmann approximation applies to carrier statistics.
3. The concept of low injection applies.
4a. The total current is a constant throughout the entire pn structure.
4b. The individual electron and hole currents are continuous functions through the pn structure.
4c. The individual electron and hole currents are constant throughout the depletion region.

### 8.1.1 Boundary Conditions

Figure 8–1 shows the conduction-band energy through the pn junction in thermal equilibrium. The n-region contains many more electrons in the conduction band than the p-region; the built-in potential barrier prevents this large density of electrons from flowing into the p-region. The built-in potential barrier maintains equilibrium between the carrier distributions on either side of the junction.

An expression for the built-in potential barrier was derived in the last chapter and was given by Equation (7–10) as

$$V_{bi} = V_t \ln \left( \frac{N_a N_d}{n_i^2} \right)$$

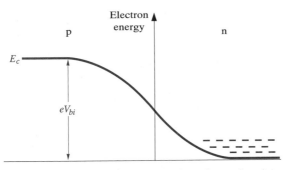

**Figure 8–1**  Conduction-band energy through a pn junction.

If we divide the equation by $V_t = kT/e$, take the exponential of both sides, and then take the reciprocal, we obtain

$$\frac{n_i^2}{N_a N_d} = \exp\left(\frac{-eV_{bi}}{kT}\right) \tag{8–1}$$

If we assume complete ionization, we can write

$$n_{n0} \cong N_d \tag{8–2}$$

where $n_{n0}$ is the thermal-equilibrium concentration of majority carrier electrons in the n-region. In the p-region, we can write

$$n_{p0} = \frac{n_i^2}{N_a} \tag{8–3}$$

where $n_{p0}$ is the thermal-equilibrium concentration of minority carrier electrons. Substituting Equations (8–2) and (8–3) into Equation (8–1) yields

$$n_{p0} = n_{n0} \exp\left(\frac{-eV_{bi}}{kT}\right) \tag{8–4}$$

This equation relates the minority carrier electron concentration on the p-side of the junction to the majority carrier electron concentration on the n-side of the junction in thermal equilibrium.

If a positive voltage is applied to the p-region with respect to the n-region, the potential barrier is reduced. Figure 8-2a shows a pn junction with an applied voltage $V_a$. The electric field in the bulk p- and n-regions is normally very small. Essentially all of the applied voltage is across the junction region. The electric field $E_{app}$ induced by the applied voltage is in the opposite direction to the thermal equilibrium space charge electric field, so the net electric field in the space charge region is reduced below the equilibrium value. The delicate balance between diffusion and the E-field force achieved at thermal equilibrium is upset. The electric field force that prevented majority carriers from crossing the space charge region is reduced; majority carrier electrons from the n-side are now injected across the depletion region

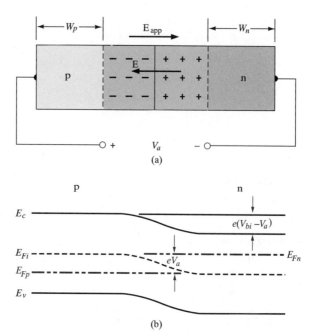

**Figure 8–2** (a) A pn junction with an applied forward-bias voltage showing the directions of the electric field induced by $V_a$ and the space charge electric field. (b) Energy-band diagram of the forward-biased pn junction.

into the p-material and majority carrier holes from the p-side are injected across the depletion region into the n-material. As long as the bias $V_a$ is applied, the injection of carriers across the space charge region continues and a current is created in the pn junction. This bias condition is known as forward bias and the energy-band diagram of the forward-biased pn junction is shown in Figure 8–2b.

The potential barrier $V_{bi}$ in Equation (8–4) can be replaced by $(V_{bi} - V_a)$ when the junction is forward biased. Equation (8–4) becomes

$$n_p = n_{n0} \exp\left(\frac{-e(V_{bi} - V_a)}{kT}\right) = n_{n0} \exp\left(\frac{-eV_{bi}}{kT}\right) \exp\left(\frac{eV_a}{kT}\right) \quad (8\text{–}5)$$

If we assume low injection, the majority carrier electron concentration $n_{n0}$, for example, does not change significantly. However, the minority carrier concentration, $n_p$, can deviate from its thermal-equilibrium value $n_{p0}$ by orders of magnitude. Using Equation (8–4), Equation (8–5) can be written as

$$n_p = n_{p0} \exp\left(\frac{eV_a}{kT}\right) \quad (8\text{–}6)$$

When a forward-bias voltage is applied to the pn junction, the junction is no longer in thermal equilibrium. The left side of Equation (8–6) is the total minority carrier electron concentration in the p-region, which is now greater than the thermal equilibrium value. The forward-bias voltage lowers the potential barrier so that majority carrier electrons from the n-region are injected across the junction into the p-region, thereby increasing the minority carrier electron concentration. We have produced excess minority carrier electrons in the p-region.

When the electrons are injected into the p-region, these excess carriers are subject to the diffusion and recombination processes we discussed in Chapter 6. Equation (8–6), then, is the expression for the minority carrier electron concentration at the edge of the space charge region in the p-region.

Exactly the same process occurs for majority carrier holes in the p-region which are injected across the space charge region into the n-region under a forward-bias voltage. We can write that

$$p_n = p_{n0} \exp \left( \frac{eV_a}{kT} \right) \tag{8–7}$$

where $p_n$ is the concentration of minority carrier holes at the edge of the space charge region in the n-region. Figure 8–3 shows these results. By applying a forward-bias voltage, we create excess minority carriers in each region of the pn junction.

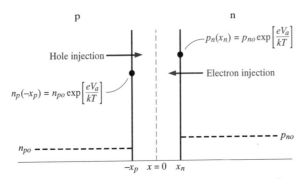

**Figure 8–3**  Excess minority carrier concentrations at the space charge edges generated by the forward-bias voltage.

---

## Example 8–1

**Objective:** To calculate the minority carrier hole concentration at the edge of the space charge region of a pn junction when a forward bias is applied.

Consider a silicon pn junction at $T = 300°K$ so that $n_i = 1.5 \times 10^{10}$ cm$^{-3}$. Assume the n-type doping is $1 \times 10^{16}$ cm$^{-3}$ and assume that a forward bias of 0.60 volt is applied to the pn junction. Calculate the minority carrier hole concentration at the edge of the space charge region.

**Solution:** From Equation (8–7), we have

$$p_n = p_{n0} \exp\left(\frac{eV_a}{kT}\right)$$

The thermal-equilibrium minority carrier hole concentration is

$$p_{n0} = \frac{n_i^2}{N_d} = \frac{(1.5 \times 10^{10})^2}{10^{16}} = 2.25 \times 10^4 \text{ cm}^{-3}$$

We then have

$$p_n = 2.25 \times 10^4 \exp\left(\frac{0.60}{0.0259}\right) = 2.59 \times 10^{14} \text{ cm}^{-3}$$

**Comment:** The minority carrier concentration can increase by many orders of magnitude when a forward-bias voltage is applied. Low injection still applies, however, since the excess-electron concentration (equal to the excess-hole concentration in order to maintain charge neutrality) is much less than the thermal-equilibrium electron concentration.

---

The minority carrier concentrations at the space charge edges, given by Equations (8–6) and (8–7), were derived assuming a forward-bias voltage ($V_a > 0$) was applied across the pn junction. However, nothing in the derivation prevents $V_a$ from being negative (reverse bias). If a reverse-bias voltage greater than a few tenths of a volt is applied to the pn junction, then we see from Equations (8–6) and (8–7) that the minority carrier concentrations at the space charge edge is essentially zero. The minority carrier concentrations for the reverse-bias condition drop below the thermal-equilibrium values.

### 8.1.2 Minority Carrier Distribution

We developed, in Chapter 6, the ambipolar transport equation for excess minority carrier holes in an n-region. This equation, in one dimension, is

$$D_p \frac{\partial^2(\delta p_n)}{\partial x^2} - \mu_p E \frac{\partial(\delta p_n)}{\partial x} + g' - \frac{\delta p_n}{\tau_{p0}} = \frac{\partial(\delta p_n)}{\partial t} \tag{8–8}$$

where $\delta p_n = p_n - p_{n0}$ is the excess minority carrier hole concentration and is the difference between the total and thermal equilibrium minority carrier concentrations. The ambipolar transport equation describes the behavior of excess carriers as a function of time and spatial coordinates.

In Chapter 5, we calculated drift current densities in a semiconductor. We determined that relatively large currents could be created with fairly small electric fields. As a first approximation, we will assume that the electric field is zero in both the neutral p- and n-regions. In the n-region for $x > x_n$, we have that $E = 0$ and $g' = 0$. If we also consider steady state so $\partial(\delta p_n)/\partial t = 0$, then Equation (8–8) reduces to

$$\frac{d^2(\delta p_n)}{dx^2} - \frac{\delta p_n}{L_p^2} = 0 \qquad (x > x_n) \qquad (8\text{--}9)$$

where $L_p^2 = D_p \tau_{p0}$. For the same set of conditions, the excess minority carrier electron concentration in the p-region is determined from

$$\frac{d^2(\delta n_p)}{dx^2} - \frac{\delta n_p}{L_n^2} = 0 \qquad (x < -x_p) \qquad (8\text{--}10)$$

where $L_n^2 = D_n \tau_{n0}$.

The boundary conditions for the total minority carrier concentrations are

$$p_n(x_n) = p_{n0} \exp\left(\frac{eV_a}{kT}\right) \qquad (8\text{--}11\text{a})$$

$$n_p(-x_p) = n_{p0} \exp\left(\frac{eV_a}{kT}\right) \qquad (8\text{--}11\text{b})$$

$$p_n(x \rightarrow +\infty) = p_{n0} \qquad (8\text{--}11\text{c})$$

$$n_p(x \rightarrow -\infty) = n_{p0} \qquad (8\text{--}11\text{d})$$

As minority carriers diffuse from the space charge edge into the neutral semiconductor regions, they will recombine with majority carriers. We will assume that the lengths $W_n$ and $W_p$ shown in Figure 8–2a are very long, meaning in particular that $W_n \gg L_p$ and $W_p \gg L_n$. The excess minority carrier concentrations must approach zero at distances far from the space charge region. This structure is referred to as a long pn junction.

The general solution to Equation (8–9) is

$$\delta p_n(x) = p_n(x) - p_{n0} = Ae^{x/L_p} + Be^{-x/L_p} \qquad (x \geq x_n) \qquad (8\text{--}12)$$

and the general solution to Equation (8–10) is

$$\delta n_p(x) = n_p(x) - n_{p0} = Ce^{x/L_n} + De^{-x/L_n} \qquad (x \leq -x_p) \qquad (8\text{--}13)$$

Applying the boundary conditions from Equations (8–11c) and (8–11d), the coefficients A and D must be zero. The coefficients B and C may be determined from the boundary conditions given by Equations (8–11a) and (8–11b). The excess carrier concentrations are then found to be, for $(x \geq x_n)$,

$$\delta p_n(x) = p_n(x) - p_{n0} = p_{n0}\left[\exp\left(\frac{eV_a}{kT}\right) - 1\right]\exp\left(\frac{x_n - x}{L_p}\right) \qquad (8\text{--}14)$$

and, for $(x \leq -x_p)$,

$$\delta n_p(x) = n_p(x) - n_{p0} = n_{p0}\left[\exp\left(\frac{eV_a}{kT}\right) - 1\right]\exp\left(\frac{x_p + x}{L_n}\right) \qquad (8\text{--}15)$$

The minority carrier concentrations decay exponentially with distance away from the junction to their thermal-equilibrium values. Figure 8–4 shows

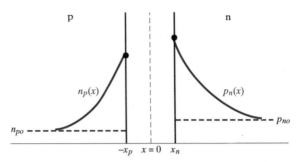

**Figure 8–4**   Steady-state minority carrier concentrations in a pn junction under forward bias.

these results. Again, we have assumed that both the n-region and the p-region lengths are long compared to the minority carrier diffusion lengths.

To review, a forward-bias voltage lowers the built-in potential barrier of a pn junction so that electrons from the n-region are injected across the space charge region creating excess minority carriers in the p-region. These excess electrons begin diffusing into the bulk p-region where they can recombine with majority carrier holes. The excess minority carrier electron concentration then decreases with distance from the junction. The same discussion applies to holes injected across the space charge region into the n-region.

### 8.1.3  Ideal pn Junction Current

The approach used to determine the current in a pn junction is based on the three parts of the fourth assumption stated earlier in this section. The total current in the junction is the sum of the individual electron and hole currents which are constant through the depletion region. Since the electron and hole currents are continuous functions through the pn junction, the total pn junction current will be the minority carrier hole diffusion current at $x = x_n$ plus the minority carrier electron diffusion current at $x = -x_p$. The gradients in the minority carrier concentrations, as shown in Figure 8–4, produce diffusion currents, and since we are assuming the electric field to be zero at the space charge edges, we can neglect any minority carrier drift current component. This approach in determining the pn junction current is shown in Figure 8–5.

We can calculate the minority carrier hole diffusion current density at $x = x_n$ from the relation

$$J_p(x_n) = -eD_p \left. \frac{dp_n(x)}{dx} \right|_{x=x_n} \tag{8–16}$$

Since we are assuming uniformly doped regions, the thermal-equilibrium carrier concentration is constant, so the hole diffusion current density may be written as

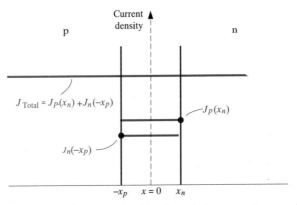

**Figure 8–5** Electron and hole current densities through the space charge region of a pn junction.

$$J_p(x_n) = -eD_p \frac{d(\delta p_n(x))}{dx}\bigg|_{x=x_n} \qquad (8\text{--}17)$$

Taking the derivative of Equation (8–14) and substituting into Equation (8–17), we obtain

$$J_p(x_n) = \frac{eD_p p_{n0}}{L_p}\left[\exp\left(\frac{eV_a}{kT}\right) - 1\right] \qquad (8\text{--}18)$$

The hole current density for this forward-bias condition is in the $+x$ direction, which is from the p- to the n-region.

Similarly, we may calculate the electron diffusion current density at $x = -x_p$. This may be written as

$$J_n(-x_p) = eD_n \frac{d(\delta n_p(x))}{dx}\bigg|_{x=-x_p} \qquad (8\text{--}19)$$

Using Equation (8–15), we obtain

$$J_n(-x_p) = \frac{eD_n n_{p0}}{L_n}\left[\exp\left(\frac{eV_a}{kT}\right) - 1\right] \qquad (8\text{--}20)$$

The electron current density is also in the $+x$ direction.

An assumption we made at the beginning was that the individual electron and hole currents were continuous functions and constant through the space charge region. The total current is the sum of the electron and hole currents and is constant through the entire junction. Figure 8–5 again shows a plot of the magnitudes of these currents.

The total current density in the pn junction is then

$$J = J_p(x_n) + J_n(-x_p) = \left[\frac{eD_p p_{n0}}{L_p} + \frac{eD_n n_{p0}}{L_n}\right]\left[\exp\left(\frac{eV_a}{kT}\right) - 1\right] \qquad (8\text{--}21)$$

Equation (8–21) is the ideal current-voltage relationship of a pn junction.
We may define a parameter $J_s$ as

$$J_s = \left[\frac{eD_pp_{n0}}{L_p} + \frac{eD_nn_{p0}}{L_n}\right]$$    (8–22)

so that Equation (8–21) may be written as

$$J = J_s\left[\exp\left(\frac{eV_a}{kT}\right) - 1\right]$$    (8–23)

Equation (8–23), known as the ideal-diode equation, gives a good description of the current-voltage characteristics of the pn junction over a wide range of currents and voltages. Although Equation (8–23) was derived assuming a forward-bias voltage ($V_a > 0$), there is nothing to prevent $V_a$ from being negative (reverse bias). Equation (8–23) is plotted in Figure 8–6 as a function of forward-bias voltage $V_a$. If the voltage $V_a$ becomes negative (reverse bias) by a few $kT/e$ volts, then the reverse-bias current density becomes independent of the reverse-bias voltage. The parameter $J_s$ is then referred to as the reverse-saturation current density. The current-voltage characteristics of the pn junction diode are obviously not bilateral.

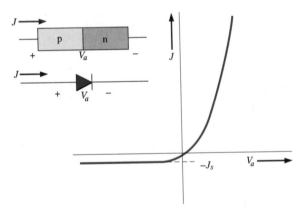

**Figure 8–6**    Ideal I-V characteristic of a pn junction diode.

---

### Example 8–2

**Objective:** To determine the ideal reverse-saturation current density in a silicon pn junction at $T = 300°K$.

Consider the following parameters in a silicon pn junction:

$N_a = N_d = 10^{16}$ cm$^{-3}$        $n_i = 1.5 \times 10^{10}$ cm$^{-3}$
$D_n = 25$ cm$^2$/sec          $\tau_{p0} = \tau_{n0} = 5 \times 10^{-7}$ sec
$D_p = 10$ cm$^2$/sec          $\varepsilon_r = 11.7$

**Solution:** The ideal reverse-saturation current density is given by

$$J_s = \frac{eD_n n_{p0}}{L_n} + \frac{eD_p p_{n0}}{L_p}$$

which may be rewritten as

$$J_s = e n_i^2 \left[ \frac{1}{N_a} \sqrt{\frac{D_n}{\tau_{n0}}} + \frac{1}{N_d} \sqrt{\frac{D_p}{\tau_{p0}}} \right]$$

Substituting the parameters, we obtain $J_s = 4.15 \times 10^{-11}$ A/cm$^2$.

**Comment:** The ideal reverse-bias saturation current density is very small. If the pn junction cross-sectional area were $A = 10^{-4}$ cm$^2$, for example, then the ideal reverse-bias diode current would be $I_s = 4.15 \times 10^{-15}$ A.

---

If the forward-bias voltage in Equation (8–23) is positive by more than a few $kT/e$ volts, then the $(-1)$ term in Equation (8–23) becomes negligible. Figure 8–7 shows the forward-bias current-voltage characteristic when the current is plotted on a log scale. Ideally, this plot yields a straight line when $V_a$ is greater than a few $kT/e$ volts. The forward-bias current is an exponential function of the forward-bias voltage.

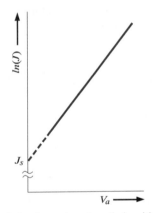

**Figure 8–7** Ideal I-V characteristic of a pn junction diode with the current plotted on a log scale.

---

## Example 8–3

**Objective:** To calculate the minority carrier electron diffusion current density in a forward-biased pn junction.

Consider a silicon pn junction at $T = 300°$K that is forward biased at $V_a = 0.60$ volt. Consider the p-region to be doped to $N_a = 3 \times 10^{15}$ cm$^{-3}$ and assume the following parameters for the minority carrier electrons: $D_n = 25$ cm²/sec; $\tau_{n0} = 10^{-7}$ sec. Determine the electron diffusion current density at the edge of the space charge region.

**Solution:** The electron diffusion current density is given by Equation (8–20) as

$$J_n = \frac{eD_n n_{p0}}{L_n} \left[ \exp\left(\frac{eV_a}{kT}\right) - 1 \right] = e \sqrt{\frac{D_n}{\tau_{n0}}} \frac{n_i^2}{N_a} \left[ \exp\left(\frac{eV_a}{kT}\right) - 1 \right]$$

Substituting the numbers, we have

$$J_n = (1.6 \times 10^{-19}) \sqrt{\frac{25}{10^{-7}}} \frac{(1.5 \times 10^{10})^2}{3 \times 10^{15}} \left[ \exp\left(\frac{0.60}{0.0259}\right) - 1 \right]$$

which yields $J_n = 2.18$ A/cm² = 2180 mA/cm².

**Comment:** Relatively large forward-bias currents can be obtained in semiconductors due to the diffusion process.

---

### 8.1.4  Summary of Physics

We have been considering the case of a forward-bias voltage being applied to a pn junction. The forward-bias voltage lowers the potential barrier so that electrons and holes are injected across the space charge region. The injected carriers become minority carriers which then diffuse from the junction and recombine with majority carriers.

We calculated the minority carrier diffusion current densities at the edge of the space charge region. We can reconsider Equations (8–14) and (8–15) and determine the minority carrier diffusion current densities as a function of distance through the p- and n-regions. These results are

$$J_p(x) = \frac{eD_p p_{n0}}{L_p} \left[ \exp\left(\frac{eV_a}{kT}\right) - 1 \right] \exp\left(\frac{x_n - x}{L_p}\right) \qquad (x \geq x_n) \qquad (8\text{-}24)$$

and

$$J_n(x) = \frac{eD_n n_{p0}}{L_n} \left[ \exp\left(\frac{eV_a}{kT}\right) - 1 \right] \exp\left(\frac{x_p + x}{L_n}\right) \qquad (x \leq -x_p) \qquad (8\text{-}25)$$

The minority carrier diffusion current densities decay exponentially in each region. However, the total current through the pn junction is constant. The difference between total current and minority carrier diffusion current is a majority carrier current. Figure 8–8 shows the various current components through the pn structure. The drift of majority carrier holes in the p-region far from the junction, for example, is to supply holes that are being injected across the space charge region into the n-region and also to supply holes that are lost by recombination with excess minority carrier electrons. The same discussion applies to the drift of electrons in the n-region.

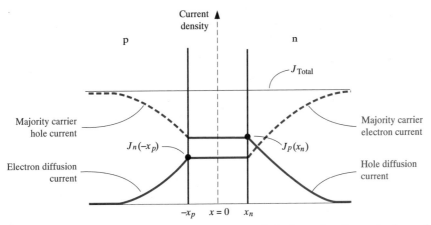

**Figure 8–8**   Ideal electron and hole current components through a pn junction under forward bias.

We have seen that excess carriers are created in a forward-biased pn junction. From the results of the ambipolar transport theory derived in Chapter 6, the behavior of the excess carriers is determined by the minority carrier parameters for low injection. In determining the current-voltage relationship of the pn junction, we consider the flow of minority carriers since we know the behavior and characteristics of these particles. It may seem strange, at times, that we concern ourselves so much with minority carriers rather than with the vast number of majority carriers, but the reason for this can be found in the results derived from the ambipolar transport theory.

The fact that we now have drift current densities in the p- and n-regions implies that the electric field in these regions is not zero as we had originally assumed. We can calculate the electric field in the neutral regions and determine the validity of our zero-field approximation.

___

**Example 8–4**

**Objective:** To calculate the electric field required to produce a given majority carrier drift current.

Consider a silicon pn junction at $T = 300°K$ with the parameters given in Example 8–2 and with an applied forward-bias voltage $V_a = 0.65$ volt.

**Solution:** The total forward-bias current density is given by

$$J = J_s \left[ \exp\left( \frac{eV}{kT} \right) - 1 \right]$$

The reverse-saturation current density was determined in Example 8–2, so we can write

$$J = (4.15 \times 10^{-11}) \left[ \exp \left( \frac{0.65}{0.0259} \right) - 1 \right] = 3.29 \text{ A/cm}^2$$

The total current far from the junction in the n-region will be majority carrier electron drift current, so we can write

$$J = J_n \cong e\mu_n N_d E$$

The doping concentration is $N_d = 10^{16} \text{ cm}^{-3}$ and if we assume $\mu_n = 1350 \text{ cm}^2/\text{volt-sec}$, then the electric field must be

$$E = \frac{J_n}{e\mu_n N_d} = \frac{3.29}{(1.6 \times 10^{-19})(1350)(10^{16})} = 1.52 \text{ V/cm}$$

**Comment:** We assumed, in the derivation of the current-voltage equation, that the electric field in the neutral p- and n-regions was zero. Although the electric field is not zero, this example shows that the magnitude is very small—thus the approximation of zero electric field is very good.

### 8.1.5 Temperature Effects

The ideal reverse-saturation current density $J_s$, given by Equation (8–22), is a function of the thermal-equilibrium minority carrier concentrations $n_{p0}$ and $p_{n0}$. These minority carrier concentrations are proportional to $n_i^2$, which is a very strong function of temperature. For a silicon pn junction, the ideal reverse-saturation current density will increase by approximately a factor of four for every 10°C increase in temperature.

The forward-bias current-voltage relation was given by Equation (8–23). This relation includes $J_s$ as well as the $\exp(eV_a/kT)$ factor, making the forward-bias current-voltage relation a function of temperature also. As temperature increases, less forward-bias voltage is required to obtain the same diode current. If the voltage is held constant, then the diode current will increase as temperature increases. The change in forward-bias current with temperature is less sensitive than the reverse-saturation current.

### Example 8–5

**Objective:** To determine the change in the forward-bias voltage on a pn junction with a change in temperature.

Consider a silicon pn junction initially biased at 0.60 volt at $T = 300°$K. Assume the temperature increases to $T = 310°$K. Calculate the change in the forward-bias voltage required to maintain a constant current through the junction.

**Solution:** The forward-bias current can be written as follows:

$$J \propto \exp \left( \frac{-E_g}{kT} \right) \exp \left( \frac{eV_a}{kT} \right)$$

If the temperature changes, we may take the ratio of the diode currents at the two temperatures. This ratio is

$$\frac{J_2}{J_1} = \frac{\exp(-E_g/kT_2)\exp(eV_{a2}/kT_2)}{\exp(-E_g/kT_1)\exp(eV_{a1}/kT_1)}$$

If current is to be held constant, then $J_1 = J_2$ and we must have

$$\frac{E_g - eV_{a2}}{kT_2} = \frac{E_g - eV_{a1}}{kT_1}$$

Let $T_1 = 300°K$, $T_2 = 310°K$, $E_g = 1.12$ eV, and $V_{a1} = 0.60$ volt. Then, solving for $V_{a2}$, we obtain $V_{a2} = 0.5827$ volt.

**Comment:** The change in the forward-bias voltage is $-17.3$ mV for a 10°C temperature change.

## 8.1.6 The "Short" Diode

We assumed in the previous analysis that both p- and n-regions were long compared with the minority carrier diffusion lengths. In many pn junction structures, one region may, in fact, be short compared with the minority carrier diffusion length. Figure 8–9 shows one such example: the length $W_n$ is assumed to be much smaller than the minority carrier hole diffusion length, $L_p$.

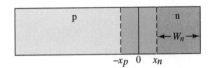

**Figure 8-9** Geometry of a "short" diode.

The steady-state excess minority carrier hole concentration in the n-region is determined from Equation (8–9), which was given as

$$\frac{d^2(\delta p_n)}{dx^2} - \frac{\delta p_n}{L_p^2} = 0$$

The original boundary condition at $x = x_n$ still applies, given by Equation (8–11a) as

$$p_n(x_n) = p_{n0}\exp\left(\frac{eV_a}{kT}\right)$$

A second boundary condition needs to be determined. In many cases we will assume that an ohmic contact exists at $x = (x_n + W_n)$, implying an infinite surface-recombination velocity and therefore an excess minority carrier concentration of zero. The second boundary condition is then written as

$$p_n(x = x_n + W_n) = p_{n0} \tag{8-26}$$

The general solution to Equation (8–9) is again given by Equation (8–12), which was

$$\delta p_n(x) = p_n(x) - p_{n0} = Ae^{x/L_p} + Be^{-x/L_p} \quad (x \geq x_n)$$

In this case, because of the finite length of the n-region, both terms of the general solution must be retained. Applying the boundary conditions of Equations (8–11b) and (8–26), the excess minority carrier concentration is given by

$$\delta p_n(x) = p_{n0} \left[ \exp\left(\frac{eV_a}{kT}\right) - 1 \right] \frac{\sinh\left[(x_n + W_n - x)/L_p\right]}{\sinh\left[W_n/L_p\right]} \tag{8-27}$$

Equation (8–27) is the general solution for the excess minority carrier hole concentration in the n-region of a forward-biased pn junction. If $W_n \gg L_p$, the assumption for the long diode, Equation (8–27) reduces to the previous result given by Equation (8–14). If $W_n \ll L_p$, we can approximate the hyberbolic sine terms by

$$\sinh\left[\frac{x_n + W_n - x}{L_p}\right] \approx \left[\frac{x_n + W_n - x}{L_p}\right] \tag{8-28a}$$

and

$$\sinh\left[\frac{W_n}{L_p}\right] \approx \left[\frac{W_n}{L_p}\right] \tag{8-28b}$$

Then Equation (8–27) becomes

$$\delta p_n(x) = p_{n0} \left[ \exp\left(\frac{eV_a}{kT}\right) - 1 \right]\left[\frac{x_n + W_n - x}{W_n}\right] \tag{8-29}$$

The minority carrier concentration becomes a linear function of distance.
   The minority carrier hole diffusion current density is given by

$$J_p = -eD_p \frac{d(\delta p_n(x))}{dx}$$

so that in the short n-region, we have

$$J_p(x) = \frac{eD_p p_{n0}}{W_n} \left[ \exp\left(\frac{eV_a}{kT}\right) - 1 \right] \tag{8-30}$$

The minority carrier hole diffusion current density now contains the length $W_n$ in the denominator, rather than the diffusion length $L_p$. The diffusion current density is larger for a short diode than for a long diode since $W_n \ll L_p$. In addition, since the minority carrier concentration is approximately a linear function of distance through the n-region, the minority carrier diffusion current density is a constant. This constant current implies that there is no recombination of minority carriers in the short region.

## 8.2 SMALL-SIGNAL MODEL OF THE PN JUNCTION

We have been considering the dc characteristics of the pn junction diode. When semiconductor devices with pn junctions are used in linear amplifier circuits, for example, sinusoidal signals are superimposed on the dc currents and voltages, so that the small-signal characteristics of the pn junction become important.

### 8.2.1 Diffusion Resistance

The ideal current-voltage relationship of the pn junction diode was given by Equation (8–23), where $J$ and $J_s$ are current densities. If we multiply both sides of the equation by the junction cross-sectional area, then we have

$$I_D = I_s \left[ \exp \left( \frac{eV_a}{kT} \right) - 1 \right] \tag{8–31}$$

where $I_D$ is the diode current and $I_s$ is the diode reverse-saturation current.

Assume that the diode is forward-biased with a dc voltage $V_0$ producing a dc diode current $I_{DQ}$. If we now superimpose a small, low-frequency sinusoidal voltage as shown in Figure 8–10, then a small sinusoidal current will be produced, superimposed on the dc current. The ratio of sinusoidal current to sinusoidal voltage is called the incremental conductance. In the limit of a very small sinusoidal current and voltage, the small-signal incremental conductance is just the slope of the dc current-voltage curve, or

$$g_d = \frac{dI_D}{dV_a} \bigg|_{V_a = V_0} \tag{8–32}$$

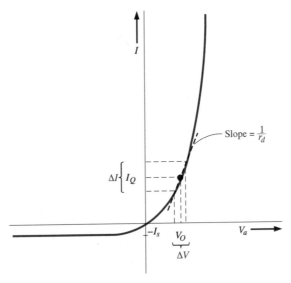

**Figure 8–10**  Curve showing the concept of the small-signal diffusion resistance.

The reciprocal of the incremental conductance is the incremental resistance, defined as

$$r_d = \frac{dV_a}{dI_D}\bigg|_{I_D=I_{DQ}} \tag{8-33}$$

where $I_{DQ}$ is the dc quiescent diode current.

If we assume that the diode is biased sufficiently far in the forward-bias region, then the $(-1)$ term can be neglected and the incremental conductance becomes

$$g_d = \frac{dI_D}{dV_a}\bigg|_{V_a=V_0} = \left(\frac{e}{kT}\right) I_s \exp\left(\frac{eV_0}{kT}\right) \cong \frac{I_{DQ}}{V_t} \tag{8-34}$$

The small-signal incremental resistance is then the reciprocal function, or

$$r_d = \frac{V_t}{I_{DQ}} \tag{8-35}$$

The incremental resistance decreases as the bias current increases, and is inversely proportional to the slope of the I-V characteristic as shown in Figure 8–10. The incremental resistance is also known as the *diffusion resistance*.

### 8.2.2 Small-Signal Admittance

In the last chapter, we considered the pn junction capacitance as a function of the reverse-bias voltage. When the pn junction diode is forward-biased, another capacitance becomes a factor in the diode admittance. The small-signal admittance, or impedance, of the pn junction under forward bias is derived using the minority carrier diffusion current relations we have already considered.

The minority carrier distribution in the pn junction will be derived for the case when a small sinusoidal voltage is superimposed on the dc junction voltage. We can then determine small signal, or ac, diffusion currents from these minority carrier functions. Figure 8–11 shows the minority carrier distribution in a pn junction when a forward-biased dc voltage is applied. The origin, $x = 0$, is set at the edge of the space charge region on the n-side for convenience. The minority carrier hole concentration at $x = 0$ is given by Equation (8–7) as $p_n(0) = p_{n0} \exp(eV_a/kT)$, where $V_a$ is the applied voltage across the junction.

Now let

$$V_a = V_0 + v_1(t) \tag{8-36}$$

where $V_0$ is the dc quiescent bias voltage and $v_1(t)$ is the ac signal voltage which is superimposed on this dc level. We may now write

$$p_n(x = 0) = p_{n0} \exp\left[\frac{e(V_0 + v_1(t))}{kT}\right] = p_n(0,t) \tag{8-37}$$

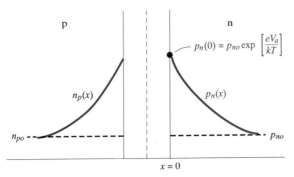

**Figure 8–11**   The dc characteristics of a forward-biased pn junction used in the small-signal admittance calculations.

Equation (8–37) may be written as

$$p_n(0,t) = p_{dc} \exp\left(\frac{ev_1(t)}{kT}\right) \tag{8–38}$$

where

$$p_{dc} = p_{n0} \exp\left(\frac{eV_0}{kT}\right) \tag{8–39}$$

If we assume that $|v_1(t)| \ll (kT/e) = V_t$, then the exponential term in Equation (8–38) may be expanded into a Taylor series retaining only the linear terms, and the minority carrier hole concentration at $x = 0$ can be written as

$$p_n(0,t) \cong p_{dc}\left(1 + \frac{v_1(t)}{V_t}\right) \tag{8–40}$$

If we assume that the time-varying voltage $v_1(t)$ is a sinusoidal signal, we can write Equation (8–40) as

$$p_n(0,t) = p_{dc}\left(1 + \frac{\hat{V}_1}{V_t}\,e^{j\omega t}\right) \tag{8–41}$$

where $\hat{V}_1$ is the phasor of the applied sinusoidal voltage. Equation (8–41) will be used as the boundary condition in the solution of the time-dependent diffusion equation for the minority carrier holes in the n-region.

   In the neutral n-region ($x > 0$), the electric field is assumed to be zero, thus the behavior of the excess minority carrier holes is determined from the equation

$$D_p\frac{\partial^2(\delta p_n)}{\partial x^2} - \frac{\delta p_n}{\tau_{p0}} = \frac{\partial(\delta p_n)}{\partial t} \tag{8–42}$$

where $\delta p_n$ is the excess hole concentration in the n-region. We are assuming that the ac signal voltage $v_1(t)$ is sinusoidal. We then expect the steady-state solution for $\delta p_n$ to be of the form of a sinusoidal solution superimposed on the dc solution, or

$$\delta p_n(x,t) = \delta p_0(x) + p_1(x)e^{j\omega t} \qquad (8\text{-}43)$$

where $\delta p_0(x)$ is the dc excess carrier concentration and $p_1(x)$ is the magnitude of the ac component of the excess carrier concentration. The expression for $\delta p_0(x)$ is the same as that given in Equation (8–14).

Substituting Equation (8–43) into the differential Equation (8–42), we obtain

$$D_p \left\{ \frac{\partial^2(\delta p_0(x))}{\partial x^2} + \frac{\partial^2 p_1(x)}{\partial x^2} e^{j\omega t} \right\} - \frac{\delta p_0(x) + p_1(x)e^{j\omega t}}{\tau_{p0}} = j\omega p_1(x)e^{j\omega t} \qquad (8\text{-}44)$$

We may rewrite this equation, combining the time-dependent and time-independent terms, as

$$\left\{ D_p \frac{\partial^2(\delta p_0(x))}{\partial x^2} - \frac{\delta p_0(x)}{\tau_{p0}} \right\} + \left\{ D_p \frac{\partial^2 p_1(x)}{\partial x^2} - \frac{p_1(x)}{\tau_{p0}} - j\omega p_1(x) \right\} e^{j\omega t} = 0 \qquad (8\text{-}45)$$

If the ac component, $p_1(x)$, is zero, then the first bracketed term is just the differential Equation (8–10) which is identically zero. Then we have, from the second bracketed term,

$$D_p \frac{d^2 p_1(x)}{dx^2} - \frac{p_1(x)}{\tau_{p0}} - j\omega p_1(x) = 0 \qquad (8\text{-}46)$$

Noting that $L_p^2 = D_p \tau_{p0}$, Equation (8–46) may be rewritten in the form

$$\frac{d^2 p_1(x)}{dx^2} - \frac{(1 + j\omega\tau_{p0})}{L_p^2} p_1(x) = 0 \qquad (8\text{-}47)$$

or

$$\frac{d^2 p_1(x)}{dx^2} - C_p^2 p_1(x) = 0 \qquad (8\text{-}48)$$

where

$$C_p^2 = \frac{(1 + j\omega\tau_{p0})}{L_p^2} \qquad (8\text{-}49)$$

The general solution to Equation (8–48) is

$$p_1(x) = K_1 e^{-C_p x} + K_2 e^{+C_p x} \qquad (8\text{-}50)$$

One boundary condition is that $p_1(x \to +\infty) = 0$, which implies that the coefficient $K_2 = 0$. Then

$$p_1(x) = K_1 e^{-C_p x} \qquad (8\text{-}51)$$

Applying the boundary condition at $x = 0$ from Equation (8–41), we obtain

$$p_1(0) = K_1 = p_{dc} \left( \frac{\hat{V}_1}{V_t} \right) e^{j\omega t} \qquad (8\text{-}52)$$

The hole diffusion current density can be calculated at $x = 0$. This will be given by

$$J_p = -eD_p \frac{\partial p_n}{\partial x} \bigg|_{x=0} \qquad (8\text{–}53)$$

If we consider a homogeneous semiconductor, the derivative of the hole concentration will be just the derivative of the excess hole concentration. Then

$$J_p = -eD_p \frac{\partial(\delta p_n)}{\partial x} \bigg|_{x=0} = -eD_p \frac{\partial(\delta p_0(x))}{\partial x} \bigg|_{x=0} -eD_p \frac{\partial p_1(x)}{\partial x} \bigg|_{x=0} \qquad (8\text{–}54)$$

We can write this equation in the form

$$J_p = J_{p0} + j_p(t) \qquad (8\text{–}55)$$

where

$$J_{p0} = -eD_p \frac{\partial(\delta p_0(x))}{\partial x} \bigg|_{x=0} = \frac{eD_p p_{n0}}{L_p} \left[ \exp\left(\frac{eV_0}{kT}\right) - 1 \right] \qquad (8\text{–}56)$$

Equation (8–56) is the dc component of the hole diffusion current density and is exactly the same as in the ideal I-V relation derived previously.

The sinusoidal component of the diffusion current density is then found from

$$j_p(t) = \hat{J}_p e^{j\omega t} = -eD_p \frac{\partial p_1(x)}{\partial x} e^{j\omega t} \bigg|_{x=0} \qquad (8\text{–}57)$$

where $\hat{J}_p$ is the current density phasor. Combining Equations (8–57), (8–51), and (8–52), we have

$$\hat{J}_p = -eD_p(-C_p) \left[ p_{dc} \left(\frac{\hat{V}_1}{V_t}\right) \right] e^{-C_p x} \bigg|_{x=0} \qquad (8\text{–}58)$$

We can write the total ac hole current phasor as

$$\hat{I}_p = A\hat{J}_p = eAD_p C_p p_{dc} \left(\frac{\hat{V}_1}{V_t}\right) \qquad (8\text{–}59)$$

where $A$ is the cross-sectional area of the pn junction. Substituting the expression for $C_p$, we obtain

$$\hat{I}_p = \frac{eAD_p p_{dc}}{L_p} \sqrt{1 + j\omega\tau_{p0}} \left(\frac{\hat{V}_1}{V_t}\right) \qquad (8\text{–}60)$$

If we define

$$I_{p0} = \frac{eAD_p p_{dc}}{L_p} = \frac{eAD_p p_{n0}}{L_p} \exp\left(\frac{eV_0}{kT}\right) \qquad (8\text{–}61)$$

then Equation (8–60) becomes

$$\hat{I}_p = I_{p0} \sqrt{1 + j\omega\tau_{p0}} \left(\frac{\hat{V}_1}{V_t}\right) \qquad (8\text{–}62)$$

We can go through the same type of analysis for the minority carrier electrons in the p-region. We will obtain

$$\hat{I}_n = I_{n0} \sqrt{1 + j\omega\tau_{n0}} \left(\frac{\hat{V}_1}{V_t}\right) \qquad (8\text{–}63)$$

where

$$I_{n0} = \frac{eAD_n n_{p0}}{L_n} \exp\left(\frac{eV_0}{kT}\right) \qquad (8\text{–}64)$$

The total ac current phasor is the sum of $\hat{I}_p$ and $\hat{I}_n$. The pn junction admittance is the total ac current phasor divided by the ac voltage phasor, or

$$Y = \frac{\hat{I}}{\hat{V}_1} = \frac{\hat{I}_p + \hat{I}_n}{\hat{V}_1} = \left(\frac{1}{V_t}\right) [I_{p0} \sqrt{1 + j\omega\tau_{p0}} + I_{n0} \sqrt{1 + j\omega\tau_{n0}}] \qquad (8\text{–}65)$$

There is not a linear, lumped, finite, passive, bilateral network that can be synthesized to give this admittance function. However, we may make the following approximations. Assume that

$$\omega\tau_{p0} \ll 1 \qquad (8\text{–}66a)$$

and

$$\omega\tau_{n0} \ll 1 \qquad (8\text{–}66b)$$

These two assumptions imply that the frequency of the ac signal is not too large. Then we may write

$$\sqrt{1 + j\omega\tau_{p0}} \approx 1 + \frac{j\omega\tau_{p0}}{2} \qquad (8\text{–}67a)$$

and

$$\sqrt{1 + j\omega\tau_{n0}} \approx 1 + \frac{j\omega\tau_{n0}}{2} \qquad (8\text{–}67b)$$

Substituting Equations (8–67a) and (8–67b) into the admittance Equation (8–65) yields

$$Y = \left(\frac{1}{V_t}\right)\left[I_{p0}\left(1 + \frac{j\omega\tau_{p0}}{2}\right) + I_{n0}\left(1 + \frac{j\omega\tau_{n0}}{2}\right)\right] \qquad (8\text{–}68)$$

If we combine the real and imaginary portions, we get

$$Y = \left(\frac{1}{V_t}\right)(I_{p0} + I_{n0}) + j\omega \left\{\left(\frac{1}{2V_t}\right) [I_{p0}\tau_{p0} + I_{n0}\tau_{n0}]\right\} \qquad (8\text{–}69)$$

Equation (8–69) may be written in the form

$$Y = g_d + j\omega C_d \qquad (8\text{–}70)$$

The parameter $g_d$ is called the *diffusion conductance* and is given by

$$g_d = \left(\frac{1}{V_t}\right)(I_{p0} + I_{n0}) = \frac{I_{DQ}}{V_t} \qquad (8\text{–}71)$$

where $I_{DQ}$ is the dc bias current. Equation (8–71) is exactly the same conductance as we obtained previously in Equation (8–34). The parameter $C_d$ is called the *diffusion capacitance* and is given by

$$C_d = \left(\frac{1}{2V_t}\right)(I_{p0}\tau_{p0} + I_{n0}\tau_{n0}) \qquad (8\text{–}72)$$

The physics of the diffusion capacitance may be seen in Figure 8–12. The dc values of the minority carrier concentrations are shown along with the changes due to the ac component of voltage. The $\Delta Q$ charge is alternately being charged and discharged through the junction as the voltage across the junction changes. The change in the stored minority carrier charge as a function of the change in voltage is the diffusion capacitance. One consequence of the approximations $\omega\tau_{p0} \ll 1$ and $\omega\tau_{n0} \ll 1$ is that there are no "wiggles" in the minority carrier curves. The sinusoidal frequency is low enough so that the exponential curves are maintained at all times.

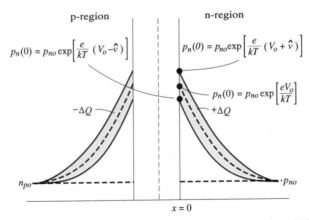

**Figure 8–12** Minority carrier concentration changes with changing forward-bias voltage.

---

## Example 8–6

**Objective:** To calculate the small-signal admittance of a pn junction diode.

This example is intended to give an indication of the magnitude of the diffusion capacitance as compared with the junction capacitance considered in the last chapter. The diffusion resistance will also be calculated. Assume that $N_a \gg N_d$ so that $p_{n0} \gg n_{p0}$. This assumption implies that $I_{p0} \gg I_{n0}$. Let $T = 300°\text{K}$, $\tau_{p0} = 10^{-7}$ sec, and $I_{p0} = I_{DQ} = 1$ mA.

**Solution:** The diffusion capacitance, with these assumptions, is given by

$$C_d \cong \left(\frac{1}{2V_t}\right)(I_{p0}\tau_{p0}) = \frac{1}{(2)(0.0259)}(10^{-3})(10^{-7}) = 1.93 \times 10^{-9} \text{ F}$$

The diffusion resistance is

$$r_d = \frac{V_t}{I_{DQ}} = \frac{0.0259 \text{ volt}}{1 \text{ mA}} = 25.9 \text{ ohms}$$

**Comment:** The value of $1.93\,nF$ for the diffusion capacitance of a forward-biased pn junction is three to four orders of magnitude larger than the junction capacitance of the reverse-biased pn junction, which we calculated in Example 7–3.

---

The diffusion capacitance tends to dominate the capacitance terms in a forward-biased pn junction. The small-signal diffusion resistance can be fairly small if the diode current is a fairly large value. As the diode current decreases, the diffusion resistance increases. The impedance of forward-biased pn junctions will be considered again when we discuss bipolar transistors.

### 8.2.3 Equivalent Circuit

The small-signal equivalent circuit of the forward-biased pn junction is derived from Equation (8–70). This circuit is shown in Figure 8–13a. We need

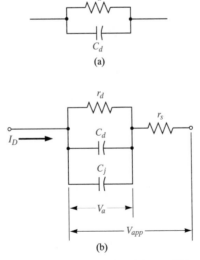

(a)

(b)

**Figure 8–13**   (a) Small-signal equivalent circuit of ideal forward-biased pn junction diode. (b) Complete small-signal equivalent circuit of pn junction.

to add the junction capacitance, which will be in parallel with the diffusion resistance and diffusion capacitance. The last element we add, to complete the equivalent circuit, is a series resistance. The neutral n- and p-regions have finite resistances so the actual pn junction will include a series resistance. The complete equivalent circuit is given in Figure 8–13b.

The voltage across the actual junction is $V_a$ and the total voltage applied to the pn diode is given by $V_{app}$. The junction voltage $V_a$ is the voltage in the ideal current-voltage expression. We can write the expression

$$V_{app} = V_a + Ir_s \qquad (8\text{–}73)$$

Figure 8–14 is a plot of the current-voltage characteristic from Equation (8–73) showing the effect of the series resistance. A larger applied voltage is required to achieve the same current value when a series resistance is included. In most diodes, the series resistance will be negligible. In some semiconductor devices with pn junctions, however, the series resistance will be in a feedback loop so that the resistance is multiplied by a gain factor and becomes non-negligible.

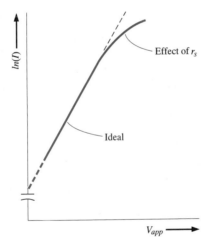

**Figure 8–14** Forward-biased I-V characteristics of a pn junction diode showing the effect of series resistance.

## 8.3 GENERATION–RECOMBINATION CURRENTS

In the derivation of the ideal current-voltage relationship, we neglected any effects occurring within the space charge region. Since other current components are generated within the space charge region, the actual I-V characteristics of a pn junction diode deviate from the ideal expression. The additional currents are generated from the recombination processes discussed in Chapter 6.

The recombination rate of excess electrons and holes, derived from the Shockley-Read-Hall recombination theory, was given by

$$R = \frac{C_n C_p N_t (np - n_i^2)}{C_n(n + n') + C_p(p + p')} \tag{8-74}$$

The parameters $n$ and $p$ are, as usual, the concentrations of electrons and holes, respectively.

### 8.3.1 Reverse-Bias Generation Current

For a pn junction under reverse bias, we have argued that the mobile electrons and holes have essentially been swept out of the space charge region. Accordingly, within the space charge region, $n \cong p \cong 0$. The recombination rate from Equation (8–74) becomes

$$R = \frac{-C_n C_p N_t n_i^2}{C_n n' + C_p p'} \tag{8-75}$$

The negative sign implies a negative recombination rate; hence we are really generating electron-hole pairs within the reverse-biased space charge region. The recombination of excess electrons and holes is the process whereby we are trying to reestablish thermal equilibrium. Since the concentration of electrons and holes is essentially zero within the reverse-biased space charge region, electrons and holes are being generated via the trap level to also try to reestablish thermal equilibrium. This generation process is schematically shown in Figure 8–15. As the electrons and holes are generated, they are swept out of the space charge region by the electric field. The flow of charge is in the direction of a reverse-bias current. This *reverse-bias*

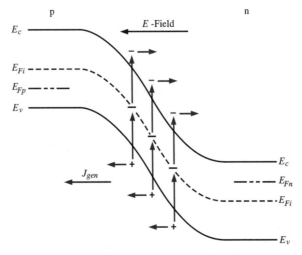

**Figure 8–15**   Generation process in a reverse-biased pn junction.

*generation current,* caused by the generation of electrons and holes in the space charge region, is in addition to the ideal reverse-bias saturation current.

We may calculate the density of the reverse-bias generation current by considering Equation (8–75). If we make a simplifying assumption and let the trap level be at the intrinsic Fermi level, then from Equations (6–84) and (6–89), we have that $n' = n_i$ and $p' = n_i$. Equation (8–75) now becomes

$$R = \frac{-n_i}{\dfrac{1}{N_t C_p} + \dfrac{1}{N_t C_n}} \qquad (8\text{–}76)$$

Using the definitions of lifetimes from Equations (6–95) and (6–96), Equation (8–76) may be written as

$$R = \frac{-n_i}{\tau_{po} + \tau_{no}} \qquad (8\text{–}77)$$

If we define a new lifetime as the average of $\tau_{po}$ and $\tau_{no}$, or

$$\tau_o = \frac{\tau_{po} + \tau_{no}}{2} \qquad (8\text{–}78)$$

then the recombination rate can be written as

$$R = \frac{-n_i}{2\tau_o} \equiv -G \qquad (8\text{–}79)$$

The negative recombination rate implies a generation rate, so $G$ is the generation rate of electrons and holes in the space charge region.

The generation current density may be determined from

$$J_{\text{gen}} = \int_0^W eG \, dx \qquad (8\text{–}80)$$

where the integral is over the space charge region. If we assume that the generation rate is constant throughout the space charge region, then we obtain

$$J_{\text{gen}} = \frac{e n_i W}{2\tau_o} \qquad (8\text{–}81)$$

The total reverse-bias current density is the sum of the ideal reverse saturation current density and the generation current density, or

$$J_R = J_s + J_{\text{gen}} \qquad (8\text{–}82)$$

The ideal reverse saturation current density $J_s$ is independent of the reverse-bias voltage. However, $J_{\text{gen}}$ is a function of the depletion width $W$ which in turn is a function of the reverse-bias voltage. The actual reverse-bias current density, then, is no longer independent of the reverse-bias voltage.

**Example 8-7**

**Objective:** To determine the relative magnitudes of the ideal reverse saturation current density and the generation current density in a silicon pn junction at $T = 300°K$.

Consider the silicon pn junction described in Example 8–2 and let $\tau_o = \tau_{po} = \tau_{no} = 5 \times 10^{-7}$ sec.

**Solution:** The ideal reverse saturation current density was calculated in Example 8–2 and was found to be $J_s = 4.15 \times 10^{-11}$ A/cm². The generation current density is again given by Equation (8–81) as

$$J_{gen} = \frac{en_i W}{2\tau_o}$$

and the depletion width is given by

$$W = \left\{ \frac{2\varepsilon_s}{e} \left( \frac{N_a + N_d}{N_a N_d} \right) (V_{bi} + V_R) \right\}^{1/2}$$

If we assume, for example, that $V_{bi} + V_R = 5$ volts, then using the parameters given in Example 8–2 we find that $W = 1.14 \times 10^{-4}$ cm, and the generation current density is then calculated to be

$$J_{gen} = 2.74 \times 10^{-7} \text{ A/cm}^2$$

**Comment:** Comparing the solutions for the two current densities, it is obvious that, for the silicon pn junction diode at room temperature, the generation current density is approximately four orders of magnitude larger than the ideal saturation current density. The generation current is the dominant reverse-bias current in a silicon pn junction diode.

### 8.3.2 Forward-Bias Recombination Current

For the reverse-biased pn junction, electrons and holes are essentially completely swept out of the space charge region so that $n \cong p \cong 0$. Under forward bias, however, electrons and holes are injected across the space charge region so we do, in fact, have some excess carriers in the space charge region. The possibility exists that some of these electrons and holes will recombine within the space charge region and not become part of the minority carrier distribution.

The recombination rate of electrons and holes is again given from Equation (8–74) as

$$R = \frac{C_n C_p N_t (np - n_i^2)}{C_n(n + n') + C_p(p + p')}$$

Dividing both numerator and denominator by $C_n C_p N_t$ and using the definitions of $\tau_{no}$ and $\tau_{po}$, we may write the recombination rate as

$$R = \frac{np - n_i^2}{\tau_{po}(n + n') + \tau_{no}(p + p')} \qquad (8\text{-}83)$$

Figure 8–16 shows the energy-band diagram of the forward-biased pn junction. Shown in the figure is the intrinsic Fermi level and the quasi-Fermi levels for electrons and holes. From the results of Chapter 6, we may write the electron concentration as

$$n = n_i \exp\left[\frac{E_{Fn} - E_{Fi}}{kT}\right] \qquad (8\text{-}84)$$

and the hole concentration as

$$p = n_i \exp\left[\frac{E_{Fi} - E_{Fp}}{kT}\right] \qquad (8\text{-}85)$$

where $E_{Fn}$ and $E_{Fp}$ are the quasi-Fermi levels for electrons and holes, respectively.

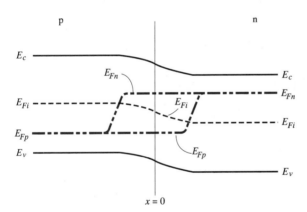

**Figure 8–16** Energy-band diagram of a forward-biased pn junction including quasi-Fermi levels.

From Figure 8–16, we may note that

$$(E_{Fn} - E_{Fi}) + (E_{Fi} - E_{Fp}) = eV_a \qquad (8\text{-}86)$$

where $V_a$ is the applied forward-bias voltage. Again, if we assume that the trap level is at the intrinsic Fermi level, then $n' = p' = n_i$. Figure 8–17 shows a plot of the relative magnitude of the recombination rate as a function of distance through the space charge region. This plot was generated using Equations (8–83), (8–84), (8–85), and (8–86). A very sharp peak occurs at the metallurgical junction ($x = 0$).

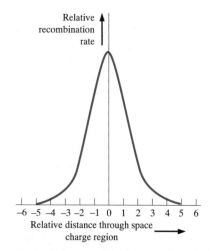

Relative recombination rate

−6 −5 −4 −3 −2 −1 0 1 2 3 4 5 6
Relative distance through space charge region

**Figure 8–17**   Relative magnitude of the recombination rate through the space charge region of a forward-biased pn junction.

At the center of the space charge region, we have

$$E_{Fn} - E_{Fi} = E_{Fi} - E_{Fp} = \frac{eV_a}{2} \qquad (8\text{–}87)$$

Equations (8–84) and (8–85) then become

$$n = n_i \exp\left(\frac{eV_a}{2kT}\right) \qquad (8\text{–}88)$$

and

$$p = n_i \exp\left(\frac{eV_a}{2kT}\right) \qquad (8\text{–}89)$$

If we assume that $n' = p' = n_i$ and that $\tau_{no} = \tau_{po} = \tau_o$, then Equation (8–83) becomes

$$R_{\max} = \frac{n_i}{2\tau_o} \frac{[\exp{(eV_a/kT)} - 1]}{[\exp{(eV_a/2kT)} + 1]} \qquad (8\text{–}90)$$

which is the maximum recombination rate for electrons and holes that occurs at the center of the forward-biased pn junction. If we assume that $V_a \gg kT/e$, the $(-1)$ term in the numerator and the $(+1)$ term in the denominator may be neglected. Equation (8–90) then becomes

$$R_{\max} = \frac{n_i}{2\tau_o} \exp\left(\frac{eV_a}{2kT}\right) \qquad (8\text{–}91)$$

The recombination current density may be calculated from

$$J_{\text{rec}} = \int_0^W eR \; dx \qquad (8\text{–}92)$$

where again the integral is over the entire space charge region. In this case, however, the recombination rate is not a constant through the space charge region. We have calculated the maximum recombination rate at the center of the space charge region, so we may write

$$J_{\text{rec}} = ex' \frac{n_i}{2\tau_o} \exp\left(\frac{eV_a}{2kT}\right) \tag{8–93}$$

where $x'$ is a length over which the maximum recombination rate is effective. However, since $\tau_o$ may not be a well-defined or known parameter, it is customary to write

$$J_{\text{rec}} = \frac{eWn_i}{2\tau_o} \exp\left(\frac{eV_a}{2kT}\right) = J_{ro} \exp\left(\frac{eV_a}{2kT}\right) \tag{8–94}$$

where $W$ is the space charge width.

### 8.3.3 Total Forward-Bias Current

The total forward-bias current density in the pn junction is the sum of the recombination and the ideal diffusion current densities. Figure 8–18 shows a plot of the minority carrier hole concentration in the neutral n-region. This distribution yields the ideal hole diffusion current density and is a function of the minority carrier hole diffusion length and the applied junction voltage. The distribution is established as a result of holes being injected across the space charge region. If, now, some of the injected holes in the space charge region are lost due to recombination, then additional holes must be injected from the p-region to make up for this loss. The flow of these additional

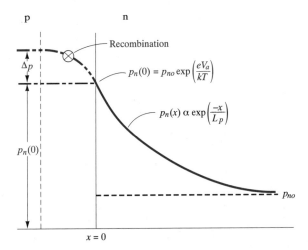

**Figure 8–18** Due to recombination, additional holes from the p-region must be injected into the space charge region to establish the minority carrier hole concentration in the n-region.

injected carriers, per unit time, results in the recombination current. This added component is schematically shown in the figure.

The total forward-bias current density is the sum of the recombination and the ideal diffusion current densities, so we can write

$$J = J_{\text{rec}} + J_D \tag{8-95}$$

where $J_{\text{rec}}$ is given by Equation (8–94) and $J_D$ is given by

$$J_D = J_s \exp\left(\frac{eV_a}{kT}\right) \tag{8-96}$$

The $(-1)$ term in Equation (8–23) has been neglected. The parameter $J_s$ is the ideal reverse-saturation current density and from previous discussion, the value of $J_{ro}$ from the recombination current is larger than the value of $J_s$.

If we take the natural log of Equations (8–94) and (8–96), we obtain

$$\ln J_{\text{rec}} = \ln J_{ro} + \frac{eV_a}{2kT} = \ln J_{ro} + \frac{V_a}{2V_t} \tag{8-97a}$$

and

$$\ln J_D = \ln J_s + \frac{eV_a}{kT} = \ln J_s + \frac{V_a}{V_t} \tag{8-97b}$$

Figure 8–19 shows the recombination and diffusion current components plotted on a log current scale as a function of $V_a/V_t$. The slopes of the two

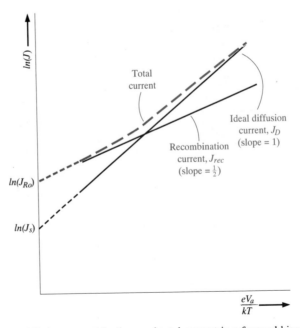

**Figure 8–19**   Ideal diffusion, recombination, and total current in a forward-biased pn junction.

curves are not the same. Also shown in the figure is the total current density—the sum of the two current components. We may notice that, at a low-current density, the recombination current dominates and, at a higher-current density, the ideal diffusion current dominates.

In general, the diode current-voltage relationship may be written as

$$I = I_s \left[ \exp\left(\frac{eV_a}{nkT}\right) - 1 \right] \tag{8–98}$$

where the parameter $n$ is called the ideality factor. For a large forward-bias voltage, $n \cong 1$ when diffusion dominates and, for low forward-bias voltage, $n \cong 2$ when recombination dominates. There is a transition region where $1 < n < 2$.

## 8.4 JUNCTION BREAKDOWN

In the ideal pn junction, a reverse-bias voltage will result in a small reverse-bias current through the device. However, the reverse-bias voltage may not increase without limit; at some particular voltage, the reverse-bias current will increase rapidly. The applied voltage at this point is called the breakdown voltage.

Two physical mechanisms give rise to the reverse-bias breakdown in a pn junction: the Zener effect and the avalanche effect. Zener breakdown occurs in highly doped pn junctions through a tunneling mechanism. In a highly doped junction, the conduction and valence bands on opposite sides of the junction are sufficiently close during reverse bias that electrons may tunnel directly from the valence band on the p-side into the conduction band on the n-side. This tunneling process is schematically shown in Figure 8–20a.

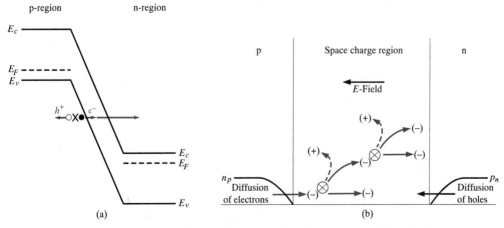

**Figure 8–20**  (a) Zener breakdown mechanism in a reverse-biased pn junction. (b) Avalanche breakdown process in a reverse-biased pn junction.

The avalanche breakdown process occurs when electrons and/or holes, moving across the space charge region, acquire sufficient energy from the electric field to create electron-hole pairs by colliding with atomic electrons within the depletion region. The avalanche process is schematically shown in Figure 8–20b. The newly created electrons and holes move in opposite directions due to the electric field and thereby add to the existing reverse-bias current. In addition, the newly generated electrons and/or holes may acquire sufficient energy to ionize other atoms, leading to the avalanche process. For most pn junctions, the predominant breakdown mechanism will be the avalanche effect.

If we assume that a reverse-bias electron current $I_{no}$ enters the depletion region at $x = 0$ as shown in Figure 8–21, the electron current $I_n$ will increase with distance through the depletion region due to the avalanche process. At $x = W$, the electron current may be written as

$$I_n(W) = M_n I_{no} \qquad (8\text{--}99)$$

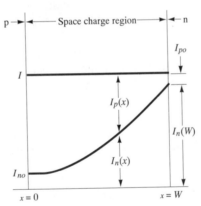

**Figure 8–21**   Electron and hole current components through the space charge region during avalanche multiplication.

where $M_n$ is a multiplication factor. The hole current is increasing through the depletion region from the n- to p-region and reaches a maximum value at $x = 0$. The total current is constant through the pn junction in steady state.

We can write an expression for the incremental electron current at some point $x$ as

$$dI_n(x) = I_n(x)\alpha_n \, dx + I_p(x)\alpha_p \, dx \qquad (8\text{--}100)$$

where $\alpha_n$ and $\alpha_p$ are the electron and hole ionization rates, respectively. The ionization rates are the number of electron-hole pairs generated per unit length by an electron ($\alpha_n$) or by a hole ($\alpha_p$). Equation (8–100) may be written as

$$\frac{dI_n(x)}{dx} = I_n(x)\alpha_n + I_p(x)\alpha_p \qquad (8\text{--}101)$$

The total current $I$ is given by

$$I = I_n(x) + I_p(x) \qquad (8\text{--}102)$$

which is a constant. Solving for $I_p(x)$ from Equation (8–102) and substituting into Equation (8–101), we obtain

$$\frac{dI_n(x)}{dx} + (\alpha_p - \alpha_n)I_n(x) = \alpha_p I \qquad (8\text{--}103)$$

If we make the assumption that the electron and hole ionization rates are equal so

$$\alpha_n = \alpha_p \equiv \alpha \qquad (8\text{--}104)$$

then Equation (8–103) may be simplified and integrated through the space charge region. We will obtain

$$I_n(W) - I_n(0) = I \int_0^W \alpha \, dx \qquad (8\text{--}105)$$

Using Equation (8–99), Equation (8–105) may be written as

$$\frac{M_n I_{no} - I_n(0)}{I} = \int_0^W \alpha \, dx \qquad (8\text{--}106)$$

Since $M_n I_{no} \approx I$ and since $I_n(0) = I_{no}$, Equation (8–106) becomes

$$1 - \frac{1}{M_n} = \int_0^W \alpha \, dx \qquad (8\text{--}107)$$

The avalanche breakdown voltage is defined to be the voltage at which $M_n$ approaches infinity. The avalanche breakdown condition is then given by

$$\int_0^W \alpha \, dx = 1 \qquad (8\text{--}108)$$

The ionization rates are strong functions of electric field and, since the electric field in not constant through the space charge region, Equation (8–108) is not easy to evaluate.

If we consider, for example, a one-sided $p^+n$ junction, the maximum electric field is given by

$$E_{max} = \frac{eN_d x_n}{\varepsilon_s} \qquad (8\text{--}109)$$

The depletion width $x_n$ is given approximately as

$$x_n \cong \left\{ \frac{2\varepsilon_s V_R}{e} \cdot \frac{1}{N_d} \right\}^{1/2} \qquad (8\text{--}110)$$

where $V_R$ is the magnitude of the applied reverse-bias voltage. We have neglected the built-in potential $V_{bi}$.

If we now define $V_R$ to be the breakdown voltage $V_B$, the maximum electric field, $E_{max}$, will be defined as a critical electric field, $E_{crit}$, at breakdown. Combining Equations (8–109) and (8–110), we may write

$$V_B = \frac{\varepsilon_s E_{crit}^2}{2eN_B} \qquad (8\text{–}111)$$

where $N_B$ is the semiconductor doping in the low-doped region of the one-sided junction. The critical electric field, plotted in Figure 8–22, is a slight function of doping.

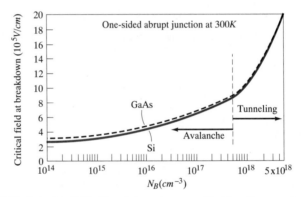

**Figure 8–22**    Critical electric field at breakdown in a one-sided junction as a function of impurity doping concentrations. (From Sze [7]).

---

### Example 8–8

**Objective:** To determine the breakdown voltage of an $n^+p$ junction.

Consider a silicon pn junction at $T = 300°K$ with doping concentrations of $N_a = 5 \times 10^{15}$ cm$^{-3}$ and $N_d = 5 \times 10^{17}$ cm$^{-3}$. Calculate the breakdown voltage assuming the abrupt junction approximation.

**Solution:** For a doping concentration in the low-doped region of $N_a = 5 \times 10^{15}$ cm$^{-3}$, the critical electric field, from Figure 8–22, is approximately $E_{crit} \approx 4 \times 10^5$ V/cm. The breakdown voltage, determined from Equation (8–111), is

$$V_B = \frac{\varepsilon_s E_{crit}^2}{2eN_B} = \frac{(11.7)(8.85 \times 10^{-14})(4 \times 10^5)^2}{2(1.6 \times 10^{-19})(5 \times 10^{15})} = 104 \text{ volts}$$

**Comment:** This breakdown voltage is for an ideal planar, abrupt junction.

---

We have been considering a uniformly doped planar junction. The breakdown voltage will decrease for a linearly graded junction. Figure 8–23 shows a plot of the breakdown voltage for a one-sided abrupt junction and a linearly graded junction. If we take into account the curvature of a diffused junction as well, the breakdown voltage will be further degraded.

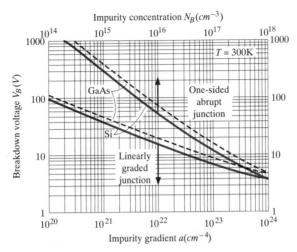

**Figure 8–23**  Breakdown voltage versus impurity concentration in uniformly doped and linearly graded junctions. (From Sze [7]).

## 8.5 CHARGE STORAGE AND DIODE TRANSIENTS

The pn junction is typically used as an electrical switch. In forward bias, referred to as the "on" state, a relatively large current can be produced by a small applied voltage; in reverse bias, referred to as the "off" state, only a very small current will exist. Of primary interest in circuit applications is the speed of the pn junction diode in switching states. We will qualitatively discuss the transients that occur and the charge storage effects. We will simply state the equations that describe the switching times without any mathematical derivations.

### 8.5.1 The Turn-Off Transient

Suppose we want to switch a diode from the forward-bias "on" state to the reverse-bias "off" state. Figure 8–24 shows a simple circuit that will switch the applied bias at $t = 0$. For $t < 0$, the forward-bias current is

$$I = I_F = \frac{V_F - V_a}{R_F} \tag{8–112}$$

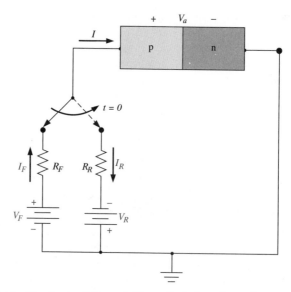

**Figure 8–24**   Simple circuit for switching a diode from forward to reverse bias.

The minority carrier concentrations in the device, for the applied forward voltage $V_F$, are shown in Figure 8–25a. There is excess minority carrier charge stored in both the p- and n-regions of the diode. The excess minority carrier concentrations at the space charge edges are supported by the forward-bias junction voltage $V_a$. When the voltage is switched from the forward- to the reverse-bias state, the excess minority carrier concentrations at the space charge edges can no longer be supported and start to decrease, as shown in Figure 8–25b.

The collapse of the minority carrier concentrations at the edges of the space charge region leads to large concentration gradients and diffusion currents in the reverse-bias direction. If we assume, for the moment, that the voltage across the diode junction is small compared with $V_R$, then the reverse-bias current is limited to approximately

$$I = -I_R \approx \frac{-V_R}{R_R} \tag{8–113}$$

The junction capacitances do not allow the junction voltage to change instantaneously. If the current $I_R$ were larger than this value, then there would be a forward-bias voltage across the junction, which violates our assumption of a reverse-bias current. If the current $I_R$ were smaller than this value, then there would be a reverse-bias voltage across the junction, which means that the junction voltage would have changed instantaneously. Since the reverse current is limited to the value given by Equation (8–113), the reverse-bias density gradient is constant; thus, the minority carrier concentrations at the space charge edge decrease with time as shown in Figure 8–25b.

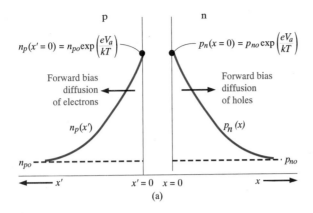

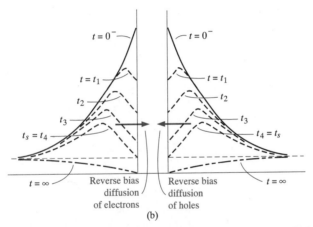

**Figure 8–25** (a) Steady-state forward-bias minority carrier concentrations. (b) Minority carrier concentrations at various times during switching.

This reverse current $I_R$ will be approximately constant for $0^+ \leq t \leq t_s$, where $t_s$ is called the *storage time*. The storage time is the length of time required for the minority carrier concentrations at the space charge edge to reach the thermal-equilibrium values. After this time, the voltage across the junction will begin to change. The current characteristic is shown in Figure 8–26. The reverse current is the flow of the stored minority carrier charge, which is the difference between the minority carrier concentrations at $t = 0^-$ and $t = \infty$, as was shown in Figure 8–25b.

The storage time $t_s$ can be determined by solving the time-dependent continuity equation. If we consider a one-sided $p^+n$ junction, the storage time is determined from the equation

$$erf \sqrt{\frac{t_s}{\tau_{po}}} = \frac{I_F}{I_F + I_R} \qquad (8–114)$$

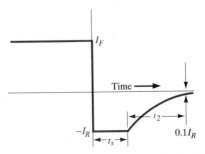

**Figure 8–26**  Current characteristic versus time during diode switching.

where $erf(x)$ is known as the error function. An approximate solution for the storage time can be obtained which is given as

$$t_s \cong \tau_{po} \ln \left[ 1 + \frac{I_F}{I_R} \right] \tag{8–115}$$

The recovery phase for $t > t_s$ is the time required for the junction to reach its steady-state reverse-bias condition. The remainder of the excess charge is being removed and the space charge width is increasing to the reverse-bias value. The decay time $t_2$ is determined from

$$erf \sqrt{\frac{t_2}{\tau_{po}}} + \frac{\exp(-t_2/\tau_{po})}{\sqrt{\pi t_2/\tau_{po}}} = 1 + 0.1 \left( \frac{I_R}{I_F} \right) \tag{8–116}$$

The total turn-off time is the sum of $t_s$ and $t_2$.

To switch the diode quickly, we need to be able to produce a large reverse current as well as have a small minority carrier lifetime. In the design of diode circuits, then, the designer must provide a path for the transient reverse-bias current pulse in order to be able to switch the diode quickly. These same effects will be considered when we discuss the switching of bipolar transistors.

### 8.5.2 The Turn-On Transient

The turn-on transient occurs when the diode is switched from its "off" state into the forward-bias "on" state. The turn-on can be accomplished by applying a forward-bias current pulse. The first stage of turn-on occurs very quickly and is the length of time required to narrow the space charge width from the reverse-bias value to its thermal-equilibrium value when $V_a = 0$. During this time, ionized donors and acceptors are neutralized as the space charge width narrows.

The second stage of the turn-on process is the time required to establish the minority-carrier distributions. During this time the voltage across the junction is increasing toward its steady-state value. A small turn-on time is achieved if the minority carrier lifetime is small and if the forward-bias current is small.

## *8.6 THE TUNNEL DIODE

The tunnel diode is a pn junction in which both the n- and p-regions are degenerately doped. As we discuss the operation of this device, we will find a region that exhibits a negative differential resistance. The tunnel diode was used in oscillator circuits in the past, but other types of solid-state devices are now used as high-frequency oscillators; thus, the tunnel diode is really only of academic interest. Nevertheless, this device does demonstrate the phenomenon of tunneling we discussed in Chapter 2.

Recall the degenerately doped semiconductors we discussed in Chapter 4: the Fermi level is in the conduction band of a degenerately doped n-type material and in the valence band of a degenerately doped p-type material. Then, even at $T = 0°K$, electrons will exist in the conduction band of the n-type material and holes (empty states) will exist in the p-type material.

The energy-band diagram of a pn junction in thermal equilibrium for the case when both the n- and p-regions are degenerately doped is shown in Figure 8–27. The depletion region width decreases as the doping increases and may be on the order of approximately 100Å for the case shown in Figure 8–27. The potential barrier at the junction can be approximated by a

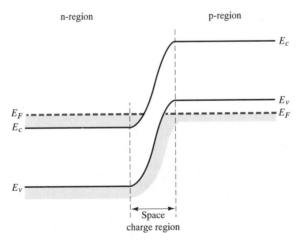

**Figure 8–27**  Energy-band diagram of a pn junction in thermal equilibrium in which both the n- and p-regions are degenerately doped.

triangular potential barrier as is shown in Figure 8–28. This potential barrier is similar to the potential barrier used in Chapter 2 to illustrate the tunneling phenomenon. The barrier width is small and the electric field in the space

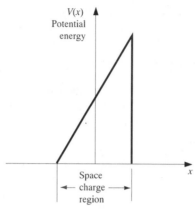

**Figure 8-28**   Triangular potential barrier approximation of the potential barrier in the tunnel diode.

charge region is quite large; thus, a finite probability exists that an electron may tunnel through the forbidden band from one side of the junction to the other.

We may qualitatively determine the current-voltage characteristics of the tunnel diode by considering the simplified energy-band diagrams in Figure 8–29. Figure 8–29a shows the energy-band diagram at zero bias, which produces zero current on the I-V diagram. If we assume, for simplicity, that we are near $0°K$, then all energy states are filled below $E_F$ on both sides of the junction.

Figure 8–29b shows the situation when a small forward-bias voltage is applied to the junction. Electrons in the conduction band of the n-region are directly opposite to empty states in the valence band of the p-region. There is a finite probability that some of these electrons will tunnel directly into the empty states, producing a forward-bias tunneling current as shown. With a slightly larger forward-bias voltage, as in Figure 8–29c, the maximum number of electrons in the n-region will be opposite the maximum number of empty states in the p-region; this will produce a maximum tunneling current.

As the forward-bias voltage continues to increase, the number of electrons on the n-side directly opposite empty states on the p-side decreases, as in Figure 8–29d, and the tunneling current will decrease. In Figure 8–29e, there are no electrons on the n-side directly opposite available empty states on the p-side. For this forward-bias voltage, the tunneling current will be zero and the normal ideal diffusion current will exist in the device as shown in the I-V characteristics.

The portion of the curve showing a decrease in current with an increase in voltage is the region of differential negative resistance. The range of voltage and current for this region is quite small; thus, any power generated from an oscillator using this negative resistance property would also be fairly small.

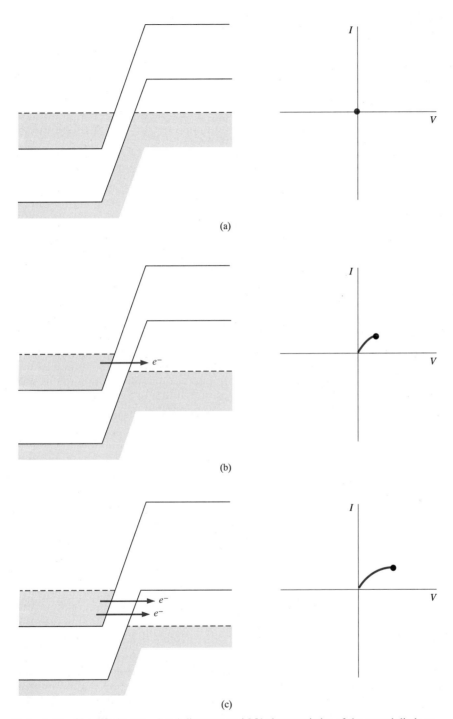

**Figure 8–29**   Simplified energy-band diagrams and I-V characteristics of the tunnel diode at (a) zero bias, (b) a slight forward bias, (c) a forward bias producing maximum tunneling current.

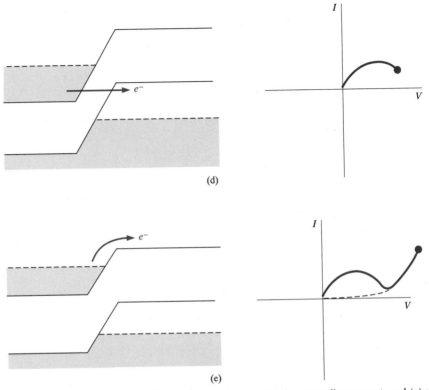

**Figure 8–29** *(concluded)* *(d)* A higher forward bias showing less tunneling current, and *(e)* a forward bias for which the diffusion current dominates.

A simplified energy-band diagram of the tunnel diode with an applied reverse-bias voltage is shown in Figure 8–30a. Electrons in the valence band on the p-side are directly opposite empty states in the conduction band on the n-side, so electrons can now tunnel directly from the p-region into the n-region, resulting in a large reverse-bias tunneling current. This tunneling current will exist for any reverse bias voltage. The reverse-bias current will increase monotonically and rapidly with reverse-bias voltage as shown in Figure 8–30b.

## 8.7 SUMMARY AND REVIEW

In this chapter, we have been dealing with currents in a pn junction. When a forward-bias voltage is applied across a pn junction (p-region positive with respect to the n-region), the potential barrier, which has been maintaining equilibrium, is lowered so that electrons from the n-region and holes from the p-region are injected across the space charge region. Electrons flow into the p-region where they become excess minority carriers and holes flow into the n-region where they too become excess minority carriers. These excess

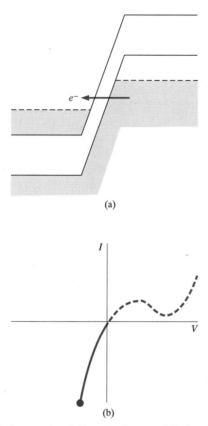

**Figure 8–30**  (a) Simplified energy-band diagram of a tunnel diode with a reverse-bias voltage. (b) I-V characteristic of a tunnel diode with a reverse-bias voltage.

carriers can now diffuse and recombine. The behavior of excess carriers is described by the ambipolar transport equation developed in Chapter 6. We saw that the behavior of excess carriers in an extrinsic semiconductor under low injection is determined by the minority-carrier parameters—for this reason, we are primarily concerned with minority carriers in a pn junction.

We initially derived boundary conditions demonstrating that the minority carrier concentrations at the edges of the space charge region are an exponential function of forward-bias voltage. The ambipolar transport equation was then used to determine the steady-state minority carrier distributions through the pn junction. Gradients in the minority carrier concentrations generate minority carrier diffusion currents. The calculation of these currents, as a function of applied forward-bias voltage, yields the ideal current-voltage relationship of the pn junction diode.

In many applications, a pn junction is used in a linear amplifier circuit in which a small time-varying voltage is superimposed on a dc voltage applied to the diode. We developed the small-signal characteristics of the pn junction in order to determine the relation between the small-signal current and voltage. Two parameters of interest are the small-signal diffusion resistance and diffusion capacitance. The diffusion resistance is the reciprocal of the slope of the dc current-voltage curve, and a function of the dc current. The diffusion capacitance of the forward-biased junction is related to minority carrier charge storage in the n- and p-regions and is usually orders of magnitude larger than the junction capacitance. These parameters determine the frequency response and switching speed of the forward-biased pn junction.

One nonideal effect in a pn junction is the generation and recombination of carriers within the space charge region. When a pn junction is reverse biased, electrons and holes are generated within the space charge region via traps, as we determined from our discussion in Chapter 6 of the Shockley-Read-Hall recombination theory. When these carriers are generated, they are swept out of the space charge region by the electric field. The flow of these charge carriers generates a current, known as the generation current, which is in the reverse-biased direction through the diode. This current is in addition to the ideal reverse-saturation current and, in silicon pn junctions, is normally orders of magnitude larger. When the diode is forward biased, electrons and holes are injected across the space charge region so that there is a finite probability of these carriers recombining within this region. This recombination gives rise to a component of forward-bias current in addition to the ideal forward-bias current. At low forward-bias voltages, the recombination current will normally dominate the diode current, whereas at higher forward-bias voltages, the ideal diffusion currents will normally dominate.

When a sufficiently large reverse-bias voltage is applied to a pn junction, the current will begin to increase drastically. This effect, called breakdown, is usually dominated by the avalanche breakdown process. Electrons and holes, crossing the space charge region, can gain sufficient energy due to the electric field that, when they collide with the lattice, valence-band electrons will be excited into the conduction band thereby creating additional electron-hole pairs. These generated electrons and holes will be swept out of the space charge region by the electric field, producing an additional reverse-bias current. At the breakdown voltage, the concentration of these generated electrons and holes becomes very large; accordingly, the reverse-bias current becomes very large. The breakdown voltage is a strong function of the impurity doping concentration in the junction—as the doping concentrations in the pn junction increase, the breakdown voltage decreases.

As we have seen, when a pn junction is forward-biased, excess minority carriers are created in the p- and n-regions. When the diode is switched to reverse bias, these excess minority carriers must be removed. The excess minority carriers will flow out of the pn junction in the reverse-bias direction, creating a large pulse of reverse-bias current in the diode. The time that it takes to get rid of the excess minority carriers is the diode turn-off time.

The turn-off time is a function both of the minority carrier lifetime and the magnitude of the reverse-bias current.

We have implicitly assumed up to this point that the p- and n-regions were nondegenerately doped. In contrast, a tunnel diode is created when both sides of the junction are degenerately doped. In both forward- and reverse-bias conditions, electrons can tunnel through the potential barrier at the junction producing tunneling currents in addition to the ideal-diode current.

The vast majority of semiconductor devices we will consider in the following chapters contain one or more pn junctions. The electrostatics of the pn junction considered in the last chapter and the current-voltage characteristics of the pn junction developed in this chapter form the basis for establishing the properties of the remaining semiconductor devices we will discuss.

## GLOSSARY OF IMPORTANT TERMS

**Avalanche breakdown:** The process whereby a large reverse-bias pn junction current is created due to the generation of electron-hole pairs by the collision of electrons and/or holes with atomic electrons within the space charge region.

**Carrier injection:** The flow of carriers across the space charge region of a pn junction when a voltage is applied.

**Critical electric field:** The peak electric field in the space charge region at breakdown.

**Diffusion capacitance:** The capacitance of a forward-biased pn junction due to minority carrier storage effects.

**Diffusion conductance:** The ratio of a low-frequency, small-signal sinusoidal current to voltage in a forward-biased pn junction.

**Diffusion resistance:** The inverse of diffusion conductance.

**Forward bias:** The condition in which a positive voltage is applied to the p-region with respect to the n-region of a pn junction so that the potential barrier between the two regions is lowered below the thermal-equilibrium value.

**Generation current:** The reverse-bias pn junction current produced by the thermal generation of electron-hole pairs within the space charge region.

**"Long" diode:** A pn junction diode in which both the neutral p- and n-regions are long compared with the respective minority carrier diffusion lengths.

**Recombination current:** The forward-bias pn junction current produced as a result of the flow of electrons and holes that recombine within the space charge region.

**Reverse-saturation current:** The ideal reverse-bias current in a pn junction.

**"Short" diode:** A pn junction diode in which at least one of the neutral p- or n-regions is short compared to the respective minority carrier diffusion length.

**Storage time:** The time required for the excess minority carrier concentrations at the space charge edge to go from their steady-state values to zero when the diode is switched from forward to reverse bias.

## PROBLEMS

### Section 8.1

1.  (a) Consider an ideal pn junction diode at $T = 300°K$ operating in the forward-bias region. Calculate the change in voltage that will cause a factor of ten increase in current. (b) Repeat part (a) for a factor of 100 increase in current.

2.  An ideal silicon pn junction at $T = 300°K$ is under forward bias. The minority carrier lifetimes are $\tau_{n0} = 10^{-6}$ sec and $\tau_{p0} = 10^{-7}$ sec. The doping on the p-side is $N_a = 10^{16}$ cm$^{-3}$. Plot the ratio of the electron current to the total current crossing the space charge region as the n-region doping varies from $N_d = 5 \times 10^{15}$ to $N_d = 5 \times 10^{17}$ cm$^{-3}$. (Use a log scale for the doping concentration.)

3.  Calculate the applied reverse-bias voltage at which the ideal reverse current in a pn junction diode at $T = 300°K$ reaches 90 percent of its reverse-saturation current value.

4.  An ideal, long silicon pn junction has a uniform cross section and constant doping on both sides of the abrupt junction. The n-region has a resistivity of 1 ohm-cm. The ratio of depletion width in the n-region to that in the p-region is 0.25. The minority carrier lifetimes are $\tau_{n0} = 10^{-7}$ sec and $\tau_{p0} = 10^{-8}$ sec. $T = 300°K$. Determine (a) $N_a$, $N_d$, (b) $V_{bi}$, (c) the density of the minority carriers at the edge of the space charge region for $V_a = 0.525$ volt, and (d) the magnitude of the electron and hole diffusion current densities at the edge of the space charge region for $V_a = 0.525$ volt.

5.  A pn junction diode is made from a material which has $L_n/L_p = 12$ and $\sigma_p/\sigma_n = 8$. Find the ratio of electron to hole current crossing the junction when the diode is forward biased by 0.525 volt. (Assume an ideal diode.)

6.  Consider an ideal silicon pn junction diode with the following parameters: $\tau_{n0} = \tau_{p0} = 0.1 \times 10^{-6}$ sec, $D_n = 25$ cm$^2$/sec, $D_p = 10$ cm$^2$/sec. What must be the ratio of $N_a/N_d$ so that 95 percent of the current in the depletion region is carried by electrons?

7.  An ideal long silicon pn junction diode at $T = 300°K$ has uniform doping in each region. The diode is made from 1 ohm-cm p-type and 0.2 ohm-cm n-type material. Let $\tau_{n0} = 10^{-6}$ sec and $\tau_{p0} = 10^{-8}$

sec. Calculate (a) $V_{bi}$, (b) $n_p(0)$ and $p_n(0)$ at $V_a = 0.585$ volt, and (c) $J_n(0)$ and $J_p(0)$ for $V_a = 0.585$ volt.

8. A silicon step junction has uniform impurity doping concentrations of $N_a = 5 \times 10^{15}$ cm$^{-3}$ and $N_d = 1 \times 10^{15}$ cm$^{-3}$, and a cross-sectional area of $A = 10^{-4}$ cm$^2$. Let $\tau_{n0} = 0.4$ $\mu$sec and $\tau_{p0} = 0.1$ $\mu$sec. Consider the geometry in Figure 8–31. Calculate (a) the ideal reverse-saturation current due to holes, (b) the ideal reverse-saturation current due to electrons, (c) the hole concentration at $x_n$ if $V_a = \frac{1}{2}V_{bi}$, and (d) the electron current at $x = x_n + \frac{1}{2}L_p$ for $V_a = \frac{1}{2}V_{bi}$.

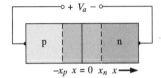

**Figure 8–31** Figure for problem 8.

9. Consider the ideal long silicon pn junction shown in Figure 8–32. $T = 300°$K. The n-region is doped with $10^{16}$ donor atoms per cm$^3$ and the p-region is doped with $5 \times 10^{16}$ acceptor atoms per cm$^3$. The minority carrier lifetimes are $\tau_{n0} = 0.05$ $\mu$sec and $\tau_{p0} = 0.01$ $\mu$sec. The minority carrier diffusion coefficients are $D_n = 23$ cm$^2$/sec and $D_p = 8$ cm$^2$/sec. The forward-bias voltage is $V_a = 0.610$ volt. Calculate (a) the excess hole concentration as a function of $x$ for $x \geq 0$, (b) the hole diffusion current density at $x = 3 \times 10^{-4}$ cm, and (c) the electron current density at $x = 3 \times 10^{-4}$ cm.

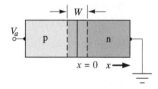

**Figure 8–32** Figure for problem 9.

10. The cross-sectional area of a silicon pn junction is $10^{-3}$ cm$^2$. The temperature of the diode is $T = 300°$K, and the doping concentrations are $N_d = 10^{16}$ cm$^{-3}$ and $N_a = 8 \times 10^{15}$ cm$^{-3}$. Assume minority carrier lifetimes of $\tau_{n0} = 10^{-6}$ sec and $\tau_{p0} = 10^{-7}$ sec. Calculate the total number of excess electrons in the p-region and the total number of excess holes in the n-region for (a) $V_a = 0.3$ volt, (b) $V_a = 0.4$ volt, and (c) $V_a = 0.5$ volt.

11. Consider two ideal pn junctions at $T = 300°K$, having exactly the same electrical and physical parameters except for the bandgap energy of the semiconductor materials. The first pn junction has a bandgap energy of 0.525 eV and a forward-bias current of 10 mA with $V_a = 0.255$ volt. For the second pn junction, "design" the bandgap energy so that a forward-bias voltage of $V_a = 0.32$ volt will produce a current of 10 $\mu$A.

12. Assume that the mobilities, diffusion coefficients, and minority carrier lifetime parameters are independent of temperature (use the $T = 300°K$ values). Assume that $\tau_{n0} = 10^{-6}$ sec, $\tau_{p0} = 10^{-7}$ sec, $N_d = 5 \times 10^{15}$ cm$^{-3}$, and $N_a = 5 \times 10^{16}$ cm$^{-3}$. Plot (on the same graph) the ideal reverse-saturation current density from $T = 200°K$ to $T = 500°K$ for (a) silicon, (b) germanium, and (c) gallium arsenide ideal pn junctions. (Use a log scale for the current density.)

13. An ideal uniformly doped silicon pn junction diode has a cross-sectional area of $10^{-4}$ cm$^2$. The p-region is doped with $5 \times 10^{18}$ acceptor atoms per cm$^3$ and the n-region is doped with $10^{15}$ donor atoms per cm$^3$. Assume that the following parameter values are independent of temperature: $E_g = 1.10$ eV, $\tau_{n0} = \tau_{p0} = 10^{-7}$ sec, $D_n = 25$ cm$^2$/sec, $D_p = 10$ cm$^2$/sec, $N_c = 2.8 \times 10^{19}$ cm$^{-3}$, and $N_v = 1.04 \times 10^{19}$ cm$^{-3}$. The ratio of the forward to reverse current is to be no less than $10^4$ with forward- and reverse-bias voltages of 0.50 volt. Also, the reverse-saturation current is to be no larger than 1 $\mu$A. What is the maximum temperature at which the diode will meet these specifications?

*14. Light is incident on a reverse-biased silicon pn junction at $T = 300°K$, producing a uniform excess carrier generation rate throughout the entire device of $g' = 10^{21}$ cm$^{-3}$ sec$^{-1}$. The donor and acceptor impurity concentrations are $N_d = N_a = 10^{17}$ cm$^{-3}$ and the minority carrier lifetimes are $\tau_{n0} = \tau_{p0} = 5 \times 10^{-7}$ sec. Consider the geometry shown in Figure 8-33. (a) What is the steady-state excess

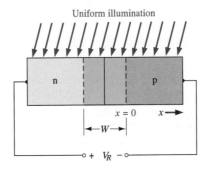

Uniform illumination

**Figure 8–33**   Figure for problem 14.

minority carrier concentration at $x = 0$? (b) What is the steady-state excess minority carrier concentration at $x = \infty$? (c) Derive the expression for the steady-state excess minority carrier concentration for $x \geq 0$.

## Section 8.2

15. Calculate the small-signal ac admittance of a pn junction biased at $V_a = 0.72$ volt and $I_{DQ} = 2.0$ mA. Assume the minority carrier lifetime is 1 $\mu$sec in both the n- and p-regions. $T = 300°K$.

16. (a) Calculate the small-signal admittance of a pn junction biased at $V_a = 0.65$ volt and $I_{DQ} = 1.0$ mA. Assume the minority carrier lifetimes are $\tau_{n0} = \tau_{p0} = 10^{-7}$ sec and let $T = 300°K$. (b) Repeat part (a) if $V_a$ is increased to 0.71 volt and if $I_{DQ}$ is increased to 10 mA.

17. A one-sided $n^+p$ silicon diode at $T = 300°K$ with a cross-sectional area of $10^{-3}$ cm$^2$ is operated under forward bias. The doping levels are $N_d = 10^{18}$ cm$^{-3}$ and $N_a = 10^{16}$ cm$^{-3}$, and the minority carrier parameters are $\tau_{p0} = 10^{-8}$ sec, $\tau_{n0} = 10^{-7}$ sec, $D_p = 10$ cm$^2$/sec, and $D_n = 25$ cm$^2$/sec. The maximum diffusion capacitance is to be 1 pF. Determine (a) the maximum current through the diode, (b) the maximum forward-bias voltage, and (c) the diffusion resistance.

18. A silicon pn junction diode at $T = 300°K$ has a cross-sectional area of $10^{-2}$ cm$^2$. The length of the p-region is 0.2 cm and the length of the n-region is 0.1 cm. The doping concentrations are $N_d = 10^{15}$ cm$^{-3}$ and $N_a = 10^{16}$ cm$^{-3}$. Determine (a) approximately the series resistance of the diode and (b) the current through the diode that will produce a 0.1 volt drop across this series resistance.

19. The small-signal diffusion resistance of an ideal forward-biased silicon pn junction diode at $T = 300°K$ is measured to be $r_d = 48$ ohms. The reverse-saturation current is $I_s = 2 \times 10^{-11}$ A. Calculate the applied forward-bias voltage required to obtain this result.

20. (a) An ideal silicon pn junction diode at $T = 300°K$ is forward biased at $V_a = +20$ mV. The reverse-saturation current is $I_s = 10^{-13}$ A. Calculate the small-signal diffusion resistance. (b) Repeat part (a) for an applied reverse-bias voltage of $V_a = -20$ mV.

## Section 8.3

21. Consider a reverse-biased gallium arsenide pn junction at $T = 300°K$. Assume that a reverse-bias voltage, $V_R = 5$ volts, is applied. Assume parameter values of: $N_a = N_d = 10^{16}$ cm$^{-3}$, $D_p = 6$ cm$^2$/sec, $D_n = 200$ cm$^2$/sec, and $\tau_{p0} = \tau_{n0} = \tau_0 = 10^{-8}$ sec. Calculate the ideal reverse-saturation current density and the reverse-biased generation current density. How does the relative value of these two currents compare to those of the silicon pn junction?

*22.  Consider Example 8–6. Assume that all parameters except $n_i$ are independent of temperature. Determine the temperature at which $J_s$ and $J_{gen}$ will be equal. What are the values of $J_s$ and $J_{gen}$ at this temperature?

23.  Consider a forward-biased silicon pn junction diode at $T = 300°K$. Assume that the ideal reverse-saturation current density is $J_s = 10^{-11}$ A/cm$^2$ and the recombination current density parameter is $J_{r0} = 10^{-7}$ A/cm$^2$. Calculate the forward-bias voltage at which the ideal diffusion current is equal to the recombination current.

*24.  Starting with Equation (8–83) and using the suitable approximations, show that the maximum recombination rate in a forward-biased pn junction is given by Equation (8–91).

25.  Consider, as shown in Figure 8–34, a uniformly doped silicon pn junction at $T = 300°K$ with impurity doping concentrations of $N_a = N_d = 5 \times 10^{15}$ cm$^{-3}$ and minority carrier lifetimes of $\tau_{n0} = \tau_{p0} = \tau_0 = 10^{-7}$ sec. A reverse-bias voltage of $V_R = 10$ volts is applied. A light source is incident only on the space charge region producing an excess carrier generation rate of $g' = 4 \times 10^{19}$ cm$^{-3}$ sec$^{-1}$. Calculate the generation current density.

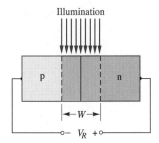

**Figure 8–34**   Figure for problem 25.

## Section 8.4

26.  A one-sided, uniformly doped, planar silicon pn junction at $T = 300°K$ is to be designed so that the minimum breakdown voltage is $V_B = 15$ volts. Using Figures 8–22 and 8–23, determine (a) the maximum impurity doping on the low-doped side, (b) the critical electric field, and (c) the maximum space charge width.

27.  Consider a one-sided abrupt silicon junction at $T = 300°K$. (a) Determine the critical electric field, using Figure 8–22, for doping concentrations of $10^{15}$, $10^{16}$, and $10^{17}$ cm$^{-3}$. (b) Using these values of critical electric field, calculate the breakdown voltage for the same doping concentrations.

28. Consider a one-sided $p^+n$, planar, uniformly doped silicon junction at $T = 300°K$. The p-region doping is $N_a = 10^{18}$ cm$^{-3}$. Breakdown will occur when $E_{max} = 3 \times 10^5$ V/cm and $V_B = 95$ volts. Determine (a) $N_d$, and (b) $x_n$ at breakdown.

29. Consider a silicon pn junction at $T = 300°K$ whose doping profile varies linearly from $N_a = 10^{18}$ cm$^{-3}$ to $N_d = 10^{18}$ cm$^{-3}$ over a distance of 2 $\mu$m. Estimate the breakdown voltage.

## Section 8.5

30. (a) In switching a pn junction from forward to reverse bias, assume that the ratio of reverse current, $I_R$, to forward current, $I_F$, is 0.2. Determine the ratio of storage time to minority carrier lifetime, $t_s/\tau_{p0}$. (b) Repeat part (a) if the ratio of $I_R$ to $I_F$ is 1.0.

31. A pn junction is switched from forward to reverse bias. We want to specify that $t_s = 0.2\,\tau_{p0}$. Determine the required ratio of $I_R$ to $I_F$ to achieve this requirement. In this case, determine $t_2/\tau_{p0}$.

## Section 8.6

32. Consider a silicon pn junction at $T = 300°K$ with doping concentrations of $N_d = N_a = 5 \times 10^{19}$ cm$^{-3}$. Assuming the abrupt junction approximation is valid, determine the space charge width at a forward-bias voltage of $V_a = 0.40$ volt.

33. Sketch the energy-band diagram of an abrupt pn junction under zero bias in which the p-region is degenerately doped and $E_C = E_F$ in the n-region. Sketch the forward- and reverse-bias current-voltage characteristics. This diode is sometimes called a backward diode. Why?

## Summary and Review

34. (a) Explain physically why the diffusion capacitance is not important in a reverse-biased pn junction. (b) Consider a silicon, germanium, gallium arsenide pn junction. If the total current density is the same in each diode under forward bias, discuss the expected relative values of electron and hole current densities.

*35. A silicon pn junction diode at $T = 300°K$ is to be designed to have a reverse-bias breakdown voltage of at least 50 volts and to handle a forward-bias current of $I_D = 100$ mA while still operating under low-injection. The minority carrier diffusion coefficients and lifetimes are $D_n = 25$ cm$^2$/sec, $D_p = 10$ cm$^2$/sec, and $\tau_{n0} = \tau_{p0} = 5 \times 10^{-7}$ sec. The diode is to be designed for minimum cross-sectional area.

*36. The donor and acceptor concentrations on either side of a silicon step junction are equal. (*a*) Derive an expression for the breakdown voltage in terms of the critical electric field and doping concentration. (*b*) If the breakdown voltage is to be $V_B = 50$ volts, specify the range of allowed doping concentrations.

*37. A forward-biased silicon diode is to be used as a temperature sensor. The diode is forward-biased with a constant current source and $V_a$ is measured as a function of temperature. (*a*) Derive an expression for $V_a(T)$ assuming that $D/L$ for electrons and holes, and $E_g$ are independent of temperature. (*b*) If the diode is biased at $I_D = 0.1$ mA and if $I_s = 10^{-15}$ A at $T = 300°$K, plot $V_a$ versus $T$ for $20°C < T < 200°C$. (c) Repeat part (*b*) if $I_D = 1$ mA. (*d*) Determine any changes in the results of parts (*a*) through (*c*) if the change in bandgap energy with temperature is taken into account.

## READING LIST

1. Muller, R. S., and T. I. Kamins. *Device Electronics for Integrated Circuits.* 2nd ed. New York: Wiley, 1986.

2. Navon, D. H. *Semiconductor Microdevices and Materials.* New York: Holt, Rinehart & Winston, 1986.

3. Neudeck, G. W. *The PN Junction Diode.* Vol. 2 of the *Modular Series on Solid State Devices.* 2nd ed. Reading, Mass.: Addison-Wesley, 1989.

*4. Shur, M. *Physics of Semiconductor Devices.* Englewood Cliffs, N.J.: Prentice Hall, 1990.

5. Streetman, B. G. *Solid State Electronic Devices.* 3rd ed. Englewood Cliffs, N.J.: Prentice Hall, 1990.

6. Sze, S. M. *Physics of Semiconductor Devices.* 2nd ed. New York: Wiley, 1981.

7. ———. *Semiconductor Devices: Physics and Technology.* New York: Wiley, 1985.

*8. Wang, S. *Fundamentals of Semiconductor Theory and Device Physics.* Englewood Cliffs, N.J.: Prentice Hall, 1989.

9. Yang, E. S. *Microelectronic Devices.* New York: McGraw-Hill, 1988.

# METAL-SEMICONDUCTOR AND SEMICONDUCTOR HETEROJUNCTIONS

## CHAPTER 9

**PREVIEW**

In the preceding two chapters, we considered the pn junction, and assumed that the semiconductor material was the same throughout the entire structure. This type of junction is referred to as a homojunction. We developed the electrostatics of the junction and derived the current-voltage relationship. In this chapter, we will consider the metal-semiconductor junction and the semiconductor heterojunction, in which the material on each side of the junction is not the same.

A metal-semiconductor rectifying junction, called a Schottky barrier diode, is initially considered. A metal in contact with a low- to moderately-doped semiconductor may form a rectifying contact, the metal taking the place of the $p^+$ region in a $p^+n$ junction. The current mechanism in the Schottky barrier diode, however, is different from that in a pn junction diode. The current in a Schottky diode is due primarily to majority carrier flow; thus there are no minority carrier storage effects. The Schottky diode, then, can switch faster than a pn junction diode. In addition, the effective turn-on voltage of the Schottky diode is smaller than that of a pn junction diode. These two characteristics make the Schottky barrier diode useful in electronic circuit applications, as we will see in later chapters.

Semiconductor devices, or integrated circuits, must make contact with the outside world. This contact is made through nonrectifying metal-semiconductor junctions, or ohmic contacts. An ohmic contact is a low-resistance junction providing current conduction in both directions. We will examine the conditions that yield metal-semiconductor ohmic contacts.

A semiconductor heterojunction is formed when two different semiconductor materials are brought into intimate contact. The "bringing together" of these different semiconductors is not a simple procedure, but involves complicated fabrication processes. Four types of heterojunction can be formed using various combinations of donor and acceptor doping in the wide- and narrow-bandgap materials. Heterojunctions in which the type of dopant changes at the junction are called anisotype and heterojunctions in which the type of dopant is the same on both sides are called isotype. Since different semiconductor materials generally do not have the same energy bandgap, there will be a discontinuity in the energy bands at the interface. Because of this energy-band discontinuity, the current-voltage characteristics of the heterojunction will differ from those of a pn homojunction. The discontinuity can also create a potential well at the interface in which an accumulation layer of electrons or holes may be confined.

Our discussion of the heterojunction in this chapter is a prelude to the study of applications for heterojunctions in later chapters. The potential-well characteristic is used in fabricating the high-mobility junction field-effect transistor that we will consider in Chapter 11. The heterojunction will also be considered in bipolar transistors and in junction LASER diodes.

## 9.1 THE SCHOTTKY BARRIER DIODE

One of the first practical semiconductor devices used in the early 1900s was the metal-semiconductor diode. This diode, also called a point contact diode, was made by touching a metallic whisker to an exposed semiconductor surface. These metal-semiconductor diodes were not easily reproduced or mechanically reliable, and were replaced by the pn junction in the 1950s. However, semiconductor and vacuum technology is now used to fabricate reproducible and reliable metal-semiconductor contacts. In this section, we will consider the metal-semiconductor rectifying contact, or Schottky barrier diode. In most cases, the rectifying contacts are made on n-type semiconductors; for this reason we will concentrate on this type of diode.

### 9.1.1 Qualitative Characteristics

The ideal energy-band diagram for a particular metal and n-type semiconductor before making contact is shown in Figure 9–1a. The vacuum level is used as a reference level. The parameter $\phi_m$ is the metal work function (measured in volts), $\phi_s$ is the semiconductor work function, and $\chi$ is known as the electron affinity. The work functions of various metals are given in Table 9–1 and the electron affinities of several semiconductors are given in Table 9–2. In Figure 9–1a, we have assumed that $\phi_m > \phi_s$. The ideal thermal-equilibrium metal-semiconductor energy band diagram, for this situation, is shown in Figure 9–1b. Before contact, the Fermi level in the semiconductor was above that in the metal. In order for the Fermi level to become a constant through the system in thermal equilibrium, electrons from the semiconductor flow into the lower energy states in the metal. Positively charged donor atoms remain in the semiconductor creating a space charge region.

The parameter $\phi_{Bn}$ is the barrier height of the semiconductor contact, the potential barrier seen by electrons in the metal trying to move into the

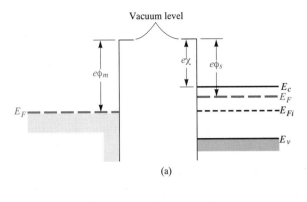

(a)

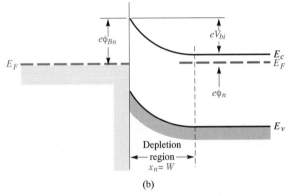

(b)

**Figure 9-1** (a) Energy-band diagram of a metal and semiconductor before contact. (b) Ideal energy-band diagram of a metal-n-semiconductor junction for $\phi_m > \phi_s$.

**Table 9-1  Work functions of some elements**

| Element | Work function, $\phi_m$ |
| --- | --- |
| Ag, silver | 4.26 |
| Al, aluminum | 4.28 |
| Au, gold | 5.1 |
| Cr, chromium | 4.5 |
| Mo, molybdenum | 4.6 |
| Ni, nickel | 5.15 |
| Pd, palladium | 5.12 |
| Pt, platinum | 5.65 |
| Ti, titanium | 4.33 |
| W, tungsten | 4.55 |

**Table 9-2  Electron affinity of some semiconductors**

| Element | Electron affinity, $\chi$ |
| --- | --- |
| Ge, germanium | 4.13 |
| Si, silicon | 4.01 |
| GaAs, gallium arsenide | 4.07 |
| AlAs, aluminum arsenide | 3.5 |

semiconductor. This barrier is known as the Schottky barrier and is given, ideally, by

$$\phi_{Bn} = (\phi_m - \chi) \tag{9-1}$$

On the semiconductor side, $V_{bi}$ is the built-in potential barrier. This barrier, similar to the case of the pn junction, is the barrier seen by electrons in the conduction band trying to move into the metal. The built-in potential barrier is given by

$$V_{bi} = \phi_{Bn} - \phi_n \tag{9-2}$$

which makes $V_{bi}$ a slight function of the semiconductor doping, as was the case in a pn junction.

If we apply a positive voltage to the semiconductor with respect to the metal, the semiconductor-to-metal barrier height increases, while $\phi_{Bn}$ remains constant in this idealized case. This bias condition is the reverse bias. If a positive voltage is applied to the metal with respect to the semiconductor, the semiconductor-to-metal barrier $V_{bi}$ is reduced while $\phi_{Bn}$ again remains essentially constant. In this situation, electrons can more easily flow from the semiconductor into the metal since the barrier has been reduced. This bias condition is the forward bias. The energy-band diagrams for the reverse and forward bias are shown in Figures 9–2a and 9–2b, where $V_R$ is the magnitude of the reverse-bias voltage and $V_a$ is the magnitude of the forward-bias voltage.

The energy-band diagrams versus voltage for the metal-semiconductor junction shown in Figure 9–2 are very similar to those of the pn junction given in the last chapter. Because of the similarity, we expect the current-voltage characteristics of the Schottky barrier junction to be similar to the exponential behavior of the pn junction diode. The current mechanism here, however, is due to the flow of majority carrier electrons. In forward bias, the barrier seen by the electrons in the semiconductor is reduced so majority carrier electrons flow more easily from the semiconductor into the metal. The forward-bias current is in the direction from metal to semiconductor; it is an exponential function of the forward-bias voltage $V_a$.

### 9.1.2 Ideal Junction Properties

We can determine the electrostatic properties of the junction in the same way as we did for the pn junction. The electric field in the space charge region is determined from Poisson's equation. We have that

$$\frac{dE}{dx} = \frac{\rho(x)}{\varepsilon_s} \tag{9-3}$$

where $\rho(x)$ is the space charge volume density and $\varepsilon_s$ is the permittivity of the semiconductor. If we assume thát the semiconductor doping is uniform, then by integrating Equation (9–3), we obtain

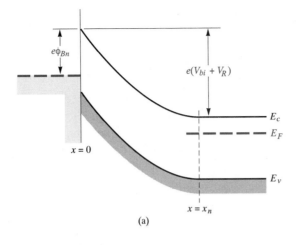

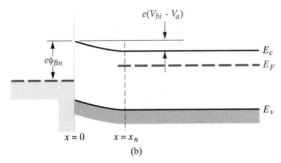

**Figure 9–2** Ideal energy-band diagram of a metal-semiconductor junction (a) under reverse bias and (b) under forward bias.

$$E = \int \frac{eN_d}{\varepsilon_s} \, dx = \frac{eN_d x}{\varepsilon_s} + C_1 \qquad (9\text{-}4)$$

where $C_1$ is a constant of integration. The electric field is zero at the space charge edge in the semiconductor, so the constant of integration can be found as

$$C_1 = -\frac{eN_d x_n}{\varepsilon_s} \qquad (9\text{-}5)$$

The electric field can then be written as

$$E = -\frac{eN_d}{\varepsilon_s} (x_n - x) \qquad (9\text{-}6)$$

which is a linear function of distance, for the uniformly doped semiconductor, and reaches a peak value at the metal-semiconductor interface. Since the E-field is zero inside the metal, a negative surface charge must exist in the metal at the metal-semiconductor junction.

The space charge region width, $W$, may be calculated as we did for the pn junction. The result is identical to that of a one-sided $p^+n$ junction. For the uniformly doped semiconductor, we have

$$W = x_n = \left[ \frac{2\varepsilon_s \, (V_{bi} + V_R)}{eN_d} \right]^{1/2} \tag{9-7}$$

where $V_R$ is the magnitude of the applied reverse-bias voltage. We are again assuming an abrupt junction approximation.

---

### Example 9-1

**Objective:** To calculate the theoretical barrier height, built-in potential barrier, and maximum electric field in a metal-semiconductor diode for zero applied bias.

Consider a contact between tungsten and n-type silicon doped to $N_d = 10^{16}$ cm$^{-3}$ at $T = 300°K$.

**Solution:** The metal work function for tungsten (W) from Table 9–1 is $\phi_m = 4.55$ volts and the electron affinity for silicon from Table 9–2 is $\chi = 4.01$ volts. The barrier height is then

$$\phi_{B0} = \phi_m - \chi = 4.55 - 4.01 = 0.54 \text{ volt}$$

where $\phi_{B0}$ is the ideal Schottky barrier height. We can calculate $\phi_n$ as

$$\phi_n = \frac{kT}{e} \ln \left( \frac{N_c}{N_d} \right) = 0.0259 \ln \left( \frac{2.8 \times 10^{19}}{10^{16}} \right) = 0.206 \text{ volt}$$

Then

$$V_{bi} = \phi_{B0} - \phi_n = 0.54 - 0.206 = 0.33 \text{ volt}$$

The space charge width at zero bias is

$$x_n = \left[ \frac{2\varepsilon_s V_{bi}}{eN_d} \right]^{1/2} = \left[ \frac{2(11.7)(8.85 \times 10^{-14})(0.33)}{(1.6 \times 10^{-19})(10^{16})} \right]^{1/2}$$

or

$$x_n = 0.207 \times 10^{-4} \text{ cm}$$

Then the maximum electric field is

$$|\mathrm{E_{max}}| = \frac{eN_d x_n}{\varepsilon_s} = \frac{(1.6 \times 10^{-19})(10^{16})(0.207 \times 10^{-4})}{(11.7)(8.85 \times 10^{-14})}$$

or finally

$$|\mathrm{E_{max}}| = 3.2 \times 10^4 \text{ V/cm}$$

**Comment:** The values of space charge width and electric field are very similar to those obtained for a pn junction.

A junction capacitance can also be determined in the same way as we did for the pn junction. We have that

$$C' = eN_d \frac{dx_n}{dV_R} = \left[\frac{e\varepsilon_s N_d}{2(V_{bi} + V_R)}\right]^{1/2} \tag{9-8}$$

where $C'$ is the capacitance per unit area. If we square the reciprocal of Equation (9-8), we obtain

$$\left(\frac{1}{C'}\right)^2 = \frac{2(V_{bi} + V_R)}{e\varepsilon_s N_d} \tag{9-9}$$

We can use Equation (9-9) to obtain, to a first approximation, the built-in potential barrier $V_{bi}$, and the slope of the curve from Equation (9-9) to yield the semiconductor doping $N_d$. We can calculate the potential $\phi_n$, and then determine the Schottky barrier $\phi_{Bn}$ from Equation (9-2).

---

**Example 9-2**

**Objective:** To calculate the semiconductor doping and Schottky barrier height from the silicon diode experimental data shown in Figure 9-3. $T = 300°K$.

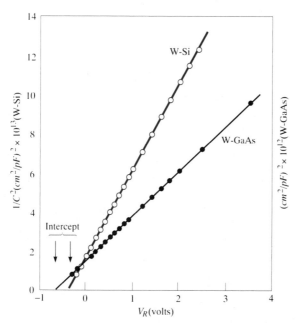

**Figure 9-3**    $1/C^2$ versus $V_R$ for W-Si and W-GaAs Schottky barrier diodes. (From Sze [9].)

**Solution:** The intercept of the tungsten-silicon curve is approximately at $V_{bi} = 0.40$ volt. From Equation (9–9), we can write

$$\frac{d(1/C')^2}{dV_R} \cong \frac{\Delta(1/C')^2}{\Delta V_R} = \frac{2}{e\varepsilon_s N_d}$$

Then, from the figure, we have

$$\frac{\Delta(1/C')^2}{\Delta V_R} \approx 4.4 \times 10^{13}$$

so that

$$N_d = \frac{2}{(1.6 \times 10^{-19})(11.7)(8.85 \times 10^{-14})(4.4 \times 10^{13})} = 2.7 \times 10^{17} \text{ cm}^{-3}$$

We can calculate

$$\phi_n = \frac{kT}{e} \ln\left(\frac{N_c}{N_d}\right) = (0.0259) \ln\left(\frac{2.8 \times 10^{19}}{2.7 \times 10^{17}}\right) = 0.12 \text{ volt}$$

so that

$$\phi_{Bn} = V_{bi} + \phi_n = 0.40 + 0.12 = 0.52 \text{ volt}$$

**Comment:** The experimental value of 0.52 volt can be compared with the ideal barrier height of 0.54 volt found in Example 9–1. These results agree fairly well. For other metals, the discrepancy between experiment and theory is larger.

---

We can see that the built-in potential barrier of the gallium arsenide Schottky diode is larger than that of the silicon diode. This experimental result is normally observed for all types of metal contacts.

### *9.1.3 Nonideal Effects on the Barrier Height

Several effects will alter the actual Schottky barrier height from the theoretical value given by Equation (9–1). The first effect we will consider is the Schottky effect, or image-force-induced lowering of the potential barrier.

An electron in a dielectric at a distance $x$ from the metal will create an electric field. The field lines must be perpendicular to the metal surface and will be the same as if an image charge, $+e$, is located at the same distance from the metal surface, but inside the metal. This image effect is shown in Figure 9–4a. The force on the electron, due to the coulomb attraction with the image charge, is

$$F = \frac{-e^2}{4\pi\varepsilon_s(2x)^2} = -e\text{E} \tag{9–10}$$

The potential can then be found as

$$-\phi(x) = +\int_x^\infty \text{E}\,dx' = +\int_x^\infty \frac{e}{4\pi\varepsilon_s \cdot 4(x')^2}\,dx' = \frac{-e}{16\pi\varepsilon_s x} \tag{9–11}$$

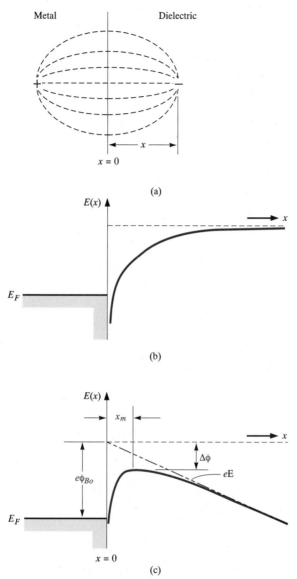

**Figure 9-4** (a) Image charge and electric field lines at a metal-dielectric interface. Distortion of the potential barrier due to image forces (b) with zero electric field and (c) with a constant electric field.

where $x'$ is the integration variable and where we have assumed that the potential is zero at $x = \infty$.

The potential energy of the electron is $-e\phi(x)$; Figure 9-4b is a plot of the potential energy assuming that no other electric fields exist. With an electric field present in the dielectric, the potential is modified and can be written as

$$-\phi(x) = \frac{-e}{16\pi\varepsilon_s x} - Ex \tag{9-12}$$

The potential energy of the electron, including the effect of a constant electric field, is plotted in Figure 9–4c. The peak potential barrier is now lowered. This lowering of the potential barrier is the Schottky effect, or image-force-induced lowering.

We can find the Schottky barrier lowering, $\Delta\phi$, and the position of the maximum barrier, $x_m$, from the condition that

$$\frac{d(e\phi(x))}{dx} = 0 \tag{9-13}$$

We find that

$$x_m = \sqrt{\frac{e}{16\pi\varepsilon_s E}} \tag{9-14}$$

and

$$\Delta\phi = \sqrt{\frac{eE}{4\pi\varepsilon_s}} \tag{9-15}$$

---

**Example 9–3**

**Objective:** To calculate the Schottky barrier lowering and the position of the maximum barrier height.

Consider a gallium arsenide metal-semiconductor contact in which the electric field in the semiconductor is assumed to be $E = 6.8 \times 10^4$ V/cm.

**Solution:** The Schottky barrier lowering is given by Equation (9–15) which then yields

$$\Delta\phi = \sqrt{\frac{eE}{4\pi\varepsilon_s}} = \sqrt{\frac{(1.6 \times 10^{-19})(6.8 \times 10^{-4})}{4\pi(13.1)(8.85 \times 10^{-14})}} = 0.0273 \text{ volt}$$

The position of the maximum barrier height is

$$x_m = \sqrt{\frac{e}{16\pi\varepsilon_s E}} = \sqrt{\frac{(1.6 \times 10^{-19})}{16\pi(13.1)(8.85 \times 10^{-14})(6.8 \times 10^4)}}$$

or

$$x_m = 2 \times 10^{-7} \text{ cm} = 20\text{Å}$$

**Comment:** Although the Schottky barrier lowering may seem like a small value, the barrier height and the barrier lowering will appear in exponential terms in the current-voltage relationship. A small change in the barrier height can then have a significant effect on the current in a Schottky barrier diode.

---

Figure 9–5 shows the measured barrier heights in gallium arsenide and silicon Schottky diodes as a function of metal work functions. There is a

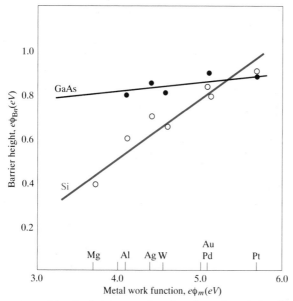

**Figure 9–5** Experimental barrier heights as a function of metal work functions for GaAs and Si. (From Pulfrey and Tarr [3].)

monotonic relation between the measured barrier height and the metal work function, but the curves do not fit the simple relation given in Equation (9–1). The barrier height of the metal-semiconductor junction is determined by both the metal work function and the semiconductor surface or interface states.

A more detailed energy-band diagram of a metal to n-type semiconductor contact in thermal equilibrium is shown in Figure 9–6. We will assume that a narrow interfacial layer of insulator exists between the metal and semicon-

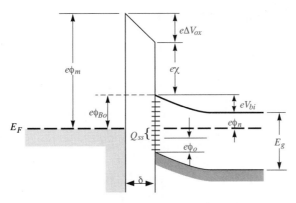

**Figure 9–6** Energy-band diagram of a metal-semiconductor junction with an interfacial layer and interface states.

ductor. The interfacial layer can support a potential difference, but will be transparent to the flow of electrons between the metal and semiconductor. The semiconductor also shows a distribution of surface states at the metal-semiconductor interface. We will assume that all states below the surface potential $\phi_o$ are donor states, which will be neutral if the state contains an electron and positively charged if the state does not contain an electron. We will also assume that all states above $\phi_o$ are acceptor states, which will be neutral if the state does not contain an electron and negatively charged if the state contains an electron.

The diagram in Figure 9–6 shows some acceptor states above $\phi_o$ and below $E_F$. These states will tend to contain electrons and will be negatively charged. If we assume that the surface state density is a constant and equal to $D_{it}$ states/cm²-eV, then the amount of surface charge density, $Q_{ss}$, can be expressed as

$$Q_{ss} = -eD_{it}(E_g - e\phi_o - e\phi_{Bo}) \text{ coul/cm}^2 \tag{9–16}$$

The space charge density, expressed as an equivalent surface charge density, can be written as

$$Q_{sc} = +eN_d x_n \text{ coul/cm}^2 \tag{9–17}$$

where $x_n$ is the space charge width. Using Equations (9–2) and (9–7) for zero applied bias, we have

$$Q_{sc} = +\sqrt{2e\varepsilon_s N_d(\phi_{Bo} - \phi_n)} \tag{9–18}$$

The potential, $\Delta V_{ox}$, across the interfacial layer can be found by applying Gauss's law to the surface charge on the metal. If $Q_M$ is the surface charge density on the metal and if $D_i$ is the electric flux density in the interfacial layer, then

$$Q_M = D_i = \varepsilon_i E_i = \varepsilon_i \left(\frac{-\Delta V_{ox}}{\delta}\right) \tag{9–19}$$

where $\varepsilon_i$ is the permittivity of the interfacial layer and $E_i$ is the electric field in the interfacial layer. With a positive gradient in the potential, $\Delta V_{ox}$, as shown in Figure 9–6, the electric field must be in the negative x-direction. From charge neutrality, we have

$$Q_M + Q_{ss} + Q_{sc} = 0 \tag{9–20}$$

The potential $\Delta V_{ox}$ can be expressed, from Equation (9–19), as

$$\Delta V_{ox} = \frac{-\delta}{\varepsilon_i} Q_M = \frac{\delta}{\varepsilon_i} [+(Q_{ss} + Q_{sc})] \tag{9–21}$$

We may also find an expression for $\Delta V_{ox}$ from the various barrier heights. Referring to Figure 9–6, we can write

$$e\phi_m = e\Delta V_{ox} + e\chi + e\phi_{Bo} \tag{9–22a}$$

or
$$\Delta V_{ox} = \phi_m - (\chi + \phi_{Bo}) \qquad (9\text{-}22b)$$

Eliminating $\Delta V_{ox}$ from Equations (9-22b) and (9-21), we have

$$\phi_m - (\chi + \phi_{Bo}) = \frac{\delta}{\varepsilon_i} (Q_{ss} + Q_{sc}) \qquad (9\text{-}23)$$

Using the relations for $Q_{ss}$ and $Q_{sc}$ from Equations (9-16) and (9-18), Equation (9-23) becomes

$$\phi_m - (\chi + \phi_{Bo}) = \frac{-eD_{it}\delta}{\varepsilon_i} (E_g - e\phi_o - e\phi_{Bo})$$

$$+ \sqrt{\frac{2e\varepsilon_s N_d \delta^2}{\varepsilon_i^2} (\phi_{Bo} - \phi_n)} \qquad (9\text{-}24)$$

If we multiply through by $\varepsilon_i / eD_{it}\delta$ and rearrange terms, we have

$$(E_g - e\phi_o - e\phi_{Bo}) = \frac{1}{eD_{it}} \sqrt{2e\varepsilon_s N_d(\phi_{Bo} - \phi_n)}$$

$$- \frac{\varepsilon_i}{eD_{it}\delta} [\phi_m - (\chi + \phi_{Bo})] \qquad (9\text{-}25)$$

We may consider several cases.

**Case 1**   Let $D_{it} \rightarrow \infty$. In this case, the right side of Equation (9-25) goes to zero. We then have

$$\phi_{Bo} = \frac{1}{e} (E_g - e\phi_o) \qquad (9\text{-}26)$$

The barrier height is now fixed by the bandgap energy and the potential $\phi_o$. The barrier height is totally independent of the metal work function and the semiconductor electron affinity. The Fermi level becomes "pinned" at the surface, at the surface potential $\phi_o$.

**Case 2**   Let $D_{it} \rightarrow \infty$ and $\delta \rightarrow 0$, but let the product $D_{it}\delta$ remain finite. Equation (9-25) now becomes

$$(E_g - e\phi_o - e\phi_{Bo}) = \frac{-\varepsilon_i}{eD_{it}\delta} [\phi_m - (\chi + \phi_{Bo})] \qquad (9\text{-}27)$$

If we define a variable as

$$\alpha^2 = \frac{e^2 D_{it}\delta}{\varepsilon_i} \qquad (9\text{-}28)$$

then, solving for $\phi_{Bo}$ from Equation (9-27), we find

$$\phi_{Bo} = \frac{(\phi_m - \chi) + \alpha^2 \left(\dfrac{E_g}{e} - \phi_o\right)}{1 + \alpha^2} \qquad (9\text{-}29)$$

In this case, the barrier height is still a monotonic function of the metal work function and semiconductor electron affinity, but the relationship is obviously modified from the original ideal expression.

**Case 3**   Let $D_{it}\delta \to 0$. Then $\alpha^2 \to 0$ and Equation (9–29) reduces to

$$\phi_{Bo} = (\phi_m - \chi)$$

which is the original ideal expression.

The Schottky barrier height is a function of the electric field in the semiconductor through the barrier lowering effect. The barrier height is also a function of the surface states in the semiconductor. The barrier height, then, is modified from the ideal theoretical value. Since the surface state density is not predictable with any degree of certainty, the barrier height must be an experimentally determined parameter.

### *9.1.4 Current-Voltage—Mathematical Derivation

The current transport in a metal-semiconductor junction is due mainly to majority carriers as opposed to minority carriers in a pn junction. The basic process in the rectifying contact with an n-type semiconductor is by transport of electrons over the potential barrier which can be described by the thermionic emission theory.

The thermionic emission characteristics are derived using the assumptions that the barrier height is much larger than $kT$, so that the Maxwell-Boltzmann approximation applies, and that thermal equilibrium is not affected by this process. Figure 9–7 shows the one-dimensional barrier with an applied forward-bias voltage $V_a$ and shows two electron current density components. The current $J_{s \to m}$ is the electron current density due to the flow of electrons from the semiconductor into the metal and the current $J_{m \to s}$ is the electron current density due to the flow of electrons from the metal into the semiconductor. The subscripts of the currents indicate the direction of electron flow. The conventional current direction is opposite to electron flow.

The current density $J_{s \to m}$ is a function of the concentration of electrons which have x-directed velocities sufficient to overcome the barrier. We may write

$$J_{s \to m} = e \int_{E_c'}^{\infty} v_x \, dn \tag{9–30}$$

where $E_c'$ is the minimum energy required for thermionic emission into the metal, $v_x$ is the carrier velocity in the direction of transport, and $e$ is the magnitude of the electronic charge. The incremental electron concentration is given by

$$dn = g_c(E)f_F(E) \, dE \tag{9–31}$$

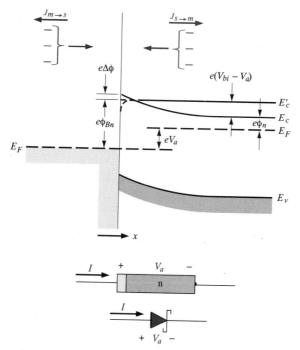

**Figure 9–7** Energy-band diagram of a forward-biased metal-semiconductor junction including the image lowering effect.

where $g_c(E)$ is the density of states in the conduction band and $f_F(E)$ is the Fermi-Dirac probability function. Assuming that the Maxwell-Boltzmann approximation applies, we may write

$$dn = \frac{4\pi(2m_n^*)^{3/2}}{h^3} \sqrt{E - E_c} \exp\left[\frac{-(E - E_F)}{kT}\right] dE \qquad (9\text{–}32)$$

If all of the electron energy above $E_c$ is assumed to be kinetic energy, then we have

$$\frac{1}{2} m_n^* v^2 = E - E_c \qquad (9\text{–}33a)$$

$$dE = m_n^* v \, dv \qquad (9\text{–}33b)$$

and

$$\sqrt{E - E_c} = v \sqrt{\frac{m_n^*}{2}} \qquad (9\text{–}33c)$$

Equation (9–32) can then be written in terms of velocity as

$$dn = 2 \left(\frac{m_n^*}{h}\right)^3 \exp\left(\frac{-e\phi_n}{kT}\right) \exp\left(\frac{-m_n^* v^2}{2kT}\right) \cdot 4\pi v^2 \, dv \qquad (9\text{–}34)$$

Equation (9–34) gives the number of electrons per unit volume that have speeds between $v$ and $v + dv$ distributed over all directions. We may decompose the speed into its components. We have

$$v^2 = v_x^2 + v_y^2 + v_z^2$$

We can also transform the differential volume $4\pi v^2\, dv$ into the differential $dv_x\, dv_y\, dv_z$.

We can then write Equation (9–30) as

$$J_{s \to m} = 2e \left(\frac{m_n^*}{h}\right)^3 \exp\left(\frac{-e\phi_n}{kT}\right) \int_{v_{ox}}^\infty v_x \exp\left(\frac{-m_n^* v_x^2}{2kT}\right) dv_x$$

$$\times \int_{-\infty}^\infty \exp\left(\frac{-m_n^* v_y^2}{2kT}\right) dv_y \times \int_{-\infty}^\infty \exp\left(\frac{-m_n^* v_z^2}{2kT}\right) dv_z \qquad (9\text{–}35)$$

The velocity $v_{ox}$ is the minimum velocity required in the $x$ direction to go over the potential barrier. We may make a change of variable in Equation (9–35) with the following definitions:

$$\frac{m_n^* v_x^2}{2kT} \equiv \alpha^2 + \frac{e(V_{bi} - V_a)}{kT} \qquad (9\text{–}36a)$$

$$\frac{m_n^* v_y^2}{2kT} \equiv \beta^2 \qquad (9\text{–}36b)$$

$$\frac{m_n^* v_z^2}{2kT} \equiv \gamma^2 \qquad (9\text{–}36c)$$

The minimum $x$-directed velocity to overcome the barrier is $v_{ox}$, which corresponds to

$$\frac{1}{2} m_n^* v_{ox}^2 = e(V_{bi} - V_a) \qquad (9\text{–}37)$$

This condition implies that the lower limit of integration in the variable $\alpha$ becomes

$$v_x \to v_{ox} \Rightarrow \alpha = 0 \qquad (9\text{–}38a)$$

We can also show that

$$v_x\, dv_x = \left(\frac{2kT}{m_n^*}\right) \alpha\, d\alpha \qquad (9\text{–}38b)$$

Then, Equation (9–35) can be written as

$$J_{s \to m} = 2e \left(\frac{m_n^*}{h}\right)^3 \left(\frac{2kT}{m_n^*}\right)^2 \exp\left(\frac{-e\phi_n}{kT}\right) \exp\left[\frac{-e(V_{bi} - V_a)}{kT}\right]$$

$$\times \int_0^\infty \alpha \exp\left(-\alpha^2\right) d\alpha \int_{-\infty}^\infty \exp\left(-\beta^2\right) d\beta \int_{-\infty}^\infty \exp\left(-\gamma^2\right) d\gamma \qquad (9\text{–}39)$$

After performing the integration, Equation (9–39) finally becomes

$$J_{s \to m} = \left(\frac{4\pi e m_n^* k^2}{h^3}\right) T^2 \exp\left[\frac{-e(\phi_n + V_{bi})}{kT}\right] \exp\left(\frac{eV_a}{kT}\right) \qquad (9\text{–}40a)$$

or

$$J_{s \to m} = \left(\frac{4\pi e m_n^* k^2}{h^3}\right) T^2 \exp\left[\frac{-e\phi_{Bn}}{kT}\right] \exp\left(\frac{eV_a}{kT}\right) \qquad (9\text{–}40b)$$

The electron current density from the metal to the semiconductor, $J_{m \to s}$, is exactly the same as $J_{s \to m}$ when the applied bias is zero. Then

$$J_{m \to s} = \left(\frac{4\pi e m_n^* k^2}{h^3}\right) T^2 \exp\left(\frac{-e\phi_{Bn}}{kT}\right) \qquad (9\text{–}41)$$

The net current density in the metal-to-semiconductor junction can be written as

$$J = J_{s \to m} - J_{m \to s} \qquad (9\text{–}42)$$

which is defined to be positive in the direction from the metal to the semiconductor. We then have

$$J = \left[A^* T^2 \exp\left(\frac{-e\phi_{Bn}}{kT}\right)\right]\left[\exp\left(\frac{eV_a}{kT}\right) - 1\right] \qquad (9\text{–}43)$$

where

$$A^* \equiv \frac{4\pi e m_n^* k^2}{h^3} \qquad (9\text{–}44)$$

The parameter $A^*$ is called the effective Richardson constant for thermionic emission.

Equation (9–43) can be written in the usual diode form as

$$J = J_{sT}\left[\exp\left(\frac{eV_a}{kT}\right) - 1\right] \qquad (9\text{–}45)$$

where $J_{sT}$ is the reverse-saturation current density and is given by

$$J_{sT} = A^* T^2 \exp\left(\frac{-e\phi_{Bn}}{kT}\right) \qquad (9\text{–}46)$$

We may recall that the Schottky barrier height $\phi_{Bn}$ changes due to the image force lowering. We have that $\phi_{Bn} = \phi_{B0} - \Delta\phi$. Then Equation (9–46) can be written as

$$J_{sT} = A^* T^2 \exp\left(\frac{-e\phi_{Bo}}{kT}\right) \exp\left(\frac{e\Delta\phi}{kT}\right) \qquad (9\text{–}47)$$

The change in barrier height, $\Delta\phi$, will increase with an increase in the electric field, or with an increase in the applied reverse-bias voltage. Figure 9–8

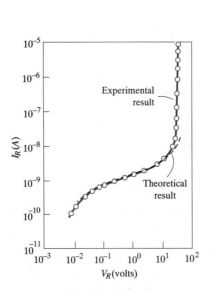

**Figure 9–8**  Experimental and theoretical reverse-bias currents in a PtSi-Si diode. (From Sze [9].)

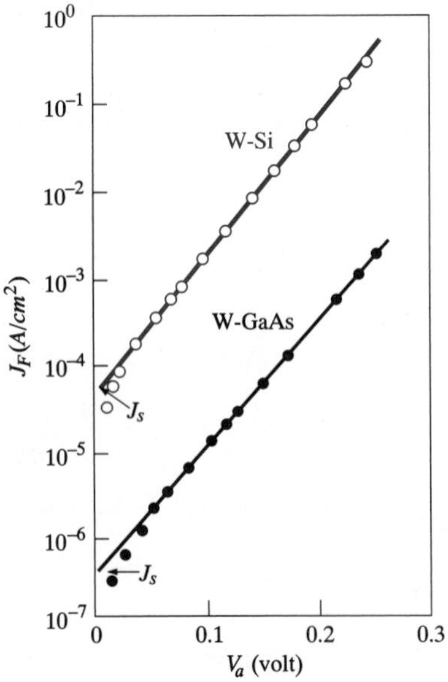

**Figure 9–9**  Forward-bias current density $J_F$ versus $V_a$ for W-Si and W-GaAs diodes. (From Sze [9].)

shows a typical reverse-bias current-voltage characteristic of a Schottky barrier diode. The reverse-bias current increases with reverse-bias voltage due to the barrier lowering effect. This figure also shows the Schottky barrier diode going into breakdown.

### 9.1.5 Current-Voltage—A Summary

The ideal current-voltage characteristics of a Schottky barrier diode, which is based on the thermionic emission of carriers over the barrier, was given by Equation (9–45) as

$$J = J_{sT} \left[ \exp \left( \frac{eV_a}{kT} \right) - 1 \right]$$

The form of the current-voltage relationship is the same as that of the pn junction. The current mechanism in a Schottky barrier diode is the flow of majority carriers rather than the flow of minority carriers as in a pn junction diode. The current $J_{sT}$ is equivalent to the reverse-saturation current density in a pn junction. We had, from Equation (9–46), that

$$J_{sT} = A^* T^2 \exp \left( \frac{-e\phi_{Bn}}{kT} \right)$$

where $\phi_{Bn}$ is the Schottky barrier height and $A^*$ is referred to as the Richardson constant.

The forward-bias current-voltage characteristics of two Schottky diodes are shown in Figure 9–9. When the forward bias voltage, $V_a$, is greater than approximately $4(kT/e)$, we can neglect the $(-1)$ term in the diode equation. We then have

$$J \cong A^*T^2 \exp\left(\frac{-e\phi_{Bn}}{kT}\right) \exp\left(\frac{eV_a}{kT}\right) \qquad (9\text{–}48)$$

The $\ln J$ versus $V_a$ curve is a straight line as shown in the figure. Figure 9–9 can be used to experimentally determine $\phi_{Bn}$ if $A^*$ is known, or to determine $A^*$ if $\phi_{Bn}$ is known.

---

**Example 9–4**

**Objective:** To calculate the effective Richardson constant from the I-V characteristics.

Consider the tungsten-silicon diode curve in Figure 9–9 and assume a barrier height of $\phi_{Bn} = 0.67$ volt. From the figure, $J_{sT} \approx 6 \times 10^{-5}$ A/cm$^2$.

**Solution:** We have that

$$J_{sT} = A^*T^2 \exp\left(\frac{-e\phi_{Bn}}{kT}\right)$$

so that

$$A^* = \frac{J_{sT}}{T^2} \exp\left(\frac{+e\phi_{Bn}}{kT}\right)$$

Then $\qquad A^* = \dfrac{6 \times 10^{-5}}{(300)^2} \exp\left(\dfrac{0.67}{0.0259}\right) = 114$ A/°K$^2$-cm$^2$

**Comment:** The experimentally determined value of $A^*$ is a very strong function of $\phi_{Bn}$, since $\phi_{Bn}$ is in the exponential term. A small change in $\phi_{Bn}$ will change the value of the Richardson constant substantially.

---

We may note that the reverse-saturation current densities of the tungsten-silicon and tungsten-gallium arsenide diodes in Figure 9–9 differ by approximately two orders of magnitude. This two-orders-of-magnitude difference will be reflected in the effective Richardson constant assuming the barrier heights in the two diodes are essentially the same. The definition of the effective Richardson constant, given by Equation (9–44), contains the electron effective mass which differs substantially between silicon and gallium arsenide. The fact that the effective mass is in the expression for the Richardson constant is a direct result of using the effective density of states function in the thermionic emission theory. The net result is that $A^*$ and $J_{sT}$ will vary widely between silicon and gallium arsenide.

### 9.1.6 Comparison of the Schottky Barrier Diode and the pn Junction Diode

Although the ideal current-voltage relationship of the Schottky barrier diode given by Equation (9–45) is of the same form as that of the pn junction diode, there are two important differences between a Schottky diode and a pn junction diode: the first is in the magnitudes of the reverse-saturation current densities and the second is in the switching characteristics.

The reverse-saturation current density of the Schottky barrier diode was given by Equation (9–46) and is

$$J_{sT} = A^*T^2 \exp\left(\frac{-e\phi_{Bn}}{kT}\right)$$

The ideal reverse-saturation current density of the pn junction diode can be written as

$$J_s = \frac{eD_n n_{p0}}{L_n} + \frac{eD_p p_{n0}}{L_p} \tag{9–49}$$

The form of the two equations is vastly different and the current mechanism in the two devices is different. The current in a pn junction is determined by the diffusion of minority carriers while the current in a Schottky barrier diode is determined by thermionic emission of majority carriers over a potential barrier.

---

**Example 9–5**

**Objective:** To calculate the reverse-saturation current densities of a Schottky barrier diode and a pn junction diode.

Consider a tungsten barrier on silicon with a measured barrier height of $e\phi_{Bn} = 0.67 \text{ eV}$. The effective Richardson constant is $A^* = 114 \text{ A}/^\circ\text{K}^2 - \text{cm}^2$. Let $T = 300^\circ\text{K}$.

**Solution:** If we neglect the barrier lowering effect, we have for the Schottky barrier diode

$$J_{sT} = A^*T^2 \exp\left(\frac{-e\phi_{Bn}}{kT}\right) = (114)(300)^2 \exp\left(\frac{-0.67}{0.0259}\right) = 5.98 \times 10^{-5} \text{ A/cm}^2$$

Consider a silicon pn junction with the following parameters at $T = 300^\circ\text{K}$.

$N_a = 10^{18} \text{ cm}^{-3}$     $N_d = 10^{16} \text{ cm}^{-3}$

$D_p = 10 \text{ cm}^2/\text{sec}$     $D_n = 25 \text{ cm}^2/\text{sec}$

$\tau_{p0} = 10^{-7} \text{ sec}$     $\tau_{n0} = 10^{-7} \text{ sec}$

We can then calculate the following parameters:

$$L_p = 1.0 \times 10^{-3} \text{ cm} \qquad L_n = 1.58 \times 10^{-3} \text{ cm}$$
$$p_{n0} = 2.25 \times 10^4 \text{ cm}^{-3} \qquad n_{p0} = 2.25 \times 10^2 \text{ cm}^{-3}$$

The ideal reverse-saturation current density of the pn junction diode can be determined from Equation (9–49) as

$$J_s = \frac{(1.6 \times 10^{-19})(25)(2.25 \times 10^2)}{(1.58 \times 10^{-3})} + \frac{(1.6 \times 10^{-19})(10)(2.25 \times 10^4)}{(1.0 \times 10^{-3})}$$
$$= 5.7 \times 10^{-13} + 3.6 \times 10^{-11} = 3.66 \times 10^{-11} \text{ A/cm}^2$$

**Comment:** The ideal reverse-saturation current density of the Schottky barrier junction is orders of magnitude larger than that of the ideal pn junction diode.

Recall that the reverse-bias current in a silicon pn junction diode is dominated by the generation current. A typical generation current density is approximately $10^{-7}$ A/cm$^2$, which is still two to three orders of magnitude less than the reverse-saturation current density of the Schottky barrier diode. A generation current also exists in the reverse-biased Schottky barrier diode. However, the generation current is negligible compared with the $J_{sT}$ value.

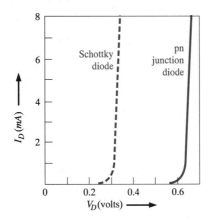

**Figure 9–10** Comparison of forward-bias I-V characteristics between a Schottky diode and a pn junction diode.

Since $J_{sT} \gg J_s$, the forward-bias characteristics of the two types of diodes will also be different. Figure 9–10 shows typical I-V characteristics of a Schottky barrier diode and a pn junction diode. The effective turn-on voltage of the Schottky diode is less than that of the pn junction diode.

**Example 9–6**

**Objective:** To calculate the forward-bias voltage required to generate a forward-bias current density of 10 A/cm$^2$ in a Schottky barrier diode and a pn junction diode.

Consider diodes with the parameters given in Example 9–5. We can assume that the pn junction diode will be sufficiently forward biased so that the ideal diffusion current will dominate. Let $T = 300°$K.

**Solution:** For the Schottky barrier diode, we have

$$J = J_{sT} \left[ \exp \left( \frac{eV_a}{kT} \right) - 1 \right]$$

Neglecting the $(-1)$ term, we can solve for the forward-bias voltage. We find

$$V_a = \left( \frac{kT}{e} \right) \ln \left( \frac{J}{J_{sT}} \right) = V_t \ln \left( \frac{J}{J_{sT}} \right) = (0.0259) \ln \left( \frac{10}{5.98 \times 10^{-5}} \right) = 0.312 \text{ volt}$$

For the pn junction diode, we have

$$V_a = V_t \ln \left( \frac{J}{J_s} \right) = (0.0259) \ln \left( \frac{10}{3.66 \times 10^{-11}} \right) = 0.682 \text{ volt}$$

**Comment:** A comparison of the two forward-bias voltages shows that the Schottky barrier diode has a turn-on voltage which, in this case, is approximately 0.37 volt smaller than the turn-on voltage of the pn junction diode.

---

The actual difference between the turn-on voltages will be a function of the barrier height of the metal-semiconductor contact and the doping concentrations in the pn junction, but the relatively large difference will always be realized. We will consider one application which utilizes the difference in turn-on voltage in the next chapter, in what is referred to as a Schottky clamped transistor.

The second major difference between a Schottky barrier diode and a pn junction diode is in the frequency response, or switching characteristics. In our discussion, we have considered the current in a Schottky diode as being due to the injection of majority carriers over a potential barrier. The energy-band diagram of Figure 9–1, for example, showed that there can be electrons in the metal directly adjacent to empty states in the semiconductor. If an electron from the valence band of the semiconductor were to flow into the metal, this effect would be equivalent to holes being injected into the semiconductor. This injection of holes would create excess minority carrier holes in the n-region. However, calculations as well as measurements have shown that the ratio of the minority carrier hole current to the total current is extremely low in most cases.

The Schottky barrier diode, then, is a majority carrier device. This fact means that there is no diffusion capacitance associated with a forward-biased Schottky diode. The elimination of the diffusion capacitance then makes the Schottky diode a higher-frequency device than the pn junction diode. Also, when switching a Schottky diode from forward to reverse bias, there is no minority-carrier stored charge to remove as was the case in the pn junction diode. Since there is no minority carrier storage time, the Schottky diodes can be used in fast switching applications. A typical switching time for a Schottky diode is in the picosecond range, while for a pn junction it is normally in the nanosecond range.

## 9.2 METAL-SEMICONDUCTOR OHMIC CONTACTS

Contacts must be made between any semiconductor device, or integrated circuit, and the outside world. These contacts are made via *ohmic contacts*. Ohmic contacts are metal-to-semiconductor contacts, but in this case they are not rectifying contacts. An ohmic contact is a low-resistance junction providing conduction in both directions between the metal and the semiconductor. Ideally, the current through the ohmic contact is a linear function of applied voltage and the applied voltage should be very small. Two general types of ohmic contacts are possible. The first type is the ideal nonrectifying barrier and the second type is the tunneling barrier. We will define a specific contact resistance which is used to characterize ohmic contacts.

### 9.2.1 Ideal Nonrectifying Barriers

We considered an ideal metal-to-n-type semiconductor contact in Figure 9–1 for the case when $\phi_m > \phi_s$. Figure 9–11 shows the same ideal contact for the opposite case of $\phi_m < \phi_s$. In Figure 9–11a we see the energy levels before contact and, in Figure 9–11b, the barrier after contact for thermal equilibrium. To achieve thermal equilibrium in this junction, electrons will flow from the metal into the lower energy states in the semiconductor, which makes the surface of the semiconductor more n-type. The excess electron charge in the n-type semiconductor exists essentially as a surface charge density. If a positive voltage is applied to the metal, there is no barrier to electrons flowing from the semiconductor into the metal. If a positive voltage is applied to the semiconductor, the effective barrier height for electrons flowing from the metal into the semiconductor will be approximately $\phi_{Bn} = \phi_n$, which is fairly small for a moderately to heavily doped semiconductor. For this bias condition, electrons can easily flow from the metal into the semiconductor.

Figure 9–12a shows the energy-band diagram when a positive voltage is applied to the metal with respect to the semiconductor. Electrons can easily flow "downhill" from the semiconductor into the metal. Figure 9–12b shows the case when a positive voltage is applied to the semiconductor with respect to the metal. Electrons can easily flow over the barrier from the metal into the semiconductor. This junction, then, is an ohmic contact.

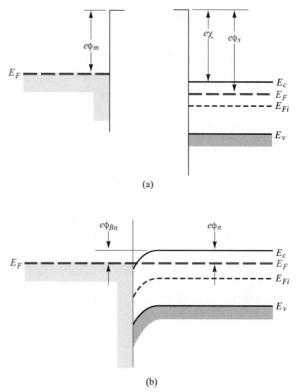

**Figure 9–11**   Ideal energy-band diagram (a) before contact and (b) after contact for a metal–n-semiconductor junction for $\phi_m < \phi_s$.

An ideal nonrectifying contact between a metal and a p-type semiconductor is shown in Figure 9–13. Figure 9–13a shows the energy levels before contact for the case when $\phi_m > \phi_s$. When contact is made, electrons from the semiconductor will flow into the metal to achieve thermal equilibrium, leaving behind more empty states, or holes. The excess concentration of holes at the surface makes the surface of the semiconductor more p-type. Electrons from the metal can readily move into the empty states in the semiconductor. This charge movement corresponds to holes flowing from the semiconductor into the metal. We can also visualize holes in the metal flowing into the semiconductor. This junction is also an ohmic contact.

The ideal energy bands shown in Figures 9–11 and 9–13 do not take into account the effect of surface states. If we assume that acceptor surface states exist in the upper half of the semiconductor bandgap, then, since all the acceptor states are below $E_F$ for the case shown in Figure 9–11b, these surface states will be negatively charged, and will alter the energy-band diagram. Similarly, if we assume that donor surface states exist in the lower half of the bandgap, then all of the donor states will be positively charged for

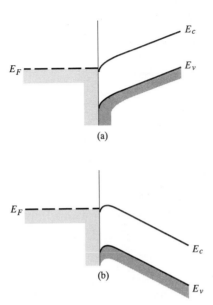

**Figure 9–12**   Ideal energy-band diagram of a metal–n-semiconductor ohmic contact (a) with a positive voltage applied to the metal and (b) with a positive voltage applied to the semiconductor.

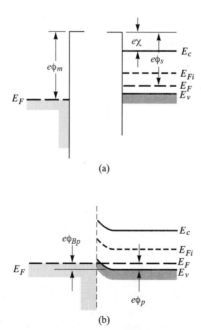

**Figure 9–13**   Ideal energy-band diagram (a) before contact and (b) after contact for a metal–p-semiconductor junction for $\phi_m > \phi_s$.

the case shown in Figure 9–13b; the positively charged surface states will also alter this energy-band diagram. Therefore, if $\phi_m < \phi_s$ for the metal–n-type semiconductor contact, and if $\phi_m > \phi_s$ for the metal–p-type semiconductor contact, we may not necessarily form a good ohmic contact.

### 9.2.2 Tunneling Barrier

The space charge width in a rectifying metal-semiconductor contact is inversely proportional to the square root of the semiconductor doping. The width of the depletion region decreases as the doping concentration in the semiconductor increases; thus, as the doping concentration increases, the probability of tunneling through the barrier increases. Figure 9–14 shows a junction in which the metal is in contact with a heavily doped n-type epitaxial layer.

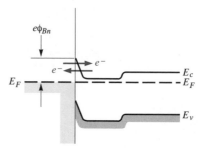

**Figure 9–14**   Energy-band diagram of a heavily doped n-semiconductor to metal junction.

---

### Example 9–7

**Objective:** To calculate the space charge width for a Schottky barrier on a heavily doped semiconductor.

Consider silicon at $T = 300°K$ doped at $N_d = 7 \times 10^{18}$ cm$^{-3}$. Assume a Schottky barrier with $\phi_{Bn} = 0.67$ volt. For this case, we can assume that $V_{bi} \approx \phi_{Bo}$. Neglect the barrier lowering effect.

**Solution:** From Equation (9–7), we have for zero applied bias

$$x_n = \left[\frac{2\varepsilon_s V_{bi}}{eN_d}\right]^{1/2} = \left[\frac{2(11.7)(8.85 \times 10^{-14})(0.67)}{(1.6 \times 10^{-19})(7 \times 10^{18})}\right]^{1/2}$$

or

$$x_n = 1.1 \times 10^{-6} \text{ cm} = 110 \text{ Å}$$

**Comment:** In a heavily doped semiconductor, the depletion width is on the order of angstroms so that tunneling is now a distinct possibility. For these types of barrier widths, tunneling may become the dominant current mechanism.

---

The tunneling current has the form

$$J_t \propto \exp \left( \frac{-e\phi_{Bn}}{E_{oo}} \right) \tag{9-50}$$

where

$$E_{oo} = \frac{e\hbar}{2} \sqrt{\frac{N_d}{\varepsilon_s m_n^*}} \tag{9-51}$$

The tunneling current increases exponentially with doping concentration.

### 9.2.3 Specific Contact Resistance

A figure of merit of ohmic contacts is the specific contact resistance, $R_c$. This parameter is defined as the reciprocal of the derivative of current density with respect to voltage evaluated at zero bias. We may write

$$R_c = \left( \frac{\partial J}{\partial V} \right)^{-1} \Bigg|_{V=0} \text{ (ohm-cm}^2\text{)} \tag{9-52}$$

We want $R_c$ to be as small as possible for an ohmic contact.

For a rectifying contact with a low to moderate semiconductor doping concentration, the current-voltage relation was given by Equation (9–43) as

$$J_n = A^* T^2 \exp \left( \frac{-e\phi_{Bn}}{kT} \right) \left[ \exp \left( \frac{eV}{kT} \right) - 1 \right]$$

The thermionic emission current is dominant in this junction. The specific contact resistance for this case is then

$$R_c = \frac{\left( \frac{kT}{e} \right) \exp \left( \frac{+e\phi_{Bn}}{kT} \right)}{A^* T^2} \tag{9-53}$$

The specific contact resistance decreases rapidly as the barrier height decreases.

For a metal-semiconductor junction with a high impurity doping concentration, the tunneling process will dominate. From Equations (9–50) and (9–51), the specific contact resistance is found to be

$$R_c \propto \exp \left[ \frac{+ 2\sqrt{\varepsilon_s m_n^*}}{\hbar} \cdot \frac{\phi_{Bn}}{\sqrt{N_d}} \right] \tag{9-54}$$

which shows that the specific contact resistance is a very strong function of semiconductor doping.

Figure 9–15 shows a plot of the theoretical values of $R_c$ as a function of semiconductor doping. For doping concentrations greater than approximately $10^{19}$ cm$^{-3}$, the tunneling process dominates and $R_c$ shows the expo-

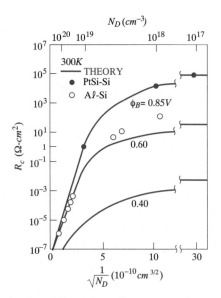

**Figure 9–15**   Theoretical and experimental specific contact resistance as a function of doping. (From Sze [9].)

nential dependence on $N_d$. For lower doping concentrations, the $R_c$ values are dependent on the barrier heights and become almost independent of the doping. Also shown in the figure are experimental data for platinum silicide-silicon and aluminum-silicon junctions.

Equation (9–54) is the specific contact resistance of the tunneling junction, which corresponds to the metal-to-$n^+$ contact shown in Figure 9–14. However, the $n^+$–$n$ junction also has a specific contact resistance since there is a barrier associated with this junction. For a fairly low doped n-region, this contact resistance may actually dominate the total resistance of the junction.

The theory of forming ohmic contacts is straightforward. To form a good ohmic contact, we need to create a low barrier and use a highly doped semiconductor at the surface. However, the actual technology of fabricating good, reliable ohmic contacts is not as easy in practice as in theory. It is also more difficult to fabricate good ohmic contacts on wide bandgap materials. In general, low barriers are not possible on these materials so a heavily doped semiconductor at the surface must be used to form a tunneling contact. The formation of a tunneling junction requires diffusion, ion implantation, or perhaps epitaxial growth. The surface doping concentration in the semiconductor may be limited to the impurity solubility, which is approximately $5 \times 10^{19}$ cm$^{-3}$ for n-type GaAs. Nonuniformities in the surface doping concentration may also prevent the theoretical limit of the specific contact resistance from being reached. In practice, a good deal of empirical processing is usually required before a good ohmic contact is obtained.

## 9.3 HETEROJUNCTIONS

In the discussion of pn junctions in the previous chapters, we assumed that the semiconductor material was homogeneous throughout the entire structure. This type of junction is called a homojunction. When two different semiconductor materials are used to form a junction, the junction is called a semiconductor heterojunction.

As with many topics in this text, our goal is to provide the basic concepts concerning the heterojunction. The complete analysis of heterojunction structures involves quantum mechanics and detailed calculations that are beyond the scope of this text. The discussion of heterojunctions will, then, be limited to the introduction of some basic concepts.

### 9.3.1 Heterojunction Materials

Since the two materials used to form a heterojunction will have different energy bandgaps, the energy band will have a discontinuity at the junction interface. We may have an abrupt junction in which the semiconductor changes abruptly from a narrow-bandgap material to a wide-bandgap material. On the other hand, if we have a GaAs–$Al_xGa_{1-x}As$ system, for example, the value of $x$ may continuously vary over a distance of several nanometers to form a graded heterojunction. Changing the value of $x$ in the $Al_xGa_{1-x}As$ system allows us to engineer, or design, the bandgap energy.

In order to have a useful heterojunction, the lattice constants of the two materials must be well matched. The lattice match is important because any lattice mismatch can introduce dislocations resulting in interface states. For example, germanium and gallium arsenide have lattice constants matched to within approximately 0.13 percent. Germanium-gallium arsenide heterojunctions have been studied quite extensively. More recently, gallium arsenide-aluminum gallium arsenide (GaAs-AlGaAs) junctions have been investigated quite thoroughly since the lattice constants of GaAs and the AlGaAs system vary by no more than 0.14 percent.

### 9.3.2 Energy-Band Diagrams

In the formation of a heterojunction with a narrow-bandgap material and a wide-bandgap material, the alignment of the bandgap energies is important in determining the characteristics of the junction. Figure 9–16 shows three possible situations. In Figure 9–16a we see the case when the forbidden bandgap of the wide-gap material completely overlaps the bandgap of the narrow-gap material. This case, called "straddling," applies to most heterojunctions. We will consider only this case here. The other possibilities are called "staggered" and "broken gap," and are shown in Figures 9–16b and 9–16c.

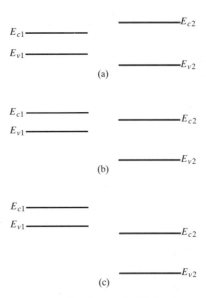

**Figure 9-16**  Relation between narrow-bandgap and wide-bandgap energies: (a) straddling, (b) staggered, and (c) broken gap.

There are four basic types of heterojunction. Those in which the dopant type changes at the junction are called anisotype. We can form n-P or N-p junctions, where the capital letter indicates the larger-bandgap material. Heterojunctions with the same dopant type on either side of the junction are called isotype. We can form n-N and p-P isotype heterojunctions.

Figure 9-17 shows the energy-band diagrams of isolated n-type and P-type materials with the vacuum level used as a reference. The electron affinity of the wide-bandgap material is less than that of the narrow-bandgap material. The difference between the two conduction band energies is denoted by $\Delta E_c$ and the difference between the two valence band energies is denoted by $\Delta E_v$. From Figure 9-17, we can see that

$$\Delta E_c = e(\chi_n - \chi_P) \qquad (9\text{-}55a)$$

and    $$\Delta E_c + \Delta E_v = E_{gP} - E_{gn} = \Delta E_g \qquad (9\text{-}55b)$$

In the ideal abrupt heterojunction using nondegenerately doped semiconductors, the vacuum level is parallel to both conduction bands and valence bands. If the vacuum level is continuous, then the same $\Delta E_c$ and $\Delta E_v$ discontinuities will exist at the heterojunction interface. This ideal situation is known as the electron affinity rule. There is still some uncertainty about the applicability of this rule, but it provides a good starting point for the discussion of heterojunctions.

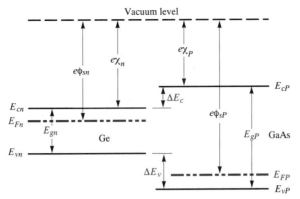

**Figure 9–17** Energy-band diagrams of a narrow-bandgap and wide-bandgap material before contact.

Figure 9–18 shows a general ideal n-P heterojunction in thermal equilibrium. In order for the Fermi levels in the two materials to become aligned, electrons from the narrow-gap n-region and holes from the wide-gap P-region must flow across the junction. As in the case of a homojunction, this flow of charge creates a space charge region in the vicinity of the metallurgical junction. The space charge width into the n-type region is denoted by $x_n$ and the space charge width into the P-type region is denoted by $x_P$. The

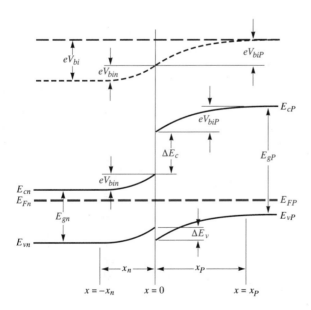

**Figure 9–18** Ideal energy-band diagram of an n-P heterojunction in thermal equilibrium.

discontinuities in the conduction and valence bands and the change in the vacuum level are shown in the figure.

### 9.3.3 Two-Dimensional Electron Gas

Before we consider the electrostatics of the heterojunction, we will discuss a unique characteristic of an isotype junction. Figure 9–19 shows the energy-

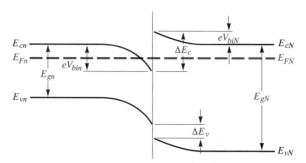

**Figure 9–19**   Ideal energy-band diagram of an n-N heterojunction in thermal equilibrium.

band diagram of an n-N GaAs–AlGaAs heterojunction in thermal equilibrium. The AlGaAs can be moderately to heavily doped n-type while the GaAs can be more lightly doped or even intrinsic. As mentioned previously, to achieve thermal equilibrium, electrons from the wide bandgap AlGaAs flow into the GaAs, forming an accumulation layer of electrons in the potential well adjacent to the interface. One basic quantum-mechanical result that we have found previously is that the energy of an electron contained in a potential well is quantized. The phrase *two-dimensional electron gas* refers to the condition in which the electrons have quantized energy levels in one spatial direction (perpendicular to the interface), but are free to move in the other two spatial directions.

The potential function near the interface can be approximated by a triangular potential well. Figure 9–20a shows the conduction band edges near the abrupt junction interface and Figure 9–20b shows the approximation of the triangular potential well. We can write

$$V(x) = e\mathrm{E}z \qquad z > 0 \qquad\qquad (9\text{–}56a)$$

$$V(z) = \infty \qquad z < 0 \qquad\qquad (9\text{–}56b)$$

Schrodinger's wave equation can be solved using this potential function. The quantized energy levels are shown in Figure 9–20b. Higher energy levels are usually not considered.

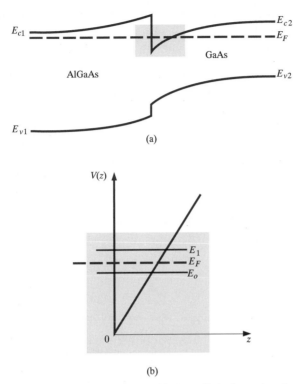

**Figure 9–20**  (a) Conduction-band edge at N-AlGaAs, n-GaAs heterojunction. (b) Triangular well approximation with discrete electron energies.

The qualitative distribution of electrons in the potential well is shown in Figure 9–21. A current parallel to the interface will be a function of this electron concentration and of the electron mobility. Since the GaAs can be lightly doped or intrinsic, the two-dimensional electron gas is in a region of low impurity doping so that impurity scattering effects are minimized. The

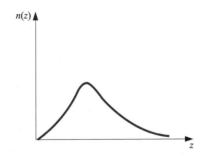

**Figure 9–21**  Electron density in triangular potential well.

electron mobility will be much larger than if the electrons were in the same region as the ionized donors.

The movement of the electrons parallel to the interface will still be influenced by the coulomb attraction of the ionized impurities in the AlGaAs. The effect of these forces can be further reduced by using a graded AlGaAs-GaAs heterojunction. The graded layer is $Al_xGa_{1-x}As$ in which the mole fraction $x$ varies with distance. In this case, an intrinsic layer of graded AlGaAs can be sandwiched between the N-type AlGaAs and the intrinsic GaAs. Figure 9–22 shows the conduction-band edges across a graded Al-GaAs-GaAs heterojunction in thermal equilibrium. The electrons in the potential well are further separated from the ionized impurities so that the electron mobility is increased above that in an abrupt heterojunction.

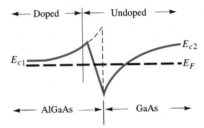

**Figure 9–22**   Conduction-band edge at a graded heterojunction.

### *9.3.4 Equilibrium Electrostatics

We will now consider the electrostatics of the n-P heterojunction that was shown in Figure 9–18. As in the case of the homojunction, potential differences exist across the space charge regions in both the n-region and the P-region. These potential differences correspond to the built-in potential barriers on either side of the junction. The built-in potential barrier for this ideal case is defined as shown in Figure 9–18 to be the potential difference across the vacuum level. The built-in potential barrier is the sum of the potential differences across each of the space charge regions. The heterojunction built-in potential barrier, however, is not equal to the difference between the conduction bands across the junction or the difference between the valence bands across the junction, as we defined for the homojunction.

Ideally, the total built-in potential barrier $V_{bi}$ can be found as the difference between the work functions, or

$$V_{bi} = \phi_{sP} - \phi_{sn} \tag{9–57}$$

Equation (9–57), using Figure 9–17, can be written as

$$eV_{bi} = [e\chi_P + E_{gP} - (E_{FP} - E_{vP})] - [e\chi_n + E_{gn} - (E_{Fn} - E_{vn})] \tag{9–58a}$$

or

$$eV_{bi} = e(\chi_P - \chi_n) + (E_{gP} - E_{gn}) + (E_{Fn} - E_{vn}) - (E_{FP} - E_{vP}) \tag{9–58b}$$

which can be expressed as

$$eV_{bi} = -\Delta E_c + \Delta E_g + kT \ln \left(\frac{N_{vn}}{p_{no}}\right) - kT \ln \left(\frac{N_{vP}}{p_{po}}\right) \qquad (9\text{–}59)$$

Finally, we can write Equation (9–59) as

$$eV_{bi} = \Delta E_v + kT \ln \left(\frac{p_{po}}{p_{no}} \cdot \frac{N_{vn}}{N_{vP}}\right) \qquad (9\text{–}60)$$

where $p_{po}$ and $p_{no}$ are the hole concentrations in the P- and n-materials, respectively, and $N_{vn}$ and $N_{vP}$ are the effective density of states functions in the n- and P-materials, respectively. We could also obtain an expression for the built-in potential barrier in terms of the conduction band shift as

$$eV_{bi} = -\Delta E_c + kT \ln \left(\frac{n_{no}}{n_{po}} \cdot \frac{N_{cP}}{N_{cn}}\right) \qquad (9\text{–}61)$$

---

**Example 9–8**

**Objective:** To determine $\Delta E_c$, $\Delta E_v$, and $V_{bi}$ for an n-Ge to P-GaAs heterojunction using the electron affinity rule.

Consider n-type Ge doped with $N_d = 10^{16}$ cm$^{-3}$ and P-type GaAs doped with $N_a = 10^{16}$ cm$^{-3}$. Let $T = 300°$K so that $n_i = 2.4 \times 10^{13}$ cm$^{-3}$ for Ge.

**Solution:** From Equation (9–55a), we have

$$\Delta E_c = e(\chi_n - \chi_P) = e(4.13 - 4.07) = 0.06 \text{ eV}$$

and from Equation (9–55b), we have

$$\Delta E_v = \Delta E_g - \Delta E_c = (1.43 - 0.67) - 0.06 = 0.70 \text{ eV}$$

To determine $V_{bi}$ using Equation (9–60), we need to determine $p_{no}$ in Ge, or

$$p_{no} = \frac{n_i^2}{N_d} = \frac{(2.4 \times 10^{13})^2}{10^{16}} = 5.76 \times 10^{10} \text{ cm}^{-3}$$

Then

$$eV_{bi} = 0.70 + (0.0259) \ln \left[\frac{(10^{16})(6 \times 10^{18})}{(5.76 \times 10^{10})(7 \times 10^{18})}\right]$$

or, finally,

$$V_{bi} \approx 1.0 \text{ volt}$$

**Comment:** There is a nonsymmetry in the $\Delta E_c$ and $\Delta E_v$ values that will tend to make the potential barriers seen by electrons and holes different. This nonsymmetry does not occur in homojunctions.

---

The electric field and potential in the junction can be determined from Poisson's equation in exactly the same way as we did for the homojunction.

For homogeneous doping on each side of the junction, we have in the n-region

$$E_n = \frac{eN_{dn}}{\varepsilon_n} (x_n + x) \qquad (-x_n \leq x < 0) \tag{9-62a}$$

and in the P-region

$$E_P = \frac{eN_{aP}}{\varepsilon_P} (x_P - x) \qquad (0 < x \leq x_P) \tag{9-62b}$$

where $\varepsilon_n$ and $\varepsilon_P$ are the permittivities of the n- and P-materials, respectively. We may note that $E_n = 0$ at $x = -x_n$ and $E_P = 0$ at $x = x_P$. The electric flux density $D$ is continuous across the junction, so

$$\varepsilon_n \, E_n(x = 0) = \varepsilon_P \, E_P(x = 0) \tag{9-63a}$$

which gives

$$N_{dn} x_n = N_{aP} x_P \tag{9-63b}$$

Equation (9–63b) simply states that the net negative charge in the P-region is equal to the net positive charge in the n-region—the same condition we had in a pn homojunction. We are neglecting any interface states that may exist at the heterojunction.

The electric potential can be found by integrating the electric field through the space charge region so that the potential difference across each region can then be determined. We find that

$$V_{bin} = \frac{eN_{dn} x_n^2}{2\varepsilon_n} \tag{9-64a}$$

and

$$V_{biP} = \frac{eN_{aP} x_P^2}{2\varepsilon_P} \tag{9-64b}$$

Equation (9–63b) can be rewritten as

$$\frac{x_n}{x_P} = \frac{N_{aP}}{N_{dn}} \tag{9-65}$$

The ratio of the built-in potential barriers can then be determined as

$$\frac{V_{bin}}{V_{biP}} = \frac{\varepsilon_P}{\varepsilon_n} \cdot \frac{N_{dn}}{N_{aP}} \cdot \frac{x_n^2}{x_P^2} = \frac{\varepsilon_P N_{aP}}{\varepsilon_n N_{dn}} \tag{9-66}$$

Assuming that $\varepsilon_n$ and $\varepsilon_P$ are of the same order of magnitude, the larger potential difference is across the lower-doped region.

The total built-in potential barrier is

$$V_{bi} = V_{bin} + V_{biP} = \frac{eN_{dn} x_n^2}{2\varepsilon_n} + \frac{eN_{aP} x_P^2}{2\varepsilon_P} \tag{9-67}$$

If we solve for $x_P$, for example, from Equation (9–63b) and substitute into Equation (9–67), we can solve for $x_n$ as

$$x_n = \left\{ \frac{2\varepsilon_n\varepsilon_P N_{aP} V_{bi}}{eN_{dn}(\varepsilon_n N_{dn} + \varepsilon_P N_{aP})} \right\}^{1/2} \qquad (9\text{–}68a)$$

We can also find

$$x_P = \left\{ \frac{2\varepsilon_n\varepsilon_P N_{dn} V_{bi}}{eN_{aP}(\varepsilon_n N_{dn} + \varepsilon_P N_{aP})} \right\}^{1/2} \qquad (9\text{–}68b)$$

The total depletion width is found to be

$$W = x_n + x_P = \left\{ \frac{2\varepsilon_n\varepsilon_P(N_{dn} + N_{aP})^2 V_{bi}}{eN_{dn}N_{aP}(\varepsilon_n N_{dn} + \varepsilon_P N_{aP})} \right\}^{1/2} \qquad (9\text{–}69)$$

If a reverse-bias voltage is applied across the heterojunction, the same equations apply if $V_{bi}$ is replaced by $V_{bi} + V_R$. Similarly, if a forward bias is applied, the same equations also apply if $V_{bi}$ is replaced by $V_{bi} - V_a$. As before, $V_R$ is the magnitude of the reverse-bias voltage and $V_a$ is the magnitude of the forward-bias voltage.

As in the case of a homojunction, a change in depletion width with a change in junction voltage yields a junction capacitance. We can find for the n-P junction

$$C_j' = \left\{ \frac{eN_{dn}N_{aP}\varepsilon_n\varepsilon_P}{2(\varepsilon_n N_{dn} + \varepsilon_P N_{aP})(V_{bi} + V_R)} \right\}^{1/2} \qquad \text{(F/cm}^2\text{)} \qquad (9\text{–}70)$$

A plot of $(1/C_j')^2$ versus $V_R$ again yields a straight line. The extrapolation of this plot to $(1/C_j')^2 = 0$ is used to find the built-in potential barrier, $V_{bi}$.

Figure 9–18 showed the ideal energy-band diagram for the n-P abrupt heterojunction. The experimentally determined values of $\Delta E_c$ and $\Delta E_v$ may differ from the ideal values determined using the electron affinity rule. One possible explanation for this difference is that most heterojunctions have interface states. If we assume that the electrostatic potential is continuous through the junction, then the electric flux density will be discontinuous at the heterojunction due to the surface charge trapped in the interface states. The interface states will then change the energy-band diagram of the semiconductor heterojunction just as they changed the energy-band diagram of the metal-semiconductor junction. Another possible explanation for the deviation from the ideal is that as the two materials are brought together to form the heterojunction, the electron orbitals of each material begin to interact with each other, resulting in a transition region of a few angstroms at the interface. The energy bandgap is then continuous through this transition region and not a characteristic of either material. However, we still have the relation that

$$\Delta E_c + \Delta E_v = \Delta E_g \qquad (9\text{–}71)$$

for the straddling type of heterojunction although the $\Delta E_c$ and $\Delta E_v$ values may differ from those determined from the electron affinity rule.

We may consider the general characteristics of the energy-band diagrams of the other types of heterojunction. Figure 9–23 shows the energy-band diagram of an N-p heterojunction. The same $\Delta E_c$ and $\Delta E_v$ discontinuities exist, although the general shape of the conduction band, for example, is different in the n-P and the N-p junctions. This difference in energy bands will influence the I-V characteristics of the two junctions.

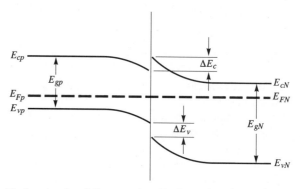

**Figure 9–23**    Ideal energy-band diagram of an N-p heterojunction in thermal equilibrium.

The other two types of heterojunctions are the n-N and the p-P isotype junctions. The energy-band diagram of the n-N junction was shown in Figure 9–19. To achieve thermal equilibrium, electrons from the wide-bandgap material will flow into the narrow-bandgap material. A positive space charge region exists in the wide-gap material and an accumulation layer of electrons now exists at the interface in the narrow-gap material. Since there are a large number of allowed energy states in the conduction band, we expect the space charge width $x_n$ and the built-in potential barrier $V_{bin}$ to be small in the narrow-gap material. The energy-band diagram of the p-P heterojunction in thermal equilibrium is shown in Figure 9–24. To achieve thermal equilibrium, holes from the wide-bandgap material will flow into the narrow-bandgap material, creating an accumulation layer of holes in the narrow-bandgap material at the interface. These types of isotype heterojunction are obviously not possible in a homojunction.

## *9.3.5 Current-Voltage Characteristics

The ideal current-voltage characteristics of a pn homojunction were developed in Chapter 8. Since the energy-band diagram of a heterojunction is more complicated than that of a homojunction, we would expect the I-V characteristics of the two junctions to differ.

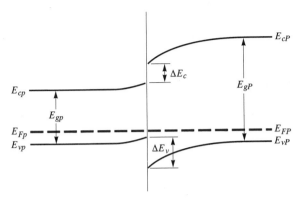

**Figure 9–24**  Ideal energy-band diagram of a p-P heterojunction in thermal equilibrium.

One immediate difference between a homojunction and a heterojunction is in the barrier heights seen by the electrons and holes. Since the built-in potential barrier for electrons and holes in a homojunction is the same, the relative magnitude of the electron and hole currents is determined by the relative doping levels. In a heterojunction, the barrier heights seen by electrons and holes are not the same. The energy-band diagrams in Figures 9–18 and 9–23 demonstrated that the barrier heights for electrons and holes in a heterojunction can be significantly different. The barrier height for electrons in Figure 9–18 is larger than for holes, so we would expect the current due to electrons to be insignificant compared to the hole current. If the barrier height for electrons is 0.2 eV larger than that for holes, the electron current will be approximately a factor of $10^4$ smaller than the hole current, assuming all other parameters are equal. The opposite situation exists for the band diagram shown in Figure 9–23.

The conduction-band edge in Figure 9–23 and the valence-band edge in Figure 9–18 are somewhat similar to that of a rectifying metal-semiconductor contact. We derive the current-voltage characteristics of a heterojunction, in general, on the basis of thermionic emission of carriers over the barrier, as we did in the metal-semiconductor junction. We can then write

$$J = A^* T^2 \exp\left(\frac{-E_w}{kT}\right) \tag{9–72}$$

where $E_w$ is an effective barrier height. The barrier height can be increased or reduced by an applied potential across the junction as in the case of a pn homojunction or a Schottky barrier junction. The heterojunction I-V characteristics, however, may need to be modified to include diffusion effects and tunneling effects. Another complicating factor is that the effective mass of a carrier changes from one side of the junction to the other. Although the actual derivation of the I-V relationship of the heterojunction is complex, the general form of the I-V equation is still similar to that of a Schottky barrier diode and is generally dominated by one type of carrier.

## 9.4 SUMMARY AND REVIEW

In this chapter, we have considered the rectifying metal-semiconductor junction known as the Schottky barrier diode; the nonrectifying metal-semiconductor junction known as an ohmic contact; and the semiconductor heterojunction. In Chapters 7 and 8, we considered a homojunction, in which the material on either side of the metallurgical junction is the same semiconductor crystal. The junctions we have considered in this chapter are heterojunctions—the materials are not the same on either side of the junction.

The Schottky barrier diode is a rectifying metal-semiconductor junction. In the ideal case, a potential barrier is formed at the interface between the metal and semiconductor if the relation between the metal and semiconductor work functions is "proper." This potential barrier can be lowered or raised by applying a voltage across the junction. The actual barrier height will deviate from the ideal case as a result of the image-force-induced lowering and also as a result of interface states. The current transport across the junction is due mainly to majority carriers; therefore, we derived the I-V characteristics by considering the thermionic emission of carriers over the barrier. Since the current in the Schottky barrier diode is due to majority carriers, there is no minority carrier storage in the junction so the switching speed of the Schottky diode is significantly faster than that of the pn junction diode. In addition, less forward-bias voltage is required in a Schottky diode than in a pn junction diode to produce a given current.

Metal-semiconductor junctions can also form ohmic contacts, which are low-resistance junctions providing conduction in both directions. Two types of ohmic contacts are possible. The first type is the ideal nonrectifying barrier in which, again if the relation between the metal and semiconductor work functions is "proper," a metal-semiconductor junction is formed that effectively has no potential barrier. Carriers are then free to flow in either direction depending on the polarity of the applied voltage. This idealized nonrectifying junction does not take into account the effect of surface states. The second type of ohmic contact is the tunneling barrier. If the semiconductor is highly doped, the width of the space charge region is reduced to the order of tens of angstroms, and tunneling through this narrow barrier becomes the dominant current mechanism. The tunneling current increases exponentially with semiconductor doping so the resistance of the junction can be made quite small. The tunneling ohmic contact is the predominant low-resistance junction used in compound semiconductor devices.

Semiconductor heterojunctions are formed between two semiconductor materials with different bandgap energies. We considered the ideal energy-band diagrams for the four basic types of heterojunction: we can form a pn junction in which the wide-bandgap material is either N-type or P-type and the narrow-bandgap material is the opposite type; and we can form n-N or p-P heterojunctions. One extremely useful property of a heterojunction is the creation of a potential well at the interface. The potential well at an n-N heterojunction, for example, contains a two-dimensional electron gas which means that, although the electrons are confined to the potential well in the direction perpendicular to the interface, they are free to move in the other two spatial dimensions. These electrons can be confined to a region with very low impurity doping so that impurity scattering is minimized and the carriers exhibit a very high mobility. The current across the junction will usually be dominated by one type of carrier since the potential barriers seen by the two types of carriers vary substantially because of the difference in bandgap energies.

The Schottky barrier diode is used in high frequency applications and in situations in which the lower turn-on voltage is useful. One such application that utilizes both the lower turn-on voltage and fast switching speed is the Schottky clamped transistor which we will consider in the next chapter. Ohmic contacts are used in every type of semiconductor device as we will see in the remaining chapters. We will usually simply indicate the existence of an ohmic contact in a device without specifying the type, so the importance of this technology may tend to be obscured while we delve into the physics of the device. We will see the heterojunction again when we discuss the bipolar transistor in Chapter 10, the junction field-effect transistor in Chapter 11, and yet again when we consider the light-emitting and laser diodes in Chapter 14.

## GLOSSARY OF IMPORTANT TERMS

**Anisotype junction:** A heterojunction in which the type of dopant changes at the metallurgical junction.

**Electron affinity rule:** The rule stating that, in an ideal heterojunction, the discontinuity at the conduction band is the difference between the electron affinities in the two semiconductors.

**Heterojunction:** The junction formed by the contact between two different semiconductor materials.

**Image-force-induced lowering:** The lowering of the peak potential barrier at the metal-semiconductor junction due to an electric field.

**Isotype junction:** A heterojunction in which the type of dopant is the same on both sides of the junction.

**Ohmic contact:** A low-resistance, metal-semiconductor contact providing conduction in both directions between the metal and semiconductor.

**Richardson constant:** The parameter $A^*$ in the current-voltage relation of a Schottky diode.

**Schottky barrier height:** The potential barrier $\phi_{Bn}$ from the metal to semiconductor in a metal-semiconductor junction.

**Schottky effect:** Another term for image-force-induced lowering.

**Specific contact resistance:** The inverse of the slope of the $J$ versus $V$ curve of a metal-semiconductor contact evaluated at $V = 0$.

**Thermionic emission:** The process by which charge flows over a potential barrier due to carriers with sufficient thermal energy.

**Tunneling barrier:** A thin potential barrier in which the current is dominated by the tunneling of carriers through the barrier.

**Two-dimensional electron gas (2-DEG):** The accumulation layer of electrons contained in a potential well at a heterojunction interface that are free to move in the "other" two spatial directions.

## PROBLEMS

(In the following problems, assume $A^* = 120$ A/°K²-cm² for silicon and $A^* = 1.12$ A/°K²-cm² for gallium arsenide Schottky diodes unless otherwise stated.)

### Section 9.1

1. Consider a contact between Al and n-Si doped at $N_d = 10^{16}$ cm⁻³. $T = 300°K$. (a) Draw the energy-band diagrams of the two materials before the junction is formed. (b) Draw the ideal energy band at zero bias after the junction is formed. (c) Calculate $\phi_{Bo}$, $x_d$, and $E_{max}$ for part (b). (d) Repeat parts (b) and (c) using the data in Figure 9–5.

2. A metal with a work function of $\phi_m = 4.95$ volts is deposited on n-type GaAs with a doping of $N_d = 5 \times 10^{15}$ cm⁻³. $T = 300°K$. Determine (a) the theoretical barrier height, $\phi_{Bo}$, (b) $\phi_n$, (c) $V_{bi}$, and (d) the zero bias space charge width.

3. Consider a gold Schottky diode at $T = 300°K$ formed on n-type GaAs doped at $N_d = 5 \times 10^{16}$ cm⁻³. Determine (a) the theoretical barrier height, $\phi_{Bo}$, (b) $\phi_n$, (c) $V_{bi}$, (d) the space charge width, $x_n$, for $V_R = 5$ volts, and (e) the electric field at the metal junction for $V_R = 5$ volts.

4. Repeat problem 3, parts (b) through (e), if the experimentally determined barrier height is found to be $\phi_{Bo} = 0.86$ volt.

5. An Au–n-Si junction with $N_d = 5 \times 10^{15}$ cm⁻³ has a cross-sectional area of $A = 5 \times 10^{-4}$ cm². $T = 300°K$. Use the data in Figure 9–5. (a) Determine the junction capacitance when $V_R = 4$ volts. (b) Repeat part (a) if the doping is increased to $N_d = 5 \times 10^{16}$ cm⁻³.

6. A Schottky diode with n-type GaAs at $T = 300°K$ yields the $1/C'^2$ versus $V_R$ plot shown in Figure 9–25, where $C'$ is the capacitance per cm². Determine (a) $V_{bi}$, (b) $N_d$, (c) $\phi_n$, and (d) $\phi_{Bo}$.

7. Consider an Al–n-Si Schottky barrier at $T = 300°K$ with $N_d = 10^{16}$ cm⁻³. Use the data in Figure 9–5 to determine the barrier height.

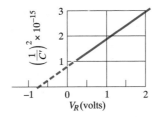

**Figure 9–25**  Figure for problem 6.

(a) Determine $V_{bi}$, $x_d$, and $E_{max}$ at zero bias. (b) Using the value of $E_{max}$ from part (a), determine $\Delta\phi$ and $x_m$ for the Schottky barrier lowering. (c) Repeat part (b) for the case when a reverse bias of $V_R = 4$ volts is applied.

8.  Starting with Equation (9–12), derive Equations (9–14) and (9–15).

9.  An Au–n–GaAs Schottky diode is at $T = 300°K$ with $N_d = 5 \times 10^{16}$ cm$^{-3}$. Use the data in Figure 9–5 to determine the barrier height. (a) Determine $V_{bi}$, $x_d$, and $E_{max}$ at zero bias. (b) Determine the reverse-bias voltage at which the Schottky barrier lowering, $\Delta\phi$, will be 7 percent of $\phi_{Bo}$. (Use the value of $E_{max}$ in the space charge region.)

*10.  The energy-band diagram of a Schottky diode is shown in Figure 9–6. Assume the following parameters:

$\phi_m = 5.2$ volt          $\chi = 4.07$ volt
$E_g = 1.43$ eV              $\phi_o = 0.60$ volt
$\varepsilon_s = (13.1)\varepsilon_0$          $\varepsilon_i = \varepsilon_0$
$\phi_n = 0.10$ volt          $N_d = 10^{16}$ cm$^{-3}$
$\delta = 25$ Å              $D_{it} = 10^{13}$ eV$^{-1}$ cm$^{-2}$

(a) Determine the theoretical barrier height $\phi_{Bo}$, without interface states. (b) Determine the barrier height with interface states. (c) Repeat parts (a) and (b) if $\phi_m$ is changed to $\phi_m = 4.5$ volts.

*11.  A Schottky barrier diode contains interface states and an interfacial layer. Assume the following parameters:

$\phi_m = 4.75$ volts          $\chi = 4.01$ volts
$E_g = 1.12$ eV               $\phi_0 = 0.230$ volt
$\varepsilon_s = (11.7)\varepsilon_0$          $\varepsilon_i = \varepsilon_0$
$\phi_n = 0.164$ volt          $N_d = 5 \times 10^{16}$ cm$^{-3}$
$\delta = 20$ Å               $\phi_{Bo} = 0.60$ volt

Determine the interface state density, $D_{it}$, in units of eV$^{-1}$ cm$^{-2}$.

12.  A PtSi Schottky diode at $T = 300°K$ is fabricated on n-type silicon with a doping of $N_d = 10^{16}$ cm$^{-3}$. From Figure 9–5, the barrier

height is 0.89 volt. Determine (a) $\phi_n$, (b) $V_{bi}$, (c) $J_{sT}$, when the barrier lowering is neglected, and (d) $V_a$ so that $J_n = 2$ A/cm$^2$.

13. (a) Consider a Schottky diode at $T = 300°$K formed with tungsten on n-type silicon. Let $N_d = 5 \times 10^{15}$ cm$^{-3}$ and assume a cross-sectional area of $A = 5 \times 10^{-4}$ cm$^2$. Determine the forward-bias voltage required to obtain a current of 1 mA. (b) Repeat part (a) if the temperature is increased to $T = 400°$K. (Neglect Schottky barrier lowering.)

14. A Schottky diode is formed by depositing Au on n-type GaAs doped at $N_d = 5 \times 10^{16}$ cm$^{-3}$. $T = 300°$K. (a) Determine the forward-bias voltage required to obtain $J_n = 5$ A/cm$^2$. (b) What is the change in forward-bias voltage necessary to double the current? (Neglect Schottky barrier lowering.)

15. A Schottky diode at $T = 300°$K is formed between tungsten and n-type silicon doped at $N_d = 10^{16}$ cm$^{-3}$. The cross-sectional area is $A = 10^{-4}$ cm$^2$. Determine the reverse-bias saturation current at (a) $V_R = 2$ volts and (b) $V_R = 4$ volts. (Take into account the Schottky barrier lowering.)

*16. Starting with the basic current equation given by Equation (9–30), derive the integral relation given by Equation (9–39).

17. Starting with the integral current relation in Equation (9–39), derive Equation (9–43).

18. A Schottky diode and a pn junction diode have cross-sectional areas of $A = 5 \times 10^{-4}$ cm$^2$. The reverse-saturation current density of the Schottky diode is $3 \times 10^{-8}$ A/cm$^2$ and the reverse-saturation current density of the pn junction diode is $3 \times 10^{-12}$ A/cm$^2$. The temperature is 300°K. Determine the forward-bias voltage in each diode required to yield diode currents of 1 mA.

19. The reverse-saturation current densities in a pn junction diode and a Schottky diode are $5 \times 10^{-12}$ A/cm$^2$ and $7 \times 10^{-8}$ A/cm$^2$, respectively, at $T = 300°$K. The cross-sectional area of the pn junction diode is $A = 8 \times 10^{-4}$ cm$^2$. Determine the cross-sectional area of the Schottky diode so that the difference in forward-bias voltages to achieve 1.2 mA is 0.265 volt.

20. (a) The reverse-saturation currents of a Schottky diode and a pn junction diode at $T = 300°$K are $5 \times 10^{-8}$ A and $10^{-12}$ A, respectively. The diodes are connected in parallel and are driven by a constant current of 0.5 mA. (i) Determine the current in each diode. (ii) Determine the voltage across each diode. (b) Repeat part (a) if the diodes are connected in series.

21. A Schottky diode and a pn junction diode have cross-sectional areas of $A = 7 \times 10^{-4}$ cm$^2$. The reverse-saturation current densities at $T = 300°$K of the Schottky diode and pn junction are $4 \times 10^{-8}$ A/cm$^2$ and $3 \times 10^{-12}$ A/cm$^2$, respectively. A forward-bias current of

0.8 mA is required in each diode. *(a)* Determine the forward-bias voltage required across each diode. *(b)* If the voltage from part *(a)* is maintained across each diode, determine the current in each diode if the temperature is increased to 400°K. (Take into account the temperature dependence of the reverse-saturation currents. Assume $E_g = 1.12$ eV for the pn junction diode and $\phi_{Bo} = 0.82$ volt for the Schottky diode.)

## Section 9.2

22.  It is possible, theoretically, to form an ohmic contact between a metal and silicon, which has a very low barrier height. Considering the specific contact resistance, determine the value of $\phi_{Bn}$ that will give a value of $R_c = 10^{-5}$ ohm–cm² at $T = 300$°K.

23.  A metal, with a work function $\phi_m = 4.2$ volts, is deposited on an n-type silicon semiconductor with $\chi_s = 4.0$ volts and $E_g = 1.12$ eV. Assume no interface states exist at the junction. Let $T = 300$°K. *(a)* Sketch the energy-band diagram for zero bias for the case when no space charge region exists at the junction. *(b)* Determine $N_d$ so that the condition in part *(a)* is satisfied. *(c)* What is the potential barrier height seen by electrons in the metal moving into the semiconductor?

24.  Consider the energy-band diagram of a silicon Schottky junction under zero bias shown in Figure 9–26. Let $\phi_{Bo} = 0.7$ volt and $T = 300$°K. Determine the doping required so that $x_d = 50$ Å at the point where the potential is $\phi_{Bo}/2$ below the peak value. (Neglect the barrier lowering effect.)

## Section 9.3

25.  Sketch the energy-band diagrams of an abrupt $Al_{0.3}Ga_{0.7}As$–GaAs heterojunction for: *(a)* N⁺–AlGaAs, intrinsic GaAs, *(b)* N⁺–

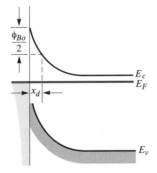

**Figure 9–26**   Figure for problem 24.

AlGaAs, $p$–GaAs, and *(c)* $P^+$–AlGaAs, $n^+$–GaAs. Assume $E_g = 1.85$ eV for $Al_{0.3}Ga_{0.7}As$ and assume $\Delta E_c = \frac{2}{3}\Delta E_g$.

26. Repeat problem 25 assuming the ideal electron affinity rule. Determine $\Delta E_c$ and $\Delta E_v$.

*27. Starting with Poisson's equation, derive Equation (9–69) for an abrupt heterojunction.

## Summary and Review

*28. *(a)* Derive an expression for $dV_a/dT$ as a function of current density in a Schottky diode. Assume the minority-carrier current is negligible. *(b)* Compare $dV_a/dT$ for a GaAs Schottky diode to that for a Si Schottky diode. *(c)* Compare $dV_a/dT$ for a Si Schottky diode to that for a Si pn junction diode.

29. The $(1/C_j)^2$ versus $V_R$ data are measured for two Schottky diodes with equal areas. One diode is fabricated with 1 ohm-cm silicon and the other diode with 5 ohm-cm silicon. The plots intersect the voltage axis at $V_R = -0.5V$ for diode A and at $V_R = -1.0V$ for diode B. The slope of the plot for diode A is $1.5 \times 10^{18}$ $(F^2 - V)^{-1}$ and that for diode B is $1.5 \times 10^{17}$ $(F^2 - V)^{-1}$. Determine which diode has the higher metal work function and which diode has the lower resistivity silicon.

*30. Both Schottky barrier diodes and ohmic contacts are to be fabricated by depositing a particular metal on a silicon integrated circuit. The work function of the metal is 4.5 volts. Considering the ideal metal-semiconductor contact, determine the allowable range of doping concentrations for each type of contact. Consider both p- and n-type silicon regions.

31. Consider an n-GaAs–p-AlGaAs heterojunction in which the bandgap offsets are $\Delta E_c = 0.3$ eV and $\Delta E_v = 0.15$ eV. Discuss the difference in the expected electron and hole currents when the junction is forward biased.

## READING LIST

1. Anderson, R. L. "Experiments on Ge–GaAs Heterojunctions." *Solid-State Electronics* 5, no. 5 (September–October 1962), pp. 341–51.

2. Michaelson, H. B. "Relation between an Atomic Electronegativity Scale and the Work Function." *IBM Journal of Research and Development* 22, no. 1 (January 1978), pp. 72–80.

3. Pulfrey, D. L., and N. G. Tarr. *Introduction to Microelectronic Devices.* Englewood Cliffs, N.J.: Prentice Hall, 1989.

4. Rideout, V. L. "A Review of the Theory, Technology and Applications of Metal-Semiconductor Rectifiers." *Thin Solid Films* 48, no. 3 (February 1, 1978), pp. 261–91.

5. Roulston, D. J. *Bipolar Semiconductor Devices*. New York: McGraw-Hill, 1990.

*6. Shur, M. *GaAs Devices and Circuits*. New York: Plenum Press, 1987.

*7. _____ . *Physics of Semiconductor Devices*. Englewood Cliffs, N.J.: Prentice Hall, 1990.

8. Streetman, B. G. *Solid State Electronic Devices*. 3rd ed. Englewood Cliffs, N.J.: Prentice Hall, 1990.

9. Sze, S. M. *Physics of Semiconductor Devices*. 2nd ed. New York: Wiley, 1981.

*10. Wang, S. *Fundamentals of Semiconductor Theory and Device Physics*. Englewood Cliffs, N.J.: Prentice Hall, 1989.

*11. Wolfe, C. M.; N. Holonyak, Jr.; and G. E. Stillman. *Physical Properties of Semiconductors*. Englewood Cliffs, N.J.: Prentice Hall, 1989.

12. Yang, E. S. *Microelectronic Devices*. New York: McGraw-Hill, 1988.

13. Zambuto, M. *Semiconductor Devices*. New York: McGraw-Hill, 1989.

# THE BIPOLAR TRANSISTOR

## CHAPTER 10

## PREVIEW

The single-junction devices we have considered, including the pn homojunction diode, can be used to obtain rectifying current-voltage characteristics, and to form electronic switching circuits. The transistor is a multi-junction semiconductor device that, in conjunction with other circuit elements, is capable of current gain, voltage gain, and signal-power gain. The transistor is therefore referred to as an active device whereas the diode is passive. The bipolar transistor has three separately doped regions and two pn junctions, sufficiently close together so that interactions occur between the two junctions. We will use much of the theory developed for the pn junction in the analysis of the bipolar transistor.

We will first discuss the basic geometry and operation of the transistor. Since there is more than one pn junction in the bipolar transistor, several combinations of reverse- and forward-bias junction voltages are possible, leading to different operating modes in the device. As with the pn junction diode, minority carrier distributions in the bipolar transistor are an important part of the physics of the device—minority carrier gradients produce diffusion currents. The minority carrier distribution in each region of the transistor will be determined as will the corresponding currents.

The bipolar transistor is a voltage controlled current source. We will consider the various factors which determine the current gain and derive its mathematical expression. As with any semiconductor device, nonideal effects influence device characteristics; a few of these effects, such as breakdown voltage, will be described.

In order to analyze or design a transistor circuit, especially using computer simulations, one needs a mathematical model or equivalent circuit of the transistor. We will develop two equivalent circuits. The first equivalent circuit, the Ebers-Moll model, can be used for a transistor biased in any of its operating modes and is especially used for transistors in switching circuits. The second equivalent circuit, the hybrid-pi model, is applied when transistors are operated in a small-single linear amplifier and takes into account frequency effects within the transistor.

Various physical factors affect the frequency response of the bipolar transistor. There are several time-delay factors within the device that determine the limiting frequency response. We will define these time delays and develop expressions for each factor. The limiting frequency is given in terms of a cutoff frequency, a figure of merit for the transistor. The frequency response generally applies to the small-signal, steady-state characteristics of the device. The switching characteristics, in contrast, determine the transient behavior of the transistor to large changes in the input signal. After describing the factors that affect the transient behavior of the bipolar transistor, we will use them to explain the fast switching characteristics of the Schottky clamped transistor.

## 10.1 THE BIPOLAR TRANSISTOR ACTION

The bipolar transistor has three separately doped regions and two pn junctions. Figure 10–1 shows the basic structure of an npn bipolar transistor and

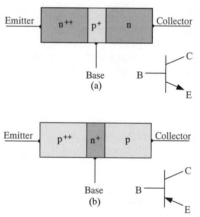

**Figure 10–1**   Simplified block diagrams and circuit symbols of (a) npn and (b) pnp bipolar transistors.

a pnp bipolar transistor, along with the circuit symbols. The three terminal connections are called the emitter, base, and collector. The width of the base region is small compared to the minority carrier diffusion length. The (++) and (+) notation indicates the relative magnitudes of the impurity doping concentrations normally used in the bipolar transistor, with (++) meaning very heavily doped and (+) meaning moderately doped. The emitter region has the largest doping concentration; the collector region has the smallest. The reasons for using these relative impurity concentrations, and for the narrow base width, will become clear as we develop the theory of the bipolar transistor. The concepts developed for the pn junction apply directly to the bipolar transistor.

The block diagrams of Figure 10–1 show the basic structure of the transistor, but in very simplified sketches. Figure 10–2a shows a cross section of a classic npn bipolar transistor fabricated in an integrated circuit configuration and Figure 10–2b shows the cross section of an npn bipolar transistor fabri-

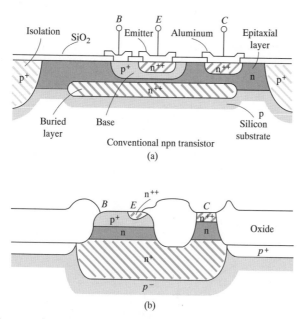

**Figure 10–2**  Cross section of (a) a conventional integrated-circuit npn bipolar transistor and (b) an oxide-isolated npn bipolar transistor. (From Muller and Kamins [1].)

cated using a more modern technology. One can immediately observe that the actual structure of the bipolar transistor is not nearly as simple as the block diagrams of Figure 10–1 might suggest. A reason for the complexity is that terminal connections are made at the surface; in order to minimize semiconductor resistances, heavily doped $n^+$ buried layers must be included. Another reason for complexity arises out of the desire to fabricate more than one bipolar transistor on a single piece of semiconductor material. Individual transistors must be isolated from each other since all collectors, for example, will not be at the same potential. This isolation is accomplished by adding $p^+$ regions so that devices are separated by reverse-biased pn junctions as shown in Figure 10–2a, or they are isolated by large oxide regions as shown in Figure 10–2b.

An important point to note from the devices shown in Figure 10–2 is that the bipolar transistor is not a symmetrical device. Although the transistor may contain two n-regions or two p-regions, the impurity doping concentrations in the emitter and collector are different and the geometry of these regions can be vastly different. The block diagrams of Figure 10–1 are highly simplified, but useful concepts in the development of the basic transistor theory.

### 10.1.1 The Basic Principle of Operation

The npn and pnp transistors are complementary devices. We will develop the bipolar transistor theory using the npn transistor, but the same basic principles and equations also apply to the pnp device. Figure 10–3 shows an idealized impurity doping profile in an npn bipolar transistor for the case when each region is uniformly doped.

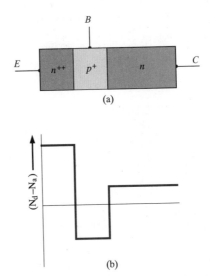

**Figure 10–3**    Idealized doping profile of a uniformly doped npn bipolar transistor.

The base-emitter (B-E) pn junction is forward-biased and the base-collector (B-C) pn junction is reverse-biased in the normal bias configuration. This configuration is called the *forward-active* operating mode: The B-E junction is forward-biased so electrons from the emitter are injected across the B-E junction into the base. These injected electrons create an excess concentration of minority carriers in the base. The B-C junction is reverse biased so the minority carrier electron concentration at the edge of the B-C junction is zero. We expect the electron concentration in the base to be like that shown in Figure 10–4. The large gradient in the electron concentration means that electrons injected from the emitter will diffuse across the base region into the B-C space charge region, where the electric field will sweep the electrons

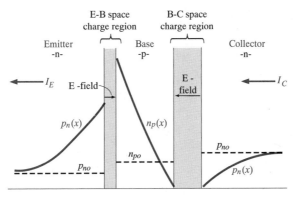

**Figure 10–4** Minority carrier distribution in an npn bipolar transistor operating in the forward-active mode.

into the collector. We want as many electrons as possible to reach the collector without recombining with any majority carrier holes in the base. For this reason, the width of the base needs to be small compared with the minority carrier diffusion length. If the base width is small, then the minority carrier electron concentration is a function of both the B-E and B-C junction voltages. The two junctions are close enough to be called interacting pn junctions.

Figure 10–5 shows a cross section of an npn transistor with the injection of electrons from the n-type emitter and the collection of the electrons in the

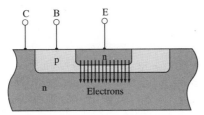

**Figure 10–5** Cross section of an npn bipolar transistor showing the injection and collection of electrons in the forward active mode.

collector. The number of electrons per unit time reaching the collector is proportional to the number of electrons injected into the base. The number of injected electrons, in turn, is a function of the B-E voltage. To a first approximation, the collector current is independent of the reverse-biased B-C voltage; thus the device looks like a constant current source. The collector current is being controlled by the B-E voltage, or the current in one part of the device is controlled by a voltage in another part of the device. This control is the basic transistor action.

Since the B-E junction is forward biased, holes from the base are injected into the emitter. However, these injected holes do not contribute to the collector current; hence, they are not part of the basic transistor action. The minority carrier distribution in each region of a bipolar transistor is shown in Figure 10–4. We will derive the mathematical expression for each minority carrier distribution and then determine the various current components in the device.

### 10.1.2 The Modes of Operation

Figure 10–6 shows the npn transistor in a simple circuit. In this configuration, the transistor may be biased in one of three modes of operation. If the

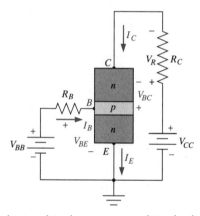

**Figure 10–6**   An npn bipolar transistor in a common emitter circuit configuration.

B-E voltage is zero or reverse biased ($V_{BE} \leq 0$), then majority carrier electrons from the emitter will not be injected into the base. The B-C junction is also reverse biased; thus, the emitter and collector currents will be zero for this case. This condition is referred to as *cutoff*—all currents in the transistor are zero.

When the B-E junction becomes forward biased, an emitter current will be generated as we have discussed, and the injection of electrons into the base results in a collector current. We may write the KVL equations around the collector-emitter loop as

$$V_{CC} = I_C R_c - V_{BC} + V_{BE} = V_R - V_{BC} + V_{BE}$$

If $V_{CC}$ is large enough and if $V_R$ is small enough, then $V_{BC} < 0$ which means that the B-C junction is reverse biased. Again, this condition is the forward-active region of operation.

As the forward-biased B-E voltage increases, the collector current and hence $V_R$ will also increase. The increase in $V_R$ means that the reverse-biased B-C voltage decreases, or $|V_{BC}|$ decreases. At some point, the collector current may become large enough so that the combination of $V_R$ and $V_{CC}$ produces zero voltage across the B-C junction. A slight increase in $I_C$ beyond this point will cause a slight increase in $V_R$ and the B-C junction will become forward-biased ($V_{BC} > 0$). This condition is called *saturation*. In the saturation mode of operation, both B-E and B-C junctions are forward biased and the collector current is no longer controlled by the B-E voltage.

A fourth mode of operation for the bipolar transistor is possible, although not with the circuit configuration shown in Figure 10–6. This fourth mode, known as *inverse active*, occurs when the B-E junction is reverse biased and the B-C junction is forward biased. In this case the transistor is operating "upside down," and the roles of the emitter and collector are reversed. We have argued that the transistor is not a symmetrical device; therefore, the inverse-active characteristics will not be the same as the forward-active characteristics.

## 10.2 MINORITY CARRIER DISTRIBUTION

We are interested in calculating currents in the bipolar transistor which, as in the simple pn junction, are determined by minority carrier diffusion. Since diffusion currents are produced by minority carrier gradients, we must determine the steady-state minority carrier distribution in each of the three transistor regions. Let us first consider the forward-active mode, and then the other modes of operation.

### 10.2.1 Forward-Active Mode

Consider a uniformly doped npn bipolar transistor with the geometry shown in Figure 10–7. When we consider the individual emitter, base, and collector regions, we will shift the origin to the edge of the space charge region and consider a positive $x$, $x'$, or $x''$ coordinate as shown in the figure.

In the forward-active mode, the B-E junction is forward biased and the B-C is reverse biased. We expect the minority carrier distributions to look like those shown in Figure 10–8. As there are two n-regions, we will have

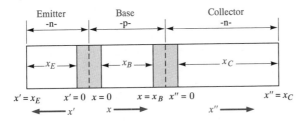

**Figure 10–7**  Geometry of the npn bipolar transistor used to calculate the minority carrier distribution.

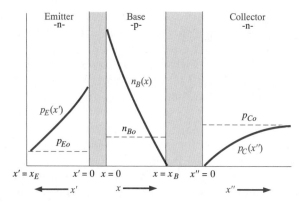

**Figure 10-8**  Minority carrier distribution in an npn bipolar transistor operating in the forward-active mode.

minority carrier holes in both emitter and collector. To distinguish between these two minority carrier hole distributions, we will use the notation shown in the figure. Keep in mind that we will be dealing only with minority carriers. The parameters $p_{E0}$, $n_{B0}$, and $p_{C0}$ denote the thermal-equilibrium minority carrier concentrations in the emitter, base, and collector, respectively. The functions $p_E(x')$, $n_B(x)$, and $p_C(x'')$ denote the steady-state minority carrier concentrations in the emitter, base, and collector, respectively. We will assume that the neutral collector length $x_C$ is long compared to the minority carrier diffusion length $L_C$ in the collector, but we will take into account a finite emitter length $x_E$. If we assume that the surface recombination velocity at $x' = x_E$ is infinite, then the excess minority carrier concentration at $x' = x_E$ is zero, or $p_E(x' = x_E) = p_{E0}$. An infinite surface recombination velocity is a good approximation when an ohmic contact is fabricated at $x' = x_E$.

**Base Region**    The steady-state excess minority carrier electron concentration is found from the ambipolar transport equation, which we discussed in detail in Chapter 6. For a zero electric field in the neutral base region, the ambipolar transport equation in steady state reduces to

$$D_B \frac{\partial^2(\delta n_B(x))}{\partial x^2} - \frac{\delta n_B(x)}{\tau_{B0}} = 0 \tag{10-1}$$

where $\delta n_B$ is the excess minority carrier electron concentration, and $D_B$ and $\tau_{B0}$ are the minority carrier diffusion coefficient and lifetime in the base region, respectively. The excess-electron concentration is defined as

$$\delta n_B(x) = n_B(x) - n_{B0} \tag{10-2}$$

The general solution to Equation (10–1) can be written as

$$\delta n_B(x) = A \exp \left( \frac{+x}{L_B} \right) + B \exp \left( \frac{-x}{L_B} \right) \tag{10-3}$$

where $L_B$ is the minority carrier diffusion length in the base, given by $L_B = \sqrt{D_B \tau_{B0}}$. The base is of finite width so both exponential terms in Equation (10-3) must be retained.

The excess minority carrier electron concentrations at the two boundaries become

$$\delta n_B(x = 0) \equiv \delta n_B(0) = A + B \tag{10-4a}$$

and

$$\delta n_B(x = x_B) \equiv \delta n_B(x_B) = A \exp \left( \frac{+x_B}{L_B} \right) + B \exp \left( \frac{-x_B}{L_B} \right) \tag{10-4b}$$

The B-E junction is forward biased, so the boundary condition at $x = 0$ is

$$\delta n_B(0) = n_B(x = 0) - n_{B0} = n_{B0} \left[ \exp \left( \frac{eV_{BE}}{kT} \right) - 1 \right] \tag{10-5a}$$

The B-C junction is reverse biased, so the second boundary condition at $x = x_B$ is

$$\delta n_B(x_B) = n_B(x = x_B) - n_{B0} = 0 - n_{B0} = -n_{B0} \tag{10-5b}$$

Using the boundary conditions given by Equations (10-5a) and (10-5b), the coefficients $A$ and $B$ from Equations (10-4a) and (10-4b) can be determined. The results are

$$A = \frac{-n_{B0} - n_{B0} \left[ \exp \left( \frac{eV_{BE}}{kT} \right) - 1 \right] \exp \left( \frac{-x_B}{L_B} \right)}{2 \sinh \left( \frac{x_B}{L_B} \right)} \tag{10-6a}$$

and

$$B = \frac{n_{B0} \left[ \exp \left( \frac{eV_{BE}}{kT} \right) - 1 \right] \exp \left( \frac{x_B}{L_B} \right) + n_{B0}}{2 \sinh \left( \frac{x_B}{L_B} \right)} \tag{10-6b}$$

Then, substituting Equations (10-6a) and (10-6b) into Equation (10-3), we can write the excess minority carrier electron concentration in the base region as

$$\delta n_B(x) = \frac{n_{B0} \left\{ \left[ \exp \left( \frac{eV_{BE}}{kT} \right) - 1 \right] \sinh \left( \frac{x_B - x}{L_B} \right) - \sinh \left( \frac{x}{L_B} \right) \right\}}{\sinh \left( \frac{x_B}{L_B} \right)} \tag{10-7}$$

Equation (10-7) may look formidable with the sinh functions. We have stressed that we want the base width $x_B$ to be small compared to the minority carrier diffusion length $L_B$. This condition may seem somewhat arbitrary at this point, but the reason will become clear as we proceed through all of the calculations. Since we want $x_B < L_B$, the argument in the sinh functions is always less than unity and in most cases will be much less than unity. Figure 10–9 shows a plot of sinh($y$) for $0 \le y \le 1$ and also shows the linear

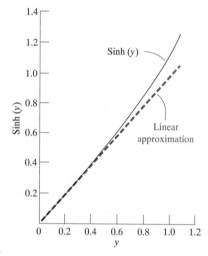

**Figure 10–9**  Hyperbolic sine function and its linear approximation.

approximation for small values of $y$. If $y < 0.4$, the sinh($y$) function differs from its linear approximation by less than 3 percent. All of this leads to the conclusion that the excess electron concentration $\delta n_B$ in Equation (10–7) is approximately a linear function of $x$ through the neutral base region. We will use the linear approximation later in some of the example calculations.

**Emitter Region**   Consider, now, the minority carrier hole concentration in the emitter. The steady-state excess-hole concentration is determined from the equation

$$D_E \frac{\partial^2(\delta p_E(x'))}{\partial x'^2} - \frac{\delta p_E(x')}{\tau_{E0}} = 0 \qquad (10\text{–}8)$$

where $D_E$ and $\tau_{E0}$ are the minority carrier diffusion coefficient and minority carrier lifetime, respectively, in the emitter. The excess-hole concentration is given by

$$\delta p_E(x') = p_E(x') - p_{E0} \qquad (10\text{–}9)$$

The general solution to Equation (10–8) can be written as

$$\delta p_E(x') = C \exp\left(\frac{+x'}{L_E}\right) + D \exp\left(\frac{-x'}{L_E}\right) \qquad (10\text{–}10)$$

where $L_E = \sqrt{D_E \tau_{E0}}$. If we assume the neutral emitter length $x_E$ is not necessarily long compared to $L_E$, then both exponential terms in Equation (10–10) must be retained.

The excess minority carrier hole concentrations at the two boundaries are

$$\delta p_E(x' = 0) \equiv \delta p_E(0) = C + D \qquad (10\text{–}11a)$$

and

$$\delta p_E(x' = x_E) \equiv \delta p_E(x_E) = C \exp\left(\frac{x_E}{L_E}\right) + D \exp\left(\frac{-x_E}{L_E}\right) \qquad (10\text{–}11b)$$

Again, the B-E junction is forward biased so

$$\delta p_E(0) = p_E(x' = 0) - p_{E0} = p_{E0}\left[\exp\left(\frac{eV_{BE}}{kT}\right) - 1\right] \qquad (10\text{–}12a)$$

An infinite surface recombination velocity at $x' = x_E$ implies that

$$\delta p_E(x_E) = 0 \qquad (10\text{–}12b)$$

Solving for $C$ and $D$ from Equations (10–11) and (10–12), the excess minority carrier hole concentration in Equation (10–10) becomes

$$\delta p_E(x') = \frac{p_{E0}\left[\exp\left(\frac{eV_{BE}}{kT}\right) - 1\right]\sinh\left(\frac{x_E - x'}{L_E}\right)}{\sinh\left(\frac{x_E}{L_E}\right)} \qquad (10\text{–}13)$$

This excess concentration will also vary approximately linearly with distance if $x_E$ is small, or will show the exponential dependence on distance if $x_E$ becomes comparable to $L_E$.

**Collector Region** The excess minority carrier hole concentration in the collector can be determined from the equation

$$D_C \frac{\partial^2(\delta p_C(x''))}{\partial x''^2} - \frac{\delta p_C(x'')}{\tau_{C0}} = 0 \qquad (10\text{–}14)$$

where $D_C$ and $\tau_{C0}$ are the minority carrier diffusion coefficient and minority carrier lifetime, respectively, in the collector. We can express the excess minority carrier hole concentration in the collector as

$$\delta p_C(x'') = p_C(x'') - p_{C0} \qquad (10\text{–}15)$$

The general solution to Equation (10–14) can be written as

$$\delta p_C(x'') = G \exp\left(\frac{x''}{L_C}\right) + H \exp\left(\frac{-x''}{L_C}\right) \qquad (10\text{–}16)$$

where $L_C = \sqrt{D_C \tau_{C0}}$. If we assume that the collector is long, then the coefficient $G$ must be zero since the excess concentration must remain finite.

The second boundary condition gives

$$\delta p_C(x'' = 0) \equiv \delta p_C(0) = p_C(x'' = 0) - p_{C0} = 0 - p_{C0} = -p_{C0} \qquad (10\text{--}17)$$

The excess minority carrier hole concentration in the collector is then given as

$$\delta p_C(x'') = -p_{C0} \exp\left(\frac{-x''}{L_C}\right) \qquad (10\text{--}18)$$

This result is exactly what we expect from the results of a reverse-biased pn junction.

### 10.2.2 Other Modes of Operation

The bipolar transistor can also operate in the cutoff, saturation, or inverse-active mode. We will qualitatively discuss the minority carrier distributions for these operating conditions and treat the actual calculations as problems at the end of the chapter.

Figure 10–10a shows the minority carrier distribution in an npn bipolar transistor in cutoff. In cutoff, both the B-E and B-C junctions are reverse biased; thus, the minority carrier concentrations are zero at each space charge edge. The emitter and collector regions are assumed to be "long" in this figure, while the base is narrow compared with the minority carrier

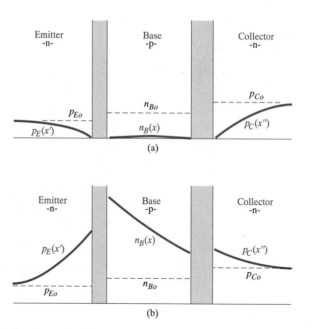

**Figure 10–10**  Minority carrier distribution in an npn bipolar transistor operating in (a) cutoff and (b) saturation.

diffusion length. Since $x_B \ll L_B$, essentially all minority carriers are swept out of the base region.

Figure 10–10b shows the minority carrier distribution in the npn bipolar transistor operating in saturation. Both the B-E and B-C junctions are forward biased; thus, excess minority carriers exist at the edge of each space charge region. However, since a collector current still exists when the transistor is in saturation, a gradient will still exist in the minority carrier electron concentration in the base.

Finally, Figure 10–11a shows the minority carrier distribution in the npn transistor for the inverse-active mode. In this case, the B-E is reverse biased and the B-C is forward biased. Electrons from the collector are now injected into the base. The gradient in the minority carrier electron concentration in the base is in the opposite direction compared with the forward-active mode, so the emitter and collector currents will change direction. Figure 10–11b

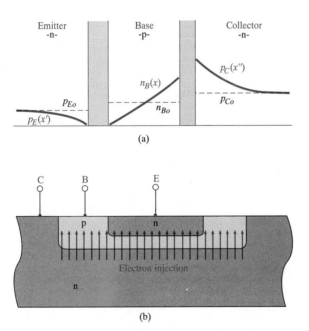

**Figure 10–11** (a) Minority carrier distribution in an npn bipolar transistor operating in the inverse-active mode. (b) Cross section of an npn bipolar transistor showing the injection and collection of electrons in the inverse-active mode.

shows the injection of electrons from the collector into the base. Since the B-C area is normally much larger than the B-E area, not all of the injected electrons will be collected by the emitter. The relative doping concentrations in the base and collector are also different compared with those in the base and emitter; thus, we see that the transistor is not symmetrical. We then expect the characteristics to be significantly different between the forward-active and inverse-active modes of operation.

## 10.3 LOW-FREQUENCY COMMON BASE CURRENT GAIN

The basic principle of operation of the bipolar transistor is the control of the collector current by the B-E voltage. The collector current is a function of the number of majority carriers reaching the collector after being injected from the emitter across the B-E junction. The *common base current gain* is defined as the ratio of collector current to emitter current. The flow of various charged carriers leads to definitions of particular currents in the device. We can use these definitions to define the current gain of the transistor in terms of several factors.

### 10.3.1  Contributing Factors

Figure 10–12 shows the various particle flux components in the npn bipolar transistor. We will define the various flux components and then consider the resulting currents. Although there seems to be a large number of flux components, we may help clarify the situation by correlating each factor with the minority carrier distributions shown in Figure 10–8.

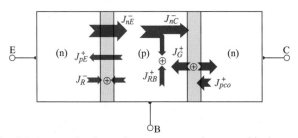

**Figure 10–12**   Particle current density or flux components in an npn bipolar transistor operating in the forward-active mode.

The factor $J_{nE}^-$ is the electron flux injected from the emitter into the base. As the electrons diffuse across the base, a few will recombine with majority carrier holes. The majority carrier holes that are lost by recombination must be replenished from the base terminal. This replacement hole flux is denoted by $J_{RB}^+$. The electron flux that reaches the collector is $J_{nC}^-$. The majority carrier holes from the base that are injected back into the emitter result in a hole flux denoted by $J_{pE}^+$. Some electrons and holes that are injected into the forward-biased B-E space charge region will recombine in this region. This recombination leads to the electron flux $J_R^-$. Generation of electrons and holes occurs in the reverse-biased B-C junction. This generation yields a hole flux $J_G^+$. Finally, the ideal reverse-saturation current in the B-C junction is denoted by the hole flux $J_{pc0}^+$.

The corresponding electric current density components in the npn transistor are shown in Figure 10–13 along with the minority carrier distributions

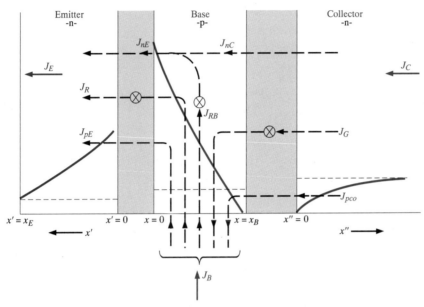

**Figure 10–13** Current density components in an npn bipolar transistor operating in the forward-active mode.

for the forward-active mode. The curves are the same as in Figure 10–8. As in the pn junction, the currents in the bipolar transistor are defined in terms of minority carrier diffusion currents. The current densities are defined as follows:

$J_{nE}$: Due to the diffusion of minority carrier electrons in the base at $x = 0$.

$J_{nC}$: Due to the diffusion of minority carrier electrons in the base at $x = x_B$.

$J_{RB}$: The difference between $J_{nE}$ and $J_{nC}$, which is due to the recombination of excess minority carrier electrons with majority carrier holes in the base. The $J_{RB}$ current is the flow of holes into the base to replace the holes lost by recombination.

$J_{pE}$: Due to the diffusion of minority carrier holes in the emitter at $x' = 0$.

$J_R$:    Due to the recombination of carriers in the forward-biased B-E junction.

$J_{pc0}$:    Due to the diffusion of minority carrier holes in the collector at $x'' = 0$.

$J_G$:    Due to the generation of carriers in the reverse-biased B-C junction.

The currents $J_{RB}$, $J_{pE}$, and $J_R$ are B-E junction currents only and do not contribute to the collector current. The currents $J_{pc0}$ and $J_G$ are B-C junction currents only. These current components do not contribute to the transistor action or the current gain.

The dc common base current gain is defined as

$$\alpha_0 = \frac{I_C}{I_E} \tag{10-19}$$

If we assume that the active cross-sectional area is the same for the collector and emitter, then we can write the current gain in terms of the current densities, or

$$\alpha_0 = \frac{J_C}{J_E} = \frac{J_{nC} + J_G + J_{pc0}}{J_{nE} + J_R + J_{pE}} \tag{10-20}$$

We are primarily interested in determining how the collector current will change with a change in emitter current. The small-signal, or sinusoidal, common base current gain is defined as

$$\alpha = \frac{\partial J_C}{\partial J_E} = \frac{J_{nC}}{J_{nE} + J_R + J_{pE}} \tag{10-21}$$

The reverse-bias B-C currents, $J_G$ and $J_{pc0}$, are not functions of the emitter current.

We can rewrite Equation (10–21) in the form

$$\alpha = \left( \frac{J_{nE}}{J_{nE} + J_{pE}} \right) \left( \frac{J_{nC}}{J_{nE}} \right) \left( \frac{J_{nE} + J_{pE}}{J_{nE} + J_R + J_{pE}} \right) \tag{10-22a}$$

or

$$\alpha = \gamma \alpha_T \delta \tag{10-22b}$$

The factors in Equation (10–22b) are defined as:

$$\gamma = \left( \frac{J_{nE}}{J_{nE} + J_{pE}} \right) \equiv \text{Emitter injection efficiency factor} \tag{10-23a}$$

$$\alpha_T = \left( \frac{J_{nC}}{J_{nE}} \right) \equiv \text{Base transport factor} \tag{10-23b}$$

$$\delta = \frac{J_{nE} + J_{pE}}{J_{nE} + J_R + J_{pE}} \equiv \text{Recombination factor} \tag{10-23c}$$

We would like to have the change in collector current be exactly the same as the change in emitter current, or ideally, to have $\alpha = 1$. However, a consideration of Equation (10–21) shows that $\alpha$ will always be less than unity. The goal is to make $\alpha$ as close to one as possible. To achieve this goal, we must make each term in Equation (10–22b) as close to one as possible, since each factor is less than unity.

The *emitter injection efficiency factor* $\gamma$ takes into account the minority carrier hole diffusion current in the emitter. This current is part of the emitter current, but does not contribute to the transistor action in that $J_{pE}$ is not part of the collector current. The *base transport factor* $\alpha_T$ takes into account any recombination of excess minority carrier electrons in the base. Ideally, we want no recombination in the base. The *recombination factor* $\delta$ takes into account the recombination in the forward-biased B-E junction. The current $J_R$ contributes to the emitter current, but does not contribute to collector current.

### 10.3.2 Mathematical Derivation of Current Gain Factors

We now wish to determine each of the gain factors in terms of the electrical and geometrical parameters of the transistor. The results of these derivations will show how the various parameters in the transistor influence the electrical properties of the device and will point the way to the design of a "good" bipolar transistor.

**Emitter Injection Efficiency Factor** Consider, initially, the emitter injection efficiency factor. We have from Equation (10–23a)

$$\gamma = \left( \frac{J_{nE}}{J_{nE} + J_{pE}} \right) = \frac{1}{\left( 1 + \dfrac{J_{pE}}{J_{nE}} \right)} \tag{10–24}$$

We derived the minority carrier distribution functions for the forward-active mode in Section 10.2.1. Noting that $J_{nE}$, as defined in Figure 10–13, is in the negative x-direction, we can write the current densities as

$$J_{pE} = -eD_E \frac{d(\delta p_E(x'))}{dx'} \Big|_{x' = 0} \tag{10–25a}$$

and

$$J_{nE} = (-)eD_B \frac{d(\delta n_B(x))}{dx} \Big|_{x = 0} \tag{10–25b}$$

where $\delta p_E(x')$ and $\delta n_B(x)$ are given by Equations (10–13) and (10–7), respectively.

Taking the appropriate derivatives of $\delta p_E(x')$ and $\delta n_B(x)$, we obtain

$$J_{pE} = \frac{eD_E p_{E0}}{L_E} \left[ \exp\left( \frac{eV_{BE}}{kT} \right) - 1 \right] \cdot \frac{1}{\tanh\left( x_E/L_E \right)} \tag{10–26a}$$

and

$$J_{nE} = \frac{eD_B n_{B0}}{L_B} \left\{ \frac{1}{\sinh\left( x_B/L_B \right)} + \frac{\left[ \exp\left( eV_{BE}/kT \right) - 1 \right]}{\tanh\left( x_B/L_B \right)} \right\} \tag{10–26b}$$

Positive $J_{pE}$ and $J_{nE}$ values imply that the currents are in the directions shown in Figure 10–13. If we assume that the B-E junction is biased sufficiently far in the forward bias so that $V_{BE} \gg kT/e$, then

$$\exp\left(\frac{eV_{BE}}{kT}\right) \gg 1$$

and also

$$\frac{\exp(eV_{BE}/kT)}{\tanh(x_B/L_B)} \gg \frac{1}{\sinh(x_B/L_B)}$$

The emitter injection efficiency, from Equation (10–24), then becomes

$$\gamma = \frac{1}{1 + \dfrac{p_{E0}D_E L_B}{n_{B0}D_B L_E} \cdot \dfrac{\tanh(x_B/L_B)}{\tanh(x_E/L_E)}} \tag{10–27}$$

If we assume that all the parameters in Equation (10–27) except $p_{E0}$ and $n_{B0}$ are fixed, then in order for $\gamma \approx 1$, we must have $p_{E0} \ll n_{B0}$. We can write

$$p_{E0} = \frac{n_i^2}{N_E} \quad \text{and} \quad n_{B0} = \frac{n_i^2}{N_B}$$

where $N_E$ and $N_B$ are the impurity doping concentrations in the emitter and base, respectively. Then the condition that $p_{E0} \ll n_{B0}$ implies that $N_E \gg N_B$. For the emitter injection efficiency to be close to unity, the emitter doping must be large compared to the base doping. This condition means that many more electrons from the n-type emitter than holes from the p-type base will be injected across the B-E space charge region.

**Base Transport Factor**    The next term to consider is the base transport factor, given by Equation (10–23b) as $\alpha_T = J_{nC}/J_{nE}$. From the definitions of the current directions shown in Figure 10–13, we can write

$$J_{nC} = (-)eD_B \frac{d(\delta n_B(x))}{dx}\Big|_{x = x_B} \tag{10–28a}$$

and

$$J_{nE} = (-)eD_B \frac{d(\delta n_B(x))}{dx}\Big|_{x = 0} \tag{10–28b}$$

Using the expression for $\delta n_B(x)$ given in Equation (10–7), we find that

$$J_{nC} = \frac{eD_B n_{B0}}{L_B}\left\{\frac{[\exp(eV_{BE}/kT) - 1]}{\sinh(x_B/L_B)} + \frac{1}{\tanh(x_B/L_B)}\right\} \tag{10–29}$$

The expression for $J_{nE}$ was given in Equation (10–26b).

If we again assume that the B-E junction is biased sufficiently far in the forward bias so that $V_{BE} \gg kT/e$, then $\exp(eV_{BE}/kT) \gg 1$. Substituting Equations (10–29) and (10–26b) into Equation (10–23b), we have

$$\alpha_T = \frac{J_{nC}}{J_{nE}} \cong \frac{\exp(eV_{BE}/kT) + \cosh(x_B/L_B)}{1 + \exp(eV_{BE}/kT)\cosh(x_B/L_B)} \tag{10–30}$$

In order for $\alpha_T$ to be close to unity, the neutral base width $x_B$ must be much smaller than the minority carrier diffusion length in the base $L_B$. If $x_B \ll L_B$, then cosh $(x_B/L_B)$ will be just slightly greater than unity. In addition, if exp $(eV_{BE}/kT) \gg 1$, then the base transport factor is approximately

$$\alpha_T \cong \frac{1}{\cosh\ (x_B/L_B)} \qquad (10\text{--}31a)$$

For $x_B \ll L_B$, we may expand the cosh function in a Taylor series, so that

$$\alpha_T = \frac{1}{\cosh\ (x_B/L_B)} \approx \frac{1}{1 + \frac{1}{2}(x_B/L_B)^2} \approx 1 - \tfrac{1}{2}(x_B/L_B)^2 \qquad (10\text{--}31b)$$

The base transport factor $\alpha_T$ will be close to one if $x_B \ll L_B$. We can now see why we indicated earlier that the neutral base width $x_B$ would be less than $L_B$.

**Recombination Factor**   The recombination factor was given by Equation (10–23c). We can write

$$\delta = \frac{J_{nE} + J_{pE}}{J_{nE} + J_R + J_{pE}} \cong \frac{J_{nE}}{J_{nE} + J_R} = \frac{1}{1 + J_R/J_{nE}} \qquad (10\text{--}32)$$

We have assumed in Equation (10–32) that $J_{pE} \ll J_{nE}$. The recombination current density, due to the recombination in a forward-biased pn junction, was discussed in Chapter 8 and can be written as

$$J_R = \frac{ex_{BE}n_i}{2\tau_0} \exp\left(\frac{eV_{BE}}{2kT}\right) = J_{r0} \exp\left(\frac{eV_{BE}}{2kT}\right) \qquad (10\text{--}33)$$

where $x_{BE}$ is the B-E space charge width.

The current $J_{nE}$ from Equation (10–26b) can be approximated as

$$J_{nE} = J_{s0} \exp\left(\frac{eV_{BE}}{kT}\right) \qquad (10\text{--}34)$$

where

$$J_{s0} = \frac{eD_B n_{B0}}{L_B \tanh\ (x_B/L_B)} \qquad (10\text{--}35)$$

The recombination factor, from Equation (10–32), can then be written as

$$\delta = \frac{1}{1 + \dfrac{J_{r0}}{J_{s0}} \exp\left(\dfrac{-eV_{BE}}{2kT}\right)} \qquad (10\text{--}36)$$

The recombination factor is a function of the B-E voltage. As $V_{BE}$ increases, the recombination current becomes less dominant and the recombination factor approaches unity.

The recombination factor must also include surface effects. The surface effects can be described by the surface recombination velocity as we discussed in Chapter 6. Figure 10–14a shows the B-E junction of an npn transis-

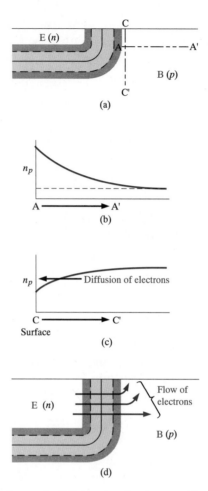

**Figure 10–14**  The surface at the E-B junction showing the diffusion of carriers toward the surface.

tor near the semiconductor surface. We will assume that the B-E junction is forward biased. Figure 10–14b shows the excess minority carrier electron concentration in the base along the cross section A-A'. This curve is the usual forward-biased junction minority carrier concentration. Figure 10–14c shows the excess minority carrier electron concentration along the cross section C-C' from the surface. We showed earlier that the excess concentration at a surface is smaller than the excess concentration in the bulk material. With this electron distribution, there is a diffusion of electrons from the bulk toward the surface where the electrons recombine with the majority carrier holes. Figure 10–14d shows the injection of electrons from the emitter into the base and the diffusion of electrons toward the surface. This diffusion generates another component of recombination current and this component of recombination current must be included in the recombination factor $\delta$. Although the actual calculation is difficult because of the two-dimensional analysis required, the form of the recombination current is the same as that of Equation (10–33).

Although we have considered an npn transistor in all of the derivations, exactly the same analysis applies to a pnp transistor; the same minority carrier distributions will be obtained except that the electron concentrations will become hole concentrations and vice versa. The current directions and voltage polarities will also change.

### 10.3.3 Example Calculations of the Gain Factors

We have been considering the common base current gain, defined in Equation (10–19) as $\alpha_0 = I_C/I_E$. The common emitter current gain is defined as $\beta_0 = I_C/I_B$. From Figure 10–6, we see that $I_E = I_B + I_C$. We can determine the relation between common emitter and common base current gains from the KCL equation. We can write

$$\frac{I_E}{I_C} = \frac{I_B}{I_C} + 1$$

Substituting the definitions of current gains, we have

$$\frac{1}{\alpha_0} = \frac{1}{\beta_0} + 1$$

Since this relation actually holds for both dc and small-signal conditions, we can drop the subscript. The common emitter current gain can now be written in terms of the common base current gain as

$$\beta = \frac{\alpha}{1 - \alpha}$$

The common base current gain, in terms of the common emitter current gain, is found to be

$$\alpha = \frac{\beta}{1 + \beta}$$

If we assume a typical value of $\beta$ to be 100, then $\alpha = 0.99$. If we also assume that $\gamma = \alpha_T = \delta$, then each factor would have to be equal to 0.9967 in order that $\beta = 100$. This calculation gives an indication of how close to unity each factor must be in order to achieve a reasonable current gain.

---

**Example 10–1**

**Objective:** To calculate the ratio of emitter doping to base doping in order to achieve an emitter injection efficiency factor equal to $\gamma = 0.9967$.

Consider an npn bipolar transistor. Assume, for simplicity, that $D_E = D_B$, $L_E = L_B$, and $x_E = x_B$.

**Solution:** Equation (10–27) reduces to

$$\gamma = \frac{1}{1 + \dfrac{p_{E0}}{n_{B0}}} = \frac{1}{1 + \dfrac{n_i^2/N_E}{n_i^2/N_B}}$$

so

$$\gamma = \frac{1}{1 + \dfrac{N_B}{N_E}} = 0.9967$$

Then

$$\frac{N_B}{N_E} = 0.00331 \quad \text{or} \quad \frac{N_E}{N_B} = 302$$

**Comment:** The emitter doping concentration must be much larger than the base doping concentration to achieve a high emitter injection efficiency.

---

### Example 10–2

**Objective:** To calculate the base width required to achieve a base transport factor equal to $\alpha_T = 0.9967$.

Consider a pnp bipolar transistor. Assume that $D_B = 10$ cm$^2$/sec and $\tau_{B0} = 10^{-7}$ sec.

**Solution:** The base transport factor applies to both pnp and npn transistors and is given by

$$\alpha_T = \frac{1}{\cosh(x_B/L_B)} = 0.9967$$

Then

$$x_B/L_B = 0.0814$$

We have

$$L_B = \sqrt{D_B \tau_{B0}} = \sqrt{(10)(10^{-7})} = 10^{-3} \text{ cm}$$

so that the base width must then be

$$x_B = 0.814 \times 10^{-4} \text{ cm} = 0.814 \ \mu\text{m}$$

**Comment:** If the base width is less than approximately 0.8 $\mu$m, then the required base transport factor will be achieved. In most cases, the base transport factor will not be the limiting factor in the bipolar transistor current gain.

---

### Example 10–3

**Objective:** To calculate the forward-biased B-E voltage required to achieve a recombination factor equal to $\delta = 0.9967$.

Consider an npn bipolar transistor at $T = 300°K$. Assume that $J_{r0} = 10^{-8}$ A/cm$^2$ and that $J_{s0} = 10^{-11}$ A/cm$^2$.

**Solution:** The recombination factor, from Equation (10–36), is

$$\delta = \frac{1}{1 + \dfrac{J_{r0}}{J_{s0}} \exp\left(\dfrac{-eV_{BE}}{2kT}\right)}$$

We then have

$$0.9967 = \frac{1}{1 + \dfrac{10^{-8}}{10^{-11}} \exp\left(\dfrac{-eV_{BE}}{2kT}\right)}$$

We can rearrange this equation and write

$$\exp\left(\frac{+eV_{BE}}{2kT}\right) = \frac{0.9967 \times 10^3}{1 - 0.9967} = 3.02 \times 10^5$$

Then $\qquad V_{BE} = 2(0.0259) \ln (3.02 \times 10^5) = 0.654$ volt

**Comment:** This example demonstrates that the recombination factor may be an important limiting factor in the bipolar current gain. In this example, if $V_{BE}$ is smaller than 0.654 volt, then the recombination factor $\delta$ will fall below the desired 0.9967 value.

---

**Example 10–4**

**Objective:** To calculate the common emitter current gain of a silicon npn bipolar transistor at $T = 300°K$ given a set of parameters.

Assume the following parameters:

$D_E = 10$ cm$^2$/sec $\qquad x_B = 0.70$ $\mu$m

$D_B = 25$ cm$^2$/sec $\qquad x_E = 0.50$ $\mu$m

$\tau_{E0} = 1 \times 10^{-7}$ sec $\qquad N_E = 1 \times 10^{18}$ cm$^{-3}$

$\tau_{B0} = 5 \times 10^{-7}$ sec $\qquad N_B = 1 \times 10^{16}$ cm$^{-3}$

$J_{r0} = 5 \times 10^{-8}$ A/cm$^2$ $\qquad V_{BE} = 0.65$ volt

The following parameters are calculated:

$$p_{E0} = \frac{(1.5 \times 10^{10})^2}{1 \times 10^{18}} = 2.25 \times 10^2 \text{ cm}^{-3}$$

$$n_{B0} = \frac{(1.5 \times 10^{10})^2}{1 \times 10^{16}} = 2.25 \times 10^4 \text{ cm}^{-3}$$

$$L_E = \sqrt{D_E \tau_{E0}} = 10^{-3} \text{ cm}$$

$$L_B = \sqrt{D_B \tau_{B0}} = 3.54 \times 10^{-3} \text{ cm}$$

**Solution:** The emitter injection efficiency factor, from Equation (10–27), is

$$\gamma = \frac{1}{1 + \dfrac{(2.25 \times 10^2)(10)(3.54 \times 10^{-3})}{(2.25 \times 10^4)(25)(10^{-3})} \cdot \dfrac{\tanh{(0.0198)}}{\tanh{(0.050)}}} = 0.9944$$

The base transport factor, from Equation (10–31a), is

$$\alpha_T = \frac{1}{\cosh\left(\dfrac{0.70 \times 10^{-4}}{3.54 \times 10^{-3}}\right)} = 0.9998$$

The recombination factor, from Equation (10–36), is

$$\delta = \frac{1}{1 + \dfrac{5 \times 10^{-8}}{J_{s0}} \exp\left(\dfrac{-0.65}{2(0.0259)}\right)}$$

where

$$J_{s0} = \frac{eD_B n_{B0}}{L_B \tanh\left(\dfrac{x_B}{L_B}\right)} = \frac{(1.6 \times 10^{-19})(25)(2.25 \times 10^4)}{3.54 \times 10^{-3} \tanh{(1.977 \times 10^{-2})}} = 1.29 \times 10^{-9} \text{ A/cm}^2$$

We can then calculate $\delta = 0.99986$. The common base current gain is then

$$\alpha = \gamma \alpha_T \delta = (0.9944)(0.9998)(0.99986) = 0.99406$$

which gives a common emitter current gain of

$$\beta = \frac{\alpha}{1 - a} = \frac{0.99406}{1 - 0.99406} = 167$$

**Comment:** In this example, the emitter injection efficiency is the limiting factor in the current gain.

## 10.4 NONIDEAL EFFECTS

In all previous discussions, we have considered a transistor with uniformly doped regions, low-injection, constant emitter and base widths, an ideal constant energy bandgap, uniform current densities, and junctions which are not in breakdown. If any of these ideal conditions are not present, then the transistor properties will deviate from the ideal characteristics we have derived.

### 10.4.1 Base Width Modulation

We have implicitly assumed that the neutral base width $x_B$ was constant. This base width, however, is a function of the B-C voltage, since the width of the space charge region extending into the base region varies with B-C voltage. As the B-C reverse-bias voltage increases, the B-C space charge region width increases, which reduces $x_B$. A change in the neutral base width will change the collector current as can be observed in Figure 10–15. A

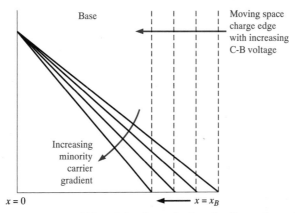

**Figure 10–15** The change in the base width and the change in the minority carrier gradient as the B-C space charge width changes.

reduction in base width will cause the gradient in the minority carrier concentration to increase, which in turn causes an increase in the diffusion current. This effect is known as *base width modulation;* it is also called the *Early effect.*

The Early effect can be seen in the current-voltage characteristics shown in Figure 10–16. In most cases, a constant base current is equivalent to a

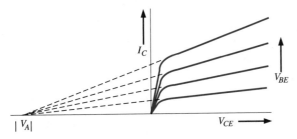

**Figure 10–16** The collector current versus collector-emitter voltage showing the Early effect and Early voltage.

constant B-E voltage. Ideally the collector current is independent of the B-C voltage so that the slope of the curves would be zero; thus the output conductance of the transistor would be zero. However, the base width modulation, or Early effect, produces a nonzero slope and gives rise to a finite output conductance. If the collector current characteristics are extrapolated to zero collector current, the curves intersect the voltage axis at a point which is defined as the Early voltage. The Early voltage is considered to be a positive value. It is a common parameter given in transistor specifications; typical values of Early voltage are in the 100- to 300-volt range.

From Figure 10–16, we can write that

$$\frac{dI_C}{dV_{CE}} \equiv g_0 = \frac{I_C}{V_{CE} + V_A} \tag{10-37a}$$

where $V_A$ and $V_{CE}$ are defined as positive quantities and $g_0$ is defined as the output conductance. Equation (10–37a) can be rewritten in the form

$$I_C = g_0(V_{CE} + V_A) \tag{10-37b}$$

showing explicitly that the collector current is now a function of the C-E voltage or the C-B voltage.

---

**Example 10–5**

**Objective:** To calculate the change in the neutral base width with a change in C-B voltage.

Consider a uniformly doped silicon bipolar transistor at $T = 300°K$ with a base doping of $N_B = 5 \times 10^{16}$ cm$^{-3}$ and a collector doping of $N_C = 2 \times 10^{15}$ cm$^{-3}$. Assume the metallurgical base width is 0.70 $\mu$m. Calculate the change in the neutral base width as the C-B voltage changes from 2 to 10 volts.

**Solution:** The space charge width extending into the base region can be written as

$$x_{dB} = \left\{ \frac{2\varepsilon_s(V_{bi} + V_{CB})}{e} \left[ \frac{N_C}{N_B} \cdot \frac{1}{(N_B + N_C)} \right] \right\}^{1/2}$$

or

$$x_{dB} = \left\{ \frac{2(11.7)(8.85 \times 10^{-14})(V_{bi} + V_{BC})}{1.6 \times 10^{-19}} \times \left[ \frac{2 \times 10^{15}}{5 \times 10^{16}} \cdot \frac{1}{(5 \times 10^{16} + 2 \times 10^{15})} \right] \right\}^{1/2}$$

which becomes

$$x_{dB} = \{(9.96 \times 10^{-12})(V_{bi} + V_{CB})\}^{1/2}$$

The built-in potential is

$$V_{bi} = \frac{kT}{e} \ln \left[ \frac{N_B N_C}{n_i^2} \right] = 0.718 \text{ volt}$$

For $V_{CB} = 2$ volts, we find $x_{dB} = 0.052$ $\mu$m, and for $V_{CB} = 10$ volts, we find $x_{dB} = 0.103$ $\mu$m. If we neglect the B-E space charge region, which will be small due to the forward-biased junction, then we can calculate the neutral base width. For $V_{CB} = 2$ volts,

$$x_B = 0.70 - 0.052 = 0.648 \ \mu m$$

and for $V_{CB} = 10$ volts,

$$x_B = 0.70 - 0.103 = 0.597 \ \mu m$$

**Comment:** This example shows that the neutral base width can easily change by approximately 8 percent as the C-B voltage changes from 2 to 10 volts.

---

**Example 10–6**

**Objective:** To calculate the change in collector current with a change in neutral base width, and to estimate the Early voltage.

Consider a uniformly doped silicon npn bipolar transistor with parameters described in Example 10–5. Assume $D_B = 25$ cm²/sec, $V_{BE} = 0.60$ volt, and also assume that $x_B \ll L_B$.

**Solution:** The excess minority carrier electron concentration in the base is given by Equation (10–7) as

$$\delta n_B(x) = \frac{n_{B0} \left\{ \left[ \exp \left( \frac{eV_{BE}}{kT} \right) - 1 \right] \sinh \left( \frac{x_B - x}{L_B} \right) - \sinh \left( \frac{x}{L_B} \right) \right\}}{\sinh \left( \frac{x_B}{L_B} \right)}$$

If $x_B \ll L_B$, then $(x_B - x) \ll L_B$ so we can write the approximations

$$\sinh \left( \frac{x_B}{L_B} \right) \cong \left( \frac{x_B}{L_B} \right) \quad \text{and} \quad \sinh \left( \frac{x_B - x}{L_B} \right) \cong \left( \frac{x_B - x}{L_B} \right)$$

The expression for $\delta n_B(x)$ can then be approximated as

$$\delta n_B(x) \cong \frac{n_{B0}}{x_B} \left\{ \left[ \exp \left( \frac{eV_{BE}}{kT} \right) - 1 \right] (x_B - x) - x \right\}$$

The collector current is now

$$|J_C| = eD_B \frac{d(\delta n_B(x))}{dx} \cong \frac{eD_B n_{B0}}{x_B} \exp \left( \frac{eV_{BE}}{kT} \right)$$

The value of $n_{B0}$ is calculated as

$$n_{B0} = \frac{n_i^2}{N_B} = \frac{(1.5 \times 10^{10})^2}{5 \times 10^{16}} = 4.5 \times 10^3 \ \text{cm}^{-3}$$

If we let $x_B = 0.648 \ \mu m$ when $V_{CB} = 2$ volts ($V_{CE} = 2.6$ volts), then

$$|J_C| = \frac{(1.6 \times 10^{-19})(25)(4.5 \times 10^3)}{0.648 \times 10^{-4}} \exp \left( \frac{0.60}{0.0259} \right) = 3.20 \ \text{A/cm}^2$$

Now let $x_B = 0.597 \ \mu m$ when $V_{CB} = 10$ volts ($V_{CE} = 10.6$ volts). In this case we have $|J_C| = 3.47$ A/cm². From Eq. (10–37a), we can write

$$\frac{dJ_C}{dV_{CE}} = \frac{J_C}{V_{CE} + V_A} \cong \frac{\Delta J_C}{\Delta V_{CE}}$$

Using the calculated values of current and voltage, we have

$$\frac{\Delta J_C}{\Delta V_{CE}} = \frac{3.47 - 3.20}{10.6 - 2.6} = \frac{J_C}{V_{CE} + V_A} = \frac{3.20}{2.6 + V_A}$$

The Early voltage is then determined to be

$$V_A \approx 92 \text{ volts}$$

**Comment:** This example indicates how much the collector current can change as the neutral base width changes with a change in the B-C space charge width, and also indicates the magnitude of the Early voltage.

---

The example demonstrates, too, that we can expect variations in transistor properties due to tolerances in transistor-fabrication processes. There will be variations, in particular, in the base width of narrow-base transistors that will cause variations in the collector current characteristics simply due to the tolerances in processing.

### 10.4.2 High Injection

The ambipolar transport equation that we have used to determine the minority carrier distributions assumed low injection. As $V_{BE}$ increases, the injected minority carrier concentration may approach, or even become larger than, the majority carrier concentration. If we assume quasi-charge neutrality, then the majority carrier hole concentration in the p-type base at $x = 0$ will increase as shown in Figure 10–17 due to the excess holes.

Two effects occur in the transistor at high injection. The first effect is a reduction in emitter injection efficiency. Since the majority carrier hole concentration at $x = 0$ increases with high injection, more holes are injected back into the emitter due to the forward-biased B-E voltage. An increase in the hole injection causes an increase in the $J_{pE}$ current and an increase in $J_{pE}$ reduces the emitter injection efficiency. The common emitter current gain decreases, then, with high injection. Figure 10–18 shows a typical common emitter current gain versus collector current curve. The low gain at low currents is due to the small recombination factor and the drop-off at the high current is due to the high-injection effect.

We will now consider the second high-injection effect. At low injection, the majority carrier hole concentration at $x = 0$ for the npn transistor is

$$p_p(0) = p_{p0} = N_a \tag{10–38a}$$

and the minority carrier electron concentration is

$$n_p(0) = n_{p0} \exp\left(\frac{eV_{BE}}{kT}\right) \tag{10–38b}$$

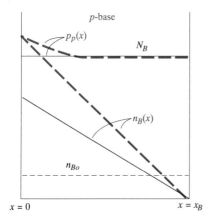

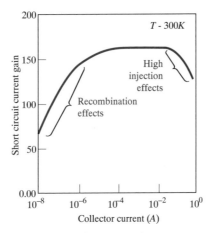

**Figure 10–17**   Minority and majority carrier concentrations in the base under low and high injection (solid line: low injection; dashed line: high injection).

**Figure 10–18**   Common emitter current gain versus collector current. (From Shur [6].)

The *pn* product is

$$p_p(0)n_p(0) = p_{p0}n_{p0} \exp\left(\frac{eV_{BE}}{kT}\right) \qquad (10\text{–}38c)$$

At high injection, Equation (10–38c) still applies. However, $p_p(0)$ will also increase, and for very high injection it will increase at nearly the same rate as $n_p(0)$. The increase in $n_p(0)$ will asymptotically approach the function

$$n_p(0) \approx n_{p0} \exp\left(\frac{eV_{BE}}{2kT}\right) \qquad (10\text{–}39)$$

The excess minority carrier concentration in the base, and hence the collector current, will increase at a slower rate with B-E voltage in high injection than low injection. This effect is shown in Figure 10–19. The high-injection effect is very similar to the effect of a series resistance in a pn junction diode.

### 10.4.3 Emitter Bandgap Narrowing

Another phenomenon affecting the emitter injection efficiency is bandgap narrowing. We implied from our previous discussion that the emitter injection efficiency factor would continue to increase and approach unity as the ratio of emitter doping to base doping continued to increase. As silicon becomes heavily doped, the discrete donor energy level in an n-type emitter splits into a band of energies. The distance between donor atoms decreases as the concentration of impurity donor atoms increases and the splitting of

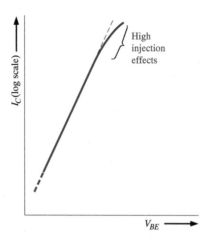

**Figure 10–19**   Collector current versus base-emitter voltage showing high-injection effects.

the donor level is caused by the interaction of donor atoms with each other. As the doping continues to increase, the donor band widens, becomes skewed, and moves up toward the conduction band, eventually merging with it. At this point, the effective bandgap energy has decreased. Figure 10–20 shows a plot of the change in the bandgap energy with impurity doping concentration.

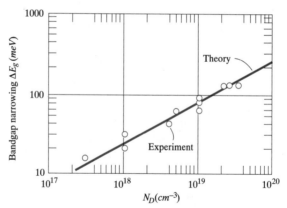

**Figure 10–20**   Bandgap-narrowing factor versus donor impurity concentration in silicon. (From Sze [8].)

A reduction in the bandgap energy increases the intrinsic carrier concentration. The intrinsic carrier concentration is given by

$$n_i^2 = N_c N_v \exp\left(\frac{-E_g}{kT}\right) \tag{10–40}$$

In a heavily doped emitter, the intrinsic carrier concentration can be written as

$$n_{iE}^2 = N_c N_v \exp\left(\frac{-(E_{g0} - \Delta E_g)}{kT}\right) = n_i^2 \exp\left(\frac{\Delta E_g}{kT}\right) \qquad (10\text{-}41)$$

where $E_{g0}$ is the bandgap energy at a low doping concentration and $\Delta E_g$ is the bandgap narrowing factor.

The emitter injection efficiency factor was given by Equation (10-27) as

$$\gamma = \frac{1}{1 + \dfrac{p_{E0}D_E L_B}{n_{B0}D_B L_E} \cdot \dfrac{\tanh(x_B/L_B)}{\tanh(x_E/L_E)}}$$

The term $p_{E0}$ is the thermal-equilibrium minority carrier concentration in the emitter and must be written as

$$p_{E0} = \frac{n_{iE}^2}{N_E} = \frac{n_i^2}{N_E} \exp\left(\frac{\Delta E_g}{kT}\right) \qquad (10\text{-}42)$$

As the emitter doping increases, $\Delta E_g$ increases; thus, $p_{E0}$ does not continue to decrease with increased emitter doping. If $p_{E0}$ starts to increase because of the bandgap narrowing, the emitter injection efficiency begins to fall off rather than continuing to increase with increased emitter doping.

---

**Example 10-7**

**Objective:** To determine the increase in $p_{E0}$ in the emitter due to bandgap narrowing.

Consider a silicon emitter at $T = 300°\text{K}$. Assume the emitter doping increases from $10^{18}$ cm$^{-3}$ to $10^{19}$ cm$^{-3}$. Calculate the change in the $p_{E0}$ value.

**Solution:** For emitter dopings of $N_E = 10^{18}$ cm$^{-3}$ and $10^{19}$ cm$^{-3}$, we have, neglecting bandgap narrowing,

$$p_{E0} = \frac{n_i^2}{N_E} = \frac{(1.5 \times 10^{10})^2}{10^{18}} = 2.25 \times 10^2 \text{ cm}^{-3}$$

and
$$p_{E0} = \frac{(1.5 \times 10^{10})^2}{10^{19}} = 2.25 \times 10^1 \text{ cm}^{-3}$$

Taking into account the bandgap narrowing, we obtain, respectively, for $N_E = 10^{18}$ cm$^{-3}$ and $N_E = 10^{19}$ cm$^{-3}$

$$p_{E0} = \frac{(1.5 \times 10^{10})^2}{10^{18}} \exp\left(\frac{0.030}{0.0259}\right) = 7.16 \times 10^2 \text{ cm}^{-3}$$

and
$$p_{E0} = \frac{(1.5 \times 10^{10})^2}{10^{19}} \exp\left(\frac{0.1}{0.0259}\right) = 1.07 \times 10^3 \text{ cm}^{-3}$$

**Comment:** If the emitter doping increases from $10^{18}$ to $10^{19}$ cm$^{-3}$, the thermal-equilibrium minority carrier concentration actually increases by a factor of 1.5 rather than decreasing by the expected factor of 10. This effect is due to bandgap narrowing.

As the emitter doping increases, the bandgap narrowing factor, $\Delta E_g$, will increase; this can actually cause $p_{E0}$ to increase. As $p_{E0}$ increases, the emitter injection efficiency decreases; this then causes the transistor gain to decrease, as in Figure 10–18. A very high emitter doping may result in a smaller current gain than we anticipate because of the bandgap-narrowing effect.

### 10.4.4 Current Crowding

It is tempting to minimize the effects of base current in a transistor, since the base current is usually much smaller than either the collector or the emitter current. Figure 10–21 is a cross section of an npn transistor showing the

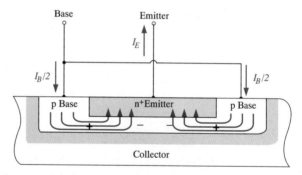

**Figure 10–21**  Cross section of an npn bipolar transistor showing the base current distribution and the lateral potential drop in the base region.

lateral distribution of base current. The base region is typically less than a micrometer thick, so there can be a sizeable base resistance. The nonzero base resistance results in a lateral potential difference under the emitter region. For the npn transistor, the potential decreases from the edge of the emitter toward the center. The emitter is highly doped, so as a first approximation the emitter can be considered an equipotential region.

The number of electrons from the emitter injected into the base is exponentially dependent on the B-E voltage. With the lateral voltage drop in the base between the edge and center of the emitter, more electrons will be injected near the emitter edges than in the center, causing the emitter current to be crowded toward the edges. This current-crowding effect is schematically shown in Figure 10–22. The larger current density near the emitter

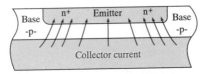

**Figure 10–22** Cross section of an npn bipolar transistor showing the emitter current crowding effect.

edge may cause localized heating effects as well as localized high-injection effects. The nonuniform emitter current also results in a nonuniform lateral base current under the emitter. A two-dimensional analysis would be required to calculate the actual potential drop versus distance because of the nonuniform base current. Another approach is to slice the transistor into a number of smaller parallel transistors and to lump the resistance of each base section into an equivalent external resistance.

Power transistors, designed to handle large currents, require large emitter areas to maintain reasonable current densities. To avoid the current-crowding effect, these transistors are usually designed with narrow emitter widths and fabricated with an interdigitated design. Figure 10–23 shows the basic

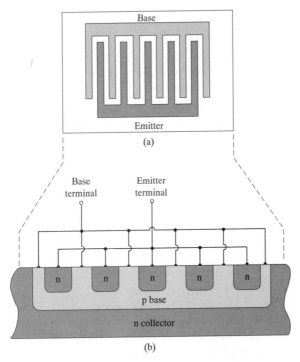

**Figure 10–23** (a) Top view and (b) cross section of an interdigitated npn bipolar transistor structure.

geometry. In effect, many narrow emitters are connected in parallel to achieve the required emitter area.

### *10.4.5 Nonuniform Base Doping

In the analysis of the bipolar transistor, we assumed uniformly doped regions. However, uniform doping rarely occurs. Figure 10–24 shows a doping

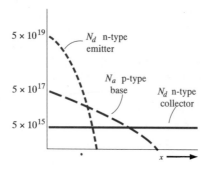

**Figure 10–24**   Impurity concentration profiles of a double diffused npn bipolar transistor.

profile in a doubly diffused npn transistor. We can start with a uniformly doped n-type substrate, diffuse acceptor atoms from the surface to form a compensated p-type base, and then diffuse donor atoms from the surface to form a doubly compensated n-type emitter. The diffusion process results in a nonuniform doping profile.

We determined in Chapter 5 that a graded impurity concentration leads to an induced electric field. For the p-type base region in thermal equilibrium, we can write

$$J_p = e\mu_p N_a E - eD_p \frac{dN_a}{dx} = 0 \qquad (10\text{–}43)$$

Then
$$E = +\left(\frac{kT}{e}\right) \frac{1}{N_a} \frac{dN_a}{dx} \qquad (10\text{–}44)$$

Using the example of Figure 10–24, $dN_a/dx$ is negative; hence the induced electric field is in the negative $x$-direction.

Electrons are injected from the n-type emitter into the base and the minority carrier base electrons begin diffusing toward the collector region. The induced electric field in the base, due to the nonuniform doping, produces a force on the electrons in the direction toward the collector. The induced electric field, then, aids the flow of minority carriers across the base region. This electric field is called an accelerating field.

The accelerating field will produce a drift component of current that is in addition to the existing diffusion current. Since the minority carrier electron concentration varies across the base, the drift current density will not be constant. The total current across the base, however, is nearly constant. The induced electric field in the base due to nonuniform base doping will alter the minority carrier distribution through the base so that the sum of drift current and diffusion current will be a constant. Calculations have shown that the uniformly doped base theory is very useful for estimating the base characteristics.

### 10.4.6 Breakdown Voltage

There are two breakdown mechanisms to consider in a bipolar transistor. The first is called punch-through. As the reverse-bias B-C voltage increases, the B-C space charge region widens and extends further into the neutral base. It is possible for the B-C depletion region to penetrate completely through the base and reach the B-E space charge region, the effect called punch-through. Figure 10–25a shows the energy-band diagram of an npn bipolar transistor in thermal equilibrium and Figure 10–25b shows the energy-band diagram for two values of reverse-bias B-C junction voltage.

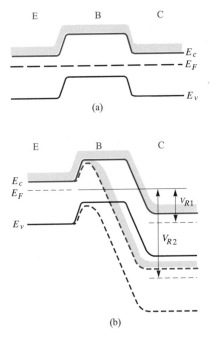

**Figure 10–25** Energy-band diagram of an npn bipolar transistor (a) in thermal equilibrium, and (b) with a reverse-bias B-C voltage before punch-through, $V_{R1}$, and after punch-through, $V_{R2}$.

When a small C-B voltage, $V_{R1}$, is applied, the B-E potential barrier is not affected; thus, the transistor current is still essentially zero. When a large reverse-bias voltage, $V_{R2}$, is applied, the depletion region extends through the base region and the B-E potential barrier is lowered due to the C-B voltage. The lowering of the potential barrier at the B-E junction produces a large increase in current with a very small increase in C-B voltage. This effect is the punch-through breakdown phenomenon.

Figure 10–26 shows the geometry for calculating the punch-through voltage. Assume that $N_B$ and $N_C$ are the uniform impurity doping concentrations

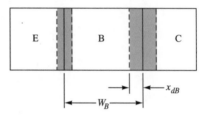

**Figure 10–26**   Geometry of a bipolar transistor to calculate the punch-through voltage.

in the base and collector, respectively. Let $W_B$ be the metallurgical width of the base and let $x_{dB}$ be the space charge width extending into the base from the B-C junction. If we neglect the narrow space charge width of a zero-biased or forward-biased B-E junction, then punch-through, assuming the abrupt junction approximation, occurs when $x_{dB} = W_B$. We can write that

$$x_{dB} = W_B = \left\{ \frac{2\varepsilon_s(V_{bi} + V_{pt})}{e} \cdot \frac{N_C}{N_B} \cdot \frac{1}{N_C + N_B} \right\}^{1/2} \qquad (10\text{–}45)$$

where $V_{pt}$ is the reverse-biased B-C voltage at punch-through. Neglecting $V_{bi}$ compared to $V_{pt}$, we can solve for $V_{pt}$ as

$$V_{pt} = \frac{eW_B^2}{2\varepsilon_s} \cdot \frac{N_B(N_C + N_B)}{N_C} \qquad (10\text{–}46)$$

---

**Example 10–8**

**Objective:** To calculate the punch-through voltage given the impurity doping concentrations and base width.

Consider a uniformly doped silicon bipolar transistor with a metallurgical base width of 0.5 $\mu$m, a base doping of $N_B = 10^{16}$ cm$^{-3}$, and a collector doping of $N_C = 10^{15}$ cm$^{-3}$.

**Solution:** Substituting the parameters into Equation (10–46), we have

$$V_{pt} = \frac{(1.6 \times 10^{-19})(0.5 \times 10^{-4})^2}{2(11.7)(8.85 \times 10^{-14})} \cdot \frac{(10^{16})(1.1 \times 10^{16})}{(10^{15})} = 21.2 \text{ volts}$$

**Comment:** We can compare this calculated punch-through voltage of $V_{pt} = 21.2$ volts to the expected breakdown voltage for a planar silicon pn junction. From Figure 8–23, the expected avalanche breakdown voltage is approximately 300 volts. Obviously punch-through will occur before the normal breakdown voltage in this case.

---

The second breakdown mechanism to consider is avalanche breakdown, but taking into account the gain of the transistor.[1] Figure 10–27a is an npn

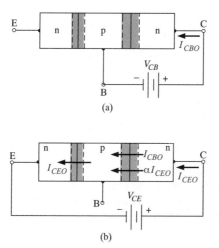

(a)

(b)

**Figure 10–27** (a) Open emitter configuration with saturation current $I_{CBO}$. (b) Open base configuration with saturation current $I_{CEO}$.

transistor with a reverse-bias voltage applied to the B-C junction and with the emitter left open. The current $I_{CBO}$ is the reverse-biased junction current. Figure 10–27b shows the transistor with an applied C-E voltage and with the base terminal left open. This bias condition also makes the B-C junction reverse biased. The current in the transistor for this bias configuration is denoted as $I_{CEO}$.

The current $I_{CBO}$ shown in Figure 10–27b is the normal reverse-biased B-C junction current. Part of this current is due to the flow of minority carrier holes from the collector across the B-C space charge region into the base. The flow of holes into the base makes the base positive with respect to the emitter, and the B-E junction becomes forward biased. The forward-biased B-E junction produces the current $I_{CEO}$, due primarily to the injection of electrons from the emitter into the base. The injected electrons diffuse

---

[1] The doping concentrations in the base and collector of the transistor are assumed to be small enough so that Zener breakdown is not a factor to be considered.

across the base toward the B-C junction. These electrons are subject to all of the recombination processes in the bipolar transistor. When the electrons reach the B-C junction, this current component is $\alpha I_{CEO}$ where $\alpha$ is the common base current gain. We therefore have

$$I_{CEO} = \alpha I_{CEO} + I_{CBO} \qquad (10\text{-}47\text{a})$$

or
$$I_{CEO} = \frac{I_{CBO}}{1 - \alpha} \cong \beta I_{CBO} \qquad (10\text{-}47\text{b})$$

where $\beta$ is the common emitter current gain. The reverse-biased junction current $I_{CBO}$ is multiplied by the current gain $\beta$ when the transistor is biased in the open-base configuration.

When the transistor is biased in the open-emitter configuration as in Figure 10-27a, the current $I_{CBO}$ at breakdown becomes $I_{CBO} \to M I_{CBO}$, where $M$ is the multiplication factor. An empirical approximation for the multiplication factor is usually written as

$$M = \frac{1}{1 - (V_{CB}/BV_{CBO})^n} \qquad (10\text{-}48)$$

where $n$ is an empirical constant, usually between 3 and 6, and $BV_{CBO}$ is the B-C breakdown voltage with the emitter left open.

When the transistor is biased with the base open circuited as shown in Figure 10-27b, the currents in the B-C junction at breakdown are multiplied so that

$$I_{CEO} = M(\alpha I_{CEO} + I_{CBO}) \qquad (10\text{-}49)$$

Solving for $I_{CEO}$, we obtain

$$I_{CEO} = \frac{M I_{CBO}}{1 - \alpha M} \qquad (10\text{-}50)$$

The condition for breakdown corresponds to

$$\alpha M = 1 \qquad (10\text{-}51)$$

Using Equation (10–48) and assuming that $V_{CB} \approx V_{CE}$, Equation (10–51) becomes

$$\frac{\alpha}{1 - (BV_{CEO}/BV_{CBO})^n} = 1 \qquad (10\text{-}52)$$

where $BV_{CEO}$ is the C-E voltage at breakdown in the open base configuration. Solving for $BV_{CEO}$, we find

$$BV_{CEO} = BV_{CBO} \sqrt[n]{1 - \alpha} \qquad (10\text{-}53)$$

where, again, $\alpha$ is the common base current gain. The common emitter and common base current gains are related by

$$\beta = \frac{\alpha}{1 - \alpha} \qquad (10\text{–}54a)$$

Normally $\alpha \approx 1$ so that

$$1 - \alpha \approx \frac{1}{\beta} \qquad (10\text{–}54b)$$

Then Equation (10–53) can be written as

$$BV_{CEO} = \frac{BV_{CBO}}{\sqrt[n]{\beta}} \qquad (10\text{–}55)$$

The breakdown voltage in the open-base configuration is smaller, by the factor $\sqrt[n]{\beta}$, than the actual avalanche junction breakdown voltage. This characteristic is shown in Figure 10–28.

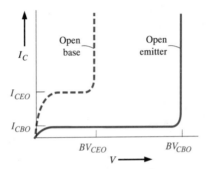

**Figure 10–28**  Relative breakdown voltages and saturation currents of the open base and open emitter configurations.

---

**Example 10–9**

**Objective:** To calculate the breakdown voltage of a bipolar transistor given the impurity doping concentrations and gain factor.

Consider a silicon transistor in which the collector and base impurity doping concentrations are $N_C = 10^{16}$ cm$^{-3}$ and $N_B = 10^{17}$ cm$^{-3}$, respectively, and the common emitter current gain is $\beta = 100$.

**Solution:** The junction breakdown voltage from Chapter 8 is given by

$$BV_{CBO} = \frac{\varepsilon_s E_{crit}^2}{2eN_C}$$

For $N_C = 10^{16}$ cm$^{-3}$, the critical electric field, from Figure 8–22, is approximately $E_{crit} = 4 \times 10^5$ V/cm.

Then $\qquad BV_{CBO} = \dfrac{(11.7)(8.85 \times 10^{-14})(4 \times 10^5)^2}{2(1.6 \times 10^{-19})(10^{16})} = 51.8$ volts

If the empirical constant $n$ in the multiplication factor is $n = 3$, then, from Equation (10–55), the C-E breakdown voltage is

$$BV_{CEO} = \frac{BV_{CBO}}{\sqrt[n]{\beta}} = \frac{51.8}{\sqrt[3]{100}} = 11.2 \text{ volts}$$

**Comment:** The breakdown voltage in the open-base configuration is significantly less than the anticipated junction breakdown voltage. This reduced breakdown voltage effect must be taken into account in the design of transistor circuits.

## 10.5 EQUIVALENT CIRCUIT MODELS

In order to analyze a transistor circuit either by hand calculations or using computer codes, a mathematical model, or equivalent circuit, of the transistor is needed. There are several possible models, each one having certain advantages and disadvantages. A detailed study of all possible models is beyond the scope of this text. However, we will consider two equivalent circuit models. Each of these follows directly from the work we have done on the pn junction diode and on the bipolar transistor. Computer analysis of electronic circuits is more commonly used than hand calculations, but it is instructive to consider the types of transistor model used in computer codes.

It is useful to divide bipolar transistors into two categories—switching and amplification—defined by their use in electronic circuits. Switching usually involves turning a transistor from its "off" state or cutoff to its "on" state, either forward-active or saturation, and then back to its "off" state. Amplification usually involves superimposing sinusoidal signals on dc values so that bias voltages and currents are only perturbed. The Ebers-Moll model is used in switching applications; the hybrid-pi model is used in amplification applications.

### *10.5.1 Ebers-Moll Model

The Ebers-Moll model, or equivalent circuit, is one of the classic models of the bipolar transistor. This particular model is based on the interacting diode junctions and is applicable in any of the transistor operating modes. Figure 10–29 shows the current directions and voltage polarities used in the Ebers-

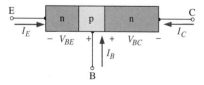

**Figure 10–29** Current direction and voltage polarity definitions for the Ebers-Moll model.

Moll model. The currents are defined as all entering the terminals so that

$$I_E + I_B + I_C = 0 \qquad (10\text{-}56)$$

The direction of the emitter current is opposite to what we have considered up to this point, but as long as we are consistent in the analysis, the defined direction does not matter.

The collector current can be written in general as

$$I_C = \alpha_F I_F - I_R \qquad (10\text{-}57a)$$

where $\alpha_F$ is the common base current gain in the forward-active mode. In this mode, Equation (10-57a) becomes

$$I_C = \alpha_F I_F + I_{CS} \qquad (10\text{-}57b)$$

where the current $I_{CS}$ is the reverse-bias B-C junction current. The current $I_F$ is given by

$$I_F = I_{ES}\left[\exp\left(\frac{eV_{BE}}{kT}\right) - 1\right] \qquad (10\text{-}58)$$

If the B-C junction becomes forward biased, such as in saturation, then we can write the current $I_R$ as

$$I_R = I_{CS}\left[\exp\left(\frac{eV_{BC}}{kT}\right) - 1\right] \qquad (10\text{-}59)$$

Using Equations (10-58) and (10-59), the collector current from Equation (10-57a) can be written as

$$I_C = \alpha_F I_{ES}\left[\exp\left(\frac{eV_{BE}}{kT}\right) - 1\right] - I_{CS}\left[\exp\left(\frac{eV_{BC}}{kT}\right) - 1\right] \qquad (10\text{-}60)$$

We can also write the emitter current as

$$I_E = \alpha_R I_R - I_F \qquad (10\text{-}61)$$

or $\qquad I_E = \alpha_R I_{CS}\left[\exp\left(\frac{eV_{BC}}{kT}\right) - 1\right] - I_{ES}\left[\exp\left(\frac{eV_{BE}}{kT}\right) - 1\right] \qquad (10\text{-}62)$

The current $I_{ES}$ is the reverse-bias B-E junction current and $\alpha_R$ is the common base current gain for the inverse-active mode. Equations (10-60) and (10-62) are the classic Ebers-Moll equations.

The equivalent circuit corresponding to Equations (10-60) and (10-62) is shown in Figure 10-30. The current sources in the equivalent circuit represent current components that depend on voltages across other junctions. The Ebers-Moll model has four parameters: $\alpha_F$, $\alpha_R$, $I_{ES}$, and $I_{CS}$. However, only three parameters are independent. The reciprocity relationship states that

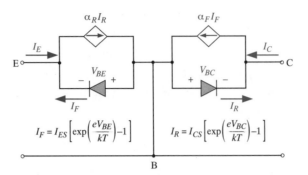

**Figure 10–30**    Basic Ebers-Moll equivalent circuit.

$$\alpha_F I_{ES} = \alpha_R I_{CS} \tag{10–63}$$

Since the Ebers-Moll model is valid in each of the four operating modes, we can, for example, use the model for the transistor in saturation. In the saturation mode, both B-E and B-C junctions are forward-biased so that $V_{BE} > 0$ and $V_{BC} > 0$. The B-E voltage will be a known parameter since we will apply a voltage across this junction. The forward-biased B-C voltage is a result of driving the transistor into saturation and is the unknown to be determined from the Ebers-Moll equations. Normally in electronic circuit applications, the collector-emitter voltage at saturation is of interest. We can define the C-E saturation voltage as

$$V_{CE}(\text{sat}) = V_{BE} - V_{BC} \tag{10–64}$$

We will find an expression for $V_{CE}(\text{sat})$ by combining the Ebers-Moll equations. In the following example we see how the Ebers-Moll equations can be used in a hand calculation, and may also see how a computer analysis would make the calculations easier.

Combining Equations (10–56) and (10–62), we have

$$-(I_B + I_C) = \alpha_R I_{CS}\left[\exp\left(\frac{eV_{BC}}{kT}\right) - 1\right] - I_{ES}\left[\exp\left(\frac{eV_{BE}}{kT}\right) - 1\right] \tag{10–65}$$

If we solve for $[\exp(eV_{BC}/kT) - 1]$ from Equation (10–65), and substitute the resulting expression into Equation (10–60), we can then find $V_{BE}$ as

$$V_{BE} = V_t \ln\left\{\frac{I_C(1 - \alpha_R) + I_B + I_{ES}(1 - \alpha_F\alpha_R)}{I_{ES}(1 - \alpha_F\alpha_R)}\right\} \tag{10–66}$$

where $V_t$ is the thermal voltage. Similarly, if we solve for $[\exp(eV_{BE}/kT) - 1]$ from Equation (10–60), and substitute this expression into Equation (10–65), we can find $V_{BC}$ as

$$V_{BC} = V_t \ln\left\{\frac{\alpha_F I_B - (1 - \alpha_F)I_C + I_{CS}(1 - \alpha_F\alpha_R)}{I_{CS}(1 - \alpha_F\alpha_R)}\right\} \tag{10–67}$$

We may neglect the $I_{ES}$ and $I_{CS}$ terms in the numerators of Equations (10–66) and (10–67). Solving for $V_{CE}(\text{sat})$, we have

$$V_{CE}(\text{sat}) = V_{BE} - V_{CB} = V_t \ln \left\{ \frac{I_C(1 - \alpha_R) + I_B}{\alpha_F I_B - (1 - \alpha_F)I_C} \cdot \frac{I_{CS}}{I_{ES}} \right\} \qquad (10\text{–}68)$$

The ratio of $I_{CS}$ to $I_{ES}$ can be written in terms of $\alpha_F$ and $\alpha_R$ from Equation (10–63). We can finally write

$$V_{CE}(\text{sat}) = V_t \ln \left\{ \frac{I_C(1 - \alpha_R) + I_B}{\alpha_F I_B - (1 - \alpha_F)I_C} \cdot \frac{\alpha_F}{\alpha_R} \right\} \qquad (10\text{–}69)$$

---

**Example 10–10**

**Objective:** To calculate the collector-emitter saturation voltage of a bipolar transistor at $T = 300°\text{K}$.

Assume that $\alpha_F = 0.99$, $\alpha_R = 0.20$, $I_C = 1$ mA, and $I_B = 50$ μA.

**Solution:** Substituting the parameters into Equation (10–69), we have

$$V_{CE}(\text{sat}) = (0.0259) \ln \left\{ \frac{(1)(1 - 0.2) + (0.05)}{(0.99)(0.05) - (1 - 0.99)(1)} \left( \frac{0.99}{0.20} \right) \right\} = 0.121 \text{ volt}$$

**Comment:** This $V_{CE}(\text{sat})$ value is typical of collector-emitter saturation voltages. Because of the log function, $V_{CE}(\text{sat})$ is not a strong function of $I_C$ or $I_B$.

---

## 10.5.2 Hybrid-Pi Model

Bipolar transistors are commonly used in circuits that amplify time varying or sinusoidal signals. In these linear amplifier circuits, the transistor is biased in the forward-active region and small sinusoidal voltages and currents are superimposed on dc voltages and currents. In these applications, the sinusoidal parameters are of interest so it is convenient to develop a small-signal equivalent circuit of the bipolar transistor using the small-signal admittance parameters of the pn junction developed in Chapter 8.

Figure 10–31a shows an npn bipolar transistor in a common emitter configuration with the small-signal terminal voltages and currents. Figure 10–31b shows the cross section of the npn transistor. The C, B, and E terminals are the external connections to the transistor, while the C', B', and E' points are the idealized internal collector, base, and emitter regions.

We can begin constructing the equivalent circuit of the transistor by considering the various terminals individually. Figure 10–32a shows the equivalent circuit between the external input base terminal and the external emitter terminal. The resistance $r_b$ is the series resistance in the base between the

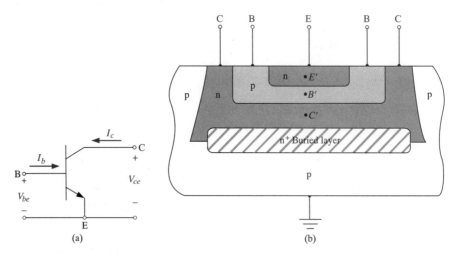

**Figure 10–31**   (a) Common emitter npn bipolar transistor with small-signal currents and voltages. (b) Cross section of an npn bipolar transistor for the hybrid-pi model.

external base terminal B and the internal base region B'. The B'-E' junction is forward biased so $C_\pi$ is the junction diffusion capacitance and $r_\pi$ is the junction diffusion resistance. The diffusion capacitance $C_\pi$ is the same as the diffusion capacitance $C_d$ given by Equation (8–72), and the diffusion resistance $r_\pi$ is the same as the diffusion resistance $r_d$ given by Equation (8–35). The values of both parameters are functions of the junction current. These two elements are in parallel with the junction capacitance, which is $C_{je}$. Finally, $r_{ex}$ is the series resistance between the external emitter terminal and the internal emitter region. This resistance is usually very small and may be on the order of 1 to 2 ohms.

Figure 10–32b shows the equivalent circuit looking into the collector terminal. The $r_c$ resistance is the series resistance between the external and internal collector connections and the capacitance $C_s$ is the junction capacitance of the reverse-biased collector-substrate junction. The dependent current source, $g_m V_{b'e'}$, is the collector current in the transistor which is controlled by the internal base-emitter voltage. The resistance $r_0$ is the inverse of the output conductance $g_0$ and is primarily due to the Early effect.

Finally, Figure 10–32c shows the equivalent circuit of the reverse-biased B'-C' junction. The $C_\mu$ parameter is the reverse-biased junction capacitance and $r_\mu$ is the reverse-biased diffusion resistance. Normally, $r_\mu$ is on the order of megohms and can be neglected. The value of $C_\mu$ is usually much smaller than $C_\pi$ but, because of the feedback effect which leads to the Miller effect and Miller capacitance, $C_\mu$ cannot be ignored in most cases. The Miller capacitance is the equivalent capacitance between B' and E' due to $C_\mu$ and the feedback effect, which includes the gain of the transistor. The Miller

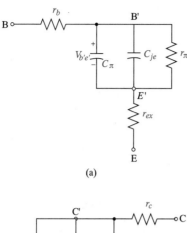

(a)

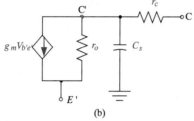

(b)

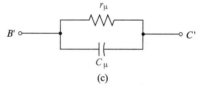

(c)

**Figure 10–32** Components of the hybrid-pi equivalent circuit between (a) the base and emitter, (b) the collector and emitter, and (c) the base and collector.

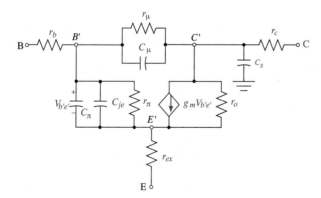

**Figure 10–33** Hybrid-pi equivalent circuit.

effect also reflects $C_\mu$ between the C' and E' terminals at the output. However, the effect on the output characteristics can usually be ignored.

The complete hybrid-pi equivalent circuit is shown in Figure 10–33. A computer simulation is usually required for this complete model because of the large number of elements. However, some simplifications can be made in order to gain an appreciation for the frequency effects of the bipolar transistor. The capacitances lead to frequency effects in the transistor which means that the gain, for example, is a function of the input signal frequency.

### Example 10–11

**Objective:** To determine, to a first approximation, the frequency at which the small-signal current gain decreases to $1/\sqrt{2}$ of its low frequency value.

Consider the simplified hybrid-pi circuit shown in Figure 10–34. We are ignoring $C_\mu$, $C_s$, $r_\mu$, $C_{je}$, $r_0$, and the series resistances. We must emphasize that this is a first order calculation and that $C_\mu$ normally cannot be neglected.

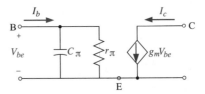

**Figure 10–34**    Simplified hybrid-pi equivalent circuit.

**Solution:** At very low frequency, we may neglect $C_\pi$ so that

$$V_{be} = I_b r_\pi \qquad \text{and} \qquad I_c = g_m V_{be} = g_m r_\pi I_b$$

We can then write

$$h_{fe0} = \frac{I_c}{I_b} = g_m r_\pi$$

where $h_{fe0}$ is the low-frequency, small-signal common emitter current gain.

Taking into account $C_\pi$, we have

$$V_{be} = I_b \left[ \frac{r_\pi}{1 + j\omega r_\pi C_\pi} \right]$$

Then
$$I_c = g_m V_{be} = I_b \left[ \frac{h_{fe0}}{1 + j\omega r_\pi C_\pi} \right]$$

or the small-signal current gain can be written as

$$A_i = \frac{I_c}{I_b} = \left[ \frac{h_{fe0}}{1 + j\omega r_\pi C_\pi} \right]$$

The magnitude of the current gain drops to $1/\sqrt{2}$ of its low-frequency value at $f = 1/2\pi r_\pi C_\pi$.

If, for example, $r_\pi = 2.6k\Omega$ and $C_\pi = 4pF$, then

$$f = 15.3 \text{ MHz}$$

**Comment:** High-frequency transistors must have small diffusion capacitances, implying the use of small devices.

---

## 10.6 FREQUENCY LIMITATIONS

The hybrid-pi equivalent circuit, developed in the last section, introduces frequency effects through the capacitor-resistor circuits. We will now discuss the various physical factors in the bipolar transistor affecting the frequency limitations of the device, and then define the transistor cutoff frequency, which is a figure of merit for a transistor.

### 10.6.1 Time-Delay Factors

The bipolar transistor is a transit time device. When the voltage across the B-E junction increases, for example, additional carriers from the emitter are injected into the base, diffuse across the base, and are collected in the collector region. As the frequency increases, this transit time can become comparable to the period of the input signal. At this point, the output response will no longer be in phase with the input and the magnitude of the current gain will decrease.

The total emitter-to-collector time constant or delay time is composed of four separate time constants. We can write

$$\tau_{ec} = \tau_e + \tau_b + \tau_d + \tau_c \tag{10-70}$$

where
$\tau_{ec}$ = emitter-to-collector time delay
$\tau_e$ = emitter-base junction capacitance charging time
$\tau_b$ = base transit time
$\tau_d$ = collector depletion region transit time
$\tau_c$ = collector capacitance charging time

The equivalent circuit of the forward-biased B-E junction was given in Figure 10–32a. The capacitance $C_{je}$ is the junction capacitance. If we ignore the series resistance, then the emitter-base junction capacitance charging time is

$$\tau_e = r_e'(C_{je} + C_p) \tag{10-71}$$

where $r'_e$ is the emitter junction or diffusion resistance. The capacitance $C_p$ includes any parasitic capacitance between the base and emitter. The resistance $r'_e$ is found as the inverse of the slope of the $I_E$ versus $V_{BE}$ curve. We obtain

$$r'_e = \frac{kT}{e} \cdot \frac{1}{I_E} \tag{10-72}$$

where $I_E$ is the dc emitter current.

The second term, $\tau_b$, is the base transit time, the time required for the minority carriers to diffuse across the neutral base region. The base transit time is related to the diffusion capacitance $C_\pi$ of the B-E junction. For the npn transistor, the electron current density in the base can be written as

$$J_n = -en_B(x)v(x) \tag{10-73}$$

where $v(x)$ is an average velocity. We can write

$$v(x) = dx/dt \qquad \text{or} \qquad dt = dx/v(x) \tag{10-74}$$

The transit time can then be found by integrating, or

$$\tau_b = \int_0^{x_B} dt = \int_0^{x_B} \frac{dx}{v(x)} = \int_0^{x_B} \frac{en_B(x)\,dx}{(-J_n)} \tag{10-75}$$

The electron concentration in the base is approximately linear (see Example 10-6) so we can write

$$n_B(x) \cong n_{B0} \left[ \exp \left( \frac{eV_{BE}}{kT} \right) \right] \left( 1 - \frac{x}{x_B} \right) \tag{10-76}$$

and the electron current density is given by

$$J_n = eD_n \frac{dn_B(x)}{dx} \tag{10-77}$$

The base transit time is then found by combining Equations (10-76) and (10-77) with Equation (10-75). We find that

$$\tau_b = \frac{x_B^2}{2D_n} \tag{10-78}$$

The third time delay factor is $\tau_d$, the collector depletion region transit time. Assuming that the electrons in the npn device travel across the B-C space charge region at their saturation velocity, we have

$$\tau_d = \frac{x_{dc}}{v_s} \tag{10-79}$$

where $x_{dc}$ is the B-C space charge width and $v_s$ is the electron saturation velocity.

The fourth time delay factor, $\tau_e$, is the collector capacitance charging time. The B-C is reverse biased so that the diffusion resistance in parallel with the junction capacitance is very large. The charging time constant is then a function of the collector series resistance $r_c$. We can write

$$\tau_c = r_c(C_\mu + C_s) \tag{10-80}$$

where $C_\mu$ is the B-C junction capacitance and $C_s$ is the collector-to-substrate capacitance. The series resistance in small epitaxial transistors is usually small; thus the time delay $\tau_c$ may be neglected in some cases.

Example calculations of the various time delay factors will be given in the next section as part of the cutoff frequency discussion.

## 10.6.2 Transistor Cutoff Frequency

The current gain as a function of frequency was developed in Example 10–11 so that we can also write the common base current gain as

$$\alpha = \frac{\alpha_0}{1 + j\dfrac{f}{f_\alpha}} \tag{10-81}$$

where $\alpha_0$ is the low-frequency common base current gain and $f_\alpha$ is defined as the alpha cutoff frequency. The frequency $f_\alpha$ is related to the emitter-to-collector time delay $\tau_{ec}$ as

$$f_\alpha = \frac{1}{2\pi\tau_{ec}} \tag{10-82}$$

When the frequency is equal to the alpha cutoff frequency, the magnitude of the common base current gain is $1/\sqrt{2}$ of its low frequency value.

We can relate the alpha cutoff frequency to the common emitter current gain by considering

$$\beta = \frac{\alpha}{1 - \alpha} \tag{10-83}$$

We may replace $\alpha$ in Equation (10–83) with the expression given by Equation (10–81). When the frequency $f$ is of the same order of magnitude as $f_\alpha$, then

$$|\beta| = \left| \frac{\alpha}{1 - \alpha} \right| \cong \frac{f_\alpha}{f} \tag{10-84}$$

where we have assumed that $\alpha_0 \approx 1$. When the signal frequency is equal to the alpha cutoff frequency, the magnitude of the common emitter current gain is equal to unity. The usual notation is to define this *cutoff frequency* as $f_T$, so we have

$$f_T = \frac{1}{2\pi\tau_{ec}} \tag{10-85}$$

From the analysis in Example 10–11, we may also write the common emitter current gain as

$$\beta = \frac{\beta_0}{1 + j(f/f_\beta)} \tag{10-86}$$

where $f_\beta$ is called the beta cutoff frequency and is the frequency at which the magnitude of the common emitter current gain $\beta$ drops to $1/\sqrt{2}$ of its low-frequency value.

Combining Equations (10–83) and (10–81), we can write

$$\beta = \frac{\alpha}{1 - \alpha} = \frac{\dfrac{\alpha_0}{1 + j(f/f_T)}}{1 - \dfrac{\alpha_0}{1 + j(f/f_T)}} = \frac{\alpha_0}{1 - \alpha_0 + j(f/f_T)} \tag{10-87}$$

or

$$\beta = \frac{\alpha_0}{(1 - \alpha_0)\left[1 + j\dfrac{f}{(1 - \alpha_0)f_T}\right]} \cong \frac{\beta_0}{1 + j\dfrac{\beta_0 f}{f_T}} \tag{10-88}$$

where

$$\beta_0 = \frac{\alpha_0}{1 - \alpha_0} \cong \frac{1}{1 - \alpha_0}$$

Comparing Equations (10–88) and (10–86), the beta cutoff frequency is related to the cutoff frequency by

$$f_\beta \cong \frac{f_T}{\beta_0} \tag{10-89}$$

Figure 10–35 shows a Bode plot of the common emitter current gain as a function of frequency and shows the relative values of the beta and cutoff frequencies. Keep in mind that the frequency is plotted on a log scale, so $f_\beta$ and $f_T$ usually have significantly different values.

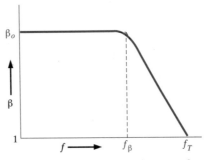

**Figure 10–35**   Bode plot of common emitter current gain versus frequency.

## Example 10–12

**Objective:** To calculate the emitter-to-collector transit time and the cutoff frequency of a bipolar transistor given the transistor parameters.

Consider a silicon npn transistor at $T = 300°K$. Assume the following parameters:

$I_E = 1$ mA     $C_{je} = 1$ pF
$x_B = 0.5$ μm     $D_n = 25$ cm²/sec
$x_{dc} = 2.4$ μm     $r_c = 20$ Ω
$C_\mu = 0.1$ pF     $C_s = 0.1$ pF

**Solution:** We will initially calculate the various time delay factors. If we neglect the parasitic capacitance, the emitter-base junction charging time is

$$\tau_e = r'_e C_{je}$$

where

$$r'_e = \frac{kT}{e} \cdot \frac{1}{I_E} = \frac{0.0259}{1 \times 10^{-3}} = 25.9 \text{ ohms}$$

Then

$$\tau_e = (25.9)(10^{-12}) = 25.9 \text{ psec}$$

The base transit time is

$$\tau_b = \frac{x_B^2}{2D_n} = \frac{(0.5 \times 10^{-4})^2}{2(25)} = 50 \text{ psec}$$

The collector depletion region transit time is

$$\tau_d = \frac{x_{dc}}{v_s} = \frac{2.4 \times 10^{-4}}{10^7} = 24 \text{ psec}$$

The collector capacitance charging time is

$$\tau_c = r_c(C_\mu + C_s) = (20)(0.2 \times 10^{-12}) = 4 \text{ psec}$$

The total emitter-to-collector time delay is then

$$\tau_{ec} = 25.9 + 50 + 24 + 4 = 103.9 \text{ psec}$$

so that the cutoff frequency is calculated as

$$f_T = \frac{1}{2\pi\tau_{ec}} = \frac{1}{2\pi(103.9 \times 10^{-12})} = 1.53 \text{ GHz}$$

If we assume a low-frequency common emitter current gain of $\beta = 100$, then the beta cutoff frequency is

$$f_\beta = \frac{f_T}{\beta_0} = \frac{1.53 \times 10^9}{100} = 15.3 \text{ MHz}$$

**Comment:** The design of high-frequency transistors requires small device geometries in order to reduce capacitances, and narrow base widths in order to reduce the base transit time.

## 10.7 LARGE-SIGNAL SWITCHING

Switching a transistor from one state to another is strongly related to the frequency characteristics just discussed. However, switching is considered to be a large-signal change whereas the frequency effects assumed only small changes in the magnitude of the signal.

### 10.7.1 Switching Characteristics

Consider an npn transistor in the circuit shown in Figure 10–36a switching from cutoff to saturation, and then switching back from saturation to cutoff. We will describe the physical processes taking place in the transistor during the switching cycle.

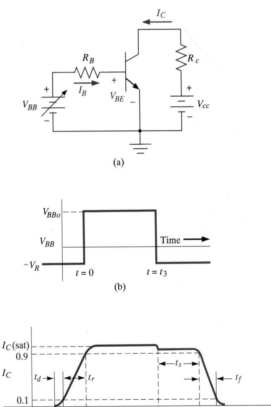

(a)

(b)

(c)

**Figure 10–36**   (a) Circuit used for transistor switching. (b) Input base drive for transistor switching. (c) Collector current versus time during transistor switching.

Consider, initially, the case of switching from cutoff to saturation. Assume that in cutoff $V_{BE} \approx V_{BB} < 0$, thus the B-E junction is reverse biased. At $t = 0$, assume that $V_{BB}$ switches to a value of $V_{BB0}$ as shown in Figure 10–36b. We will assume that $V_{BB0}$ is sufficiently positive to eventually drive the transistor into saturation. For $0 \leq t \leq t_1$, the base current supplies charge to bring the B-E junction from reverse bias to a slight forward bias. The space charge width of the B-E junction is narrowing and ionized donors and acceptors are being neutralized. A small amount of charge is also injected into the base during this time. The collector current increases from zero to 10 percent of its final value during this time period, referred to as the delay time.

During the next time period, $t_1 \leq t \leq t_2$, the base current is supplying charge which increases the B-E junction voltage from near cutoff to near saturation. During this time, additional carriers are being injected into the base so that the gradient of the minority carrier electron concentration in the base increases, causing the collector current to increase. We refer to this time period as the rise time, during which the collector current increases from 10 percent to 90 percent of the final value. For $t > t_2$, the base drive continues to supply base current, driving the transistor into saturation and establishing the final minority carrier distribution in the device.

The switching of the transistor from saturation to cutoff involves removing all of the excess minority carriers stored in the emitter, base, and collector regions. Figure 10–37 shows the charge storage in the base and collector

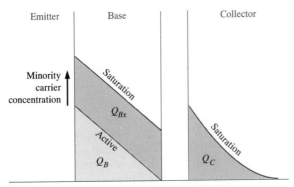

**Figure 10–37** Charge storage in the base and collector at saturation and in the active mode.

when the transistor is in saturation. The charge $Q_B$ is the excess charge stored in a forward-active transistor and $Q_{BX}$ and $Q_C$ are the extra charges stored when the transistor is biased in saturation. At $t = t_3$, the base voltage $V_{BB}$ switches to a negative value of $(-V_R)$. The base current in the transistor reverses direction as was the case in switching a pn junction diode from forward to reverse bias. The reverse base current pulls the excess stored carriers from the emitter and base regions. Initially, the collector current does not change significantly since the gradient of the minority carrier concentration in the base does not change instantaneously. Recall that when the transistor is biased in saturation, both the B-E and B-C junctions are forward biased. The charge $Q_{BX}$ in the base must be removed to reduce the forward-biased B-C voltage to zero volts before the collector current can change. This time delay is called the *storage time* and is denoted by $t_s$. The storage time is the time between the point at which $V_{BB}$ switches to the time when the collector current is reduced to 90 percent of its maximum saturation value. The storage time is usually the most important parameter in the switching speed of the bipolar transistor.

The final switching delay time is the fall time $t_f$ during which the collector current decreases from the 90 percent to the 10 percent value. During this time, the B-C junction is reverse biased but excess carriers in the base are still being removed and the B-E junction voltage is decreasing.

The switching-time response of the transistor can be determined using the Ebers-Moll model. The frequency-dependent gain parameters must be used and normally the Laplace transform technique is used to obtain the time response. The details of this analysis are quite tedious and will not be presented here.

### 10.7.2 The Schottky Clamped Transistor

One method frequently employed to reduce the storage time and increase the switching speed is the use of a Schottky clamped transistor. This is a normal npn bipolar device with a Schottky diode connected between base and collector, as shown in Figure 10–38a. The circuit symbol for the Schottky clamped transistor is shown in Figure 10–38b. When the transistor is biased in the forward-active mode, the B-C junction is reverse biased; hence the Schottky diode is reverse-biased and effectively out of the circuit. The characteristics of the Schottky clamped transistor—or simply the Schottky transistor—are those of the normal npn bipolar device.

When the transistor is driven into saturation, the B-C junction becomes forward biased; hence the Schottky diode also becomes forward biased. We may recall from our discussion in the previous chapter that the effective turn-on voltage of the Schottky diode is approximately half that of the pn junction. The difference in turn-on voltage means that most of the excess base current will be shunted through the Schottky diode and away from the base so that the amount of excess stored charge in the base and collector is drastically reduced. The excess minority carrier concentration in the base and collector at the B-C junction is an exponential function of $V_{BC}$. If $V_{BC}$ is reduced from 0.5 volt to 0.3 volt, for example, the excess minority carrier concentration is reduced by over three orders of magnitude. The reduced excess stored charge in the base of the Schottky transistor greatly reduces

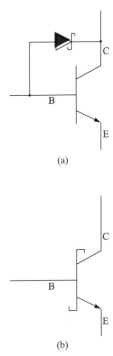

(a)

(b)

**Figure 10–38**   (a) The Schottky clamped transistor. (b) Circuit symbol of the Schottky clamped transistor.

the storage time—storage times on the order of 1 ns or less are common in Schottky transistors.

## *10.8 HETEROJUNCTION BIPOLAR TRANSISTORS

One of the basic limitations of the current gain in the bipolar transistor is the emitter injection efficiency. The emitter injection efficiency $\gamma$ can be increased by reducing the value of the thermal-equilibrium minority carrier concentration $p_{E0}$. However, as the emitter doping increases, the bandgap-narrowing effect offsets any improvement in the emitter injection efficiency. One possible solution is to use a wide-bandgap material for the emitter which will minimize the injection of carriers from the base back into the emitter.

Figure 10–39a shows a discrete aluminum gallium arsenide/gallium arsenide heterojunction bipolar transistor, and Figure 10–39b shows the band diagram of the N-AlGaAs emitter to p-GaAs base junction. The large potential barrier $V_h$ limits the number of holes that will be injected back from the base into the emitter.

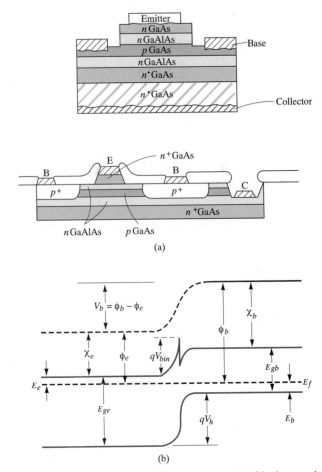

Figure 10–39   (a) Cross section of AlGaAs/GaAs heterojunction bipolar transistor showing a discrete and integrated structure. (b) Energy-band diagram of the N-AlGaAs emitter and p-GaAs base junction. (From Tiwari et al. [9].)

The intrinsic carrier concentration is a function of bandgap energy as

$$n_i^2 \propto \exp\left(\frac{-E_g}{kT}\right)$$

For a given emitter doping, the number of minority carrier holes injected into the emitter is reduced by a factor of

$$\exp\left(\frac{\Delta E_g}{kT}\right)$$

in changing from a narrow- to wide-bandgap emitter. If $\Delta E_g = 0.30$ eV, for example, $n_i^2$ would be reduced by approximately $10^5$ at $T = 300°$K. The drastic reduction in $n_i^2$ for the wide-bandgap emitter means that the requirements of a very high emitter doping can be relaxed and a high emitter injection efficiency can still be obtained. A lower emitter doping reduces the bandgap-narrowing effect.

The heterojunction GaAs bipolar transistor has the potential of being a very high frequency device. A lower emitter doping in the wide-bandgap emitter leads to a smaller junction capacitance, increasing the speed of the device. Also, for the GaAs npn device, the minority carriers in the base are electrons with a high mobility. The electron mobility in GaAs is approximately five times that in silicon; thus, the base transit time in the GaAs base is very short. Experimental AlGaAs/GaAs heterojunction transistors with base widths on the order of 0.1 $\mu$m have shown cutoff frequencies on the order of 40 GHz.

One disadvantage of GaAs is the low minority carrier lifetime. The small lifetime is not a factor in the base of a narrow-base device, but results in a larger B-E recombination current, which decreases the recombination factor and reduces the current gain. A current gain of 150 has been reported.

## 10.9 SUMMARY AND REVIEW

The physics of the bipolar transistor was considered in this chapter. The bipolar transistor is a multijunction semiconductor device that, in conjunction with a circuit, is capable of current gain, voltage gain, and signal-power gain.

In the forward-active mode, the B-E junction is forward biased and the B-C junction is reverse biased. Majority carriers from the emitter are injected into the base by the forward-biased B-E voltage where they become minority carriers. A large gradient in the minority carrier concentration is created in the base since the B-C is reverse biased and since the base width is narrow. Minority carriers diffuse across the base into the B-C space charge region, where they are swept into the collector region by the electric field. The collector current, then, is a function of the number of carriers being injected from the emitter, which in turn is a function of the B-E voltage. The basic bipolar transistor action is the control of a current in one part of the device, the collector current, by a voltage in another part of the device, the B-E voltage.

The various current components in the device are determined by the diffusion of minority carriers, whose distributions in the device we determined as functions of electrical and geometrical parameters. We then calculated diffusion currents using these minority carrier distributions.

The common base current gain, which leads to the common emitter current gain, is a primary parameter characterizing the bipolar transistor. The common base current gain is a function of several factors including emitter injection efficiency, base transport factor, and the recombination factor. Emitter injection efficiency takes into account carriers from the base that are injected back into the emitter by the forward-biased B-E voltage. These carriers do not contribute to the collector current and so are not part of the transistor action. The base transport factor takes into account carriers that do not reach the collector due to recombination within the base region. The recombination factor takes into account carriers that are injected from the emitter and are lost due to recombination within the forward-biased B-E junction. The recombination factor should also take into account surface recombination effects.

Several nonideal effects in the bipolar transistor were described. Base width modulation, or Early effect, is the change in collector current with a change in B-C voltage, producing a change in the neutral base width. As the reverse-bias B-C voltage increases, the neutral base width decreases and the collector current increases. Since the collector current is a function of the B-C voltage, the transistor has a finite output impedance. If the n-type emitter doping concentration, for example, is increased to improve the emitter injection efficiency factor, the donor impurity energy level can split into a band of energies and merge into the conduction band. The bandgap energy is effectively reduced, lowering the emitter injection efficiency factor and reducing current gain. The heterojunction bipolar transistor, discussed later in the chapter, is one method of increasing the emitter injection efficiency without encountering the bandgap-narrowing effect.

A lateral voltage drop in the base under the emitter is due to a finite base resistance and causes a nonuniform emitter current density, largest at the emitter edges. The emitter current crowding effect may cause localized heating within the device and may result in localized high-injection effects. For power transistors, especially, the emitter width must be minimized to prevent current crowding.

We considered two voltage breakdown mechanisms in the bipolar transistor. The first is punch-through, in which the B-C space charge region expands completely through the base region as the reverse-bias B-C voltage increases. The second is the avalanche multiplication process. If the base terminal is left open, the avalanche multiplication in the reverse-biased B-C region will be further multiplied by the gain of the transistor. The breakdown voltage in this configuration can be significantly smaller than that of a planar pn junction and will need to be accounted for in a circuit design.

In order to analyze a transistor circuit, one must develop a mathematical model or equivalent circuit of the transistor. The Ebers-Moll model and equivalent circuit was initially developed. This model is based on the interacting junctions and is applicable in any of the transistor operating modes. The hybrid-pi equivalent circuit was developed using the small-signal admittance parameters of a pn junction. This equivalent circuit applies to transistors used in linear amplifier circuits, and includes capacitances so that the small-signal frequency characteristics of the transistor and circuit can be analyzed.

The frequency limitations of the transistor include the B-E and B-C capacitance charging times and the base transit time as the primary limiting factors. The design of high-frequency devices implies the use of small geometries to minimize capacitance, and the use of narrow base widths to minimize the base transit time. The cutoff frequency, a parameter that is a figure of merit for the transistor, is the frequency at which the magnitude of the common emitter current gain becomes equal to unity.

The switching characteristics are closely related to the frequency limitations although switching involves large changes in currents and voltages. We followed the switching of a transistor from cutoff to saturation and back to cutoff. An important parameter in switching is the charge storage time, which can be minimized by using a Schottky clamped transistor.

## GLOSSARY OF IMPORTANT TERMS

**Alpha cutoff frequency:** The frequency at which the magnitude of the common base current is $1/\sqrt{2}$ of its low-frequency value; also equal to the cutoff frequency.

**Bandgap narrowing:** The reduction in the forbidden-energy bandgap with high emitter doping concentration.

**Base transit time:** The time that it takes a minority carrier to cross the neutral base region.

**Base transport factor:** The factor in the common base current gain that accounts for recombination in the neutral base width.

**Base width modulation:** The change in the neutral base width with C-E or C-B voltage.

**Beta cutoff frequency:** The frequency at which the magnitude of the common emitter current gain is $1/\sqrt{2}$ of its low frequency value.

**Collector capacitance charging time:** The time constant that describes the time required for the B-C and collector-substrate space charge widths to change with a change in emitter current.

**Collector depletion region transit time:** The time that it takes a carrier to be swept across the B-C space charge region.

**Common base current gain:** The ratio of collector current to emitter current.

**Common emitter current gain:** The ratio of collector current to base current.

**Current crowding:** The nonuniform current density across the emitter junction area created by a lateral voltage drop in the base region due to a finite base current and base resistance.

**Cutoff:** The bias condition in which zero- or reverse-bias voltages are applied to both transistor junctions resulting in zero transistor currents.

**Cutoff frequency:** The frequency at which the magnitude of the common emitter current gain is unity.

**Early effect:** Another term for base width modulation.

**Early voltage:** The value of voltage (magnitude) at the intercept on the voltage axis obtained by extrapolating the $I_C$ versus $V_{CE}$ curves to zero current.

**Emitter-base junction capacitance charging time:** The time constant describing the time for the B-E space charge width to change with a change in emitter current.

**Emitter injection efficiency factor:** The factor in the common base current gain that takes into account the injection of carriers from the base into the emitter.

**Forward active:** The bias condition in which the B-E junction is forward biased and the B-C junction is reverse biased.

**Inverse active:** The bias condition in which the B-E junction is reverse biased and the B-C junction is forward biased.

**Output conductance:** The ratio of a differential change in collector current to the corresponding differential change in C-E voltage.

**Punch-through:** The bias condition such that the B-C space charge region extends completely through the neutral base region to the B-E depletion region.

**Recombination factor:** The factor in the common base current gain that accounts for recombination in the forward-biased B-E junction.

**Saturation:** The bias condition in which both B-E and B-C junctions are forward biased and the collector current is no longer controlled by the B-E voltage.

**Storage time:** The time duration from the point when the input base signal switches from forward to reverse bias, to the point when the collector current is reduced to 90 percent of its saturation value.


## PROBLEMS

(Note: In the following problems, use the transistor geometry shown in Figure 10–7.)


### Section 10.1

1. For a uniformly doped $n^{++}p^{+}n$ bipolar transistor in thermal equilibrium, (a) sketch the energy-band diagram, (b) sketch the electric field through the device, and (c) repeat parts (a) and (b) for the transistor biased in the forward-active region.

2. Consider a $p^{++}n^{+}p$ bipolar transistor, uniformly doped in each region. Sketch the energy-band diagram for the case when the transistor is (a) in thermal equilibrium, (b) biased in the forward-active mode, (c) biased in the inverse-active region, and (d) biased in cutoff with both the B-E and B-C junctions reverse biased.

## Section 10.2

3. A uniformly doped silicon npn bipolar transistor is to be biased in the forward-active mode. The transistor dopings are $N_E = 10^{17}$ cm$^{-3}$, $N_B = 10^{16}$ cm$^{-3}$, and $N_C = 10^{15}$ cm$^{-3}$. (a) For $T = 300°$K, calculate the B-E voltage at which the minority carrier electron concentration at $x = 0$ is 10 percent of the majority carrier hole concentration. (b) At this bias, determine the minority carrier hole concentration at $x' = 0$. (c) Repeat parts (a) and (b) if the temperature is increased to $T = 435°$K.

4. Derive the expressions for the coefficients given by Equations (10–6a) and (10–6b).

5. Derive the expression for the excess minority carrier hole concentration in the base region of a uniformly doped pnp bipolar transistor operating in the forward-active region.

6. The idealized excess minority carrier concentration $\delta n_{B0}(x)$ in the base of an npn transistor is a linear function of distance through the base when the transistor is biased in the forward-active region (see Example 10–6). Using Equation (10–7) for the uniformly doped npn device, determine

$$\frac{\delta n_{B0}(x) - \delta n_B(x)}{\delta n_{B0}(x)} \times 100 \text{ percent}$$

at $x = x_B/2$ for (a) $x_B/L_B = 0.1$ and (b) $x_B/L_B = 1.0$. Assume $V_{BE} \gg kT/e$.

7. Consider a pnp bipolar transistor. Assume that the excess minority carrier hole concentrations at the edges of the B-E and B-C space charge regions are $\delta p_B(0) = 8 \times 10^{14}$ cm$^{-3}$ and $\delta p_B(x_B) = -2.25 \times 10^4$ cm$^{-3}$, respectively. Plot, on the same graph, $\delta p_B(x)$ for (a) the ideal case when no recombination occurs in the base, and (b) the case when $x_B = L_B = 10$ $\mu$m. What can be said about the slopes of the two curves at $x = 0$ and $x = x_B$?

*8. (a) A uniformly doped npn bipolar transistor at $T = 300°$K is biased in saturation. Starting with the continuity equation for minority carriers, show that the excess electron concentration in the base region can be expressed as

$$\delta n_B(x) = n_{B0}\left\{\left[\exp\left(\frac{eV_{BE}}{kT}\right) - 1\right]\left[1 - \frac{x}{x_B}\right] + \left[\exp\left(\frac{eV_{BC}}{kT}\right) - 1\right]\left[\frac{x}{x_B}\right]\right\}$$

for $x_B/L_B \ll 1$ where $x_B$ is the neutral base width. (b) Show that the minority carrier diffusion current in the base is then given by

$$J_n = -\frac{eD_B n_{B0}}{x_B}\left\{\exp\left(\frac{eV_{BE}}{kT}\right) - \exp\left(\frac{eV_{BC}}{kT}\right)\right\}$$

(c) Show that the total excess minority carrier charge (coul/cm$^2$) in the base region is given by

$$\delta Q_{nB} = \frac{-en_{B0}x_B}{2}\left\{\left[\exp\left(\frac{eV_{BE}}{kT}\right) - 1\right] + \left[\exp\left(\frac{eV_{BC}}{kT}\right) - 1\right]\right\}$$

*9.  Consider a silicon pnp bipolar transistor at $T = 300°$K with uniform dopings of $N_E = 5 \times 10^{18}$ cm$^{-3}$, $N_B = 10^{17}$ cm$^{-3}$, and $N_C = 5 \times 10^{15}$ cm$^{-3}$. Let $D_B = 10$ cm$^2$/sec, $x_B = 0.7$ $\mu$m, and assume $x_B \ll L_B$. The transistor is operating in saturation with $J_p = 165$ A/cm$^2$ and $V_{EB} = 0.75$ volt. Determine (a) $V_{CB}$, (b) $V_{EC}$(sat), (c) the #/cm$^2$ of excess minority carrier holes in the base, and (d) the #/cm$^2$ of excess minority carrier electrons in the long collector. Let $L_c = 35$ $\mu$m.

10.  An npn silicon bipolar transistor at $T = 300°$K has uniform dopings of $N_E = 10^{19}$ cm$^{-3}$, $N_B = 10^{17}$ cm$^{-3}$, and $N_C = 7 \times 10^{15}$ cm$^{-3}$. The transistor is operating in the inverse-active mode with $V_{BE} = -2$ volts and $V_{BC} = 0.565$ volt. (a) Sketch the minority carrier distribution through the device. (b) Determine the minority carrier concentrations at $x = x_B$ and $x'' = 0$.

11.  A uniformly doped silicon pnp bipolar transistor at $T = 300°$K with dopings of $N_E = 5 \times 10^{17}$ cm$^{-3}$, $N_B = 10^{16}$ cm$^{-3}$, and $N_C = 5 \times 10^{14}$ cm$^{-3}$ is biased in the inverse-active mode. What is the maximum B-C voltage so that the low-injection condition applies?

12.  Consider a pnp silicon transistor at $T = 300°$K. Assume that the base doping is $N_B = 5 \times 10^{16}$ cm$^{-3}$, the neutral base width is $x_B = 1.2$ $\mu$m, and the minority carrier diffusion length in the base is $L_B = 10$ $\mu$m. (a) For $V_{EB} = V_{CB} = 0$, calculate the minority carrier charge (coul/cm$^2$) in the base. (Assume the abrupt junction approximation applies.) (b) If the B-C and B-E junctions are reverse biased at 1 volt, calculate the minority carrier charge (coul/cm$^2$) in the base. (Assume the neutral base width is unchanged.)

## Section 10.3

13.  The following currents are measured in a uniformly doped npn bipolar transistor:

$$I_{nE} = 1.20 \text{ mA} \qquad I_{pE} = 0.10 \text{ mA}$$
$$I_{nC} = 1.18 \text{ mA} \qquad I_R = 0.20 \text{ mA}$$
$$I_G = 0.001 \text{ mA} \qquad I_{pc0} = 0.001 \text{ mA}$$

Determine (a) $\alpha$, (b) $\gamma$, (c) $\alpha_T$, (d) $\delta$, and (e) $\beta$.

14.  Consider a uniformly doped npn bipolar transistor at $T = 300°$K with the following parameters:

$$N_E = 10^{18} \text{ cm}^{-3} \qquad N_B = 5 \times 10^{16} \text{ cm}^{-3} \qquad N_C = 10^{15} \text{ cm}^{-3}$$
$$D_E = 8 \text{ cm}^2/\text{sec} \qquad D_B = 15 \text{ cm}^2/\text{sec} \qquad D_C = 12 \text{ cm}^2/\text{sec}$$
$$\tau_{E0} = 10^{-8} \text{ sec} \qquad \tau_{B0} = 5 \times 10^{-8} \text{ sec} \qquad \tau_{C0} = 10^{-7} \text{ sec}$$
$$x_E = 0.8 \ \mu\text{m} \qquad x_B = 0.7 \ \mu\text{m} \qquad J_{r0} = 3 \times 10^{-8} \text{ A/} \atop \text{cm}^2$$

For $V_{BE} = 0.60$ volt and $V_{CE} = 5$ volts, calculate ($a$) the currents $J_{nE}$, $J_{pE}$, $J_{nC}$, and $J_R$ and ($b$) the current gain factors $\gamma$, $\alpha_T$, $\delta$, $\alpha$, and $\beta$.

15. ($a$) Calculate and plot, for a bipolar transistor, the base transport factor, $\alpha_T$, as a function of $(x_B/L_B)$ over the range $0.01 \le (x_B/L_B) \le 10$. (Use a log scale on the horizontal axis.) ($b$) Assuming that the emitter injection efficiency and recombination factors are unity, plot the common emitter current gain for the conditions in part ($a$). ($c$) Considering the results of part ($b$), what can be said about the base transport factor being the limiting factor in the common emitter current gain?

16. ($a$) Calculate and plot the emitter injection efficiency as a function of the doping ratio, $N_B/N_E$, over the range $0.01 \le N_B/N_E \le 10$. Assume that $D_E = D_B$, $L_B = L_E$, and $x_B = x_E$. (Use a log scale on the horizontal axis.) Neglect bandgap narrowing effects. ($b$) Assuming that the base transport factor and recombination factors are unity, plot the common emitter current gain for the conditions in part ($a$). ($c$) Considering the results of part ($b$), what can be said about the emitter injection efficiency being the limiting factor in the common emitter current gain.

17. ($a$) Calculate and plot the recombination factor as a function of the forward-bias B-E voltage for $0.1 \le V_{BE} \le 0.6$. Assume the following parameters:

$$D_B = 25 \text{ cm}^2/\text{sec} \qquad D_E = 10 \text{ cm}^2/\text{sec}$$
$$N_E = 5 \times 10^{18} \text{ cm}^{-3} \qquad N_B = 1 \times 10^{17} \text{ cm}^{-3}$$
$$N_C = 5 \times 10^{15} \text{ cm}^{-3} \qquad x_B = 0.7 \ \mu\text{m}$$
$$\tau_{B0} = \tau_{E0} = 10^{-7} \text{ sec} \qquad J_{r0} = 2 \times 10^{-9} \text{ A/cm}^2$$
$$n_i = 1.5 \times 10^{10} \text{ cm}^{-3}$$

($b$) Assuming the base transport and emitter injection efficiency factors are unity, plot the common emitter current gain for the conditions in part ($a$). ($c$) Considering the results of part ($b$), what can be said about the recombination factor being the limiting factor in the common emitter current gain.

18. Consider an npn silicon bipolar transistor at $T = 300°K$ with the following parameters:

$$D_B = 25 \text{ cm}^2/\text{sec} \qquad D_E = 10 \text{ cm}^2/\text{sec}$$
$$\tau_{B0} = 10^{-7} \text{ sec} \qquad \tau_{E0} = 5 \times 10^{-8} \text{ sec}$$
$$N_B = 10^{16} \text{ cm}^{-3} \qquad x_E = 0.5 \ \mu\text{m}$$

The recombination factor, $\delta$, has been determined to be $\delta = 0.998$. We need a common emitter current gain of $\beta = 120$. Assuming that $\alpha_T = \gamma$, determine the maximum base width, $x_B$, and the minimum emitter doping, $N_E$, to achieve this specification.

*19.   (a) The recombination current density, $J_{r0}$, in an npn silicon bipolar transistor at $T = 300°K$ is $J_{r0} = 5 \times 10^{-8}$ A/cm$^2$. The uniform dopings are $N_E = 10^{18}$ cm$^{-3}$, $N_B = 5 \times 10^{16}$ cm$^{-3}$, and $N_C = 10^{15}$ cm$^{-3}$. Other parameters are $D_E = 10$ cm$^2$/sec, $D_B = 25$ cm$^2$/sec, $\tau_{E0} = 10^{-8}$ sec, and $\tau_{B0} = 10^{-7}$ sec. Determine the neutral base width so that the recombination factor is $\delta = 0.995$ when $V_{BE} = 0.55$ volt. (b) If $J_{r0}$ remains constant with temperature, what is the value of $\delta$ when $V_{BE} = 0.55$ volt for the case when the temperature is $T = 400°K$? Use the value of $x_B$ determined in part (a).

### Section 10.4

20.   A silicon pnp bipolar transistor at $T = 300°K$ has uniform dopings of $N_E = 10^{18}$ cm$^{-3}$, $N_B = 10^{16}$ cm$^{-3}$, and $N_C = 10^{15}$ cm$^{-3}$. The metallurgical base width is 1.2 $\mu$m. Let $D_B = 10$ cm$^2$/sec and $\tau_{B0} = 5 \times 10^{-7}$ sec. Assume that the minority carrier hole concentration in the base can be approximated by a linear distribution. Let $V_{EB} = 0.625$ volt. Determine the hole diffusion current density in the base for (a) $V_{CB} = 5$ volts, (b) $V_{CB} = 10$ volts, and (c) $V_{CB} = 15$ volts.

*21.   In a silicon npn bipolar transistor at $T = 300°K$, the doping concentrations in the emitter and collector are $N_E = 10^{18}$ cm$^{-3}$ and $N_C = 10^{15}$ cm$^{-3}$, respectively. The netural base width is 0.65 $\mu$m when $V_{BE} = 0.70$ volt and $V_{CB} = 5$ volts. Let $D_B = 25$ cm$^2$/sec and $\tau_{B0} = 5 \times 10^{-7}$ sec. The minority carrier diffusion current in the base is to increase by no more than 5 percent when $V_{CB}$ increases to 10 volts. (a) Determine the minimum base doping. (b) Estimate the Early voltage.

22.   Consider a uniformly doped silicon npn bipolar transistor in which $x_E = x_B$, $L_E = L_B$, and $D_E = D_B$. Assume that $\alpha_T = \delta = 0.995$ and let $N_B = 10^{17}$ cm$^{-3}$. Calculate and plot the common emitter current gain $\beta$ for $N_E = 10^{17}$, $10^{18}$, $10^{19}$, and $10^{20}$ cm$^{-3}$, and for the case (a) when the bandgap narrowing effect is neglected, and (b) when the bandgap narrowing effect is taken into account.

23.   A silicon pnp bipolar transistor at $T = 300°K$ is to be designed so that the emitter injection efficiency is $\gamma = 0.996$. Assume that $x_E = x_B$, $L_E = L_B$, $D_E = D_B$, and let $N_E = 10^{19}$ cm$^{-3}$. (a) Determine the maximum base doping, taking into account bandgap narrowing. (b) If bandgap narrowing were neglected, what would be the maximum base doping required?

24.   A first-approximation type calculation of the current crowding effect can be made using the geometry shown in Figure 10–40. As-

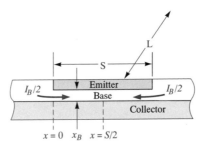

**Figure 10–40**  Figure for problems 24 and 25.

sume that one half of the base current enters from each side of the
emitter strip and flows uniformly to the center of the emitter. Assume the base is p-type with the following parameters:

$$N_B = 10^{16} \text{ cm}^{-3} \qquad\qquad x_B = 0.70 \ \mu m$$
$$\mu_p = 400 \text{ cm}^2/\text{V-sec} \qquad S = 8 \ \mu m$$
$$Emitter \ Length, \qquad\qquad L = 100 \ \mu m$$

(a) Calculate the resistance between $x = 0$ and $x = S/2$. (b) If $(\frac{1}{2})I_B$
$= 10 \ \mu A$, calculate the voltage drop between $x = 0$ and $x = S/2$. (c)
If $V_{BE} = 0.6$ volt at $x = 0$, estimate in percent the number of electrons being injected into the base at $x = S/2$ compared to $x = 0$.

25.  Consider the geometry shown in Figure 10–40 and the device parameters in problem 24 except the emitter width, S. The emitter
width S is to be changed so that the number of electrons injected
into the base at $x = S/2$ is no more than 10 percent less than the
number of electrons injected into the base at $x = 0$. Calculate S.

*26.  The base doping in a diffused $n^+pn$ bipolar transistor can be ap-
proximated by an exponential as

$$N_B = N_B(0) \exp\left(\frac{-ax}{x_B}\right)$$

where $a$ is a constant and is given by

$$a = \ln\left(\frac{N_B(0)}{N_B(x_B)}\right)$$

(a) Show that, in thermal equilibrium, the electric field in the neutral base region is a constant. (b) Indicate the direction of the electric field. Does this electric field aid or retard the flow of minority
carrier electrons across the base? (c) Derive an expression for the
steady-state minority carrier electron concentration in the base under forward bias. Assume no recombination occurs in the base.
(Express the electron concentration in terms of the electron current
density.)

27.  Consider a silicon npn bipolar transistor with uniform dopings of $N_E = 5 \times 10^{18}$ cm$^{-3}$, $N_B = 10^{17}$ cm$^{-3}$, and $N_C = 5 \times 10^{15}$ cm$^{-3}$. Assume the common base current gain is $\alpha = 0.9920$. Determine (a) $BV_{CBO}$, (b) $BV_{CEO}$, and (c) the base-emitter breakdown voltage. (Assume $n = 3$ for the empirical constant.)

28.  A high-voltage silicon npn bipolar transistor is to be fabricated with a uniform base doping of $N_B = 10^{16}$ cm$^{-3}$ and a common emitter current gain of $\beta = 50$. The breakdown voltage $BV_{CEO}$ is to be at least 60 volts. Determine the maximum collector doping and the minimum collector length to support this voltage. (Assume $n = 3$.)

29.  A uniformly doped silicon epitaxial npn bipolar transistor is fabricated with a base doping of $N_B = 3 \times 10^{16}$ cm$^{-3}$ and a heavily doped collector region with $N_C = 5 \times 10^{17}$ cm$^{-3}$. The neutral base width is $x_B = 0.70$ $\mu$m when $V_{BE} = V_{BC} = 0$. Determine $V_{BC}$ at which punch-through occurs. Compare this value to the expected avalanche breakdown voltage of the junction.

30.  A uniformly doped silicon pnp bipolar transistor is fabricated with $N_E = 10^{19}$ cm$^{-3}$ and $N_C = 10^{16}$ cm$^{-3}$. The metallurgical base width is 0.75 $\mu$m. Determine the minimum base doping so that the punch-through voltage is no less than $V_{pt} = 25$ volts.

**Section 10.5**

31.  The $V_{CE}$(sat) voltage of a npn transistor in saturation continues to decrease slowly as the base current increases. In the Ebers-Moll model, assume $\alpha_F = 0.99$, $\alpha_R = 0.20$, and $I_C = 1$ mA. For $T = 300°$K, determine the base current, $I_B$, necessary to give $V_{CE}$(sat) = 0.10 volt.

32.  Consider an npn bipolar transistor biased in the active mode. Using the Ebers-Moll model, derive the equation for the base current, $I_B$, in terms of $\alpha_F$, $\alpha_R$, $I_{ES}$, $I_{CS}$, and $V_{BE}$.

33.  Consider the Ebers-Moll model and let the base terminal be open so $I_B = 0$. Show that, when a collector-emitter voltage is applied, we have

$$I_C \equiv I_{CEO} = I_{CS} \frac{(1 - \alpha_F \alpha_R)}{(1 - \alpha_F)}$$

34.  In the Ebers-Moll model, let $\alpha_F = 0.98$, $I_{ES} = 10^{-13}$ A, and $I_{CS} = 5 \times 10^{-13}$ A at $T = 300°$K. Plot $I_C$ versus $V_{CB}$ for $-V_{BE} < V_{CB} < 3$ volts and for $V_{BE} = 0.2$, 0.4, and 0.6 volt. (Note that $V_{CB} = -V_{BC}$.) What can be said about the base width modulation effect using this model?

## Section 10.6

35. Consider a silicon npn transistor at $T = 300°K$. Assume the following parameters:

$$I_E = 0.5 \text{ mA} \qquad C_{je} = 0.8 \text{ pF}$$
$$x_B = 0.7 \ \mu m \qquad D_n = 25 \text{ cm}^2/\text{sec}$$
$$x_{dc} = 2.0 \ \mu m \qquad r_c = 30 \ \Omega$$
$$C_s = C_\mu = 0.08 \text{ pF} \qquad \beta = 50$$

(a) Calculate the transit time factors. (b) Calculate the cutoff and beta cutoff frequencies, $f_T$ and $f_\beta$, respectively.

36. A silicon npn bipolar transistor has a beta cutoff frequency of $f_\beta = 100$ MHz and a common emitter current gain of $\beta = 125$. Assume that the base transit time and the emitter base junction charging time are equal limiting factors. If $I_E = 1$ mA, determine $C_{je}$, $x_B$, and $f_T$.

37. A silicon npn bipolar transistor at $T = 300°K$ has a uniformly doped base region. The cutoff frequency is $f_T = 500$ MHz and is limited by the base transit time. Estimate the base width. Discuss the variation in $f_T$ expected as the temperature increases to $T = 400°K$.

## Summary and Review

*38. (a) A silicon npn bipolar transistor at $T = 300°K$ is to be designed with an Early voltage of at least 200 volts and a current gain of at least $\beta = 80$. (b) Repeat part (a) for a pnp bipolar transistor.

*39. Design a uniformly doped silicon npn bipolar transistor so that $\beta = 100$ at $T = 300°K$. The maximum CE voltage is to be 15 volts and any breakdown voltage is to be at least three times this value. Assume the recombination factor is constant at $\delta = 0.995$. The transistor is to be operated in low-injection with a maximum collector current of $I_C = 5$ mA. Bandgap narrowing effects and base width modulation effects are to be minimized. Let $D_E = 6$ cm²/sec, $D_B = 25$ cm²/sec, $\tau_{E0} = 10^{-8}$ sec, and $\tau_{B0} = 10^{-7}$ sec. Determine doping concentrations, the metallurgical base width, the active area, and the maximum allowable $V_{BE}$.

*40. Design a pair of complementary npn and pnp bipolar transistors. The transistors are to have the same metallurgical base and emitter widths of $W_B = 0.75 \ \mu m$ and $x_E = 0.5 \ \mu m$. Assume that the following minority carrier parameters apply to each device.

$$D_n = 23 \text{ cm}^2/\text{sec} \quad \tau_{n0} = 10^{-7} \text{ sec}$$
$$D_p = 8 \text{ cm}^2/\text{sec} \quad \tau_{p0} = 5 \times 10^{-8} \text{ sec}$$

The collector doping concentration in each device is $5 \times 10^{15} \text{ cm}^{-3}$ and the recombination factor in each device is constant at $\delta =$

0.9950. (*a*) Design, if possible, the devices so that $\beta = 100$ in each device. If this is not possible, how close a match can be obtained? (*b*) With equal forward-bias base-emitter voltages applied, the collector currents are to be $I_C = 5$ mA with each device operating in low-injection. Determine the active cross-sectional areas.

## READING LIST

1. Muller, R. S., and T. I. Kamins. *Device Electronics for Integrated Circuits.* 2nd ed. New York: Wiley, 1986.

2. Navon, D. H. *Semiconductor Microdevices and Materials.* New York: Holt, Rinehart & Winston, 1986.

3. Neudeck, G. W. *The Bipolar Junction Transistor.* Vol. 3 of the *Modular Series on Solid State Devices.* 2nd ed. Reading, Mass.: Addison-Wesley, 1989.

4. Roulston, D. J. *Bipolar Semiconductor Devices.* New York: McGraw-Hill, 1990.

*5. Shur, M. *GaAs Devices and Circuits.* New York: Plenum Press, 1987.

*6. ———. *Physics of Semiconductor Devices.* Englewood Cliffs, N.J.: Prentice Hall, 1990.

7. Streetman, B. G. *Solid State Electronic Devices.* 3rd ed. Englewood Cliffs, N.J.: Prentice Hall, 1990.

8. Sze, S. M. *Physics of Semiconductor Devices.* 2nd ed. New York: Wiley, 1981.

9. Tiwari, S.; S. L. Wright; and A. W. Kleinsasser. "Transport and related properties of (Ga, Al)As/GaAs double heterojunction bipolar junction transistors." *IEEE Transactions on Electron Devices* ED-34 (February 1987), pp. 185–187.

*10. Wang, S. *Fundamentals of Semiconductor Theory and Device Physics.* Englewood Cliffs, N.J.: Prentice Hall, 1989.

*11. Warner, R. M., Jr., and B. L. Grung. *Transistors: Fundamentals for the Integrated-Circuit Engineer.* New York: Wiley, 1983.

12. Yang, E. S. *Microelectronic Devices.* New York: McGraw-Hill, 1988.

# THE JUNCTION FIELD-EFFECT TRANSISTOR

## CHAPTER 11

## PREVIEW

In this chapter we will cover the physics of the junction field-effect transistor. Although we discussed the bipolar transistor in the previous chapter, the material in this chapter presumes a knowledge only of semiconductor material properties and the characteristics of pn and Schottky barrier junctions.

The Junction Field-Effect Transistor (JFET), in the conjunction with other circuit elements, is capable of voltage gain and signal-power gain. There are two general categories of junction field-effect transistors. The first is the pn junction FET, or pn JFET, and the second is the MEtal-Semiconductor Field-Effect Transistor (MESFET). The pn JFET is fabricated with a pn junction, and the MESFET, with a Schottky barrier rectifying junction.

The current in a junction field-effect transistor is through a semiconductor region known as the channel, with ohmic contacts at each end. The basic transistor action is the modulation of the channel conductance by an electric field that is perpendicular to the channel. The modulating electric field is induced in the space charge region of a reverse-biased pn junction or a reverse-biased Schottky barrier junction and is, therefore, a function of a gate voltage. The modulation of the channel conductance by the gate voltage modulates the channel current. The ideal current-voltage characteristics of the pn JFET will be initially derived in terms of the semiconductor and geometrical properties of the device. We will then consider the transistor gain, or transconductance, of the pn JFET, develop the special properties of the MESFET, and describe some nonideal effects of the JFETs.

To examine the frequency response and limitations of the JFET, a small-signal equivalent circuit is initially developed and the various physical factors affecting the frequency response of the JFET are analyzed. The transistor cutoff frequency is a figure of merit for the frequency response of the transistor; we will derive an expression for this factor.

A specialized JFET structure is considered. This device is called a modulation-doped field-effect transistor (MODFET) and utilizes a heterojunction.

Electrons, confined in a potential well at the heterojunction interface, can easily move in the channel parallel to the junction. The electrons in the channel are separated from the ionized donors so that ionized impurity scattering is minimized; thus electron mobility is not degraded. A large mobility results in a large transistor gain and good high-frequency characteristics.

## 11.1  JFET CONCEPTS

The concept of the field-effect phenomenon was the basis for the first proposed solid-state transistor. Patents filed in the 1920s and 1930s conceived and investigated the transistor shown in Figure 11–1. A voltage applied to

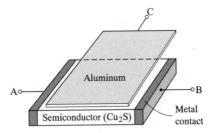

**Figure 11–1**    Idealization of the Lilienfeld transistor. (From Pierret [7].)

the metal plate modulated the conductance of the semiconductor under the metal and controlled the current between the ohmic contacts. Good semiconductor materials and processing technology were not available at that time so the device was not seriously considered again until the 1950s.

The phenomenon of modulating the conductance of a semiconductor by an electric field applied perpendicular to the surface of a semiconductor is called field effect. This type of transistor has also been called the unipolar transistor, to emphasize that only one type of carrier, the majority carrier, is involved in the operation. We will qualitatively discuss the basic operation of the two types of JFETs in this section, and introduce some of the JFET terminology.

### 11.1.1  Basic pn JFET Operation

The first type of field-effect transistor is the pn junction field-effect transistor, or pn JFET. A simplified cross-section of a symmetrical device is shown in Figure 11–2. The n-region between the two p-regions is known as the channel and, in this n-channel device, majority carrier electrons flow between the source and drain terminals. The source is the terminal from which carriers enter the channel from the external circuit, the drain is the terminal

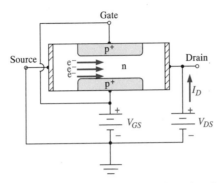

**Figure 11–2**   Cross section of a symmetrical n-channel pn junction FET.

where carriers leave, or are drained from, the device, and the gate is the control terminal. The two gate terminals shown in Figure 11–2 are tied together to form a single gate connection. Since majority carrier electrons are primarily involved in the conduction in this n-channel transistor, the JFET is a majority-carrier device.

A complementary p-channel JFET can also be fabricated in which the p- and n-regions are reversed from those of the n-channel device. Holes will flow in the p-type channel between source and drain and the source terminal will now be the source of the holes. The current direction and voltage polarities in the p-channel JFET are the reverse of those in the n-channel device. The p-channel JFET is generally a lower frequency device than the n-channel JFET due to the lower hole mobility.

Figure 11–3a shows an n-channel pn JFET with zero volts applied to the gate. If the source is at ground potential, and if a small positive drain voltage is applied, a drain current $I_D$ is produced between the source and drain terminals. The n-channel is essentially a resistance so the $I_D$ versus $V_{DS}$ characteristic, for small $V_{DS}$ values, is approximately linear as shown in the figure.

When we apply a voltage to the gate of a pn JFET with respect to the source and drain, we alter the channel conductance. If a negative voltage is applied to the gate of the n-channel pn JFET shown in Figure 11–3, the gate-to-channel pn junction becomes reverse biased. The space charge region now widens so the channel region becomes narrower and the resistance of the n-channel increases. The slope of the $I_D$ versus $V_{DS}$ curve, for small $V_{DS}$, decreases. These effects are shown in Figure 11–3b. If a larger negative gate voltage is applied, the condition shown in Figure 11–3c can be achieved. The reverse-biased gate-to-channel space charge region has completely filled the channel region. This condition is known as *pinchoff*. The drain current at pinchoff is essentially zero, since the depletion region isolates the source and drain terminals. The $I_D$ versus $V_{DS}$ curve for this case, as well as the other two cases, is shown in Figure 11–3c.

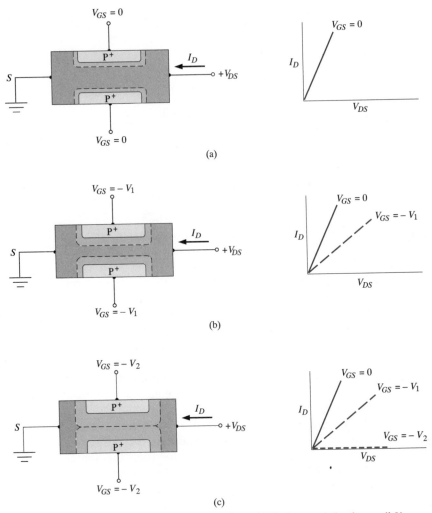

**Figure 11–3** Gate-to-channel space charge regions and I-V characteristics for small $V_{DS}$ values and for (a) zero gate voltage, (b) small reverse-biased gate voltage, and (c) a gate voltage to achieve pinchoff.

The current in the channel is controlled by the gate voltage. The control of the current in one part of the device by a voltage in another part of the device is the basic transistor action. This device is a normally on or *depletion mode* device, which means that a voltage must be applied to the gate terminal to turn the device off.

Now consider the situation in which the gate voltage is held at zero volts, $V_{GS} = 0$, and the drain voltage changes. Figure 11–4a is a replica of Figure 11–3a for zero gate voltage and a small drain voltage. As the drain voltage

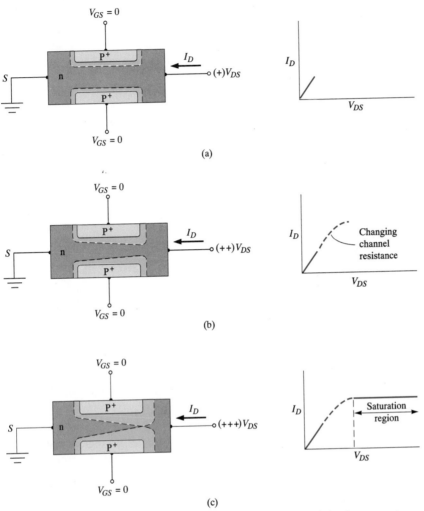

**Figure 11–4**  Gate-to-channel space charge regions and I-V characteristics for zero gate voltage and for (a) a small drain voltage, (b) a larger drain voltage, and (c) a drain voltage to achieve pinchoff at the drain terminal.

increases (positive), the gate-to-channel pn junction becomes reverse biased near the drain terminal so that the space charge region extends further into the channel. The channel is essentially a resistor and the effective channel resistance increases as the space charge region widens; therefore the slope of the $I_D$ versus $V_{DS}$ characteristic decreases as shown in Figure 11–4b. The effective channel resistance now varies along the channel length and, since the channel current must be constant, the voltage drop through the channel becomes dependent on position.

If the drain voltage further increases, the condition shown in Figure 11–4c can result. The channel has been pinched off at the drain terminal. Any further increase in drain voltage will not cause an increase in drain current. The I-V characteristic for this condition is also shown in this figure. The drain voltage at pinchoff is referred to as $V_{DS}$(sat). For $V_{DS} > V_{DS}$(sat), the transistor is said to be in the saturation region and the drain current, for this ideal case, is independent of $V_{DS}$. At first glance, we might expect the drain current to go to zero when the channel becomes pinched off at the drain terminal but we will show why this does not happen.

Figure 11–5 shows an expanded view of the pinchoff region in the channel. The n-channel and drain terminal are now separated by a space charge

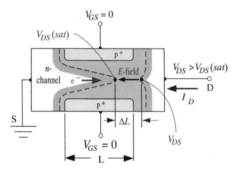

**Figure 11–5** Expanded view of the space charge region in the channel for $V_{DS} > V_{DS}$(sat).

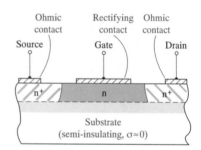

**Figure 11–6** Cross section of an n-channel MESFET with a semi-insulating substrate.

region which has a length $\Delta L$. The electrons move through the n-channel from the source and are injected into the space charge region where, subjected to the E-field force, they are swept through into the drain contact area. If we assume that $\Delta L \ll L$, then the electric field in the n-channel region remains unchanged from the $V_{DS}$(sat) case; the drain current will remain constant as $V_{DS}$ changes. Once the carriers are in the drain region, the drain current will be independent of $V_{DS}$; thus the device looks like a constant current source.

### 11.1.2 Basic MESFET Operation

The second type of junction field-effect transistor is the MESFET. The gate junction in the pn junction FET is replaced by a Schottky barrier rectifying contact. Although MESFETs can be fabricated in silicon, they are usually associated with gallium arsenide or other compound semiconductor materials. A simplified cross section of a GaAs MESFET is shown in Figure 11–6. A thin epitaxial layer of GaAs is used for the active region; the substrate is a very high resistivity GaAs material referred to as a semi-insulating substrate. GaAs is intentionally doped with chromium, which behaves as a single acceptor close to the center of the energy bandgap, to make it semi-insulating with a resistivity as high as $10^9$ ohm-cm. The advantages of these devices include higher electron mobility, hence smaller transit time and faster response; and decreased parasitic capacitance and a simplified fabrication process, resulting from the semi-insulating GaAs substrate.

In the MESFET shown in Figure 11–6, a reverse-bias gate-to-source voltage induces a space charge region under the metal gate which modulates the channel conductance as in the case of the pn junction FET. The space charge region will eventually reach the substrate if the applied negative gate voltage is sufficiently large. This condition, again, is known as pinchoff. The device shown in this figure is also a depletion mode device since a gate voltage must be applied to pinch off the channel.

If we treat the semi-insulating substrate as an intrinsic material, then the energy-band diagram of the substrate-channel-metal structure is shown in Figure 11–7 for the case of zero bias applied to the gate. Because there is a

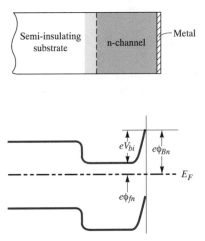

**Figure 11–7**   Idealized energy-band diagram of the substrate-channel-metal in the n-channel MESFET.

potential barrier between the channel and substrate and between the channel and metal, the majority carrier electrons are confined to the channel region.

Consider, now, another type of MESFET in which the channel is pinched off even at $V_{GS} = 0$. Figure 11–8a shows this condition, in which the channel thickness is smaller than the zero-biased space charge width. To open a channel, the depletion region must be reduced: a forward-bias voltage must be applied to the gate–semiconductor junction. When a slightly forward-bias voltage is applied, the depletion region just extends through the channel—a condition known as threshold, shown in Figure 11–8b. The threshold voltage is the gate-to-source voltage that must be applied to create the pinchoff

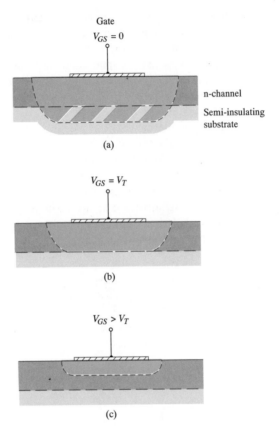

**Figure 11–8** Channel space charge region of an enhancement mode MESFET for (a) $V_{GS} = 0$, (b) $V_{GS} = V_T$, and (c) $V_{GS} > V_T$.

condition. The threshold voltage for this n-channel MESFET is positive, in contrast to the negative voltage for the n-channel depletion mode device. If a larger forward bias is applied, the channel region opens as shown in Figure 11–8c. The applied forward-bias gate voltage is limited to a few tenths of a volt before there is significant gate current. This device is known as an n-channel enhancement mode MESFET. Enhancement mode p-channel MESFETs and enhancement mode pn junction FETs have also been fabricated. The advantage of enhancement mode MESFETs is that circuits can be designed in which the voltage polarity on the gate and drain is the same. However, the output voltage swing will be quite small with these devices.

## 11.2 THE DEVICE CHARACTERISTICS

To describe the basic electrical characteristics of the JFET, we will initially consider a uniformly doped depletion mode pn JFET and then later discuss the enhancement mode device. The pinchoff voltage and drain-to-source saturation voltage will be defined and expressions for these parameters derived in terms of geometry and electrical properties. The ideal current-voltage relationship will be developed, and then the transconductance, or transistor gain.

Figure 11–9a shows a symmetrical, two-sided pn JFET and Figure 11–9b shows a MESFET with the semi-insulating substrate. One can derive the

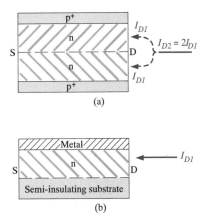

**Figure 11–9** Drain currents of (a) a symmetrical, two-sided pn JFET, and (b) a one-sided MESFET.

ideal DC current-voltage relationship for both devices by simply considering the two-sided device to be two JFETs in parallel. We will derive the I-V characteristics in terms of $I_{D1}$ so that the drain current in the two-sided device will then be $I_{D2} = 2I_{D1}$. We will ignore any depletion region at the substrate of the one-sided device in the ideal case.

## 11.2.1 Internal Pinchoff Voltage, Pinchoff Voltage, and Drain-to-Source Saturation Voltage

Figure 11–10a shows a simplified one-sided n-channel pn JFET. The metallurgical channel thickness between the $p^+$ gate region and the substrate is $a$, and the induced depletion region width for the one-sided $p^+n$ junction is $h$. Assume the drain-to-source voltage is zero. If we assume the abrupt depletion approximation, then the space charge width is given by

$$h = \left\{ \frac{2\varepsilon_s (V_{bi} - V_{GS})}{eN_d} \right\}^{1/2}$$

(11–1)

where $V_{GS}$ is the gate-to-source voltage and $V_{bi}$ is the built-in potential barrier. For a reverse-biased $p^+n$ junction, $V_{GS}$ must be a negative voltage.

At pinchoff, $h = a$ and the total potential across the $p^+n$ junction is called the *internal pinchoff voltage*, denoted by $V_{p0}$. We now have

$$a = \left\{ \frac{2\varepsilon_s V_{p0}}{eN_d} \right\}^{1/2} \qquad (11-2)$$

or

$$V_{p0} = \frac{ea^2 N_d}{2\varepsilon_s} \qquad (11-3)$$

Note that the internal pinchoff voltage is defined as a positive quantity.

The internal pinchoff voltage $V_{p0}$ is not the gate-to-source voltage to achieve pinchoff. The gate-to-source voltage that must be applied to achieve pinchoff is described as the *pinchoff voltage* and is also variously called the *turn-off voltage* or *threshold voltage*. The pinchoff voltage is denoted by $V_p$ and is defined from Equations (11–1) and (11–2) as

$$V_{bi} - V_p = V_{p0} \qquad \text{or} \qquad V_p = V_{bi} - V_{p0} \qquad (11-4)$$

The gate-to-source voltage to achieve pinchoff in an n-channel depletion mode JFET is negative; thus, $V_{p0} > V_{bi}$.

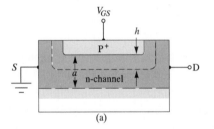

(a)

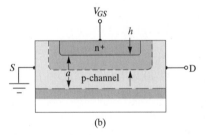

(b)

**Figure 11–10**   Geometries of simplified (a) n-channel and (b) p-channel pn JFETs.

---

### Example 11–1

**Objective:** To calculate the internal pinchoff voltage and pinchoff voltage of an n-channel JFET.

Assume that the $p^+n$ junction of a uniformly doped silicon n-channel JFET at $T = 300°K$ has doping concentrations of $N_a = 10^{18}$ cm$^{-3}$ and $N_d = 10^{16}$ cm$^{-3}$. Assume that the metallurgical channel thickness, $a$, is 0.75 $\mu$m = 0.75 $\times$ 10$^{-4}$ cm.

**Solution:** The internal pinchoff voltage is given by Equation (11–3), so we have

$$V_{p0} = \frac{ea^2N_d}{2\varepsilon_s} = \frac{(1.6 \times 10^{-19})(0.75 \times 10^{-4})^2(10^{16})}{2(11.7)(8.85 \times 10^{-14})} = 4.35 \text{ volts}$$

The built-in potential barrier is

$$V_{bi} = V_t \ln\left(\frac{N_aN_d}{n_i^2}\right) = (0.0259) \ln\left[\frac{(10^{18})(10^{16})}{(1.5 \times 10^{10})^2}\right] = 0.814 \text{ volt}$$

The pinchoff voltage, from Equation (11–4), is then found as

$$V_p = V_{bi} - V_{p0} = 0.814 - 4.35 = -3.54 \text{ volts}$$

**Comment:** The pinchoff voltage, or gate-to-source voltage to achieve pinchoff, for the n-channel depletion mode device is a negative quantity as we have said.

---

The pinchoff voltage is the gate-to-source voltage that must be applied to turn the JFET off and so must be within the voltage range of the circuit design. The magnitude of the pinchoff voltage must also be less than the breakdown voltage of the junction.

Figure 11–10b shows a p-channel JFET with the same basic geometry as the n-channel JFET we considered. The induced depletion region for the one-sided $n^+p$ junction is again denoted by $h$ and is given by

$$h = \left\{\frac{2\varepsilon_s(V_{bi} + V_{GS})}{eN_a}\right\}^{1/2} \tag{11–5}$$

For a reverse-biased $n^+p$ junction, $V_{GS}$ must be positive. The internal pinchoff voltage is again defined to be the total pn junction voltage to achieve pinchoff, so that when $h = a$ we have

$$a = \left\{\frac{2\varepsilon_s V_{p0}}{eN_a}\right\}^{1/2} \tag{11–6}$$

or

$$V_{p0} = \frac{ea^2N_a}{2\varepsilon_s} \tag{11–7}$$

The internal pinchoff voltage for the p-channel device is also defined to be a positive quantity.

The pinchoff voltage is again defined as the gate-to-source voltage to achieve the pinchoff condition. For the p-channel depletion mode device, we have, from Equation (11–5), at pinchoff

$$V_{bi} + V_p = V_{p0} \quad \text{or} \quad V_p = V_{p0} - V_{bi} \tag{11–8}$$

The pinchoff voltage for a p-channel depletion mode JFET is a positive quantity.

**Example 11–2**

**Objective:** To design the channel doping concentration and metallurgical channel thickness to achieve a given pinchoff voltage.

Consider a silicon p-channel pn JFET at $T = 300°K$. Assume that the gate doping concentration is $N_d = 10^{18}$ cm$^{-3}$. Determine the channel doping concentration and channel thickness so that the pinchoff voltage is $V_p = 2.25$ volts.

**Solution:** There is not a unique solution to this design problem. We will pick a channel doping concentration of $N_a = 2 \times 10^{16}$ cm$^{-3}$ and determine the channel thickness. The built-in potential barrier is

$$V_{bi} = V_t \ln \left( \frac{N_a N_d}{n_i^2} \right) = (0.0259) \ln \left[ \frac{(2 \times 10^{16})(10^{18})}{(1.5 \times 10^{10})^2} \right] = 0.832 \text{ volt}$$

From Equation (11–8), the internal pinchoff voltage must be

$$V_{p0} = V_{bi} + V_p = 0.832 + 2.25 = 3.08 \text{ volts}$$

and from Equation (11–6), the channel thickness can be determined as

$$a = \left\{ \frac{2\varepsilon_s V_{p0}}{eN_a} \right\}^{1/2} = \left\{ \frac{2(11.7)(8.85 \times 10^{-14})(3.08)}{(1.6 \times 10^{-19})(2 \times 10^{16})} \right\}^{1/2} = 0.446 \; \mu\text{m}$$

**Comment:** If the channel doping concentration chosen were larger, the required channel thickness would decrease; a very small value of channel thickness would be difficult to fabricate within reasonable tolerance limits.

Also, we will see later that if the channel doping concentration were smaller, the current capability of the device would decrease. There are definite trade-offs to be considered in any design problem.

We have determined the pinchoff voltage for both n-channel and p-channel JFETs when the drain-to-source voltage is zero. Now consider the case when both gate and drain voltages are applied. The depletion region width will vary with distance through the channel. Figure 11–11 shows the simpli-

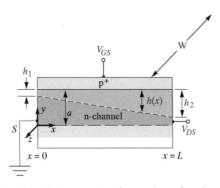

**Figure 11–11**  Simplified geometry of an n-channel pn junction FET.

fied geometry for an n-channel device. The depletion width $h_1$ at the source end is a function of $V_{bi}$ and $V_{GS}$ but is not a function of drain voltage. The depletion width at the drain terminal is given by

$$h_2 = \left\{ \frac{2\varepsilon_s(V_{bi} + V_{DS} - V_{GS})}{eN_d} \right\}^{1/2} \tag{11-9}$$

Again, we must keep in mind that $V_{GS}$ is a negative quantity for the n-channel device.

Pinchoff at the drain terminal occurs when $h_2 = a$. At this point we reach what is known as the saturation condition; thus we can write that $V_{DS} = V_{DS}(\text{sat})$. Then

$$a = \left\{ \frac{2\varepsilon_s(V_{bi} + V_{DS}(\text{sat}) - V_{GS})}{eN_d} \right\}^{1/2} \tag{11-10}$$

This can be rewritten as

$$V_{bi} + V_{DS}(\text{sat}) - V_{GS} = \frac{ea^2N_d}{2\varepsilon_s} = V_{p0} \tag{11-11}$$

or

$$V_{DS}(\text{sat}) = V_{p0} - (V_{bi} - V_{GS}) \tag{11-12}$$

Equation (11–12) gives the drain-to-source voltage to cause pinchoff at the drain terminal. The drain-to-source saturation voltage decreases with increasing reverse-bias gate-to-source voltage. We may note that Equation (11–12) has no meaning if $|V_{GS}| > |V_p|$.

In a p-channel JFET, the voltage polarities are the reverse of those in the n-channel device. We can show that, in the p-channel JFET at saturation,

$$V_{SD}(\text{sat}) = V_{p0} - (V_{bi} + V_{GS}) \tag{11-13}$$

where now the source is positive with respect to the drain.

### 11.2.2 Ideal DC Current-Voltage Relationship—Depletion Mode JFET

The derivation of the ideal current-voltage relation of the JFET is somewhat tedious and the resulting equations are cumbersome in hand calculations. Before we go through this derivation, consider the following expression, which is a good approximation to the I-V characteristics when the JFET is biased in the saturation region. This equation is used extensively in JFET applications and is given by

$$I_D = I_{DSS} \left( 1 - \frac{V_{GS}}{V_p} \right)^2 \tag{11-14}$$

where $I_{DSS}$ is the saturation current when $V_{GS} = 0$. At the end of this section, we will compare the approximation given by Equation (11–14) and the ideal current-voltage equation which we have derived.

**I-V Derivation**    The ideal current-voltage relationship of the JFET is derived by starting with Ohm's law. Consider an n-channel JFET with the geometry shown in Figure 11–11. We are considering half of the two-sided symmetrical geometry. The differential resistance of the channel at a point $x$ in the channel is

$$dR = \frac{\rho dx}{A(x)} \qquad (11\text{--}15)$$

where $\rho$ is the resistivity and $A(x)$ is the cross-sectional area. If we neglect the minority carrier holes in the n-channel, the channel resistivity is

$$\rho = \frac{1}{e\mu_n N_d} \qquad (11\text{--}16)$$

The cross-sectional area is given by

$$A(x) = (a - h(x))W \qquad (11\text{--}17)$$

where $W$ is the channel width. Equation (11–15) can now be written as

$$dR = \frac{dx}{e\mu_n N_d(a - h(x))W} \qquad (11\text{--}18)$$

The differential voltage across a differential length $dx$ can be written as

$$dV(x) = I_{D1}dR(x) \qquad (11\text{--}19)$$

where the drain current $I_{D1}$ is constant through the channel. Substituting Equation (11–18) into Equation (11–19), we have

$$dV(x) = \frac{I_{D1}dx}{e\mu_n N_d W(a - h(x))} \qquad (11\text{--}20a)$$

or

$$I_{D1}dx = e\mu_n N_d W(a - h(x))dV(x) \qquad (11\text{--}20b)$$

The depletion width $h(x)$ is given by

$$h(x) = \left\{\frac{2\varepsilon_s(V(x) + V_{bi} - V_{GS})}{eN_d}\right\}^{1/2} \qquad (11\text{--}21)$$

where $V(x)$ is the potential in the channel due to the drain-to-source voltage. Solving for $V(x)$ in Equation (11–21) and taking the differential, we have

$$dV(x) = \frac{eN_d h(x)dh(x)}{\varepsilon_s} \qquad (11\text{--}22)$$

Then Equation (11–20b) becomes

$$I_{D1}dx = \frac{\mu_n(eN_d)^2W}{\varepsilon_s} [ah(x)dh(x) - h(x)^2dh(x)] \qquad (11\text{--}23)$$

The drain current $I_{D1}$ is found by integrating Equation (11–23) along the channel length. Assuming the current and mobility are constant through the channel, we obtain

$$I_{D1} \overset{\zeta}{=} \frac{\mu_n(eN_d)^2W}{\varepsilon_s L} \left\{ \int_{h_1}^{h_2} ah\,dh - \int_{h_1}^{h_2} h^2 dh \right\} \qquad (11\text{--}24)$$

or

$$I_{D1} = \frac{\mu_n(eN_d)^2W}{\varepsilon_s L} \left\{ \frac{a}{2}(h_2^2 - h_1^2) - \frac{1}{3}(h_2^3 - h_1^3) \right\} \qquad (11\text{--}25)$$

Noting that

$$h_2^2 = \frac{2\varepsilon_s(V_{DS} + V_{bi} - V_{GS})}{eN_d} \qquad (11\text{--}26a)$$

$$h_1^2 = \frac{2\varepsilon_s(V_{bi} - V_{GS})}{eN_d} \qquad (11\text{--}26b)$$

and

$$V_{p0} = \frac{ea^2N_d}{2\varepsilon_s} \qquad (11\text{--}26c)$$

Equation (11–25) can be written as

$$I_{D1} = \frac{\mu_n(eN_d)^2Wa^3}{2\varepsilon_s L} \left\{ \frac{V_{DS}}{V_{p0}} - \frac{2}{3}\left[\frac{V_{DS} + V_{bi} - V_{GS}}{V_{p0}}\right]^{3/2} \right.$$
$$\left. + \frac{2}{3}\left[\frac{V_{bi} - V_{GS}}{V_{p0}}\right]^{3/2} \right\} \qquad (11\text{--}27)$$

We may define

$$I_{P1} \equiv \frac{\mu_n(eN_d)^2Wa^3}{6\varepsilon_s L} \qquad (11\text{--}28)$$

where $I_{P1}$ is called the pinchoff current. Equation (11–27) becomes

$$I_{D1} = I_{P1}\left\{ 3\left(\frac{V_{DS}}{V_{p0}}\right) - 2\left(\frac{V_{DS} + V_{bi} - V_{GS}}{V_{p0}}\right)^{3/2} + 2\left(\frac{V_{bi} - V_{GS}}{V_{p0}}\right)^{3/2} \right\} \qquad (11\text{--}29)$$

Equation (11–29) is valid for $0 \le |V_{GS}| \le |V_p|$ and $0 \le V_{DS} \le V_{DS}(\text{sat})$. The pinchoff current $I_{P1}$ would be the maximum drain current in the JFET if the zero-biased depletion regions could be ignored, or if $V_{GS}$ and $V_{bi}$ were both zero.

Equation (11–29) is the current-voltage relationship for the one-sided n-channel JFET in the nonsaturation region. For the two-sided symmetrical JFET shown in Figure 11–9a, the total drain current would be $I_{D2} = 2I_{D1}$. Equation (11–27) can also be written as

$$I_{D1} = G_{01} \left\{ V_{DS} - \frac{2}{3} \sqrt{\frac{1}{V_{p0}}} \left[ (V_{DS} + V_{bi} - V_{GS})^{3/2} - (V_{bi} - V_{GS})^{3/2} \right] \right\} \qquad (11\text{--}30)$$

where

$$G_{01} = \frac{\mu_n (eN_d)^2 Wa^3}{2\varepsilon_s L V_{p0}} = \frac{e\mu_n N_d Wa}{L} = \frac{3I_{P1}}{V_{p0}} \qquad (11\text{--}31)$$

The channel conductance is defined as

$$g_d = \frac{\partial I_{D1}}{\partial V_{DS}} \bigg|_{V_{DS} \to 0} \qquad (11\text{--}32)$$

Taking the derivative of Equation (11–30) with respect to $V_{DS}$, we obtain

$$g_d = \frac{\partial I_{D1}}{\partial V_{DS}} \bigg|_{V_{DS} \to 0} = G_{01} \left\{ 1 - \left( \frac{V_{bi} - V_{GS}}{V_{p0}} \right)^{1/2} \right\} \qquad (11\text{--}33)$$

We may note from Equation (11–33) that $G_{01}$ would be the conductance of the channel if both $V_{bi}$ and $V_{GS}$ were zero. This condition would exist if no space charge regions existed in the channel. We may also note, from Equation (11–33), that the channel conductance is modulated or controlled by the gate voltage. This channel conductance modulation is the basis of the field-effect phenomenon.

We have shown that the drain becomes pinched off, for the n-channel JFET, when

$$V_{DS} = V_{DS}(\text{sat}) = V_{p0} - (V_{bi} - V_{GS}) \qquad (11\text{--}34)$$

In the saturation region, the saturation drain current is determined by setting $V_{DS} = V_{DS}(\text{sat})$ in Equation (11–29) so that

$$I_{D1} = I_{D1}(\text{sat}) = I_{P1} \left\{ 1 - 3 \left( \frac{V_{bi} - V_{GS}}{V_{p0}} \right) \left[ 1 - \frac{2}{3} \sqrt{\frac{V_{bi} - V_{GS}}{V_{p0}}} \right] \right\} \qquad (11\text{--}35)$$

The ideal saturation drain current is independent of the drain-to-source voltage. Figure 11–12 shows the ideal current-voltage characteristics of a silicon n-channel JFET.

---

### Example 11–3

**Objective:** To calculate the maximum current in an n-channel JFET.

Consider a silicon n-channel JFET at $T = 300°K$ with the following parameters: $N_a = 10^{18} \text{ cm}^{-3}$, $N_d = 10^{16} \text{ cm}^{-3}$, $a = 0.75 \ \mu\text{m}$, $L = 10 \ \mu\text{m}$, $W = 30 \ \mu\text{m}$, and $\mu_n = 1000 \text{ cm}^2/\text{V-sec}$.

**Solution:** The pinchoff current from Equation (11–28) becomes

$$I_{P1} = \frac{(1000)[(1.6 \times 10^{-19})(10^{16})]^2(30 \times 10^{-4})(0.75 \times 10^{-4})^3}{6(11.7)(8.85 \times 10^{-14})(10 \times 10^{-4})} = 0.522 \text{ mA}$$

We also have from Example 11–1 that $V_{bi} = 0.814$ volt and $V_{p0} = 4.35$ volts. The maximum current occurs when $V_{GS} = 0$ so from Equation (11–35)

$$I_{D1}(\text{max}) = I_{P1} \left\{ 1 - 3 \left( \frac{V_{bi}}{V_{p0}} \right) \left[ 1 - \frac{2}{3} \sqrt{\frac{V_{bi}}{V_{p0}}} \right] \right\} \qquad (11\text{–}36)$$

or

$$I_{D1}(\text{max}) = (0.522) \left\{ 1 - 3 \left( \frac{0.814}{4.35} \right) \left[ 1 - \frac{2}{3} \sqrt{\frac{0.814}{4.35}} \right] \right\} = 0.313 \text{ mA}$$

**Comment:** The maximum current through the JFET is less than the pinchoff current $I_{P1}$.

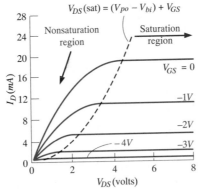

**Figure 11–12**   Ideal current-voltage characteristics of a silicon n-channel JFET with $a = 1.5$ $\mu$m, $W/L = 170$, and $N_d = 2.5 \times 10^{15}$ cm$^{-3}$. (From Yang [13].)

The maximum saturation current calculated in this example is considerably less than that shown in Figure 11–12 because of the big difference in the width-to-length ratios. Once the pinchoff voltage of a JFET has been designed, the channel width $W$ is the primary design variable for determining the current capability of a device.

**Summary**   Equations (11–29) and (11–35) are rather cumbersome to use in any hand calculations. We may show that, in the saturation region, the drain current is given to a good approximation by Equation (11–14), stated at the beginning of this section as

$$I_D = I_{DSS} \left( 1 - \frac{V_{GS}}{V_p} \right)^2$$

The current $I_{DSS}$ is the maximum drain current and is the same as $I_{D1}(\text{max})$ in Equation (11–36). The parameter $V_{GS}$ is the gate-to-source voltage and $V_p$ is the pinchoff voltage. We may note that, for the n-channel depletion mode JFET, both $V_{GS}$ and $V_p$ are negative and, for the p-channel depletion mode device, both are positive. Figure 11–13 shows the comparison between Equations (11–14) and (11–35).

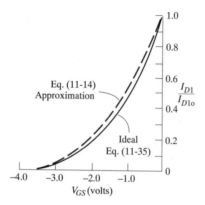

**Figure 11–13**   Comparison of Equations (11–14) and (11–35) for the $I_D$ versus $V_{GS}$ characteristics of a JFET biased in the saturation region.

### 11.2.3 Transconductance

The transconductance is the transistor gain of the JFET; it indicates the amount of control the gate voltage has on the drain current. The transconductance is defined as

$$g_m = \frac{\partial I_D}{\partial V_{GS}} \tag{11–37}$$

Using the expressions for the ideal drain current derived in the last section, we can write the expressions for the transconductance.

The drain current for an n-channel depletion mode device in the non-saturation region was given by Equation (11–29). We can then determine the transconductance of the transistor in the same region as

$$g_{mL} = \frac{\partial I_{D1}}{\partial V_{GS}} = \frac{3I_{P1}}{V_{p0}} \sqrt{\frac{V_{bi} - V_{GS}}{V_{p0}}} \left\{ \sqrt{\left(\frac{V_{DS}}{V_{bi} - V_{GS}}\right) + 1} - 1 \right\} \tag{11–38}$$

Taking the limit as $V_{DS}$ becomes small, the transconductance becomes

$$g_{mL} \cong \frac{3I_{P1}}{2V_{p0}} \cdot \frac{V_{DS}}{\sqrt{V_{p0}(V_{bi} - V_{GS})}} \tag{11–39}$$

We can also write Equation (11–39) in terms of the conductance parameter $G_{01}$ as

$$g_{mL} = \frac{G_{01}}{2} \cdot \frac{V_{DS}}{\sqrt{V_{p0}(V_{bi} - V_{GS})}} \qquad (11\text{–}40)$$

The ideal drain current in the saturation region for the JFET was given by Equation (11–35). The transconductance in the saturation region is then found to be

$$g_{ms} = \frac{\partial I_{D1}(\text{sat})}{\partial V_{GS}} = \frac{3I_{P1}}{V_{p0}}\left[1 - \sqrt{\frac{V_{bi} - V_{GS}}{V_{p0}}}\right] = G_{01}\left[1 - \sqrt{\frac{V_{bi} - V_{GS}}{V_{p0}}}\right]$$

$$(11\text{–}41a)$$

Using the current-voltage approximation given by Equation (11–14), we can also write the transconductance as

$$g_{ms} = \frac{-2I_{DSS}}{V_p}\left(1 - \frac{V_{GS}}{V_p}\right) \qquad (11\text{–}41b)$$

Since $V_p$ is negative for the n-channel JFET, $g_{ms}$ is positive.

---

**Example 11–4**

**Objective:** To determine the maximum transconductance of an n-channel depletion mode JFET biased in the saturation region.

Consider the silicon JFET described in Example 11–3. We had calculated $I_{P1} = 0.522$ mA, $V_{bi} = 0.814$ volt, and $V_{p0} = 4.35$ volts.

**Solution:** The maximum transconductance occurs when $V_{GS} = 0$. Then Equation (11–41a) can be written as

$$g_{ms}(\text{max}) = \frac{3I_{p1}}{V_{p0}}\left[1 - \sqrt{\frac{V_{bi}}{V_{p0}}}\right] = \frac{3(0.522)}{4.35}\left[1 - \sqrt{\frac{0.814}{4.35}}\right] = 0.204 \text{ mA/volt}$$

**Comment:** The saturation transconductance is a function of $V_{GS}$ and becomes zero when $V_{GS} = V_p$.

---

The experimental transconductance may deviate from this ideal expression due to a source series resistance. This effect will be considered later in the discussion of the small signal model of the JFET.

## 11.2.4 The MESFET

So far in our discussion, we have explicitly considered the pn JFET. The MESFET is the same basic device except that the pn junction is replaced by a Schottky barrier rectifying junction. The simplified MESFET geometry was shown in Figure 11–9b. MESFETs are usually fabricated in gallium arsenide. We will neglect any depletion region that may exist between the n-channel and the substrate. We have also limited our discussion to depletion mode devices, wherein a gate-to-source voltage is applied to turn the transistor off. Enhancement mode GaAs MESFETs can be fabricated—their basic operation was discussed in Section 11.1.2. We can also consider enhancement mode GaAs pn JFETs.

Since the electron mobility in GaAs is much larger than the hole mobility, we will concentrate our discussion on n-channel GaAs MESFETs or JFETs. The definition of internal pinchoff voltage, given by Equation (11–3), also applies to these devices. In considering the enhancement mode JFET, the term threshold voltage is commonly used in place of pinchoff voltage. For this reason, we shall use the term threshold voltage in our discussion of MESFETs.

For the n-channel MESFET, the threshold voltage is defined from Equation (11–4) as

$$V_{bi} - V_T = V_{p0} \qquad \text{or} \qquad V_T = V_{bi} - V_{p0} \qquad (11\text{–}42)$$

For an n-channel depletion mode JFET, $V_T < 0$, and for the enhancement mode device, $V_T > 0$. We can see from Equation (11–42) that $V_{bi} > V_{p0}$ for an enhancement mode n-channel JFET.

---

### Example 11–5

**Objective:** To calculate the internal pinchoff voltage and threshold voltage of a GaAs MESFET.

Consider an n-channel GaAs MESFET at $T = 300°K$ with a gold Schottky barrier contact. Assume the barrier height is $\phi_{Bn} = 0.89$ volt. The n-channel doping is $N_d = 2 \times 10^{15}$ cm$^{-3}$ and the channel thickness is $a = 0.6$ $\mu$m.

**Solution:** The pinchoff voltage is

$$V_{p0} = \frac{ea^2 N_d}{2\varepsilon_s} = \frac{(1.6 \times 10^{-19})(0.6 \times 10^{-4})^2(2 \times 10^{15})}{2(13.1)(8.85 \times 10^{-14})} = 0.497 \text{ volt}$$

The difference between the conduction band and the Fermi level is given by

$$\phi_n = V_t \ln\left(\frac{N_c}{N_d}\right) = (0.0259) \ln\left(\frac{4.7 \times 10^{17}}{2 \times 10^{15}}\right) = 0.141 \text{ volt}$$

The built-in potential barrier is then

$$V_{bi} = \phi_{Bn} - \phi_n = 0.89 - 0.141 = 0.749 \text{ volt}$$

Finally, the threshold voltage, from Equation (11–42), is

$$V_T = V_{bi} - V_{p0} = 0.749 - 0.497 = +0.252 \text{ volt}$$

**Comment:** A positive threshold voltage indicates an enhancement mode n-channel JFET, as does the fact that the internal pinchoff voltage is less than the built-in potential barrier.

The design of enhancement mode JFETs implies the use of narrow channel thicknesses and low channel doping concentrations to achieve this condition. The precise control of the channel thickness and doping concentration necessary to achieve internal pinchoff voltages of a few tenths of a volt makes the fabrication of enhancement mode MESFETs difficult.

---

**Example 11–6**

**Objective:** To calculate the forward-bias voltage required in an n-channel GaAs enhancement mode pn JFET to open up a channel.

Consider a GaAs n-channel pn junction FET at $T = 300°K$ with $N_a = 10^{18}$ cm$^{-3}$, $N_d = 3 \times 10^{15}$ cm$^{-3}$, and $a = 0.70$ $\mu$m. Determine the forward-bias gate voltage required to open a channel region that is 0.10 $\mu$m thick with zero drain voltage.

**Solution:** The built-in potential barrier is

$$V_{bi} = V_t \ln \left( \frac{N_a N_d}{n_i^2} \right) = (0.0259) \ln \left[ \frac{(10^{18})(3 \times 10^{15})}{(1.8 \times 10^6)^2} \right] = 1.25 \text{ volts}$$

The internal pinchoff voltage is

$$V_{p0} = \frac{ea^2 N_d}{2\varepsilon_s} = \frac{(1.6 \times 10^{-19})(0.7 \times 10^{-4})^2(3 \times 10^{15})}{2(13.1)(8.85 \times 10^{-14})} = 1.01 \text{ volts}$$

which gives a threshold voltage of

$$V_T = V_{bi} - V_{p0} = 0.24 \text{ volt}$$

The channel depletion width is given by Equation (11–1). Setting $h = 0.60$ $\mu$m will yield an undepleted channel thickness of 0.1 $\mu$m. Solving for $V_{GS}$, we obtain

$$V_{GS} = V_{bi} - \frac{eh^2 N_d}{2\varepsilon_s} = 1.25 - \frac{(1.6 \times 10^{-19})(0.6 \times 10^{-4})^2(3 \times 10^{15})}{2(13.1)(8.85 \times 10^{-14})}$$

$$= 1.25 - 0.745 = 0.50 \text{ volt}$$

**Comment:** An applied gate voltage of 0.50 volt is greater than the threshold voltage so the induced depletion region will be smaller than the metallurgical channel thickness. An n-channel region is then formed between the source and drain contacts. The forward-bias gate voltage must not be too large or an undesirable gate current will be present in the device.

---

Ideally, the I-V characteristics of the enhancement mode device are the same as the depletion mode device—the only real difference is the relative values of the internal pinchoff voltage. The current in the saturation region was given by Equation (11–35) as

$$I_{D1} = I_{D1}(\text{sat}) = I_{P1} \left\{ 1 - 3 \left( \frac{V_{bi} - V_{GS}}{V_{p0}} \right) \left[ 1 - \frac{2}{3} \sqrt{\frac{V_{bi} - V_{GS}}{V_{p0}}} \right] \right\}$$

The threshold voltage for the n-channel device was defined in Equation (11–42) as $V_T = V_{bi} - V_{p0}$, so we can also write

$$V_{bi} = V_T + V_{p0} \tag{11–43}$$

Substituting this expression for $V_{bi}$ into Equation (11–35), we obtain

$$I_{D1}(\text{sat}) = I_{P1}\left\{1 - 3\left[1 - \left(\frac{V_{GS} - V_T}{V_{p0}}\right)\right] + 2\left[1 - \left(\frac{V_{GS} - V_T}{V_{p0}}\right)\right]^{3/2}\right\} \tag{11–44}$$

Equation (11–44) is valid for $V_{GS} \geq V_T$.

When the transistor first turns on, we have $(V_{GS} - V_T) \ll V_{p0}$. Equation (11–44) can then be expanded into a Taylor series and we obtain

$$I_{D1}(\text{sat}) \cong I_{P1}\left[\frac{3}{4}\left(\frac{V_{GS} - V_T}{V_{p0}}\right)\right]^2 \tag{11–45}$$

Substituting the expressions for $I_{P1}$ and $V_{p0}$, Equation (11–45) becomes

$$I_{D1}(\text{sat}) = \frac{\mu_n \varepsilon_s W}{2aL}(V_{GS} - V_T)^2 \qquad \text{for } V_{GS} \geq V_T \tag{11–46}$$

We can now write Equation (11–46) as

$$I_{D1}(\text{sat}) = k_n(V_{GS} - V_T)^2 \tag{11–47}$$

where

$$k_n = \frac{\mu_n \varepsilon_s W}{2aL} \tag{11–48}$$

The factor $k_n$ is called a conduction parameter. The form of Equation (11–47) is the same for a MOSFET, as we will see in the next chapter.

The square root of Equation (11–47), or $\sqrt{I_{D1}(\text{sat})}$ versus $V_{GS}$, is plotted as the ideal dotted curve shown in Figure 11–14. The ideal curve

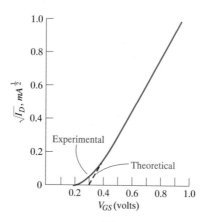

**Figure 11–14**  Experimental and theoretical $\sqrt{I_D}$ versus $V_{GS}$ characteristics of an enhancement mode JFET.

intersects the voltage axis at the threshold voltage, $V_T$. The solid line shows an experimental plot. Equation (11–46) does not describe the experimental results well near the threshold voltage. The ideal current-voltage relationship was derived assuming an abrupt depletion approximation for the pn junction. However, when the depletion region extends almost through the channel, a more accurate model of the space charge region must be used to more accurately predict the drain current characteristics near threshold. The subthreshold conduction will be considered in Section 11.3.3.

---

**Example 11–7**

**Objective:** To calculate the conduction parameter and $I_{D1}(sat)$ value for an n-channel gallium arsenide enhancement mode pn JFET.

Consider the GaAs JFET in Example 11–6. We calculated $V_T = 0.24$ volt. In addition, let $\mu_n = 8000$ cm$^2$/volt-sec, $L = 1.2$ $\mu$m, and $W = 20$ $\mu$m.

**Solution:** From the conduction parameter given by Equation (11–48) we can write

$$k_n = \frac{\mu_n \varepsilon_s W}{2aL} = \frac{(8000)(13.1)(8.85 \times 10^{-14})(20 \times 10^{-4})}{2(0.7 \times 10^{-4})(1.2 \times 10^{-4})} = 1.1 \text{ mA/V}^2$$

Assume that $V_{GS} = 0.5$ volt. The saturation current is then

$$I_{D1}(sat) = k_n(V_{GS} - V_T)^2 = (1.1)(0.5 - 0.24)^2 = 74.4 \text{ } \mu\text{A}$$

**Comment:** The saturation current will obviously increase if $V_{GS}$ increases or if we increase the size of the transistor by increasing $W$.

---

The transconductance of the enhancement mode device operating in the saturation region can also be derived. Using Equation (11–47), we can write

$$g_{ms} = \frac{\partial I_{D1}(sat)}{\partial V_{GS}} = 2k_n(V_{GS} - V_T) \qquad (11–49)$$

The transconductance increases as $V_{GS}$ increases for the enhancement mode device as it did for the depletion mode device.

## *11.3 NONIDEAL EFFECTS

As with any semiconductor device, there are nonideal effects that will change the ideal device characteristics. In all of the previous discussion, we have considered an ideal transistor with a constant channel length and constant mobility; we have also neglected gate currents. However, when a JFET is biased in the saturation region, the effective electrical channel length is a function of $V_{DS}$. This nonideal effect is called channel length modulation. In addition, when a transistor is biased near or in the saturation region, the electric field in the channel can become large enough so that the majority carriers reach their saturation velocity. At this point, the mobility is no longer a constant. The magnitude of the gate current will affect the input impedance which may need to be taken into account in a circuit design.

### 11.3.1 Channel Length Modulation

The expression for the drain current is inversely proportional to the channel length $L$ as given, for example, by Equation (11–27). In deriving the current equations, we implicitly assumed that the channel length was constant. However, the effective channel length can change. Figure 11–5 showed the space charge region in the channel when the transistor is biased in the saturation region. The neutral n-channel length decreases as $V_{DS}$ increases; thus the drain current will increase. The change in the effective channel length and the corresponding change in drain current is called channel length modulation.

The pinchoff current, Equation (11–28), is modified by the channel length modulation and can be written as

$$I'_{P1} = \frac{\mu_n(eN_d)^2 Wa^3}{6\varepsilon_s L'} \tag{11–50}$$

where

$$L' \cong L - \frac{1}{2}\Delta L \tag{11–51}$$

If we assume the channel depletion region, shown in Figure 11–5 extends equally into the channel and drain regions, then as a first approximation, we will include the factor $\frac{1}{2}$ in the expression for $L'$.

The drain current can be written as

$$I'_{D_1} = I_{D1} \cdot \frac{I'_{P1}}{I_{P1}} = I_{D1}\left(\frac{L}{L - \frac{1}{2}\Delta L}\right) \tag{11–52}$$

where $I_{D1}$ is the ideal drain current predicted by Equation (11–35). Another form of the current-voltage characteristic in the saturation region is given by

$$I'_{D_1}(\text{sat}) = I_{D1}(\text{sat})(1 + \lambda V_{DS}) \tag{11–53}$$

The effective channel length $L'$ supports the $V_{DS}(\text{sat})$ voltage and the space charge region length $\Delta L$ in the channel supports the drain voltage beyond the saturation value. This depletion length is given by

$$\Delta L = \left[\frac{2\varepsilon_s(V_{DS} - V_{DS}(\text{sat}))}{eN_d}\right]^{1/2} \tag{11–54}$$

Since the effective channel length changes with $V_{DS}$, the drain current is now a function of $V_{DS}$. The small-signal output impedance at the drain terminal can be defined as

$$r_{ds} = \frac{\partial V_{DS}}{\partial I'_D} \cong \frac{\Delta V_{DS}}{\Delta I'_D} \qquad (11\text{-}55)$$

## Example 11-8

**Objective:** To calculate the small-signal output resistance at the drain terminal due to channel length modulation effects.

Consider an n-channel depletion mode silicon JFET with a channel doping of $N_d = 3 \times 10^{15}$ cm$^{-3}$. Calculate $r_{ds}$ for the case when $V_{DS}$ changes from $V_{DS}(1) = V_{DS}(\text{sat}) + 2.0$ to $V_{DS}(2) = V_{DS}(\text{sat}) + 2.5$. Assume $L = 10$ $\mu$m and $I_{P1} = 4.0$ mA

**Solution:** We have that

$$r_{ds} = \frac{\Delta V_{DS}}{\Delta I'_{D1}} = \frac{V_{DS}(2) - V_{DS}(1)}{I'_{D1}(2) - I'_{D1}(1)}$$

We can calculate the change in the channel length for the two voltages as

$$\Delta L(2) = \left[ \frac{2\varepsilon_s(V_{DS}(2) - V_{DS}(\text{sat}))}{eN_d} \right]^{1/2}$$

$$= \left[ \frac{2(11.7)(8.85 \times 10^{-14})(2.5)}{(1.6 \times 10^{-19})(3 \times 10^{15})} \right]^{1/2} = 1.04 \ \mu m$$

and

$$\Delta L(1) = \left[ \frac{2(11.7)(8.85 \times 10^{-14})(2.0)}{(1.6 \times 10^{-19})(3 \times 10^{15})} \right]^{1/2} = 0.929 \ \mu m$$

The drain currents are then

$$I'_{D1}(2) = I_{D1} \left( \frac{L}{L - \frac{1}{2} \Delta L(2)} \right) = 4.0 \left( \frac{10}{9.48} \right)$$

and

$$I'_{D1}(1) = I_{D1} \left( \frac{L}{L - \frac{1}{2} \Delta L(1)} \right) = 4.0 \left( \frac{10}{9.54} \right)$$

The output resistance can be calculated as

$$r_{ds} = \frac{2.5 - 2.0}{4 \left( \frac{10}{9.48} \right) - 4 \left( \frac{10}{9.54} \right)} = 18.9 \ k\Omega$$

**Comment:** This value of output resistance is significantly less than the ideal value of infinity.

For high-frequency MESFETs, typical channel lengths are on the order of 1 $\mu$m. Channel length modulation and other effects become very important in short-channel devices.

## 11.3.2 Velocity Saturation Effects

We have seen that the drift velocity of a carrier in silicon saturates with increasing electric field. This velocity saturation effect implies that the mobility is not a constant. For very short channels, the carriers can easily reach their saturation velocity, which changes the I-V characteristics of the JFET.

Figure 11–15 shows the channel region with an applied drain voltage. As the channel narrows at the drain terminal, the velocity of the carriers in-

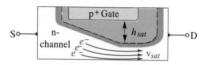

**Figure 11–15**   Cross section of JFET showing carrier velocity and space charge width saturation effects.

creases since the current through the channel is constant. The carriers first saturate at the drain end of the channel. The depletion region will reach a saturation thickness, so we can write

$$I_{D1}(\text{sat}) = eN_d v_{\text{sat}}(a - h_{\text{sat}})W \tag{11–56}$$

where $v_{\text{sat}}$ is the saturation velocity and $h_{\text{sat}}$ is the saturation depletion width. This saturation effect occurs at a drain voltage smaller than the $V_{DS}(\text{sat})$ value determined previously. Both $I_{DS}(\text{sat})$ and $V_{DS}(sat)$ will be smaller than previously calculated.

Figure 11–16 shows normalized plots of $I_D$ versus $V_{DS}$. Figure 11–16a is for the case of a constant mobility and Figure 11–16b is for the case of velocity saturation. Since the I-V characteristics change when velocity saturation occurs, the transconductance will also change—the transconductance will become smaller; hence, the effective gain of the transistor decreases when velocity saturation occurs.

## 11.3.3 Subthreshold and Gate Current Effects

The subthreshold current is the drain current in the JFET that exists when the gate voltage is below the pinchoff or threshold value. The subthreshold conduction was shown in Figure 11–14. When the JFET is biased in the saturation region, the drain current varies quadratically with gate-to-source voltage. When $V_{GS}$ is below the threshold value, the drain current varies exponentially with gate-to-source voltage. Near threshold, the abrupt depletion approximation does not accurately model the channel region: a more detailed potential profile in the space charge region must be used. However, these calculations are beyond the scope of this text.

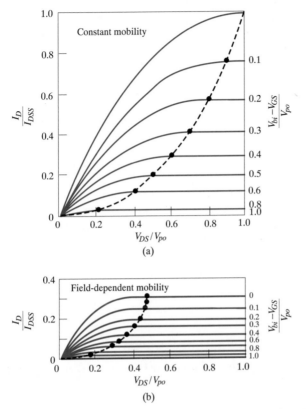

Figure 11–16  Normalized $I_D$ versus $V_{DS}$ plots for a constant mobility and field-dependent mobility. (From Sze [10].)

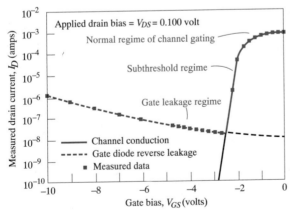

Figure 11–17  Measured drain current versus $V_{GS}$ for a GaAs MESFET showing the normal drain current, subthreshold current, and gate leakage current. (From Daring [2].)

When the gate voltage is approximately 0.5 to 1.0 volt below threshold in an n-channel MESFET, the drain current reaches a minimum value and then slowly increases as the gate voltage decreases. The drain current in this region is the gate leakage current. Figure 11–17 is a plot of the drain current versus $V_{GS}$ for the three regions of gate voltage. The curve illustrates that the drain current becomes small below threshold, but is not zero. The minimum drain current may need to be accounted for in low-power circuit applications.

## *11.4 EQUIVALENT CIRCUIT AND FREQUENCY LIMITATIONS

In order to analyze a transistor circuit, one needs a mathematical model or equivalent circuit of the transistor. One of the most useful models is the small-signal equivalent circuit, which applies to transistors used in linear amplifier circuits. This equivalent circuit will introduce frequency effects in the transistor through the equivalent capacitor-resistor circuits. The various physical factors in the JFET affecting the frequency limitations will be considered here and a transistor cutoff frequency, which is a figure of merit, will then be defined.

### 11.4.1 Small-Signal Equivalent Circuit

The cross section of an n-channel pn JFET is shown in Figure 11–18, including source and drain series resistances. The substrate may be semi-insulating gallium arsenide or it may be a $p^+$ type substrate.

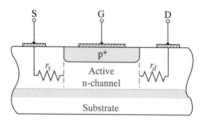

**Figure 11–18**   Cross section of JFET with source and drain series resistance.

Figure 11–19 shows a small-signal equivalent circuit for the JFET. The voltage $V_{g's'}$ is the internal gate-to-source voltage that controls the drain current. The $r_{gs}$ and $C_{gs}$ parameters are the gate-to-source diffusion resistance and junction capacitance, respectively. The gate-to-source junction is reverse biased for depletion mode devices and has only a small forward-bias voltage for enhancement mode devices so that normally $r_{gs}$ is large. The parameters $r_{gd}$ and $C_{gd}$ are the gate-to-drain resistance and capacitance, respectively. The resistance $r_{ds}$ is the finite drain resistance, which is a function of the channel length modulation effect. The $C_{ds}$ capacitance is

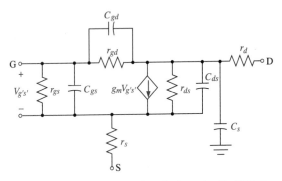

**Figure 11–19**   Small-signal equivalent circuit of JFET.

mainly a drain-to-source parasitic capacitance and $C_s$ is the drain-to-sub-strate capacitance.

The ideal small-signal equivalent circuit is shown in Figure 11–20a. All diffusion resistances are infinite, the series resistances are zero, and at low frequency the capacitances become open circuits. The small-signal drain current is now

$$I_{ds} = g_m V_{gs} \tag{11–57}$$

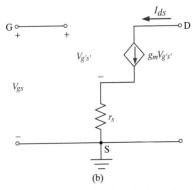

**Figure 11–20**   (a) Ideal low-frequency small-signal equivalent circuit. (b) Ideal equivalent circuit including $r_s$.

which is a function only of the transconductance and the input-signal voltage.

The effect of the source series resistance can be determined using Figure 11–20b. We have

$$I_{ds} = g_m V_{g's'} \tag{11-58}$$

The relation between $V_{gs}$ and $V_{g's'}$ can be found from

$$V_{gs} = V_{g's'} + (g_m V_{g's'})r_s = (1 + g_m r_s)V_{g's'} \tag{11-59}$$

Equation (11–58) can then be written as

$$I_{ds} = \left(\frac{g_m}{1 + g_m r_s}\right) V_{gs} = g_m' V_{gs} \tag{11-60}$$

The effect of the source resistance is to reduce the effective transconductance or transistor gain.

Recall that $g_m$ is a function of the dc gate-to-source voltage so $g_m'$ will also be a function of $V_{GS}$. Equation (11–41b) is the relation between $g_m$ and $V_{GS}$ when the transistor is biased in the saturation region. Figure 11–21 shows a

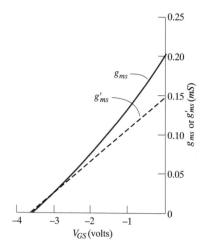

**Figure 11–21**    JFET transconductance versus $V_{GS}$ (a) without, and (b) with a source series resistance.

comparison between the theoretical and experimental transconductance values using the parameters from Example 11–4 and letting $r_s = 2000$ ohms. (A value of $r_s = 2000$ ohms may seem excessive, but keep in mind that the active thickness of the semiconductor may be on the order of 1 $\mu$m or less; thus, a large series resistance may result if special care is not taken.)

## 11.4.2 Frequency Limitation Factors and Cutoff Frequency

There are two frequency limitation factors in a JFET. The first is the channel transit time. If we assume a channel length of 1 $\mu$m and assume carriers are traveling at their saturation velocity, then the transit time is on the order of

$$\tau_t = \frac{L}{v_s} = \frac{1 \times 10^{-4}}{1 \times 10^{+7}} = 10 \text{ ps} \tag{11-61}$$

The channel transit time is normally not the limiting factor except in very high frequency devices.

The second frequency limitation factor is the capacitance charging time. Figure 11–22 is a simplified equivalent circuit that includes the primary

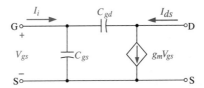

**Figure 11–22** A small-signal equivalent circuit with capacitance.

capacitances and ignores the diffusion resistances. The output current will be the short-circuit current. As the frequency of the input-signal voltage $V_{gs}$ increases, the impedance of $C_{gd}$ and $C_{gs}$ decreases so the current through $C_{gd}$ will increase. For a constant $g_m V_{gs}$, the $I_{ds}$ current will then decrease. The output current then becomes a function of frequency.

If the capacitance charging time is the limiting factor, then the cutoff frequency $f_T$ is defined as the frequency at which the magnitude of the input current $I_i$ is equal to the magnitude of the ideal output current $g_m V_{gs}$ of the intrinsic transistor. We have, when the output is short-circuited,

$$I_i = j\omega(C_{gs} + C_{gd})V_{gs} \tag{11-62}$$

If we let $C_G = C_{gs} + C_{gd}$, then at the cutoff frequency

$$|I_i| = 2\pi f_T C_G V_{gs} = g_m V_{gs} \tag{11-63}$$

or

$$f_T = \frac{g_m}{2\pi C_G} \tag{11-64}$$

From Equation (11–41b), the maximum possible transconductance is

$$g_{ms}(\text{max}) = G_{01} = \frac{e\mu_n N_d W a}{L} \tag{11-65}$$

and the minimum gate capacitance is

$$C_G(\text{min}) = \frac{\varepsilon_s WL}{a} \tag{11–66}$$

where $a$ is the maximum space charge width. The maximum cutoff frequency can be written as

$$f_T = \frac{e\mu_n N_d a^2}{2\pi\varepsilon_s L^2} \tag{11–67}$$

---

**Example 11–9**

**Objective:** To calculate the cutoff frequency of a silicon JFET.
Consider a silicon JFET with the following parameters:

$\mu_n = 1000 \text{ cm}^2/\text{V-sec}$ $\qquad a = 0.60 \ \mu m$

$N_d = 10^{16} \text{ cm}^{-3}$ $\qquad L = 5 \ \mu m$

**Solution:** Substituting the parameters into Equation (11–67), we have

$$f_T = \frac{e\mu_n N_d a^2}{2\pi\varepsilon_s L^2} = \frac{(1.6 \times 10^{-19})(1000)(10^{16})(0.6 \times 10^{-4})^2}{2\pi(11.7)(8.85 \times 10^{-14})(5 \times 10^{-4})^2} = 3.54 \ GHz$$

**Comment:** This example shows that even silicon JFETs can have relatively large cutoff frequencies.

---

For gallium arsenide JFETs or MESFETs with very small geometries, the cutoff frequency is even larger. The channel transit time may also become a factor in very-high-frequency devices, in which case the expression for cutoff frequency would need to be modified.

One application of GaAs FETs is in ultrafast digital integrated circuits. Conventional GaAs MESFET logic gates can achieve propagation delay times in the sub-nanosecond range. These delay times are at least comparable with, if not shorter than, fast ECL, but the power dissipation is three orders of magnitude smaller than in the ECL circuits. Enhancement mode GaAs JFETs have been used as drivers in logic circuits and depletion mode devices may be used as loads. Propagation delay times of as low as 45 ps have been observed. Special JFET structures may be used to further increase the speed. These structures include the modulation-doped field-effect transistor which is discussed in the following section.

## *11.5 MODULATION-DOPED FIELD-EFFECT TRANSISTOR

As frequency needs, power capacity, and low noise performance requirements increase, the gallium arsenide MESFET is being pushed to its limit of design and performance. These requirements imply a very small FET with a short channel length, large saturation current, and large transconductance. These requirements are generally achieved by increasing the channel doping under the gate. In all of the devices we have considered, the channel region is in a doped layer of bulk semiconductor with the majority carriers and doping impurities in the same region. The majority carriers experience ionized impurity scattering, which reduces carrier mobility and degrades device performance.

The degradation in mobility and peak velocity in GaAs due to increased doping can be minimized by separating the majority carriers from the ionized impurities. This separation can be achieved in a heterostructure that has an abrupt discontinuity in conduction and valence bands. We considered the basic heterojunction properties in Chapter 9. Figure 11–23 shows the con-

**Figure 11–23**   Conduction-band edges for N-AlGaAs–intrinsic GaAs abrupt heterojunction.

duction-band energy relative to the Fermi energy of an N-AlGaAs–intrinsic GaAs heterojunction in thermal equilibrium. Thermal equilibrium is achieved when electrons from the wide-bandgap AlGaAs flow into the GaAs and are confined to the potential well. However, the electrons are free to move parallel to the heterojunction interface. In this structure, the majority carrier electrons in the potential well are now separated from the impurity dopant atoms in the AlGaAs; thus, impurity scattering tends to be minimized.

The FETs fabricated from these heterojunctions are known by several names. The term used here is the MOdulation-Doped Field-Effect Transistor (MODFET). Other names include the high-electron mobility transistor (HEMT), selectively doped heterojunction field-effect transistor (SDHT), and two-dimensional electron gas field-effect transistor (TEGFET).

### 11.5.1 Quantum Well Structures

Figure 11–23 showed the conduction-band energy of an N-AlGaAs–intrinsic GaAs heterojunction. A two-dimensional surface channel layer of electrons is formed in the thin potential well ($\approx 80$ Å) in the undoped GaAs. Electron sheet carrier densities on the order of $10^{12}$ cm$^{-2}$ have been obtained. An improvement in the low-field mobility of the carriers moving parallel to the heterojunction is observed since the impurity-scattering effects are reduced. At 300°K, mobilities have been reported in the range of 8500–9000 cm$^2$/V-sec whereas GaAs MESFETs doped to $N_d = 10^{17}$ cm$^{-3}$ have low-field mobilities of less than 5000 cm²/V-sec. The electron mobility in the heterojunction now tends to be dominated by lattice or phonon scattering, so as the temperature is reduced, the mobility increases rapidly.

Impurity-scattering effects can be further reduced by increasing the separation of the electrons and ionized donor impurities. The electrons in the potential well of the abrupt heterojunction shown in Figure 11–23 are separated from the donor atoms, but are still close enough to be subjected to a coulomb attraction. A thin spacer layer of undoped AlGaAs can be placed between the doped AlGaAs and the undoped GaAs. The energy-band diagram for this structure is shown in Figure 11–24. Increasing the separation between the carriers and ionized donors increases further the electron mobility, since there is even less coulomb interaction. One disadvantage of this

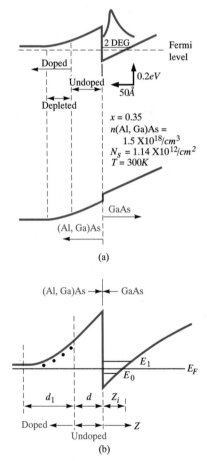

Figure 11–24    Conduction-band edges for N-AlGaAs–undoped AlGaAs–undoped GaAs heterojunction. (From Shur [9].)

graded heterojunction is that the electron density in the potential well tends to be smaller than in the abrupt junction.

The molecular beam epitaxial process allows the growth of very thin layers of specific semiconductor materials with specific dopings. In particular, a multilayer modulation-doped heterostructure can be formed, as shown in Figure 11–25. Several surface channel layers of electrons are formed in

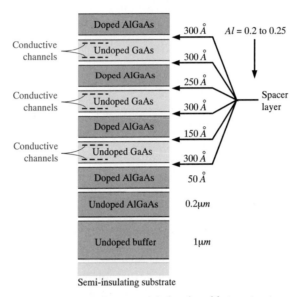

**Figure 11–25**   Multilayer modulation-doped heterostructure.

parallel. This structure would be equivalent to increasing the channel electron density, which would increase the current capability of the FET.

## 11.5.2 Transistor Performance

A typical MODFET structure is shown in Figure 11–26. The N-AlGaAs is separated from the undoped GaAs by an undoped AlGaAs spacer. A

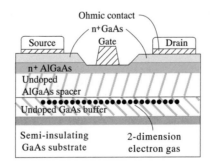

**Figure 11–26**   A "normal" AlGaAs-GaAs MODFET.

Schottky contact to the N-AlGaAs forms the gate of the transistor. This structure is a "normal" MODFET. An "inverted" structure is shown in Figure 11–27. In this case the Schottky contact is made to the undoped GaAs layer. The inverted MODFET has been investigated less than the normal structure because the normal structure has yielded superior results.

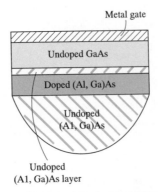

**Figure 11–27**   An "inverted" GaAs-AlGaAs MODFET. (From Shur [9].)

The density of electrons in the two-dimensional electron gas layer in the potential well can be controlled by the gate voltage. The electric field of the Schottky gate depletes the two-dimensional electron gas layer in the potential well when a sufficiently large negative voltage is applied to the gate. Figure 11–28 shows the energy-band diagrams of the metal-AlGaAs-GaAs

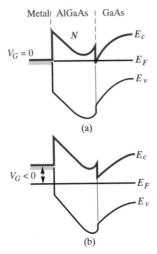

**Figure 11–28**   Energy-band diagram of a normal MODFET (a) with zero gate bias, and (b) with a negative gate bias.

structure under zero bias and with a reverse bias applied to the gate. With zero bias, the conduction-band edge in the GaAs is below the Fermi level, implying a large density of the two-dimensional electron gas. With a negative voltage applied to the gate, the conduction-band edge in the GaAs is above the Fermi level, implying that the density of the two-dimensional electron gas is very small and the current in an FET would be essentially zero.

The Schottky barrier depletes the AlGaAs layer from the surface and the heterojunction depletes the AlGaAs layer from the heterojunction interface. Ideally the device should be designed so that the two depletion regions just overlap to prevent electron conduction through the AlGaAs layer. For depletion mode devices, the depletion layer from the Schottky gate should extend only to the heterojunction depletion layer. For enhancement mode devices, the thickness of the doped AlGaAs layer is smaller, and the Schottky gate built-in potential barrier will completely deplete the AlGaAs layer and the two-dimensional electron gas channel. A positive voltage applied to the gate of the enhancement mode device will turn on the device.

The density of the two-dimensional electron gas in a normal structure can be described using a charge control model. We may write

$$n_s = \frac{\varepsilon_N}{q(d + \Delta d)} (V_g - V_{\text{off}}) \tag{11-68}$$

where $\varepsilon_N$ is the permittivity of the N-AlGaAs, $d = d_d + d_i$ is the thickness of the doped-plus-undoped AlGaAs layer, and $\Delta d$ is a correction factor given by

$$\Delta d = \frac{\varepsilon_N a}{q} \approx 80 \mathring{A} \tag{11-69}$$

The threshold voltage $V_{\text{off}}$ is given by

$$V_{\text{off}} = \phi_B - \frac{\Delta E_c}{q} - V_{p2} \tag{11-70}$$

where $\phi_B$ is the Schottky barrier height and $V_{p2}$ is

$$V_{p2} = \frac{qN_d d_d^2}{2\varepsilon_N} \tag{11-71}$$

A negative gate bias will reduce the two-dimensional electron gas concentration. If a positive gate voltage is applied, the density of the two-dimensional electron gas will increase. Increasing the gate voltage will increase the two-dimensional electron gas density until the conduction band of the AlGaAs crosses the Fermi level of the electron gas. Figure 11–29 shows this effect. At this point the gate loses control over the electron gas since a parallel conduction path in the AlGaAs has been formed.

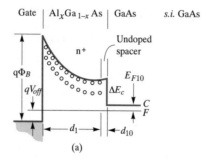

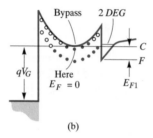

**Figure 11–29**   Energy-band diagram of an enhancement mode MODFET (a) with a slight forward gate voltage, and (b) with a larger forward gate voltage which creates a conduction channel in the AlGaAs. (From Fritzsche [4].)

### Example 11–10

**Objective:** To determine the two-dimensional electron concentration for an n-AlGaAs-intrinsic GaAs heterojunction.

Consider $N - Al_{0.3}Ga_{0.7}As$ doped to $10^{18}$ cm$^{-3}$ and having a thickness of 500Å. Assume an undoped spacer layer of 20Å. Let $\phi_B = 0.85$ volt and $\Delta E_c/q = 0.22$ volt. The relative dielectric constant of $Al_{0.3}Ga_{0.7}As$ is $\varepsilon_N = 12.2$.

**Solution:** The parameter $V_{p2}$ is found as

$$V_{p2} = \frac{qN_d d_d^2}{2\varepsilon_N} = \frac{(1.6 \times 10^{-19})(10^{18})(500 \times 10^{-8})^2}{2(12.2)(8.85 \times 10^{-14})} = 1.85 \text{ volts}$$

Then the threshold voltage is

$$V_{off} = \phi_B - \frac{\Delta E_c}{q} - V_{p2} = 0.85 - 0.22 - 1.85 = -1.22 \text{ volts}$$

The channel electron concentration for $V_g = 0$ is found from Equation (11–68) to be

$$n_s = \frac{(12.2)(8.85 \times 10^{-14})}{(1.6 \times 10^{-19})(500 + 20 + 80) \times 10^{-8}} [-(-1.22)] = 1.37 \times 10^{12} \text{ cm}^{-2}$$

**Comment:** The threshold voltage $V_{off}$ is negative, making this device a depletion mode MODFET: applying a negative gate voltage will turn off the device. A value of $N_s \approx 10^{12}$ cm$^{-2}$ is a typical channel concentration.

The current-voltage characteristics of the MODFET can be found using the charge control model and the gradual channel approximation. The channel carrier concentration can be written as

$$n_s(x) = \frac{\varepsilon_N}{q(d + \Delta d)}(V_g - V_{off} - V(x)) \tag{11-72}$$

where $V(x)$ is the potential along the channel due to the drain-to-source voltage. The drain current is

$$I_D = qn_s v(E)W \tag{11-73}$$

where $v(E)$ is the carrier drift velocity and $W$ is the channel width. This analysis is very similar to that for the pn JFET in Section 11.2.2.

If we assume a constant mobility, then for low $V_{DS}$ values, we have

$$I_D = \frac{\varepsilon_N \mu W}{2L(d + \Delta d)}[2(V_g - V_{off})V_{DS} - V_{DS}^2] \tag{11-74}$$

The form of this equation is the same as that for the pn JFET or MESFET operating in the nonsaturation region. If $V_{DS}$ increases so that the carriers reach the saturation velocity, then

$$I_D(\text{sat}) = \frac{\varepsilon_N W}{(d + \Delta d)}(V_g - V_{off} - V_0)v_{sat} \tag{11-75}$$

where $v_{sat}$ is the saturation velocity and $V_0 = E_s L$ with $E_s$ being the electric field in the channel which produces the saturation velocity.

Various velocity versus electric field models can be used to derive different I-V expressions. However, Equations (11-74) and (11-75) yield satisfactory results for most situations. Figure 11-30 shows a comparison between experimental and calculated I-V characteristics. As observed in the figure, the current in these heterojunction devices can be quite large. The transconductance of the MODFET is defined as it was for the pn JFET and MESFET. Typical measured values at $T = 300K°$ are in the range of 250 mS/mm.

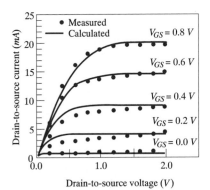

**Figure 11-30** Current-voltage characteristics of an enhancement mode MODFET. Solid curves are numerical calculations and dots are measured points. (From Shur [9].)

Higher values have been reported. These transconductance values are significantly larger than for either the pn JFET or MESFET.

MODFETs may also be fabricated with multiple heterojunction layers. This device type is shown in Figure 11–31. A single heterojunction for an

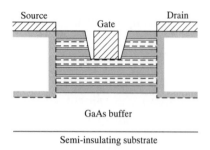

GaAs buffer

Semi-insulating substrate

**Figure 11–31**   A multilayer MODFET.

AlGaAs–GaAs interface has a maximum two-dimensional electron sheet density on the order of $1 \times 10^{12}$ cm$^{-2}$. This concentration can be increased by fabricating two or more AlGaAs-GaAs interfaces in the same epitaxial layer. The device current capacity is increased, and power performance is improved. The multichannel MODFET behaves as multiple single-channel MODFETs connected in parallel and modulated by the same gate but with slightly different threshold voltages. The maximum transconductance will not scale directly with the number of channels because of the change in threshold voltage with each channel. In addition, the effective channel length increases as the distance between the gate and channel increases.

MODFETs can be used in high-speed logic circuits. They have been used in flip-flop circuits operating at clock frequencies of 5.5 GHz at $T = 300°$K; the clock frequency can be increased at lower temperatures. Small-signal, high-frequency amplifiers have also been investigated. MODFETs showing low noise and reasonable gains have been operated at 35 GHz. The maximum frequency increases as the channel length decreases. Cutoff frequencies on the order of 100 GHz have been measured with channel lengths of 0.25 $\mu$m.

It seems clear that MODFETs are inherently superior to other FET technologies in terms of achieving higher speeds of operation, lower power dissipation, and lower noise. These advantages derive directly from the superior transport properties obtained by using undoped GaAs as the channel layer for the FET. One way to achieve an adequate carrier concentration in an undoped channel is to accumulate the carriers at a semiconductor heterojunction interface, as we have seen. The disadvantage of the MODFET is that the fabrication processes for the heterojunction are more complicated.

## 11.6 SUMMARY AND REVIEW

In this chapter we considered the physics and characteristics of the junction field-effect transistor. The three general types of JFETs we discussed are the pn JFET, the MESFET, and the MODFET. Each of these transistors, in conjunction with a circuit, is capable of voltage gain and signal-power gain.

The current in a JFET is controlled by an electric field applied perpendicular to the direction of current. The current is in the channel region between the source and drain contacts. In a pn JFET, the channel forms one side of a pn junction that is used to modulate the channel conductance. The gate voltage controls the reverse-biased pn junction space charge width, which extends into the channel region. The MESFET is similar to the pn JFET except that the gate is a Schottky barrier junction rather than a pn junction. The basic JFET transistor action is the control of a current in one part of the device, the drain current, by a voltage in another part of the device, the gate voltage.

Two primary parameters of the JFET are the internal pinchoff voltage, $V_{p0}$, and the pinchoff voltage, $V_p$. The internal pinchoff voltage is defined as a positive quantity and is the total gate-to-channel potential that causes the junction space charge layer to completely fill the channel region. The drain current is essentially zero for this case since the source and drain terminals are isolated by this space charge region. The pinchoff voltage is defined as the gate voltage that must be applied to achieve pinchoff and includes the sign of the applied voltage. The difference between the internal pinchoff and pinchoff voltages is the built-in potential barrier of the junction.

We derived the ideal current-voltage relationship by considering the resistance of the channel and using Ohm's law. The drain current is a function of both gate-to-source and drain-to-source voltages in the nonsaturation region but, for the ideal transistor, is a function of only the gate-to-source voltage in the saturation region. The transconductance, or transistor gain, is the rate of change of drain current with respect to the corresponding change in gate-to-source voltage. This parameter is a function of the semiconductor properties and the transistor geometry.

Three nonideal effects were considered. First we discussed channel length modulation. The JFET drain current is inversely proportional to the channel length, which we assumed to be constant in the ideal derivation. However, the effective length of the channel decreases with drain voltage when the drain voltage becomes larger than the $V_{DS}(\text{sat})$ value. The drain current, then, is no longer completely independent of the drain-to-source voltage in the saturation region so that channel length modulation results in a finite output impedance at the drain terminal. The ideal current-voltage relationship was derived, also assuming that carrier mobility was constant. For short-channel JFETs in particular, the electric field in the channel can be large enough so that carriers will be traveling at their saturation velocity. In this case, the drain current will saturate at a lower value than predicted in the ideal case. This was the second nonideal effect we considered. The third was the subthreshold drain current. The JFET drain current is controlled by the space charge width of a reverse-biased junction. Near pinchoff or threshold, the abrupt depletion approximation does not accurately model the channel region; a more detailed potential profile is required. At pinchoff and slightly below, the drain current is not zero but varies exponentially with gate-to-source voltage. In low-power applications, this current may need to be taken into account to fully determine the circuit power consumption.

We developed a small-signal equivalent circuit of the JFET. The equivalent circuit includes capacitance effects which introduce frequency effects in the transistor. Various physical factors in the JFET affect the frequency limitations; the two main factors are channel transit time and capacitance charging time. The channel transit time is normally not the limiting factor for short-channel devices. The capacitance charging time is a function of the gate-to-source and gate-to-drain capacitances. The design of a high-frequency JFET implies small geometries and the minimization of any parasitic capacitance. The cutoff frequency is a figure of merit for the frequency response of a transistor. Typical values of $f_T$ for high-frequency MESFETs are in the tens of GHz range. The MESFET, in general, has the highest frequency capability of any of the three basic transistor types.

The modulation-doped field-effect transistor, MODFET, structure utilizes a heterojunction. A two-dimensional electron gas is confined to a potential well at the heterojunction interface. However, the electrons are free to move parallel to the interface. These electrons are separated from the ionized donors so that ionized impurity scattering effects are minimized, resulting in a high mobility. The high mobility results in large saturation currents, large transconductance values, and very high cutoff frequencies.

## GLOSSARY OF IMPORTANT TERMS

**Capacitance charging time:** The time associated with charging or discharging the input gate capacitance with a change in the input gate signal.

**Channel conductance:** The ratio of a differential change in drain current to the corresponding differential change in drain-to-source voltage in the limit as the drain-to-source voltage approaches zero.

**Channel conductance modulation:** The process whereby the channel conductance changes with gate voltage; this is the basic field-effect transistor action.

**Channel length modulation:** The change in effective channel length with drain-to-source voltage with the JFET biased in the saturation region.

**Conduction parameter:** The multiplying factor $k_n$ in the expression for drain current versus gate-to-source voltage for the enhancement mode MESFET.

**Cutoff frequency:** A figure of merit for the transistor defined to be the frequency at which the ratio of the small-signal input gate current to small-signal drain current is equal to unity.

**Depletion mode JFET:** A JFET in which a gate-to-source voltage must be applied to create pinchoff and turn the device off.

**Enhancement mode JFET:** A JFET in which pinchoff exists at zero gate voltage and a gate-to-source voltage must be applied to induce a channel, turning the device on.

**Internal pinchoff voltage:** The total potential drop across the gate junction at pinchoff.

**Output resistance:** The ratio of a differential change in drain-to-source voltage to the corresponding differential change in drain current at a constant gate-to-source voltage.

**Pinchoff:** The condition whereby the gate junction space charge region extends completely through the channel so that the channel is completely depleted of free carriers.

**Pinchoff voltage:** The gate-to-source voltage that must be applied to produce pinchoff throughout the channel.

**Saturation:** The condition whereby pinchoff occurs at the drain terminal and the drain current is no longer a function of drain-to-source voltage.

**Semi-insulating substrate:** A very high resistivity gallium arsenide substrate.

**Threshold voltage:** Another term for pinchoff voltage, primarily used for enhancement mode MESFETs.

**Transconductance:** The ratio of a differential change in drain current to the corresponding differential change in gate-to-source voltage: also referred to as the transistor gain.

## PROBLEMS

### Section 11.1

1. *(a)* Draw the structure of a p-channel JFET similar to the structure shown in Figure 11–2. *(b)* Qualitatively discuss the I-V characteristics, including current directions and voltage polarities, similar to those shown in Figures 11–3 and 11–4.

2. Consider the n-channel JFET in Figure 11–32. The p-type substrate is connected to the n-type source terminal. Sketch the space charge regions for various $V_{GS}$ values when $V_{DS} = 0$ and for various $V_{DS}$ values when $V_{GS} = 0$.

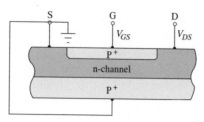

**Figure 11–32**   Figure for problem 2.

## Section 11.2

3. A p-channel silicon JFET at $T = 300°K$ has doping concentrations of $N_d = 5 \times 10^{18}$ cm$^{-3}$ and $N_a = 3 \times 10^{16}$ cm$^{-3}$. The channel thickness dimension is $a = 0.5$ $\mu$m. *(a)* Compute the internal pinchoff voltage $V_{p0}$ and the pinchoff voltage $V_p$. *(b)* Determine the minimum undepleted channel thickness, $a - h$, for $V_{GS} = 1$ volt and for *(i)* $V_{DS} = 0$, *(ii)* $V_{DS} = -2.5$ volts, and *(iii)* $V_{DS} = -5$ volts.

4. Repeat problem 3 for a GaAs JFET with the same electrical and geometrical parameters.

5. Consider an n-channel silicon JFET at $T = 300°K$ with impurity doping concentrations of $N_d = 4 \times 10^{16}$ cm$^{-3}$ and $N_a = 5 \times 10^{18}$ cm$^{-3}$. The channel thickness dimension is $a = 0.35$ $\mu$m. *(a)* Compute the internal pinchoff voltage $V_{p0}$ and the pinchoff voltage $V_p$. *(b)* Determine the undepleted channel thickness, $a - h$, at the drain terminal for *(i)* $V_{GS} = 0$, $V_{DS} = 1$ volt; *(ii)* $V_{GS} = -1.0$ volt, $V_{DS} = 1$ volt; and *(iii)* $V_{GS} = -1.0$ volt, $V_{DS} = 2$ volts.

6. Repeat problem 5 for a GaAs JFET with the same electrical and geometrical parameters.

7. An n-channel silicon JFET at $T = 300°K$ has the following parameters:

   $$N_a = 10^{19} \text{ cm}^{-3} \qquad N_d = 10^{16} \text{ cm}^{-3}$$
   $$a = 0.50 \text{ } \mu\text{m} \qquad L = 20 \text{ } \mu\text{m}$$
   $$W = 400 \text{ } \mu\text{m} \qquad \mu_n = 1000 \text{ cm}^2/\text{V-sec}$$

   Ignoring velocity saturation effects, calculate *(a)* $I_{p1}$, *(b)* $V_{DS}$(sat) for *(i)* $V_{GS} = 0$, *(ii)* $V_{GS} = V_p/4$, *(iii)* $V_{GS} = V_p/2$, and *(iv)* $V_{GS} = 3V_p/4$; and *(c)* $I_{D1}$(sat) for the same $V_{GS}$ values in part *(b)*. *(d)* Using the results from parts *(b)* and *(c)*, plot the I-V characteristics.

8. Consider the JFET described in problem 7. Compute and plot the channel conductance, $g_d$, as a function of $V_{GS}$ for $0 < |V_{GS}| < |V_p|$.

9. Consider an n-channel GaAs JFET at $T = 300°K$ with the following parameters:

$$N_a = 5 \times 10^{18} \text{ cm}^{-3} \qquad N_d = 2 \times 10^{16} \text{ cm}^{-3}$$
$$a = 0.35 \ \mu\text{m} \qquad L = 10 \ \mu\text{m}$$
$$W = 30 \ \mu\text{m} \qquad \mu_n = 8000 \text{ cm}^2/\text{V-sec}$$

Ignoring velocity saturation effects, calculate *(a)* $G_{01}$; *(b)* $V_{DS}(\text{sat})$ for $V_{GS} = 0$ and $V_{GS} = V_p/2$; *(c)* $I_{D1}(\text{sat})$ for $V_{GS} = 0$ and $V_{GS} = V_p/2$. *(d)* Sketch the I-V characteristics using the results from parts *(b)* and *(c)*.

10. Using the parameters from problem 7, calculate the maximum transconductance in the saturation region. Normalize this transconductance to mS per unit width, or mS/mm.

11. *(a)* Calculate the maximum transconductance for the transistor described in problem 9. *(b)* Determine the maximum transconductance if the channel length is reduced to 2 $\mu$m.

12. The Schottky barrier height, $\phi_{Bn}$, of a metal–n-GaAs MESFET is 0.90 volt. The channel doping is $N_d = 1.5 \times 10^{16} \text{ cm}^{-3}$ and the channel thickness is $a = 0.5 \ \mu$m. $T = 300°$K. *(a)* Calculate the internal pinchoff voltage $V_{p0}$ and the threshold voltage $V_T$. *(b)* Determine whether the MESFET is depletion type or enhancement type.

13. Consider an n-channel GaAs MESFET at $T = 300°$K with a gold Schottky barrier contact. Assume $\phi_{Bn} = 0.89$ volt. The channel thickness is $a = 0.35 \ \mu$m. *(a)* Determine the uniform channel doping so that the threshold voltage is $V_T = 0.10$ volt. *(b)* Using the results of part *(a)*, determine the threshold voltage at $T = 400°$K.

14. The barrier height, $\phi_{Bn}$, of a metal–n-GaAs MESFET is 0.89 volt. The channel doping is $N_d = 10^{16} \text{ cm}^{-3}$ and the temperature is $T = 300°$K. *(a)* Determine the channel thickness so that $V_T = +0.15$ volt. *(b)* Using the results of part *(a)*, determine $V_T$ at $T = 350°$K and at $T = 400°$K.

15. An n-channel silicon MESFET is fabricated using a gold contact. The n-channel doping is $N_d = 10^{16} \text{ cm}^{-3}$ and the temperature is $T = 300°$K. When a gate voltage of $V_{GS} = 0.35$ volt is applied with $V_{DS} = 0$, the undepleted channel thickness is 0.075 $\mu$m. *(a)* Determine the channel thickness dimension $a$ and the threshold voltage $V_T$. *(b)* Determine the value of $V_{DS}(\text{sat})$ for $V_{GS} = 0.35$ volt.

16. The Schottky barrier height, $\phi_{Bn}$, of a metal–n-silicon MESFET is 0.80 volt. The channel thickness dimension is $a = 0.40 \ \mu$m and the channel doping is $N_d = 2 \times 10^{16} \text{ cm}^{-3}$. $T = 300°$K. *(a)* For this device, calculate $V_{bi}$, $V_{p0}$, $V_T$, and $V_{DS}(\text{sat})$ for $V_{GS} = -1$ volt. *(b)* What concentration of impurity atoms must be added to the channel so that $V_{p0} = 4.5$ volts? Repeat the calculations in part *(a)* for this new channel doping.

17. A GaAs n-channel MESFET at $T = 300°$K has a threshold voltage of $V_T = +0.12$ volt. Let $\mu_n = 7800 \text{ cm}^2/\text{V-sec}$, $W = 20 \ \mu$m, $L = 1.2$

$\mu$m, and $a = 0.30$ $\mu$m. *(a)* Calculate the conduction parameter $k_n$.
*(b)* Determine $V_{DS}(\text{sat})$ for $V_{GS} = 1.5$ $V_T$ and for $V_{GS} = 2.0$ $V_T$.
*(c)* Determine $I_{D1}(\text{sat})$ for $V_{GS} = 1.5$ $V_T$ and for $V_{GS} = 2.0$ $V_T$.

18.   Consider a GaAs n-channel MESFET at $T = 300°$K with $V_T = +0.25$ volt. Let $\mu_n = 8000$ cm$^2$/V-sec, $L = 1.0$ $\mu$m, and $a = 0.35$ $\mu$m. The transconductance in the saturation region is to be $g_m = 1.75$ mA/V at $V_{GS} = 0.50$ volt. *(a)* Determine the channel width $W$. *(b)* Determine $I_{D1}(\text{sat})$ for $V_{GS} = 0.40$ volt and $V_{GS} = 0.65$ volt.

## Section 11.3

19.   Consider a uniformly doped n-channel silicon depletion mode JFET with the following parameters:

$$N_a = 10^{19} \text{ cm}^{-3} \qquad N_d = 3 \times 10^{16} \text{ cm}^{-3}$$
$$a = 0.40 \text{ } \mu\text{m} \qquad \mu_n = 1000 \text{ cm}^2/\text{V-sec}$$

The maximum drain-to-source voltage is to be 5 volts. When $V_{GS} = 0$, the effective channel length, $L'$, is to be 90 percent of the original channel length. Determine $L$.

*20.   If the change in the channel length, $\Delta L$, is assumed small, derive an approximate expression in terms of channel parameters for $\lambda$ given in Equation (11–53). (Note: The parameter $\lambda$ may not be a constant. However, justify using Equation (11–53) by plotting the expression for $\lambda$ over the range $1.5V_{DS}(\text{sat}) \leq V_{DS} \leq 3.0V_{DS}(\text{sat})$. Use typical parameter values.)

*21.   As a first approximation, assume that the electric field in the channel of an n-channel silicon JFET is uniform through the channel. Also, assume that the drift velocity versus electric field for the electrons is given by the piecewise linear approximation given in Figure 11–33. Let:

$$N_a = 5 \times 10^{18} \text{ cm}^{-3} \qquad N_d = 4 \times 10^{16} \text{ cm}^{-3}$$
$$L = 2 \text{ } \mu\text{m} \qquad W = 30 \text{ } \mu\text{m}$$
$$a = 0.50 \text{ } \mu\text{m}$$

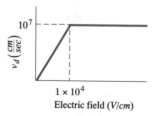

**Figure 11–33**   Figure for problem 21.

(a) Determine $V_{DS}$ at which velocity saturation occurs. Let $V_{GS} = 0$. (b) For $V_{GS} = 0$, determine $h_{sat}$. (c) Calculate $I_{D1}$(sat) if velocity saturation occurs. (d) If the electron mobility is a constant and equal to $\mu_n = 1000$ cm$^2$/V-sec, calculate $I_{D1}$(sat) if velocity saturation did not occur.

*22. (a) Repeat problem 21 if $L = 1$ $\mu$m and all other parameters remain the same. (b) If velocity saturation occurs, does the relation $I_{D1}$(sat) $\propto L^{-1}$ still apply? Explain.

23. Consider a one-sided silicon n-channel JFET at $T = 300°$K, pinched off as shown in Figure 11–34. The source-to-gate and drain-to-gate reverse-bias currents are split geometrically as shown. Assume that the reverse-bias currents are dominated by the generation current. Assume the following parameters:

$$N_a = 5 \times 10^{18} \text{ cm}^{-3} \quad\quad N_d = 3 \times 10^{16} \text{ cm}^{-3}$$
$$\tau_0 = 5 \times 10^{-8} \text{ sec} \quad\quad a = 0.30 \ \mu\text{m}$$
$$W = 30 \ \mu\text{m} \quad\quad L = 2.4 \ \mu\text{m}$$

Calculate $I_{DG}$ for (a) $V_{DS} = 0$, (b) $V_{DS} = 1$ volt, and (c) $V_{DS} = 5$ volts. (Use Equation (8–81) and consider the volume of the depletion region.)

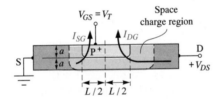

Figure 11–34   Figure for problem 23.

## Section 11.4

24. The source series resistance of a MESFET will reduce the value of transconductance, $g_{ms}$. Assume the doping in the source region of a GaAs MESFET is $N_d = 7 \times 10^{16}$ cm$^{-3}$ and the dimensions are $a = 0.3$ $\mu$m, $L = 1.5$ $\mu$m, and $W = 5.0$ $\mu$m. Let $\mu_n = 4500$ cm$^2$/V-sec and $\phi_{Bn} = 0.89$ volt. (a) Determine the ideal value of $g_{ms}$ for $V_{GS} = 0$. (b) Determine the value of $r_s$ for which the value of $g'_{ms}$ is 80 percent of the ideal value. (c) Determine the maximum effective distance from the edge of the channel to the source terminal so that $r_s$ is no larger than the value determined in part (b).

25. Estimate the cutoff frequency of the MESFET in problem 24.

## Section 11.5

26. Consider an N-$Al_{0.3}Ga_{0.7}As$–intrinsic GaAs abrupt heterojunction. Assume that the AlGaAs is doped to $N_d = 3 \times 10^{18}$ cm$^{-3}$ and has a thickness of 350 Å. Let $\phi_{Bn} = 0.89$ volt and assume that $\Delta E_c = 0.24$ eV. (a) Calculate $V_{off}$ and (b) calculate $n_s$ for $V_g = 0$.

27. If the electrons in the channel of the JFET in problem 26 are traveling at a saturation velocity of $2 \times 10^7$ cm/sec, determine (a) the transconductance per unit width at $V_g = 0$ and (b) the saturation current per unit width at $V_g = 0$. (Assume $V_0 = 1$ volt.)

28. Consider an abrupt N-$Al_{0.3}Ga_{0.7}As$–intrinsic GaAs heterojunction. The N-AlGaAs is doped to $N_d = 2 \times 10^{18}$ cm$^{-3}$. The Schottky barrier height is 0.85 volt and the heterojunction conduction-band edge discontinuity is $\Delta E_c = 0.22$ eV. Determine the thickness of the AlGaAs layer so that $V_{off} = -0.3$ volt.

## Summary and Review

*29. Design a one-sided silicon n-channel JFET so that $V_p = -5.0$ V, $I_{DSS} = I_{DS}$ (sat) = 2.0 mA at $V_{GS} = 0$ V, and $f_T = 3$ GHz. Neglect velocity saturation effects. Determine channel thickness, width, and doping concentration.

*30. Design a one-sided GaAs n-channel MESFET with a gold Schottky contact so that $V_T = +0.15$V, $I_D$(sat) = 0.70 $\mu$A at $V_{GS} = 0.50$V, and $f_T = 30$ GHz.

*31. Design a pair of complementary n-channel and p-channel silicon JFETs so that $I_{DSS} = 1$ mA and $|V_p| = 3.2$ V for each device at $T = 300°$K. If the devices are to operate for $0 \le V_{DS} \le 5$ V, comment on velocity saturation and channel length modulation effects in your design.

## READING LIST

1. Chang, C. S., and D. Y. S. Day. "Analytic Theory for Current-Voltage Characteristics and Field Distribution of GaAs MESFETs." *IEEE Transactions on Electron Devices* 36, no. 2 (February 1989), pp. 269–80.

2. Daring, R. B. "Subthreshold Conduction in Uniformly Doped Epitaxial GaAs MESFETs." *IEEE Transactions on Electron Devices* 36, no. 7 (July 1989), pp. 1264–73.

3. Drummond, T. J.; W. T. Masselink; and H. Morkoc. "Modulation-Doped GaAs/(Al,Ga)As Heterojunction Field-Effect Transistors: MODFETs." *Proceedings of the IEEE* 74, no. 6 (June 1986), pp. 773–812.

4. Fritzsche, D. "Heterostructures in MODFETs." *Solid-State Electronics* 30, no. 11 (November 1987), pp. 1183–95.

5.  Ghandhi, S. K. *VLSI Fabrication Principles: Silicon and Gallium Arsenide*. New York: Wiley, 1983.

6.  Liao, S. Y. *Microwave Solid-State Devices*. Englewood Cliffs, N.J.: Prentice Hall, 1985.

7.  Pierret, R. F. *Field Effect Devices*. Vol. 4 of the *Modular Series on Solid State Devices*. 2nd ed. Reading, Mass.: Addison-Wesley, 1990.

8.  Pulfrey, D. L., and N. G. Tarr. *Introduction to Microelectronic Devices*. Englewood Cliffs, N.J.: Prentice Hall, 1989.

*9.  Shur, M. *GaAs Devices and Circuits*. New York: Plenum Press, 1987.

10.  Sze, S. M. *Physics of Semiconductor Devices*. 2nd ed. New York: Wiley, 1981.

11.  _____. *Semiconductor Devices: Physics and Technology*. New York: Wiley, 1985.

12.  Turner, J. A.; R. S. Butlin; D. Parker; R. Bennet; A. Peake; and A. Hughes. "The Noise and Gain Performance of Submicron Gate Length GaAs FETs." *GaAs FET Principles and Technology*. Edited by J. V. DiLorenzo and D. D. Khandelwal. Dedham, Mass.: Artech House, 1982.

13.  Yang, E. S. *Microelectronic Devices*. New York: McGraw-Hill, 1988.

14.  Zambuto, M. *Semiconductor Devices*. New York: McGraw-Hill, 1989.

# FUNDAMENTALS OF THE METAL-OXIDE-SEMICONDUCTOR FIELD-EFFECT TRANSISTOR

## CHAPTER 12

## PREVIEW

The fundamental physics of the **M**etal-**O**xide-**S**emiconductor **F**ield-**E**ffect **T**ransistor (MOSFET) is developed in this chapter. Although the bipolar and junction field-effect transistors were discussed in Chapters 10 and 11, respectively, the material in this chapter presumes a knowledge only of the semiconductor material properties and characteristics of the pn junction.

The MOSFET, in conjunction with other circuit elements, is capable of voltage gain and signal-power gain. The MOSFET is also used extensively in digital circuit applications where, because of its relatively small size, thousands of devices can be fabricated in a single integrated circuit. The MOSFET is, without doubt, the core of integrated circuit design at the present time.

The MOS designation is implicitly used only for the metal-silicon dioxide (SiO$_2$)-silicon system. The more general terminology is metal-insulator-semiconductor (MIS), where the insulator is not necessarily silicon dioxide and the semiconductor is not necessarily silicon. We will use the MOS system throughout this chapter although the same basic physics applies to the MIS system.

The heart of the MOSFET is a metal-oxide-semiconductor structure known as a MOS capacitor. The energy bands in the semiconductor near the oxide–semiconductor interface bend as a voltage is applied across the MOS capacitor. The position of the conduction and valence bands relative to the Fermi level at the oxide–semiconductor interface is a function of the MOS capacitor voltage, so that the characteristics of the semiconductor surface can be inverted from p-type to n-type, or from n-type to p-type, by applying the proper voltage. The operation and characteristics of the MOSFET are dependent on this inversion and the creation of an inversion charge density at the semiconductor surface. The threshold voltage is defined as the applied gate voltage required to create the inversion layer charge and is one of the important parameters of the MOSFET. We will derive a mathematical expression for the threshold voltage and consider the electrical and geometrical factors affecting it.

The various types of MOSFETs are examined and a qualitative discussion of the current-voltage characteristics is initially presented. A mathematical derivation of the current-voltage relation is then covered in detail followed by a discussion of the transconductance or transistor gain. The frequency response and limitations of the MOSFET are then considered. A small-signal equivalent circuit is developed and the various physical factors that affect the frequency response of the MOSFET are analyzed. The transistor cutoff frequency is a figure of merit for the frequency response of the transistor; an expression for this factor is derived.

Although we have not discussed fabrication processes in any detail in this text, there are some MOS technologies that should be considered since these fabrication techniques directly influence the characteristics and properties of the MOS devices and circuits. We will consider the complementary MOS (CMOS) process and power MOSFETs. These technologies will be discussed only briefly, but should provide a good base for further in-depth study.

## 12.1 THE TWO-TERMINAL MOS STRUCTURE

The heart of the MOSFET is the metal-oxide-semiconductor capacitor shown in Figure 12–1. The metal may be aluminum or some other type of

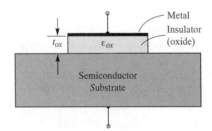

**Figure 12–1**    The basic MOS capacitor structure.

metal although in many cases, it is actually a high-conductivity polycrystalline silicon that has been deposited on the oxide; however, the term metal is usually still used. The parameter $t_{ox}$ in the figure is the thickness of the oxide and $\varepsilon_{ox}$ is the permittivity of the oxide.

### 12.1.1 Energy-Band Diagrams

The physics of the MOS structure can be more easily explained with the aid of the simple parallel-plate capacitor. Figure 12–2a shows a parallel plate capacitor with the top plate at a negative voltage with respect to the bottom plate. An insulator material separates the two plates. With this bias, a negative charge exists on the top plate, a positive charge exists on the bottom plate, and an electric field is induced between the two plates as shown. The capacitance per unit area for this geometry is

$$C' = \frac{\varepsilon}{d} \tag{12-1}$$

where $\varepsilon$ is the permittivity of the insulator and $d$ is the distance between the two plates. The magnitude of the charge per unit area on either plate is

$$Q' = C'V \tag{12-2}$$

where the prime indicates charge or capacitance per unit area. The magnitude of the electric field is

$$E = \frac{V}{d} \tag{12-3}$$

A MOS capacitor with a p-type semiconductor substrate is shown in Figure 12-2b. The top metal gate is at a negative voltage with respect to the semiconductor substrate. From the example of the parallel-plate capacitor, we can see that a negative charge will exist on the top metal plate and an electric field will be induced with the direction shown in the figure. If the electric field were to penetrate into the semiconductor, the majority carrier holes would experience a force toward the oxide–semiconductor interface. The equilibrium distribution of charge in the MOS capacitor with this particular applied voltage is shown in Figure 12-2c. An *accumulation layer* of holes in the oxide–semiconductor junction corresponds to the positive charge on the bottom "plate" of the MOS capacitor.

Figure 12-3a shows the same MOS capacitor in which the polarity of the applied voltage is reversed. A positive charge now exists on the top metal plate and the induced electric field is in the opposite direction as shown. If the electric field penetrates the semiconductor in this case, majority carrier holes will experience a force away from the oxide–semiconductor interface. As the holes are pushed away from the interface, a negative space charge region is created due to the fixed ionized acceptor atoms. The negative charge in the induced depletion region corresponds to the negative charge on the bottom "plate" of the MOS capacitor. Figure 12-3b shows the equilibrium distribution of charge in the MOS capacitor with this applied voltage.

The energy-band diagram of the MOS capacitor with the p-type substrate, for the case when a negative voltage is applied to the top metal gate, is shown in Figure 12-4a. The valence-band edge is closer to the Fermi level at the oxide–semiconductor interface than in the bulk material, which implies that there is an accumulation of holes. The semiconductor surface appears to be more p-type than the bulk material. The Fermi level is a constant in the semiconductor since the MOS system is in thermal equilibrium and there is no current through the oxide.

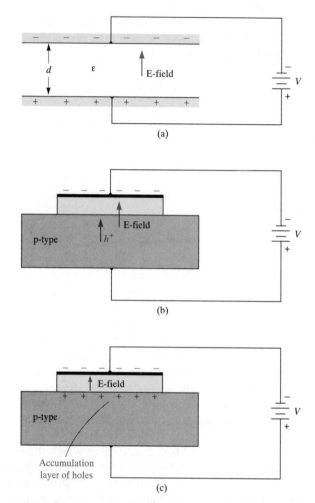

**Figure 12–2** (a) A parallel-plate capacitor showing the electric field and conductor charges. (b) A corresponding MOS capacitor with a negative gate bias showing the electric field and charge flow. (c) The MOS capacitor with an accumulation layer of holes.

Figure 12–4b shows the energy-band diagram of the MOS system when a positive voltage is applied to the gate. The conduction and valence band edges bend as shown in the figure, indicating a space charge region similar to that in a pn junction. The conduction band and intrinsic Fermi levels move closer to the Fermi level. The induced space charge width is $x_d$.

Now consider the case when a still larger positive voltage is applied to the top metal gate of the MOS capacitor. We expect the induced electric field to increase in magnitude and the corresponding positive and negative charges on the MOS capacitor to increase. A larger negative charge in the MOS capacitor implies a larger induced space charge region and more band bend-

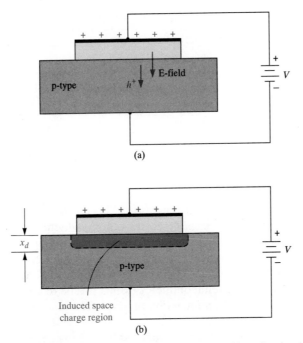

(a)

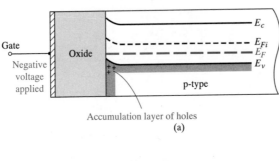

(b)

**Figure 12–3**   The MOS capacitor with a moderate positive gate bias, showing (a) the electric field and charge flow, and (b) the induced space charge region.

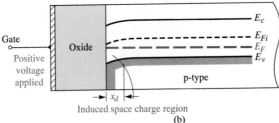

**Figure 12–4**   The energy-band diagram of a MOS capacitor with a p-type substrate for (a) a negative gate bias and (b) a moderate positive gate bias.

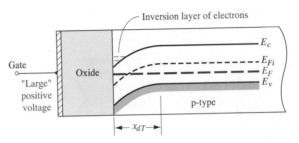

**Figure 12–5**  The energy-band diagram of the MOS capacitor with a p-type substrate for a "large" positive gate bias.

ing. Figure 12–5 shows such a condition. The intrinsic Fermi level at the surface is now below the Fermi level, thus the conduction band is closer to the Fermi level than the valence band is. This result implies that the surface in the semiconductor adjacent to the oxide–semiconductor interface is n-type. By applying a sufficiently large positive gate voltage, we have inverted the surface of the semiconductor from a p-type to an n-type semiconductor. We have created an *inversion layer* of electrons at the oxide–semiconductor interface.

In the MOS capacitor structure that we have considered, we assumed a p-type semiconductor substrate. The same type of energy-band diagrams

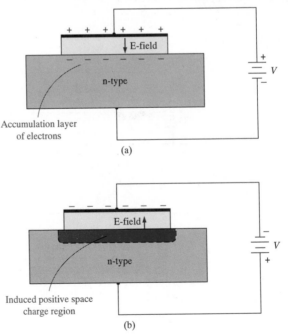

**Figure 12–6**  The MOS capacitor with an n-type substrate for (a) a positive gate bias, and (b) a moderate negative gate bias.

can be constructed for a MOS capacitor with an n-type semiconductor substrate. Figure 12–6a shows the MOS capacitor structure with a positive voltage applied to the top gate terminal. A positive charge exists on the top gate and an electric field is induced with the direction shown in the figure. An accumulation layer of electrons will be induced in the n-type substrate. The case when a negative voltage is applied to the top gate is shown in Figure 12–6b. A positive space charge region is induced in the n-type semiconductor in this situation.

The energy-band diagrams for this MOS capacitor with the n-type substrate are shown in Figure 12–7. Figure 12–7a shows the case when a positive voltage is applied to the gate and an accumulation layer of electrons is formed. The positive space charge region induced by an applied negative

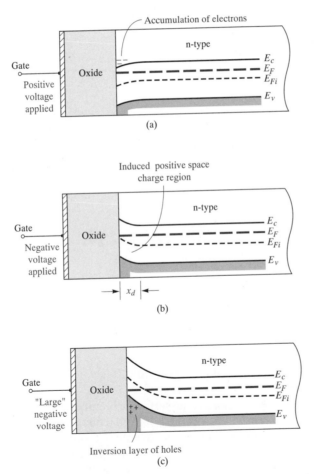

**Figure 12–7**   The energy-band diagram of the MOS capacitor with an n-type substrate for (a) a positive gate bias, (b) a moderate negative gate bias, and (c) a "large" negative gate bias.

gate voltage is shown in Figure 12–7b in which the conduction and valence band energies bend upward. Figure 12–7c shows the energy bands when a larger negative voltage is applied to the gate. The conduction and valence bands are bent even more and the intrinsic Fermi level has moved above the Fermi level so that the valence band is closer to the Fermi level than the conduction band is. This result implies that the semiconductor surface adjacent to the oxide–semiconductor interface is p-type. By applying a sufficiently large negative voltage to the gate of the MOS capacitor, the semiconductor surface has been inverted from n-type to p-type. An inversion layer of holes has been induced at the oxide–semiconductor interface.

### 12.1.2 Depletion Layer Thickness

We may calculate the width of the induced space change region adjacent to the oxide-semiconductor interface. Figure 12–8 shows the space charge re-

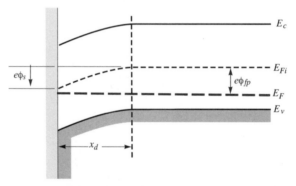

**Figure 12–8**   The energy-band diagram in the p-type semiconductor, indicating surface potential.

gion in a p-type semiconductor substrate. The potential $\phi_{fp}$ is the difference (in volts) between $E_{Fi}$ and $E_F$ and is given by

$$\phi_{fp} = V_t \ln \left( \frac{N_a}{n_i} \right) \tag{12–4}$$

where $N_a$ is the acceptor doping concentration and $n_i$ is the intrinsic carrier concentration.

The potential $\phi_s$ is called the surface potential; it is the difference (in volts) between $E_{Fi}$ measured in the bulk semiconductor and $E_{Fi}$ measured at the surface. The surface potential is the potential difference across the space charge layer. The space charge width can now be written in a form similar to that of a one-sided pn junction. We can write that

$$x_d = \left\{ \frac{2\varepsilon_s \phi_s}{eN_a} \right\}^{1/2} \tag{12–5}$$

where $\varepsilon_s$ is the permittivity of the semiconductor. Equation (12–5) assumes that the abrupt depletion approximation is valid.

Figure 12–9 shows the energy bands for the case in which $\phi_s = 2\phi_{fp}$. The Fermi level at the surface is as far above the intrinsic level as the Fermi level

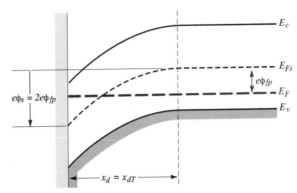

**Figure 12–9**   The energy-band diagram in the p-type semiconductor at the threshold inversion point.

is below the intrinsic level in the bulk semiconductor. The electron concentration at the surface is the same as the hole concentration in the bulk material. This condition is known as the *threshold inversion point*. The applied gate voltage creating this condition is known as the *threshold voltage*. If the gate voltage increases above this threshold value, the conduction band will bend slightly closer to the Fermi level, but the change in the conduction band at the surface is now only a slight function of gate voltage. The electron concentration at the surface, however, is an exponential function of the surface potential. The surface potential may increase by a few $(kT/e)$ volts, which will change the electron concentration by orders of magnitude, but the space charge width changes only slightly. In this case, then, the space charge region has essentially reached a maximum width.

The maximum space charge width, $x_{dT}$, at this inversion transition point can be calculated from Equation (12–5) by setting $\phi_s = 2\phi_{fp}$. Then

$$x_{dT} = \left\{ \frac{4\varepsilon_s \phi_{fp}}{eN_a} \right\}^{1/2} \tag{12–6}$$

**Example 12–1**

**Objective:**  To calculate the maximum space charge width given a particular semiconductor doping concentration.

Consider silicon at $T = 300°K$ doped to $N_a = 10^{16}$ cm$^{-3}$. The intrinsic carrier concentration is $n_i = 1.5 \times 10^{10}$ cm$^{-3}$.

**Solution:** From Equation (12–4), we have

$$\phi_{fp} = V_t \ln \left( \frac{N_a}{n_i} \right) = (0.0259) \ln \left( \frac{10^{16}}{1.5 \times 10^{10}} \right) = 0.347 \text{ volt}$$

Then the maximum space charge width is

$$x_{dT} = \left\{ \frac{4\varepsilon_s \phi_{fp}}{eN_a} \right\}^{1/2} = \left\{ \frac{4(11.7)(8.85 \times 10^{-14})(0.347)}{(1.6 \times 10^{-19})(10^{16})} \right\}^{1/2}$$

or

$$x_{dT} = 0.30 \times 10^{-4} \text{ cm} = 0.30 \ \mu\text{m}$$

**Comment:** The maximum induced space charge width is on the same order of magnitude as pn junction space charge widths.

---

We have been considering a p-type semiconductor substrate. The same maximum induced space charge region width occurs in an n-type substrate. Figure 12–10 is the energy-band diagram at the threshold voltage with an n-type substrate. We can write

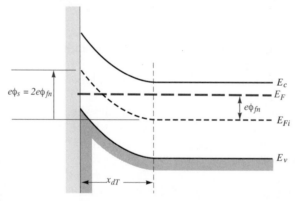

**Figure 12–10** The energy-band diagram in the n-type semiconductor at the threshold inversion point.

$$\phi_{fn} = V_t \ln \left( \frac{N_d}{n_i} \right) \tag{12–7}$$

and

$$x_{dT} = \left\{ \frac{4\varepsilon_s \phi_{fn}}{eN_d} \right\}^{1/2} \tag{12–8}$$

We may note that we are always assuming the parameters $\phi_{fp}$ and $\phi_{fn}$ to be positive quantities.

Figure 12–11 is a plot of $x_{dT}$ at $T = 300°K$ as a function of doping concentration in silicon. The semiconductor doping can be either n-type or p-type.

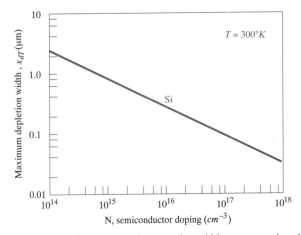

**Figure 12–11** Maximum induced space charge region width versus semiconductor doping.

### 12.1.3 Work Function Differences

We have been concerned, so far, with the energy-band diagrams of the semiconductor material. Figure 12–12a shows the energy levels in the metal, silicon dioxide, and silicon relative to the vacuum level. The metal work function is $\phi_m$ and the electron affinity is $\chi$. The parameter $\chi_i$ is the oxide electron affinity and, for silicon dioxide, $\chi_i = 0.9$ volt.

Figure 12–12b shows the energy-band diagram of the entire metal-oxide-semiconductor structure with zero gate voltage applied. The Fermi level is a constant through the entire system at thermal equilibrium. We may define $\phi'_m$ as a modified metal work function—the potential required to inject an electron from the metal into the conduction band of the oxide. Similarly, $\chi'$ is defined as a modified electron affinity. The voltage $V_{ox0}$ is the potential drop across the oxide for zero applied gate voltage and is not necessarily zero because of the difference between $\phi_m$ and $\chi$. The potential $\phi_{s0}$ is the surface potential for this case.

If we sum the energies from the Fermi level on the metal side to the Fermi level on the semiconductor side, we have

$$e\phi'_m + eV_{ox0} = e\chi' + \frac{E_g}{2} - e\phi_{s0} + e\phi_{fp} \tag{12–9}$$

Equation (12–9) can be rewritten as

$$V_{ox0} + \phi_{s0} = -\left[\phi'_m - \left(\chi' + \frac{E_g}{2e} + \phi_{fp}\right)\right] \tag{12–10}$$

We can define a potential $\phi_{ms}$ as

$$\phi_{ms} \equiv \left[\phi'_m - \left(\chi' + \frac{E_g}{2e} + \phi_{fp}\right)\right] \tag{12–11}$$

which is known as the metal–semiconductor work function difference.

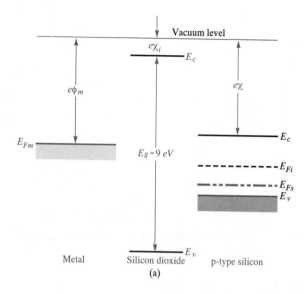

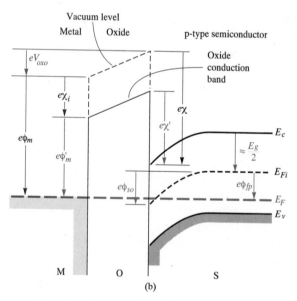

**Figure 12–12** (a) Energy levels in a MOS system prior to contact. (b) Energy-band diagram through the MOS structure in thermal equilibrium after contact.

## Example 12–2

**Objective:** To calculate the metal–semiconductor work function difference $\phi_{ms}$ for a given MOS system and semiconductor doping.

For an aluminum–silicon dioxide junction, $\phi'_m = 3.20$ volts and for a silicon–silicon dioxide junction, $\chi' = 3.25$ volts. We may assume that $E_g = 1.11$ eV. Let the p-type doping be $N_a = 10^{14}$ cm$^{-3}$.

**Solution:** For silicon at $T = 300°$K, we may calculate $\phi_{fp}$ as

$$\phi_{fp} = V_t \ln\left(\frac{N_a}{n_i}\right) = (0.0259) \ln\left(\frac{10^{14}}{1.5 \times 10^{10}}\right) = 0.228 \text{ volt}$$

Then the work function difference is

$$\phi_{ms} = \phi'_m - \left(\chi' + \frac{E_g}{2e} + \phi_{fp}\right) = 3.20 - (3.25 + 0.555 + 0.228)$$

or
$$\phi_{ms} = -0.83 \text{ volt}$$

**Comment:** The value of $\phi_{ms}$ will become more negative as the doping of the p-type substrate increases.

Degenerately doped polysilicon deposited on the oxide is also often used as the metal gate. Figure 12–13a shows the energy-band diagram of a MOS capacitor with an $n^+$ polysilicon gate and a p-type substrate. Figure 12–13b

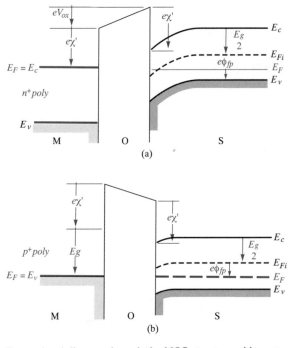

**Figure 12–13**   Energy-band diagram through the MOS structure with a p-type substrate at zero gate bias for (a) an $n^+$ polysilicon gate and (b) a $p^+$ polysilicon gate.

shows the energy-band diagram for the case of a $p^+$ polysilicon gate and the p-type silicon substrate. In the degenerately doped polysilicon, we will initially assume that $E_F = E_c$ for the $n^+$ case and $E_F = E_v$ for the $p^+$ case.

For the $n^+$ polysilicon gate, the metal–semiconductor work function difference can be written as

$$\phi_{ms} = \left[ \chi' - \left( \chi' + \frac{E_g}{2e} + \phi_{fp} \right) \right] = - \left( \frac{Eg}{2e} + \phi_{fp} \right) \qquad (12\text{–}12)$$

and for the $p^+$ polysilicon gate, we have

$$\phi_{ms} = \left[ \left( \chi' + \frac{E_g}{e} \right) - \left( \chi' + \frac{E_g}{2e} + \phi_{fp} \right) \right] = \left( \frac{E_g}{2e} - \phi_{fp} \right) \qquad (12\text{–}13)$$

However, for degenerately doped $n^+$ polysilicon and $p^+$ polysilicon, the Fermi level can be above $E_c$ and below $E_v$, respectively, by 0.1 to 0.2 volt. The experimental $\phi_{ms}$ values will then be slightly different from the values calculated using Equations (12–12) and (12–13).

We have been considering a p-type semiconductor substrate. We may also have an n-type semiconductor substrate in a MOS capacitor. Figure 12–14 shows the energy-band diagram of the MOS capacitor with a metal gate and the n-type semiconductor substrate, for the case when a negative voltage is applied to the gate. The metal–semiconductor work function difference for this case is defined as

$$\phi_{ms} = \phi'_m - \left( \chi' + \frac{E_g}{2e} - \phi_{fn} \right) \qquad (12\text{–}14)$$

where $\phi_{fn}$ is assumed to be a positive value. We will have similar expressions for $n^+$ and $p^+$ polysilicon gates.

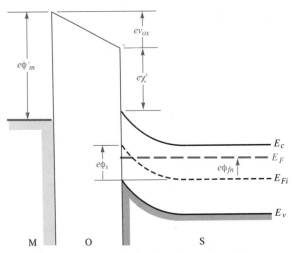

**Figure 12–14**  Energy-band diagram through the MOS structure with an n-type substrate for a negative applied gate bias.

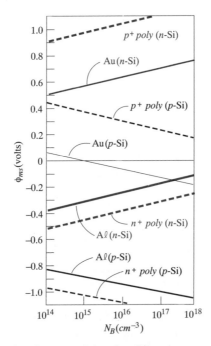

**Figure 12-15** Metal–semiconductor work function difference versus doping for aluminum, gold, $n^+$-, and $p^+$-polysilicon gates. (From Sze [10] and Werner [12].)

Figure 12–15 shows the work function differences as a function of semiconductor doping for the various types of gates. We may note that the magnitude of $\phi_{ms}$ for the polysilicon gates are somewhat larger than Equations (12–12) and (12–13) predict. This difference again is because the Fermi level is not equal to the conduction band energy for the $n^+$ gate and is not equal to the valence band energy for the $p^+$ gate. The metal–semiconductor work function difference becomes important in the flat-band and threshold voltage parameters discussed next.

## 12.1.4 Flat-Band Voltage

The flat-band voltage is defined as the applied gate voltage such that there is no band bending in the semiconductor and, as a result, zero net space charge in this region. Figure 12–16 shows this flat-band condition. Because of the work function difference and possible trapped charge in the oxide, the voltage across the oxide for this case is not necessarily zero.

We have implicitly been assuming that there is zero net charge density in the oxide material. This assumption may not be valid—a net fixed charge density, usually positive, may exist in the insulator. The positive charge has been identified with broken or dangling covalent bonds near the oxide–semiconductor interface. During the thermal formation of $SiO_2$, oxygen dif-

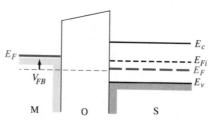

**Figure 12–16**   Energy-band diagram of a MOS capacitor at flat-band.

fuses through the oxide and reacts near the Si–SiO$_2$ interface to form the SiO$_2$. Silicon atoms may also break away from the silicon material just prior to reacting to form SiO$_2$. When the oxidation process is terminated, excess silicon may exist in the oxide near the interface, resulting in the dangling bonds. The magnitude of this oxide charge seems, in general, to be a strong function of the oxidizing conditions such as oxidizing ambient and temperature. The charge density can be altered to some degree by annealing the oxide in an argon or nitrogen atmosphere. However, the charge is rarely zero.

The net fixed charge in the oxide appears to be located fairly close to the oxide–semiconductor interface. We will assume in the analysis of the MOS structure that an equivalent trapped charge per unit area, $Q'_{ss}$, is located in the oxide directly adjacent to the oxide–semiconductor interface. For the moment, we will ignore any other oxide-type charges that may exist in the device. The parameter $Q'_{ss}$ is usually given in terms of number of electronic charges per unit area.

Equation (12–10), for zero applied gate voltage, can be written as

$$V_{ox0} + \phi_{s0} = -\phi_{ms} \tag{12–15}$$

If a gate voltage is applied, the potential drop across the oxide and the surface potential will change. We can then write

$$V_G = \Delta V_{ox} + \Delta \phi_s = (V_{ox} - V_{ox0}) + (\phi_s - \phi_{s0}) \tag{12–16}$$

Using Equation (12–15), we have

$$V_G = V_{ox} + \phi_s + \phi_{ms} \tag{12–17}$$

Figure 12–17 shows the charge distribution in the MOS structure for the flat-band condition. There is zero net charge in the semiconductor and we can assume that an equivalent fixed surface charge density exists in the oxide. The charge density on the metal is $Q'_m$ and from charge neutrality, we have

$$Q'_m + Q'_{ss} = 0 \tag{12–18}$$

We can relate $Q'_m$ to the voltage across the oxide by

$$V_{ox} = \frac{Q'_m}{C_{ox}} \tag{12–19}$$

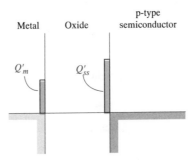

**Figure 12–17** Charge distribution in a MOS capacitor at flat-band.

where $C_{ox}$ is the oxide capacitance per unit area.[1] Substituting Equation (12–18) into Equation (12–19), we have

$$V_{ox} = \frac{-Q'_{ss}}{C_{ox}} \qquad (12\text{–}20)$$

In the flat-band condition, the surface potential is zero, or $\phi_s = 0$. Then from Equation (12–17), we have

$$V_G = V_{FB} = \phi_{ms} - \frac{Q'_{ss}}{C_{ox}} \qquad (12\text{–}21)$$

Equation (12–21) is the flat-band voltage for this MOS device.

---

**Example 12–3**

**Objective:** To calculate the flat-band voltage for a MOS capacitor with a p-type semiconductor substrate.

Consider a MOS structure with a p-type semiconductor substrate doped to $N_a = 10^{16}$ cm$^{-3}$, a silicon-dioxide insulator with a thickness of $t_{ox} = 500$ Å, and an $n^+$ polysilicon gate. Assume that $Q'_{ss} = 10^{11}$ electronic charges per cm$^2$.

**Solution:** The work function difference from Figure 12–15 is $\phi_{ms} = -1.1$ volts. The oxide capacitance can be found as

$$C_{ox} = \frac{\varepsilon_{ox}}{t_{ox}} = \frac{(3.9)(8.85 \times 10^{-14})}{500 \times 10^{-8}} = 6.9 \times 10^{-8} \text{ F/cm}^2$$

The equivalent oxide surface charge density is

$$Q'_{ss} = (10^{11})(1.6 \times 10^{-19}) = 1.6 \times 10^{-8} \text{ coul/cm}^2$$

---

[1] Although we will, in general, use the primed notation for capacitance per unit area or charge per unit area, we will omit, for convenience, the prime on the oxide capacitance per unit area parameter.

The flat-band voltage is then calculated as

$$V_{FB} = \phi_{ms} - \frac{Q'_{ss}}{C_{ox}} = -1.1 - \left(\frac{1.6 \times 10^{-8}}{6.9 \times 10^{-8}}\right) = -1.33 \text{ volts}$$

**Comment:** The applied gate voltage required to achieve the flat-band condition for this p-type substrate is negative. If the amount of fixed oxide charge increases, the flat-band voltage becomes even more negative.

### 12.1.5 Threshold Voltage

The threshold voltage was defined as the applied gate voltage required to achieve the threshold inversion point. The threshold inversion point, in turn, is defined as the condition when the surface potential is $\phi_s = 2\phi_{fp}$ for the p-type semiconductor and $\phi_s = 2\phi_{fn}$ for the n-type semiconductor. These conditions were shown in Figures 12–9a and 12–10. The threshold voltage will be derived in terms of the electrical and geometrical properties of the MOS capacitor.

Figure 12–18 shows the charge distribution through the MOS device at the threshold inversion point for a p-type semiconductor substrate. The

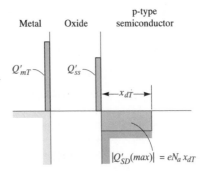

**Figure 12–18**   Charge distribution in a MOS capacitor with a p-type substrate at the threshold inversion point.

space charge width has reached its maximum value. We will assume that there is an equivalent oxide charge $Q'_{ss}$ and the positive charge on the metal gate at threshold is $Q'_{mT}$. The prime on the charge terms indicates charge per unit area. Even though we are assuming that the surface has been inverted, we will ignore the inversion layer charge at this threshold inversion point. From conservation of charge, we can write

$$Q'_{mT} + Q'_{ss} = |Q'_{SD}(\text{max})| \tag{12–22}$$

where
$$|Q'_{SD}(\text{max})| = eN_a x_{dT} \qquad (12\text{–}23)$$

and is the magnitude of the maximum space charge density per unit area of the depletion region.

The energy-band diagram of the MOS system with an applied positive gate voltage is shown in Figure 12–19. As we mentioned, an applied gate

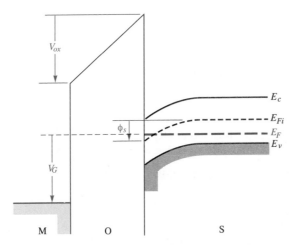

**Figure 12–19**  Energy-band diagram through the MOS structure with a positive applied gate bias.

voltage will change the voltage across the oxide and will change the surface potential. We had from Equation (12–16) that

$$V_G = \Delta V_{\text{ox}} + \Delta \phi_s = V_{\text{ox}} + \phi_s + \phi_{ms}$$

At threshold, we can define $V_G = V_{TN}$, where $V_{TN}$ is the threshold voltage that creates the electron inversion layer charge. The surface potential is $\phi_s = 2\phi_{fp}$ at threshold so Equation (12–16) can be written as

$$V_{TN} = V_{\text{ox}T} + 2\phi_{fp} + \phi_{ms} \qquad (12\text{–}24)$$

where $V_{\text{ox}T}$ is the voltage across the oxide at this threshold inversion point.

The voltage $V_{\text{ox}T}$ can be related to the charge on the metal and to the oxide capacitance by

$$V_{\text{ox}T} = \frac{Q'_{mT}}{C_{\text{ox}}} \qquad (12\text{–}25)$$

where again $C_{\text{ox}}$ is the oxide capacitance per unit area. Using Equation (12–22), we can write

$$V_{\text{ox}T} = \frac{Q'_{mT}}{C_{\text{ox}}} = \frac{1}{C_{\text{ox}}} (|Q'_{SD}(\text{max})| - Q'_{ss}) \qquad (12\text{–}26)$$

Finally, the threshold voltage can be written as

$$V_{TN} = (|Q'_{SD}(\text{max})| - Q'_{ss}) \left(\frac{t_{ox}}{\varepsilon_{ox}}\right) + \phi_{ms} + 2\phi_{fp} \qquad (12\text{–}27)$$

For a given semiconductor material, oxide material, and gate metal, the threshold voltage is a function of semiconductor doping, oxide charge $Q'_{ss}$, and oxide thickness.

---

**Example 12–4**

**Objective:** To calculate the threshold voltage of an MOS system using an $n^+$ polysilicon gate.

Consider a p-type silicon substrate doped to $N_a = 3 \times 10^{16}$ cm$^{-3}$. Assume that the oxide is silicon dioxide which has a thickness of $t_{ox} = 500$ Å and assume that $Q'_{ss} = 10^{11}$ cm$^{-2}$. The work function difference from Figure 12–15 is $\phi_{ms} = -1.13$ volts.

**Solution:** The various parameters can be calculated as

$$\phi_{fp} = V_t \ln\left(\frac{N_a}{n_i}\right) = (0.0259) \ln\left(\frac{3 \times 10^{16}}{1.5 \times 10^{10}}\right) = 0.376 \text{ volt}$$

and

$$x_{dT} = \left\{\frac{4\varepsilon_s \phi_{fp}}{eN_a}\right\}^{1/2} = \left\{\frac{4(11.7)(8.85 \times 10^{-14})(0.376)}{(1.6 \times 10^{-19})(3 \times 10^{16})}\right\}^{1/2} = 0.18 \text{ } \mu\text{m}$$

Then

$$|Q'_{SD}(\text{max})| = eN_a x_{dT} = (1.6 \times 10^{-19})(3 \times 10^{16})(0.18 \times 10^{-4})$$

or

$$|Q'_{SD}(\text{max})| = 8.64 \times 10^{-8} \text{ coul/cm}^2$$

The threshold voltage can now be calculated as

$$V_{TN} = (|Q'_{SD}(\text{max})| - Q'_{ss}) \left(\frac{t_{ox}}{\varepsilon_{ox}}\right) + \phi_{ms} + 2\phi_{fp}$$

$$= [(8.64 \times 10^{-8}) - (10^{11})(1.6 \times 10^{-19})] \left[\frac{500 \times 10^{-8}}{(3.9)(8.85 \times 10^{-14})}\right]$$

$$- 1.13 + 2(0.376)$$

$$= +0.642 \text{ volt}$$

**Comment:** The threshold voltage for this case is a positive quantity which means that this MOS device is an enhancement mode device: a gate voltage must be applied to create the inversion layer charge, which is zero for zero applied gate voltage.

---

The threshold voltage must be within the voltage range of a circuit design. Although we have not yet considered the current in a MOS transistor, the threshold voltage is the point at which the transistor turns on. If a circuit is to operate between 0 and 5 volts and the threshold voltage of a MOSFET is

10 volts, for example, the device and circuit cannot be turned "on" and "off." The threshold voltage, then, is one of the important parameters of the MOSFET.

---

**Example 12–5**

**Objective:** To calculate the threshold voltage of a MOS system using an aluminum gate.

Consider a p-type silicon substrate at $T = 300°K$ doped to $N_a = 10^{14}$ cm$^{-3}$. Let $Q'_{ss} = 10^{10}$ cm$^{-2}$, $t_{ox} = 500$Å, and assume the oxide is silicon dioxide. From Figure 12–15, we have that $\phi_{ms} = -0.83$ volt.

**Solution:** We can start calculating the various parameters as

$$\phi_{fp} = V_t \ln \left(\frac{N_a}{n_i}\right) = (0.0259) \ln \left(\frac{10^{14}}{1.5 \times 10^{10}}\right) = 0.228 \text{ volt}$$

and      $$x_{dT} = \left\{\frac{4\varepsilon_s \phi_{fp}}{eN_a}\right\}^{1/2} = \left\{\frac{4(11.7)(8.85 \times 10^{-14})(0.228)}{(1.6 \times 10^{-19})(10^{14})}\right\}^{1/2} = 2.43 \ \mu m$$

Then

$$|Q'_{SD}(\max)| = eN_a x_{dT} = (1.6 \times 10^{-19})(10^{14})(2.43 \times 10^{-4}) = 3.89 \times 10^{-9} \text{ coul/cm}^2$$

We can now calculate the threshold voltage as

$$V_{TN} = (|Q'_{SD}(\max)| - Q'_{ss}) \left(\frac{t_{ox}}{\varepsilon_{ox}}\right) + \phi_{ms} + 2\phi_{fp}$$

$$= [(3.89 \times 10^{-9}) - (10^{10})(1.6 \times 10^{-19})] \left[\frac{500 \times 10^{-8}}{(3.9)(8.85 \times 10^{-14})}\right]$$

$$- 0.83 + 2(0.228)$$

$$= -0.341 \text{ volt}$$

**Comment:** In this example, the semiconductor is very lightly doped which, in conjunction with the positive charge in the oxide and the work function potential difference, is sufficient to induce an electron inversion layer charge even with zero applied gate voltage. This condition makes the threshold voltage negative.

---

A negative threshold voltage for a p-type substrate implies a depletion mode device. A negative voltage on the gate must be applied in order to make the inversion layer charge equal to zero, whereas a positive gate voltage will induce a larger inversion layer charge.

Figure 12–20 is a plot of the threshold voltage $V_{TN}$ as a function of the acceptor doping concentration for various positive oxide charge values. We

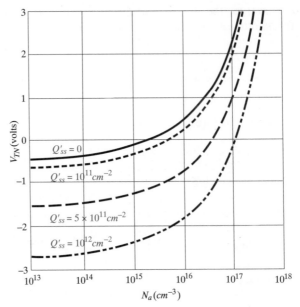

**Figure 12–20**  Threshold voltage of an n-channel MOSFET versus the p-type substrate doping concentration for various values of oxide trapped charge ($t_{ox}$ = 500 Å, aluminum gate).

may note that the p-type semiconductor must be somewhat heavily doped in order to obtain an enhancement mode device.

The previous derivation of the threshold voltage assumed a p-type semiconductor substrate. The same type of derivation can be done with an n-type semiconductor substrate, where a negative gate voltage can induce an inversion layer of holes at the oxide–semiconductor interface.

Figure 12–14 showed the energy-band diagram of the MOS structure with an n-type substrate and with an applied negative gate voltage. The threshold voltage for this case can be derived and is given by

$$V_{TP} = (-|Q'_{SD}(\text{max})| - Q'_{ss})\left(\frac{t_{ox}}{\varepsilon_{ox}}\right) + \phi_{ms} - 2\phi_{fn} \qquad (12\text{–}28)$$

where

$$\phi_{ms} = \phi'_m - \left(\chi' + \frac{E_g}{2e} - \phi_{fn}\right) \qquad (12\text{–}14)$$

$$|Q'_{SD}(\text{max})| = eN_d x_{dT} \qquad (12\text{–}29a)$$

$$x_{dT} = \left\{\frac{4\varepsilon_s \phi_{fn}}{eN_d}\right\}^{1/2} \qquad (12\text{–}29b)$$

and

$$\phi_{fn} = V_t \ln\left(\frac{N_d}{n_i}\right) \qquad (12\text{–}29c)$$

We may note that $x_{dT}$ and $\phi_{fn}$ are defined as positive quantities. We may also note that the notation of $V_{TP}$ is the threshold voltage which will induce an inversion layer of holes. We will later drop the $N$ and $P$ subscript notation on the threshold voltage but, for the moment, the notation may be useful for clarity.

---

### Example 12–6

**Objective:** To calculate the threshold voltage of a MOS structure with an n-type semiconductor substrate.

   Consider an aluminum-silicon dioxide-silicon MOS structure. Assume that the semiconductor is doped to $N_d = 10^{14}$ cm$^{-3}$. Let $t_{ox} = 650$Å and $Q'_{ss} = 10^{10}$ cm$^{-2}$. We will still assume that the oxide charge is positive. From Figure 12–15, we find that $\phi_{ms} = -0.38$ volt.

**Solution:** The various parameters can be calculated as

$$\phi_{fn} = V_t \ln \left( \frac{N_d}{n_i} \right) = 0.228 \text{ volt}$$

and

$$x_{dT} = \left\{ \frac{4\varepsilon_s \phi_{fn}}{eN_d} \right\}^{1/2} = 2.43 \ \mu\text{m}$$

Then

$$|Q'_{SD}(\text{max})| = eN_d x_{dT} = 3.89 \times 10^{-9} \text{ coul/cm}^2$$

We can now calculate the threshold voltage as

$$V_{TP} = (-|Q'_{SD}(\text{max})| - Q'_{ss}) \left( \frac{t_{ox}}{\varepsilon_{ox}} \right) + \phi_{ms} - 2\phi_{fn}$$

$$= [-(3.89 \times 10^{-9}) - (10^{10})(1.6 \times 10^{-19})] \left[ \frac{650 \times 10^{-8}}{(3.9)(8.85 \times 10^{-14})} \right]$$

$$- 0.38 - 2(0.228)$$

$$= -0.939 \text{ volt}$$

**Comment:** The threshold voltage is negative, implying that this MOS capacitor, with the n-type substrate, is an enhancement mode device. The inversion layer charge is zero with zero gate voltage and a negative gate voltage must be applied to induce the hole inversion layer.

---

   Figure 12–21 is a plot of $V_{TP}$ versus doping concentration for several values of $Q'_{ss}$. We may note that, for all values of positive oxide charge, this MOS capacitor is always an enhancement mode device. As the $Q'_{ss}$ charge increases, the threshold voltage becomes more negative which means that it takes a larger applied gate voltage to create the inversion layer of holes at the oxide–semiconductor interface.

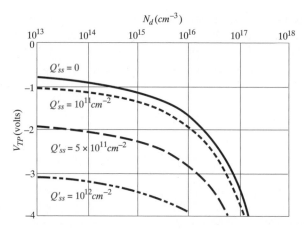

**Figure 12–21**    Threshold voltage of a p-channel MOSFET versus the n-type substrate doping concentration for various values of oxide trapped charge ($t_{ox}$ = 500Å, aluminum gate).

## 12.2 CAPACITANCE-VOLTAGE CHARACTERISTICS

The MOS capacitor structure is the heart of the MOSFET. A great deal of information about the MOS device and the oxide–semiconductor interface can be obtained from the capacitance versus voltage or C-V characteristics of the device. The capacitance of a device is defined as

$$C = \frac{dQ}{dV} \tag{12–30}$$

where $dQ$ is the magnitude of the differential change in charge on one plate as a function of the differential change in voltage $dV$ across the capacitor. The capacitance is a small-signal or ac parameter and is measured by super-imposing a small ac voltage on an applied dc gate voltage. The capacitance, then, is measured as a function of the applied dc gate voltage.

### 12.2.1 Ideal C-V Characteristics

First we will consider the ideal C-V characteristics of the MOS capacitor and then discuss some of the deviations that occur from these idealized results. We will initially assume that there is zero charge trapped in the oxide and also that there is no charge trapped at the oxide–semiconductor interface.

There are three operating conditions of interest in the MOS capacitor: accumulation, depletion, and inversion. Figure 12–22a shows the energy-band diagram of a MOS capacitor with a p-type substrate for the case when a negative voltage is applied to the gate, inducing an accumulation layer of holes in the semiconductor at the oxide–semiconductor interface. A small differential change in voltage across the MOS structure will cause a differen-

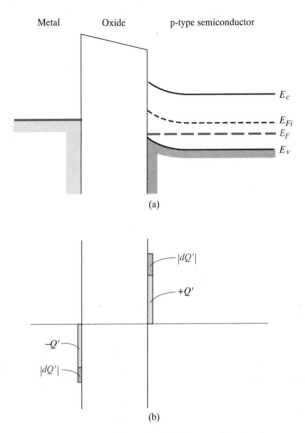

Metal            Oxide            p-type semiconductor

$E_c$

$E_{Fi}$
$E_F$

$E_v$

(a)

$|dQ'|$

$+Q'$

$-Q'$

$|dQ'|$

(b)

**Figure 12–22** (a) Energy-band diagram through a MOS capacitor for the accumulation mode. (b) Differential charge distribution at accumulation for a differential change in gate voltage.

tial change in charge on the metal gate and also in the hole accumulation charge, as shown in Figure 12–22b. The differential changes in charge density occur at the edges of the oxide, as in a parallel-plate capacitor. The capacitance $C'$ per unit area of the MOS capacitor for this accumulation mode is just the oxide capacitance, or

$$C'(\text{acc}) = C_{\text{ox}} = \frac{\varepsilon_{\text{ox}}}{t_{\text{ox}}} \qquad (12\text{–}31)$$

Figure 12–23a shows the energy-band diagram of the MOS device when a small positive voltage is applied to the gate inducing a space charge region in the semiconductor; Figure 12–23b shows the charge distribution through the device for this condition. The oxide capacitance and the capacitance of the depletion region are in series. A small differential change in voltage across the capacitor will cause a differential change in the space charge width. The

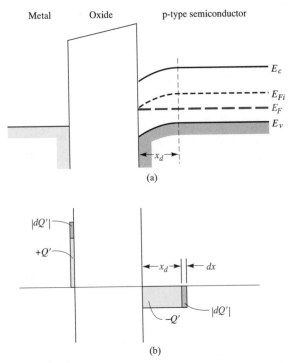

**Figure 12–23**   (a) Energy-band diagram through a MOS capacitor for the depletion mode. (b) Differential charge distribution at depletion for a differential change in gate voltage.

corresponding differential changes in charge densities are shown in the figure. The total capacitance of the series combination is

$$\frac{1}{C'(\text{depl})} = \frac{1}{C_{\text{ox}}} + \frac{1}{C'_{SD}} \tag{12–32a}$$

or

$$C'(\text{depl}) = \frac{C_{\text{ox}}C'_{SD}}{C_{\text{ox}} + C'_{SD}} \tag{12–32b}$$

Since $C_{\text{ox}} = \dfrac{\varepsilon_{\text{ox}}}{t_{\text{ox}}}$ and $C'_{SD} = \dfrac{\varepsilon_s}{x_d}$, Equation (12–32b) can be written as

$$C'(\text{depl}) = \frac{C_{\text{ox}}}{1 + \dfrac{C_{\text{ox}}}{C_{SD}}} = \frac{\varepsilon_{\text{ox}}}{t_{\text{ox}} + \left(\dfrac{\varepsilon_{\text{ox}}}{\varepsilon_s}\right) x_d} \tag{12–33}$$

As the space charge width increases, the total capacitance $C'(\text{depl})$ decreases.

We had defined the threshold inversion point to be the condition when the maximum depletion width is reached but there is essentially zero inversion charge density. This condition will yield a minimum capacitance $C'_{\text{min}}$ which is given by

$$C'_{min} = \frac{\varepsilon_{ox}}{t_{ox} + \left(\dfrac{\varepsilon_{ox}}{\varepsilon_s}\right) x_{dT}} \tag{12–34}$$

Figure 12–24a shows the energy-band diagram of this MOS device for the inversion condition. In the ideal case, a small incremental change in the voltage across the MOS capacitor will cause a differential change in the inversion layer charge density. The space charge width does not change. If the inversion charge can respond to the change in capacitor voltage as indicated in Figure 12–24b, then the capacitance is again just the oxide capacitance, or

$$C'(\text{inv}) = C_{ox} = \frac{\varepsilon_{ox}}{t_{ox}} \tag{12–35}$$

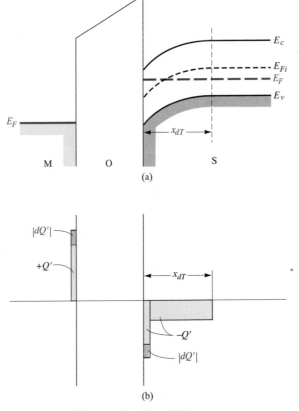

**Figure 12–24** (a) Energy-band diagram through a MOS capacitor for the inversion mode. (b) Differential charge distribution at inversion for a low-frequency differential change in gate voltage.

The ideal capacitance versus gate voltage, or C-V, characteristics of the MOS capacitor with a p-type substrate are shown in Figure 12–25. The three dashed segments correspond to the three components $C_{ox}$, $C'_{SD}$, and $C'_{min}$. The solid curve is the ideal net capacitance of the MOS capacitor. Moderate inversion, which is indicated in the figure, is the transition region between the point when only the space charge density changes with gate voltage and when only the inversion charge density changes with gate voltage.

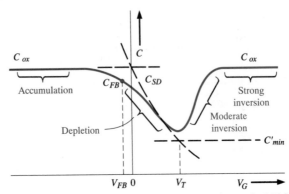

**Figure 12–25**   Ideal low-frequency capacitance versus gate voltage of a MOS capacitor with a p-type substrate. Individual capacitance components are also shown.

The point on the curve that corresponds to the flat-band condition is of interest. The flat-band condition occurs between the accumulation and depletion conditions. The capacitance at flat-band is given by

$$C'_{FB} = \frac{\varepsilon_{ox}}{t_{ox} + \left(\dfrac{\varepsilon_{ox}}{\varepsilon_s}\right)\sqrt{\left(\dfrac{kT}{e}\right)\left(\dfrac{\varepsilon_s}{eN_a}\right)}} \tag{12–36}$$

We may note that the flat-band capacitance is a function of oxide thickness as well as semiconductor doping. The general location of this point on the C-V plot is shown in Figure 12–25.

---

**Example 12–7**

**Objective:** To calculate $C_{ox}$, $C'_{min}$, and $C'_{FB}$ for a MOS capacitor.

Consider a p-type silicon substrate at $T = 300°K$ doped to $N_a = 10^{16}$ cm$^{-3}$. The oxide is silicon dioxide with a thickness of 550 Å and the gate is aluminum.

**Solution:** The oxide capacitance is

$$C_{ox} = \frac{\varepsilon_{ox}}{t_{ox}} = \frac{(3.9)(8.85 \times 10^{-14})}{550 \times 10^{-8}} = 6.28 \times 10^{-8} \text{ F/cm}^2$$

To find the minimum capacitance, we need to calculate

$$\phi_{fp} = V_t \ln \left(\frac{N_a}{n_i}\right) = (0.0259) \ln \left(\frac{10^{16}}{1.5 \times 10^{10}}\right) = 0.347 \text{ volt}$$

and

$$x_{dT} = \left\{\frac{4\varepsilon_s \phi_{fp}}{eN_a}\right\}^{1/2} = \left\{\frac{4(11.7)(8.85 \times 10^{-14})(0.347)}{(1.6 \times 10^{-19})(10^{16})}\right\}^{1/2} = 0.30 \times 10^{-4} \text{ cm}$$

Then

$$C'_{min} = \frac{\varepsilon_{ox}}{t_{ox} + \left(\frac{\varepsilon_{ox}}{\varepsilon_s}\right) x_{dT}} = \frac{(3.9)(8.85 \times 10^{-14})}{(550 \times 10^{-8}) + \left(\frac{3.9}{11.7}\right)(0.3 \times 10^{-4})} = 2.23 \times 10^{-8} \text{ F/cm}^2$$

We may note that

$$\frac{C'_{min}}{C_{ox}} = \frac{2.23 \times 10^{-8}}{6.28 \times 10^{-8}} = 0.355$$

The flat-band capacitance is

$$C'_{FB} = \frac{\varepsilon_{ox}}{t_{ox} + \left(\frac{\varepsilon_{ox}}{\varepsilon_s}\right)\sqrt{\left(\frac{kT}{e}\right)\left(\frac{\varepsilon_s}{eN_a}\right)}}$$

$$= \frac{(3.9)(8.85 \times 10^{-14})}{(550 \times 10^{-8}) + \left(\frac{3.9}{11.7}\right)\sqrt{(0.0259)\frac{(11.7)(8.85 \times 10^{-14})}{(1.6 \times 10^{-19})(10^{16})}}}$$

$$= 5.03 \times 10^{-8} \text{ F/cm}^2$$

We may also note that

$$\frac{C'_{FB}}{C_{ox}} = \frac{5.03 \times 10^{-8}}{6.28 \times 10^{-8}} = 0.80$$

**Comment:** The ratios of $C'_{min}$ to $C_{ox}$ and of $C'_{FB}$ to $C_{ox}$ are typical values obtained in C-V plots.

---

Typical values of channel length and width are 2 $\mu$m and 20 $\mu$m, respectively. The total gate oxide capacitance for this example is then

$$C_{oxT} = (6.28 \times 10^{-8})(2 \times 10^{-4})(20 \times 10^{-4}) = 0.025 \times 10^{-12} \text{ F} = 0.025 \text{ pF}$$

The total oxide capacitance in a typical MOS device is quite small.

The same type of ideal C-V characteristics are obtained for a MOS capacitor with an n-type substrate by changing the sign of the voltage axis. The accumulation condition is obtained for a positive gate bias and the inversion condition is obtained for a negative gate bias. This ideal curve is shown in Figure 12–26.

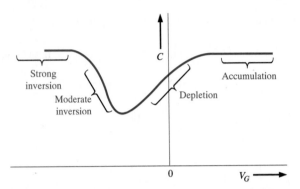

**Figure 12-26**  Ideal low-frequency capacitance versus gate voltage of a MOS capacitor with an n-type substrate.

### 12.2.2 Frequency Effects

The MOS capacitor with a p-type substrate and biased in the inversion condition was shown in Figure 12–24a. We have argued that a differential change in the capacitor voltage in the ideal case causes a differential change in the inversion layer charge density. However, we must consider the source of electrons that produces a change in the inversion charge density.

There are two sources of electrons that can change the charge density of the inversion layer. The first source is by diffusion of minority carrier electrons from the p-type substrate across the space charge region. This diffusion process is the same as that in a reverse-biased pn junction which generates the ideal reverse saturation current. The second source of electrons is by thermal generation of electron-hole pairs within the space charge region. This process is again the same as that in a reverse-biased pn junction which generates the reverse-biased generation current. Both of these processes

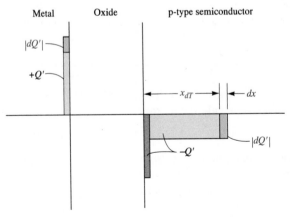

**Figure 12-27**  Differential charge distribution at inversion for a high-frequency differential change in gate voltage.

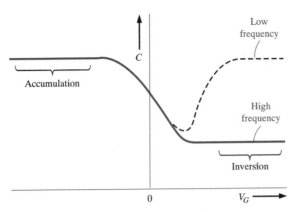

**Figure 12–28** Low-frequency and high-frequency capacitance versus gate voltage of a MOS capacitor with a p-type substrate.

generate electrons at a particular rate. The electron concentration in the inversion layer, then, cannot change instantaneously. If the ac voltage across the MOS capacitor changes rapidly, the change in the inversion layer charge will not be able to respond. The C-V characteristics will then be a function of the frequency of the ac signal used to measure the capacitance.

In the limit of a very high frequency, the inversion layer charge will not respond to a differential change in capacitor voltage. Figure 12–27 shows the charge distribution in the MOS capacitor with a p-type substrate. At a high signal frequency, the differential change in charge occurs at the metal and in the space charge width in the semiconductor. The capacitance of the MOS capacitor is then $C'_{min}$, which we discussed earlier.

The high-frequency and low-frequency limits of the C-V characteristics are shown in Figure 12–28. In general, high frequency corresponds to a value on the order of 1 MHz and low frequency corresponds to values in the range of 5 to 100 Hz. Typically, the high-frequency characteristics of the MOS capacitor are measured.

### 12.2.3 Fixed Oxide and Interface Charge Effects

In all of the discussion concerning C-V characteristics so far, we have assumed an ideal oxide in which there are no fixed oxide or oxide–semiconductor interface charges. These two types of charges will change the C-V characteristics.

We previously discussed how the fixed oxide charge affects the threshold voltage. This charge will also affect the flat-band voltage. The flat-band voltage from Equation (12–21) was given by

$$V_{FB} = \phi_{ms} - \frac{Q'_{ss}}{C_{ox}}$$

where $Q'_{ss}$ is the equivalent fixed oxide charge and $\phi_{ms}$ is the metal–semiconductor work function difference. The flat-band voltage shifts to more nega-

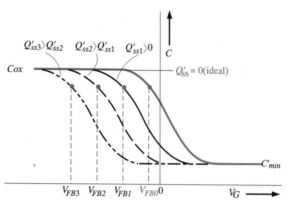

**Figure 12–29** High-frequency capacitance versus gate voltage of a MOS capacitor with a p-type substrate for several values of effective trapped oxide charge.

tive voltages for a positive fixed oxide charge. Since the oxide charge is not a function of gate voltage, the curves show a parallel shift with oxide charge and the shape of the C-V curves remains the same as the ideal characteristics. Figure 12–29 shows the high-frequency characteristics of a MOS capacitor with a p-type substrate for several values of fixed positive oxide charge.

The C-V characteristics can be used to determine the equivalent fixed oxide charge. For a given MOS structure, $\phi_{ms}$ and $C_{ox}$ are known so the ideal flat-band voltage and flat-band capacitance can be calculated. The experimental value of flat-band voltage can be measured from the C-V curve and the value of fixed oxide charge can then be determined. The C-V measurements are a valuable diagnostic tool to characterize a MOS device. This characterization is especially useful in the study of radiation effects on MOS devices, for example, which we will discuss in the next chapter.

We first encountered oxide–semiconductor interface states in Chapter 9 in the discussion of Schottky barrier diodes. Figure 12–30 shows the energy-band diagram of a semiconductor at the oxide–semiconductor interface. The periodic nature of the semiconductor is abruptly terminated at the interface

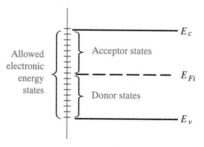

**Figure 12–30** Schematic diagram showing interface states at the oxide–semiconductor interface.

so that allowed electronic energy levels will exist within the forbidden bandgap. These allowed energy states are referred to as interface states. Charge can flow between the semiconductor and interface states, in contrast to the fixed oxide charge. The net charge in these interface states is a function of the position of the Fermi level in the bandgap.

In general, acceptor states exist in the upper half of the bandgap and donor states exist in the lower half of the bandgap. An acceptor state is neutral if the Fermi level is below the state, and becomes negatively charged if the Fermi level is above the state. A donor state is neutral if the Fermi level is above the state and becomes positively charged if the Fermi level is below the state. The charge of the interface states is then a function of the gate voltage applied across the MOS capacitor.

Figure 12–31a shows the energy-band diagram in a p-type semiconductor of a MOS capacitor biased in the accumulation condition. In this case, there

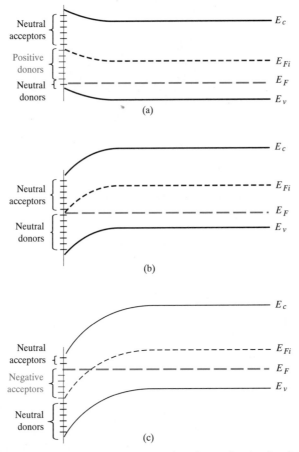

**Figure 12–31**   Energy-band diagram in a p-type semiconductor showing the charge trapped in the interface states when the MOS capacitor is biased (a) in accumulation, (b) at midgap, and (c) at inversion.

is a net positive charge trapped in the donor states. Now let the gate voltage change to produce the energy-band diagram shown in Figure 12–31b. The Fermi level corresponds to the intrinsic Fermi level at the surface; thus, all interface states are neutral. This particular bias condition is known as *midgap*. Figure 12–31c shows the condition at inversion in which there is now a net negative charge in the acceptor states.

The net charge in the interface states changes from positive to negative as the gate voltage sweeps from the accumulation, depletion, to the inversion condition. We noted that the C-V curves shifted in the negative gate voltage direction due to positive fixed oxide charge. When interface states are present, the amount and direction of the shift changes as we sweep through the gate voltage, since the amount and sign of the interface trapped charge changes. The C-V curves now become "smeared out" as shown in Figure 12–32.

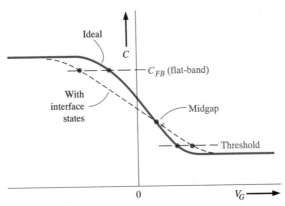

**Figure 12–32**    High-frequency C-V characteristics of a MOS capacitor showing effects of interface states.

Again, the C-V measurements can be used as a diagnostic tool in semiconductor device process control. For a given MOS device, the ideal C-V curve can be determined. Any "smearing out" in the experimental curve indicates the presence of interface states and any parallel shift indicates the presence of fixed oxide charge. The amount of smearing out can be used to determine the density of interface states. These types of measurement are extremely useful in the study of radiation effects on MOS devices, which we will consider in the next chapter.

## 12.3 THE BASIC MOSFET OPERATION

The current in a MOS field-effect transistor is due to the flow of charge in the inversion layer or channel region adjacent to the oxide–semiconductor interface. We have discussed the creation of the inversion layer charge in enhancement-type MOS capacitors. We may also have depletion-type devices in which a channel already exists at zero gate voltage.

### 12.3.1 MOSFET Structures

There are four basic MOSFET device types. Figure 12–33 shows an n-channel enhancement mode MOSFET. Implicit in the enhancement mode notation is the idea that the semiconductor substrate is not inverted directly under the oxide with zero gate voltage. A positive gate voltage induces the

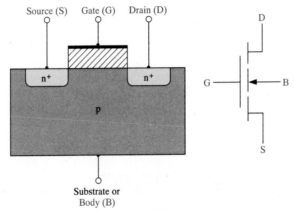

**Figure 12–33** Cross section and circuit symbol for an n-channel enhancement mode MOSFET.

electron inversion layer, which then "connects" the n-type source and the n-type drain regions. The source terminal is the source of carriers that flow through the channel to the drain terminal. For this n-channel device, electrons flow from the source to the drain so the conventional current will enter the drain and leave the source. The conventional circuit symbol for this n-channel enhancement mode device is also shown in this figure.

Figure 12–34 shows an n-channel depletion mode MOSFET. An n-channel region exists under the oxide with zero volts applied to the gate. However, we have shown that the threshold voltage of a MOS device with a p-type substrate may be negative; this means that an electron inversion layer already exists with zero gate voltage applied. Such a device is also consid-

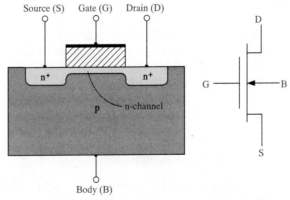

**Figure 12–34** Cross section and circuit symbol for an n-channel depletion mode MOSFET.

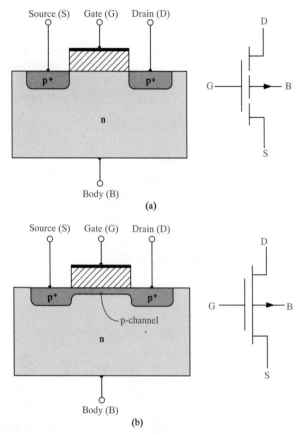

**Figure 12–35**   Cross section and circuit symbol for (a) a p-channel enhancement mode MOSFET, and (b) a p-channel depletion mode MOSFET.

ered to be a depletion mode device. The n-channel shown in this figure can be an electron inversion layer or an intentionally doped n-region. The conventional circuit symbol for the n-channel depletion mode MOSFET is also shown in the figure.

Figures 12–35a and 12–35b show a p-channel enhancement mode MOSFET and a p-channel depletion mode MOSFET. In the p-channel enhancement mode device, a negative gate voltage must be applied to create an inversion layer of holes that will "connect" the p-type source and drain regions. Holes flow from the source to the drain so the conventional current will enter the source and leave the drain. A p-channel region exists in the depletion mode device even with zero gate voltage. The conventional circuit symbols are shown in the figure.

### 12.3.2 Current-Voltage Relationship—Concepts

Figure 12–36a shows an n-channel enhancement mode MOSFET with a gate-to-source voltage that is less than the threshold voltage and with only a

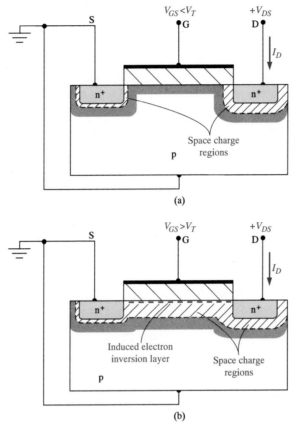

**Figure 12–36**   The n-channel enhancement mode MOSFET (a) with an applied gate voltage $V_{GS} < V_T$, and (b) with an applied gate voltage $V_{GS} > V_T$.

very small drain-to-source voltage. The source and substrate, or body, terminals are held at ground potential. With this bias configuration there is no electron inversion layer, the drain-to-substrate pn junction is reverse biased, and the drain current is zero (disregarding pn junction leakage currents).

Figure 12–36b shows the same MOSFET with an applied gate voltage such that $V_{GS} > V_T$. An electron inversion layer has been created so that, when a small drain voltage is applied, the electrons in the inversion layer will flow from the source to the positive drain terminal. The conventional current enters the drain terminal and leaves the source terminal. In this ideal case, there is no current through the oxide to the gate terminal.

For small $V_{DS}$ values, the channel region has the characteristics of a resistor, so we can write

$$I_D = g_d V_{DS} \qquad (12\text{--}37)$$

where $g_d$ is defined as the channel conductance in the limit as $V_{DS} \rightarrow 0$. The channel conductance is given by

$$g_d = \frac{W}{L} \cdot \mu_n |Q_n'| \qquad (12\text{--}38)$$

where $\mu_n$ is the mobility of the electrons in the inversion layer and $|Q_n'|$ is the magnitude of the inversion layer charge per unit area. The inversion layer charge is a function of the gate voltage; thus the basic MOS transistor action is the modulation of the channel conductance by the gate voltage. The channel conductance, in turn, determines the drain current. We will initially assume that the mobility is a constant; we will discuss mobility effects and variations in the next chapter.

The $I_D$ versus $V_{DS}$ characteristics, for small values of $V_{DS}$, are shown in Figure 12–37. When $V_{GS} < V_T$, the drain current is zero. As $V_{GS}$ becomes larger than $V_T$, the channel inversion charge density increases, which increases the channel conductance. A larger value of $g_d$ produces a larger initial slope of the $I_D$ versus $V_{DS}$ characteristic as shown in the figure.

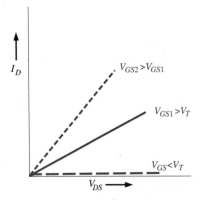

**Figure 12–37**   $I_D$ versus $V_{DS}$ characteristics for small values of $V_{DS}$ at three $V_{GS}$ voltages.

Figure 12–38a shows the basic MOS structure for the case when $V_{GS} > V_T$ and the applied $V_{DS}$ voltage is small. The thickness of the inversion channel layer in the figure qualitatively indicates the relative charge density, which is essentially constant along the entire channel length for this case. The corresponding $I_D$ versus $V_{DS}$ curve is shown in the figure.

Figure 12–38b shows the situation when the $V_{DS}$ value increases. As the drain voltage increases, the voltage drop across the oxide near the drain terminal decreases, which means that the induced inversion charge density near the drain also decreases. The incremental conductance of the channel at the drain decreases which then means that the slope of the $I_D$ versus $V_{DS}$ curve will decrease. This effect is shown in the $I_D$ versus $V_{DS}$ curve in the figure.

When $V_{DS}$ increases to the point where the potential drop across the oxide at the drain terminal is equal to $V_T$, the induced inversion charge density is zero at the drain terminal. This effect is schematically shown in Figure 12–38c. At this point, the incremental conductance at the drain is zero which

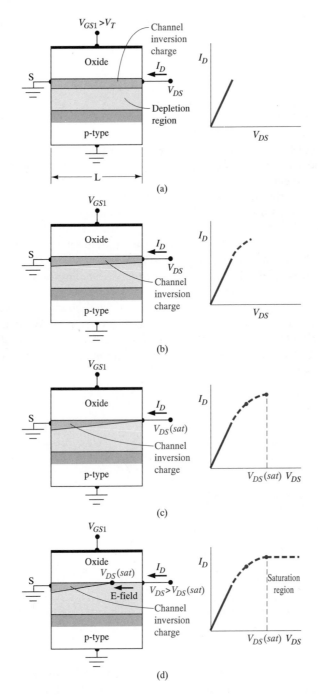

**Figure 12–38** Cross section and $I_D$ versus $V_{DS}$ curve when $V_{GS} > V_T$ for (a) a small $V_{DS}$ value, (b) a larger $V_{DS}$ value, (c) a value of $V_{DS} = V_{DS}(\text{sat})$, and (d) a value of $V_{DS} > V_{DS}(\text{sat})$.

means that the slope of the $I_D$ versus $V_{DS}$ curve is zero. We can write

$$V_{GS} - V_{DS}(\text{sat}) = V_T \tag{12–39a}$$

or

$$V_{DS}(\text{sat}) = V_{GS} - V_T \tag{12–39b}$$

where $V_{DS}(\text{sat})$ is the drain-to-source voltage producing zero inversion charge density at the drain terminal.

When $V_{DS}$ becomes larger than the $V_{DS}(\text{sat})$ value, the point in the channel at which the inversion charge is just zero moves toward the source terminal. In this case, electrons enter the channel at the source, travel through the channel toward the drain, and then, at the point where the charge goes to zero, the electrons are injected into the space charge region where they are swept by the E-field to the drain contact. If we assume that the change in channel length $\Delta L$ is small compared to the original length $L$, then the drain current will be a constant for $V_{DS} > V_{DS}(\text{sat})$. The region of the $I_D$ versus $V_{DS}$ characteristic is referred to as the *saturation region*. This region of operation is shown in Figure 12–38d.

When $V_{GS}$ changes, the $I_D$ versus $V_{DS}$ curve will change. We saw that, if $V_{GS}$ increases, the initial slope of $I_D$ versus $V_{DS}$ increases. We can also note from Equation (12–39b) that the value of $V_{DS}(\text{sat})$ is a function of $V_{GS}$. We can generate the family of curves for this n-channel enhancement mode MOSFET as shown in Figure 12–39.

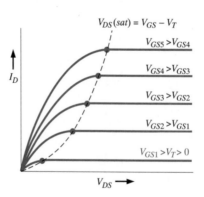

**Figure 12–39**  Family of $I_D$ versus $V_{DS}$ curves for an n-channel enhancement mode MOSFET.

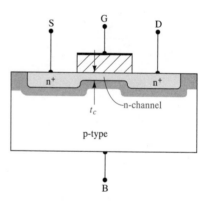

**Figure 12–40**  Cross section of an n-channel depletion mode MOSFET.

Figure 12–40 shows an n-channel depletion mode MOSFET. If the n-channel region is actually an induced electron inversion layer created by the metal–semiconductor work function difference and fixed charge in the oxide, the current-voltage characteristics are exactly the same as we have discussed except that $V_T$ is a negative quantity. We may also consider the case when the n-channel region is actually an n-type semiconductor region. In this type of device, a negative gate voltage will induce a space charge region under the oxide, reducing the thickness of the n-channel region. The reduced thickness decreases the channel conductance, which reduces the drain current. A positive gate voltage will create an electron accumulation layer, which increases the drain current. One basic requirement for this device is that the channel thickness $t_c$ must be less than the maximum induced space charge width in order to be able to turn the device off. The general $I_D$ versus $V_{DS}$ family of curves for an n-channel depletion mode MOSFET is shown in Figure 12–41.

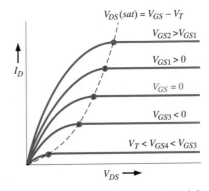

**Figure 12–41** Family of $I_D$ versus $V_{DS}$ curves for an n-channel depletion mode MOSFET.

In the next section we will derive the ideal current-voltage relation for the n-channel MOSFET. In the nonsaturation region, we will obtain

$$I_D = \frac{W\mu_n C_{ox}}{2L}\left[2(V_{GS} - V_T)V_{DS} - V_{DS}^2\right] \qquad (12\text{–}40)$$

and, in the saturation region, we will have

$$I_D = \frac{W\mu_n C_{ox}}{2L}(V_{GS} - V_T)^2 \qquad (12\text{–}41)$$

The operation of a p-channel device is the same as the n-channel device except the charge carrier is the hole and the conventional current direction and voltage polarities are reversed.

### *12.3.3 Current-Voltage Relationship—Mathematical Derivation

In the previous section, we qualitatively discussed the current-voltage characteristics. In this section, we will derive the mathematical relation between the drain current, the gate-to-source voltage, and the drain-to-source volt-

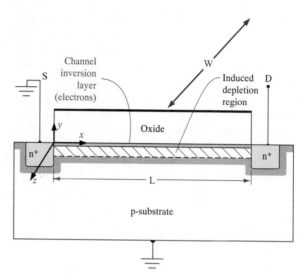

**Figure 12–42**    Geometry of a MOSFET for $I_D$ versus $V_{DS}$ derivation.

age. Figure 12–42 shows the geometry of the device that we will use in this derivation.

In this analysis, we will make the following assumptions:

1.  The current in the channel is due to drift rather than diffusion.
2.  There is no current through the gate oxide.
3.  A gradual channel approximation is used in which $\partial E_y/\partial y \gg \partial E_x/\partial x$. This approximation means that $E_x$ is essentially a constant.
4.  Any fixed oxide charge is assumed to be an equivalent charge density at the oxide-semiconductor interface.
5.  The carrier mobility in the channel is assumed to be constant.

We start the analysis with Ohm's law, which can be written as

$$J_x = \sigma E_x \qquad (12\text{--}42)$$

where $\sigma$ is the channel conductivity and $E_x$ is the electric field along the channel created by the drain-to-source voltage. The channel conductivity is given by $\sigma = e\mu_n n(y)$ where $\mu_n$ is the electron mobility and $n(y)$ is the electron concentration in the inversion layer.

The total channel current is found by integrating $J_x$ over the cross-sectional area in the $y$- and $z$-directions. Then

$$I_x = \int_y \int_z J_x \, dy \, dz \qquad (12\text{--}43)$$

We may write that

$$Q'_n = -\int en(y) \, dy \qquad (12\text{--}44)$$

where $Q'_n$ is the inversion layer charge per unit area and is a negative quantity for this case.

Equation (12–43) then becomes

$$I_x = -W\mu_n Q'_n E_x \qquad (12\text{–}45)$$

where $W$ is the channel width, the result of integrating over $z$.

Two concepts we will use in the current-voltage derivation are charge neutrality and Gauss's law. Figure 12–43 shows the charge densities through

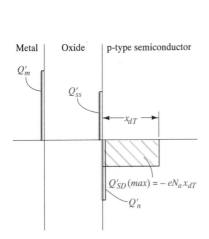

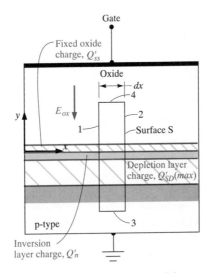

**Figure 12–43** Charge distribution in the n-channel enhancement mode MOSFET for $V_{GS} > V_T$.

**Figure 12–44** Geometry for applying Gauss's law.

the device for $V_{GS} > V_T$. The charges are all given in terms of charge per unit area. Using the concept of charge neutrality, we can write

$$Q'_m + Q'_{ss} + Q'_n + Q'_{SD}(\text{max}) = 0 \qquad (12\text{–}46)$$

The inversion layer charge and induced space charge will be negative for this n-channel device.

Gauss's law can be written as

$$\oint_S \varepsilon E_n \, dS = Q_T \qquad (12\text{–}47)$$

where the integral is over a closed surface, $Q_T$ is the total charge enclosed by the surface, and $E_n$ is the outward directed normal component of the electric field crossing the surface $S$. Gauss's law will be applied to the surface defined in Figure 12–44. Since the surface must be enclosed, we must take into account the two end surfaces in the $x$-$y$ plane. However, there is no z-component of the electric field so these two end surfaces do not contribute to the integral of Equation (12–47).

Now consider the surfaces labeled 1 and 2 in Figure 12–44. From the gradual channel approximation, we will assume that $E_x$ is essentially a constant along the channel length. This assumption means that $E_x$ into surface 2 is the same as $E_x$ out of surface 1. Since the integral in Equation (12–47) involves the outward component of the E-field, the contributions of surfaces 1 and 2 cancel each other. Surface 3 is in the neutral p-region so the electric field is zero at this surface.

Surface 4 is the only surface that contributes to Equation (12–47). Taking into account the direction of the electric field in the oxide, Equation (12–47) becomes

$$\oint_S \varepsilon E_n \, dS = -\varepsilon_{ox} E_{ox} W \, dx = Q_T \tag{12–48}$$

where $\varepsilon_{ox}$ is the permittivity of the oxide. The total charge enclosed is

$$Q_T = (Q'_{ss} + Q'_n + Q'_{SD}(max)) W \, dx \tag{12–49}$$

Combining Equations (12–48) and (12–49), we have

$$-\varepsilon_{ox} E_{ox} = Q'_{ss} + Q'_n + Q'_{SD}(max) \tag{12–50}$$

We now need an expression for $E_{ox}$. Figure 12–45a shows the oxide and channel. We will assume that the source is at ground potential. The voltage $V_x$ is the potential in the channel at a point $x$ along the channel length. The potential difference across the oxide at $x$ is a function of $V_{GS}$, $V_x$, and the metal–semiconductor work function difference.

The energy-band diagram through the MOS structure at point $x$ is shown in Figure 12–45b. The Fermi level in the p-type semiconductor is $E_{Fp}$ and the Fermi level in the metal is $E_{Fm}$. We have

$$E_{Fp} - E_{Fm} = e(V_{GS} - V_x) \tag{12–51}$$

Considering the potential barriers, we can write

$$V_{GS} - V_x = (\phi'_m + V_{ox}) - \left(\chi' + \frac{E_g}{2e} - \phi_s + \phi_{fp}\right) \tag{12–52}$$

which can also be written as

$$V_{GS} - V_x = V_{ox} + 2\phi_{fp} + \phi_{ms} \tag{12–53}$$

where $\phi_{ms}$ is the metal–semiconductor work function difference, and $\phi_s = 2\phi_{fp}$ for the inversion condition.

The electric field in the oxide is

$$E_{ox} = \frac{V_{ox}}{t_{ox}} \tag{12–54}$$

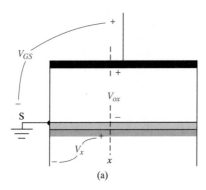

(a)

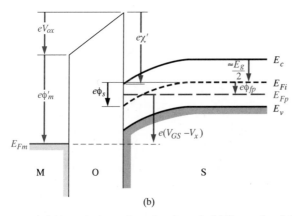

(b)

**Figure 12-45** (a) Potentials at a point $x$ along the channel. (b) Energy-band diagram through the MOS structure at the point $x$.

Combining Equations (12–50), (12–53), and (12–54), we find that

$$-\varepsilon_{ox}\mathrm{E}_{ox} = -\frac{\varepsilon_{ox}}{t_{ox}}[(V_{GS} - V_x) - (\phi_{ms} + 2\phi_{fp})]$$

$$= Q'_{ss} + Q'_n + Q'_{SD}(\max) \qquad (12\text{–}55)$$

The inversion charge density, $Q'_n$, from Equation (12–55) can be substituted into Equation (12–45) and we obtain

$$I_x = -W\mu_n C_{ox} \frac{dV_x}{dx}\{(V_{GS} - V_x) - V_T\} \qquad (12\text{–}56)$$

where $\varepsilon_x = -dV_x/dx$ and $V_T$ is the threshold voltage defined by Equation (12–27).

We can now integrate Equation (12–56) over the length of the channel. We have

$$\int_0^L I_x \, dx = -W\mu_n C_{\text{ox}} \int_{V_x(0)}^{V_x(L)} [(V_{GS} - V_T) - V_x] \, dV_x \qquad (12\text{–}57)$$

We are assuming a constant mobility $\mu_n$. For the n-channel device, the drain current enters the drain terminal and is a constant along the entire channel length. Letting $I_D = -I_x$, Equation (12–57) becomes

$$I_D = \frac{W\mu_n C_{\text{ox}}}{2L} [2(V_{GS} - V_T)V_{DS} - V_{DS}^2] \qquad (12\text{–}58)$$

Equation (12–58) is valid for $V_{GS} \geq V_T$ and for $0 \leq V_{DS} \leq V_{DS}(\text{sat})$.

Figure 12–46 shows plots of Equation (12–58) as a function of $V_{DS}$ for several values of $V_{GS}$. We can find the value of $V_{DS}$ at the peak current value

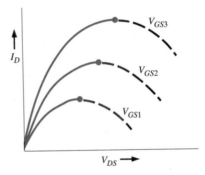

**Figure 12–46**    Plots of $I_D$ versus $V_{DS}$ from Equation (12–58).

from $\partial I_D / \partial V_{DS} = 0$. Then, using Equation (12–58), the peak current occurs when

$$V_{DS} = V_{GS} - V_T \qquad (12\text{–}59)$$

This value of $V_{DS}$ is just $V_{DS}(\text{sat})$, the point at which saturation occurs. For $V_{DS} > V_{DS}(\text{sat})$, the ideal drain current is a constant and is equal to

$$I_D(\text{sat}) = \frac{W\mu_n C_{\text{ox}}}{2L} [2(V_{GS} - V_T)V_{DS}(\text{sat}) - V_{DS}^2(\text{sat})] \qquad (12\text{–}60)$$

Using Equation (12–59) for $V_{DS}(\text{sat})$, Equation (12–60) becomes

$$I_D(\text{sat}) = \frac{W\mu_n C_{\text{ox}}}{2L} (V_{GS} - V_T)^2 \qquad (12\text{–}61)$$

Equation (12–58) is the ideal current-voltage relationship of the n-channel MOSFET in the nonsaturation region for $0 \leq V_{DS} \leq V_{DS}(\text{sat})$ and Equation (12–61) is the ideal current-voltage relationship of the n-channel MOSFET in the saturation region for $V_{DS} \geq V_{DS}(\text{sat})$. These I-V expressions were explicitly derived for an n-channel enhancement mode device. However, these same equations apply to an n-channel depletion mode MOSFET in which the threshold voltage $V_T$ is a negative quantity.

---

**Example 12–8**

**Objective:** To calculate $I_D(\text{sat})$ for an n-channel MOSFET.

Consider an n-channel MOSFET with the same parameters as given in Example 12–4. In addition, assume that $L = 2$ $\mu$m, $W = 20$ $\mu$m, and $\mu_n = 650$ cm²/V-sec.

**Solution:** We have from Equation (12–61), that

$$I_D(\text{sat}) = \frac{W\mu_n C_{\text{ox}}}{2L}(V_{GS} - V_T)^2 = \frac{(20 \times 10^{-4})(650)(6.9 \times 10^{-8})}{2(2 \times 10^{-4})}(V_{GS} - V_T)^2$$

$$= 2.24 \times 10^{-4}(V_{GS} - V_T)^2 \text{ amps}$$

If $V_{GS} = 5$ volts, then

$$I_D(\text{sat}) = 0.224(5 - 0.642)^2 \text{ mA} = 4.25 \text{ mA}$$

**Comment:** The current capability of a MOSFET is directly proportional to the channel width $W$. The current handling capability can be increased by increasing $W$.

---

We can use the I-V relations to experimentally determine the mobility and threshold voltage parameters. From Equation (12–58), we can write, for very small values of $V_{DS}$,

$$I_D = \frac{W\mu_n C_{\text{ox}}}{L}(V_{GS} - V_T)V_{DS} \tag{12–62a}$$

Figure 12–47a shows a plot of Equation (12–62a) as a function of $V_{GS}$ for constant $V_{DS}$. A straight line is fitted through the points. The deviation from the straight line at low values of $V_{GS}$ is due to subthreshold conduction and the deviation at higher values of $V_{GS}$ is due to mobility being a function of gate voltage. Both of these effects will be considered in the next chapter. The extrapolation of the straight line to zero current gives the threshold voltage and the slope is proportional to the inversion carrier mobility.

If we take the square root of Equation (12–61), we obtain

$$\sqrt{I_D(\text{sat})} = \sqrt{\frac{W\mu_n C_{\text{ox}}}{2L}}(V_{GS} - V_T) \tag{12–62b}$$

Figure 12–47b is a plot of Equation (12–62b). In the ideal case, we can obtain the same information from both curves. However, as we will see in the next chapter, the threshold voltage may be a function of $V_{DS}$ in short channel devices. Since Equation (12–62b) applies to devices biased in the saturation region, the $V_T$ parameter in this equation may differ from the extrapolated value determined in Figure 12–47a. In general, the nonsaturation current-voltage characteristics will produce the more reliable data.

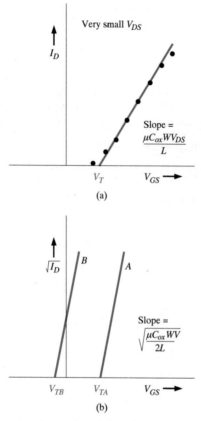

Figure 12–47    (a) $I_D$ versus $V_{GS}$ (for small $V_{DS}$) for enhancement mode MOSFET. (b) Ideal $\sqrt{I_D}$ versus $V_{GS}$ in saturation region for enhancement mode (curve A) and depletion mode (curve B) n-channel MOSFETs.

---

**Example 12–9**

**Objective:** To determine the inversion carrier mobility from experimental results.

Consider an n-channel MOSFET with $W = 15 \ \mu m$, $L = 2 \ \mu m$, and $C_{ox} = 6.9 \times 10^{-8} \ F/cm^2$. Assume that the drain current in the nonsaturation region for $V_{DS} = 0.10$ V is $I_D = 35 \ \mu A$ at $V_{GS} = 1.5$ volts and $I_D = 75 \ \mu A$ at $V_{GS} = 2.5$ volts.

**Solution:** From Equation (12–62a), we can write

$$I_{D2} - I_{D1} = \frac{W \mu_n C_{ox}}{L} (V_{GS2} - V_{GS1}) V_{DS}$$

so that

$$75 \times 10^{-6} - 35 \times 10^{-6} = \left(\frac{15}{2}\right) \mu_n (6.9 \times 10^{-8})(2.5 - 1.5)(0.10)$$

which yields

$$\mu_n = 773 \text{ cm}^2/\text{V-sec}$$

We can then determine

$$V_T = 0.625 \text{ V}$$

**Comment:** The mobility of carriers in the inversion layer is less than that in the bulk semiconductor due to the surface scattering effect. We will discuss this effect in the next chapter.

---

The current-voltage relationship of a p-channel device can be obtained using the same type of analysis. Figure 12–48 shows a p-channel enhance-

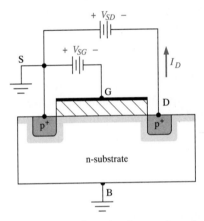

**Figure 12–48**   Cross section and bias configuration for a p-channel enhancement mode MOSFET.

ment mode MOSFET. The voltage polarities and current direction are the reverse of those in the n-channel device. We may note the change in the subscript notation for this device. For the current direction shown in the figure, the I-V relations for the p-channel MOSFET are

$$I_D = \frac{W\mu_p C_{ox}}{2L} [2(V_{SG} + V_T)V_{SD} - V_{SD}^2] \qquad (12\text{–}63)$$

for $0 \le V_{SD} \le V_{SD}(\text{sat})$, and

$$I_D(\text{sat}) = \frac{W\mu_p C_{ox}}{2L} (V_{SG} + V_T)^2 \qquad (12\text{–}64)$$

for $V_{SD} \ge V_{SD}(\text{sat})$, where

$$V_{SD}(\text{sat}) = V_{SG} + V_T \qquad (12\text{–}65)$$

Note the change in the sign in front of $V_T$ and note that the mobility is now the mobility of the holes in the hole inversion layer charge. Keep in mind that $V_T$ is negative for a p-channel enhancement mode MOSFET and positive for a depletion mode p-channel device.

One assumption we made in the derivation of the current-voltage relationship was that the charge neutrality condition given by Equation (12–46) was valid over the entire length of the channel. We implicitly assumed that $Q'_{SD}(\text{max})$ was constant along the length of the channel. The space charge width, however, varies between source and drain due to the drain-to-source voltage; it is widest at the drain when $V_{DS} > 0$. A change in the space charge density along the channel length must be balanced by a corresponding change in the inversion layer charge. An increase in the space charge width means that the inversion layer charge is reduced, implying that the drain current and drain-to-source saturation voltage are less than the ideal values. The actual saturation drain current may be as much as 20 percent less than the predicted value due to this bulk charge effect.

### 12.3.4 Transconductance

The MOSFET transconductance is defined as the change in drain current with respect to the corresponding change in gate voltage, or

$$g_m = \frac{\partial I_D}{\partial V_{GS}} \qquad (12-66)$$

The transconductance is sometimes referred to as the transistor gain.

If we consider an n-channel MOSFET operating in the nonsaturation region, then, using Equation (12–58), we have

$$g_{mL} = \frac{\partial I_D}{\partial V_{GS}} = \frac{W \mu_n C_{ox}}{L} \cdot V_{DS} \qquad (12-67)$$

The transconductance increases linearly with $V_{DS}$ but is independent of $V_{GS}$ in the nonsaturation region.

The I-V characteristics of an n-channel MOSFET in the saturation region were given by Equation (12–61). The transconductance in this region of operation is given by

$$g_{ms} = \frac{\partial I_D(\text{sat})}{\partial V_{GS}} = \frac{W \mu_n C_{ox}}{L} (V_{GS} - V_T) \qquad (12-68)$$

In the saturation region, the transconductance is a linear function of $V_{GS}$ and is independent of $V_{DS}$.

The transconductance is a function of the geometry of the device as well as carrier mobility and threshold voltage. The transconductance increases as the width of the device increases, and also increases as the channel length and oxide thickness decrease. In the design of MOSFET circuits, the size of the transistor, in particular the channel width $W$, is an important engineering design parameter.

### 12.3.5 Substrate Bias Effects

In all of our analyses so far, the substrate, or body, has been connected to the source and held at ground potential. In MOSFET circuits, the source and body may not be at the same potential. Figure 12–49a shows an n-channel MOSFET and the associated double subscripted voltage variables. The source-to-substrate pn junction must always be zero or reverse biased so that $V_{SB}$ must always be greater than or equal to zero.

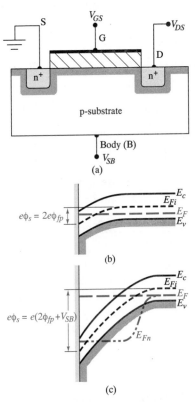

**Figure 12–49** (a) Applied voltages on an n-channel MOSFET. (b) Energy-band diagram at inversion point when $V_{SB} = 0$. (c) Energy-band diagram at inversion point when $V_{SB} > 0$ is applied.

If $V_{SB} = 0$, threshold is defined as the condition when $\phi_s = 2\phi_{fp}$ as we discussed previously and as shown in Figure 12–49b. When $V_{SB} > 0$, the surface will still try to invert when $\phi_s = 2\phi_{fp}$. However, these electrons are at a higher potential energy than are the electrons in the source. The newly created electrons will move laterally and flow out of the source terminal. When $\phi_s = 2\phi_{fp} + V_{SB}$, the surface reaches an equilibrium inversion condition. The energy-band diagram for this condition is shown in Figure 12–49c. The curve represented as $E_{Fn}$ is the Fermi level from the p-substrate through the reverse-biased source-substrate junction to the source contact.

The space charge region width under the oxide increases from the original $x_{dT}$ value when a reverse-biased source-substrate junction voltage is applied. With an applied $V_{SB} > 0$, there is more charge associated with this region. Considering the charge neutrality condition through the MOS structure, the positive charge on the top metal gate must increase to compensate for the increased negative space charge in order to reach the threshold inversion point. So when $V_{SB} > 0$, the threshold voltage of the n-channel MOSFET increases.

When $V_{SB} = 0$, we had

$$Q'_{SD}(\text{max}) = -eN_a x_{dT} = -\sqrt{2e\varepsilon_s N_a (2\phi_{fp})} \tag{12–69}$$

When $V_{SB} > 0$, the space charge width increases and we now have

$$Q'_{SD} = -eN_a x_d = -\sqrt{2e\varepsilon_s N_a (2\phi_{fp} + V_{SB})} \tag{12–70}$$

The change in the space charge density is then

$$\Delta Q'_{SD} = -\sqrt{2e\varepsilon_s N_a} \left[\sqrt{2\phi_{fp} + V_{SB}} - \sqrt{2\phi_{fp}}\right] \tag{12–71}$$

To reach the threshold condition, the applied gate voltage must be increased. The change in threshold voltage can be written as

$$\Delta V_T = -\frac{\Delta Q'_{SD}}{C_{ox}} = \frac{\sqrt{2e\varepsilon_s N_a}}{C_{ox}} \left[\sqrt{2\phi_{fp} + V_{SB}} - \sqrt{2\phi_{fp}}\right] \tag{12–72}$$

where $\Delta V_T = V_T(V_{SB} > 0) - V_T(V_{SB} = 0)$. We may note that $V_{SB}$ must always be positive so that, for the n-channel device, $\Delta V_T$ is always positive. The threshold voltage of the n-channel MOSFET will increase as a function of the source-substrate junction voltage.

---

**Example 12–10**

**Objective:** To calculate the change in the threshold voltage due to an applied source-to-body voltage.

Consider an n-channel silicon MOSFET at $T = 300°K$. Assume the substrate is doped to $N_a = 3 \times 10^{16}$ cm$^{-3}$ and assume the oxide is silicon dioxide with a thickness of $t_{ox} = 500$ Å. Let $V_{SB} = 1$ volt.

**Solution:** We can calculate that

$$\phi_{fp} = V_t \ln\left(\frac{N_a}{n_i}\right) = (0.0259) \ln\left(\frac{3 \times 10^{16}}{1.5 \times 10^{10}}\right) = 0.376 \text{ volt}$$

We can also find

$$C_{\text{ox}} = \frac{\varepsilon_{\text{ox}}}{t_{\text{ox}}} = \frac{(3.9)(8.85 \times 10^{-14})}{500 \times 10^{-8}} = 6.9 \times 10^{-8} \ \text{F/cm}^2$$

Then from Equation (12–72), we can obtain

$$\Delta V_T = \frac{[2(1.6 \times 10^{-19})(11.7)(8.85 \times 10^{-14})(3 \times 10^{16})]^{1/2}}{6.9 \times 10^{-8}}$$

$$\times \{[2(0.376) + 1]^{1/2} - [2(0.376)]^{1/2}\}$$

or

$$\Delta V_T = 1.445(1.324 - 0.867) = 0.66 \ \text{volt}$$

**Comment:** Figure 12–50 shows plots of $\sqrt{I_D}(\text{sat})$ versus $V_{GS}$ for various values of applied $V_{SB}$. The original threshold voltage, $V_{T0}$, is 0.64 volt.

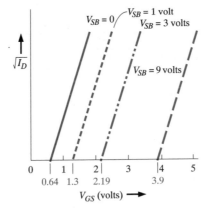

**Figure 12–50**  Plots of $\sqrt{I_D}$ versus $V_{GS}$ at several values of $V_{SB}$ for an n-channel MOSFET.

If a body or substrate bias is applied to a p-channel device, the threshold voltage is shifted to more negative values. Because the threshold voltage of a p-channel enhancement mode MOSFET is negative, a body voltage will increase the applied negative gate voltage required to create inversion. The same general observation was made for the n-channel MOSFET.

## 12.4 FREQUENCY LIMITATIONS

In many applications, the MOSFET is used in a linear amplifier circuit. A small-signal equivalent circuit for the MOSFET is needed in order to mathematically analyze the electronic circuit. The equivalent circuit contains capacitances and resistances that introduce frequency effects. We will initially develop a small-signal equivalent circuit and then discuss the physical factors that limit the frequency response of the MOSFET. A transistor cutoff frequency, which is a figure of merit, will then be defined and an expression derived for this factor.

### 12.4.1 Small-Signal Equivalent Circuit

The small-signal equivalent circuit of the MOSFET is constructed from the basic MOSFET geometry. A model based on the inherent capacitances and resistances within the transistor structure, along with elements that represent the basic device equations, is shown in Figure 12–51. One simplifying

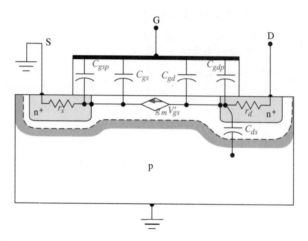

**Figure 12–51**    Inherent resistances and capacitances in the n-channel MOSFET structure.

assumption we will make in the equivalent circuit is that the source and substrate are both tied to ground potential.

Two of the capacitances connected to the gate are inherent in the device. These capacitances are $C_{gs}$ and $C_{gd}$, which represent the interaction between the gate and the channel charge near the source and drain terminals, respectively. The remaining two gate capacitances, $C_{gsp}$ and $C_{gdp}$, are parasitic or overlap capacitances. In real devices, the gate oxide will overlap the source and drain contacts because of tolerance or fabrication factors. As we will see, the drain overlap capacitance $C_{gdp}$, in particular, will lower the frequency response of the device. The parameter $C_{ds}$ is the drain-to-substrate pn junction capacitance and $r_s$ and $r_d$ are the series resistances associated with the source and drain terminals. The small-signal channel current is controlled by the internal gate-to-source voltage through the transconductance.

The small-signal equivalent circuit for the n-channel common-source MOSFET is shown in Figure 12–52. The voltage $V'_{gs}$ is the internal gate-to-source voltage that controls the channel current. The parameters $C_{gsT}$ and $C_{gdT}$ are the total gate-to-source and total gate-to-drain capacitances. One

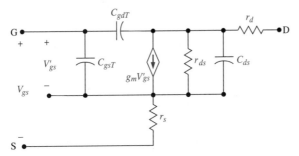

**Figure 12–52** Small-signal equivalent circuit of a common-source n-channel MOSFET.

parameter, $r_{ds}$, shown in Figure 12–52, is not shown in Figure 12–51. This resistance is associated with the slope of $I_D$ versus $V_{DS}$. In the ideal MOSFET biased in the saturation region, $I_D$ is independent of $V_{DS}$ so that $r_{ds}$ would be infinite. In short channel length devices, in particular, $r_{ds}$ is finite because of channel length modulation, which we will consider in the next chapter.

A simplified small-signal equivalent circuit valid at low frequency is shown in Figure 12–53. The series resistances, $r_s$ and $r_d$, have been ne-

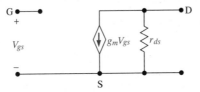

**Figure 12–53** Simplified, low-frequency small-signal equivalent circuit of a common-source n-channel MOSFET.

glected, so the drain current is essentially only a function of the gate-to-source voltage through the transconductance. The input gate impedance is infinite in this simplified model.

The source resistance $r_s$ can have a significant effect on the transistor characteristics. Figure 12–54 shows a simplified, low-frequency equivalent circuit including $r_s$ but neglecting $r_{ds}$. The drain current is given by

$$I_d = g_m V'_{gs} \tag{12–73}$$

and the relation between $V_{gs}$ and $V'_{gs}$ can be found from

$$V_{gs} = V'_{gs} + (g_m V'_{gs})r_s = (1 + g_m r_s)V'_{gs} \tag{12–74}$$

The drain current from Equation (12–73) can now be written as

$$I_d = \left(\frac{g_m}{1 + g_m r_s}\right) V_{gs} = g'_m V_{gs} \tag{12–75}$$

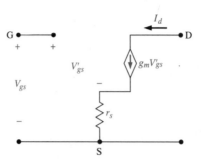

**Figure 12–54** Simplified, low-frequency small-signal equivalent circuit of common-source n-channel MOSFET including source resistance $r_s$.

The source resistance reduces the effective transconductance or transistor gain.

The equivalent circuit of the p-channel MOSFET is exactly the same as that of the n-channel except that all voltage polarities and current directions are reversed. The same capacitances and resistances that are in the n-channel model apply to the p-channel model.

### 12.4.2 Frequency Limitation Factors and Cutoff Frequency

There are two basic frequency limitation factors in the MOSFET. The first factor is the channel transit time. If we assume that carriers are traveling at their saturation drift velocity $v_{sat}$, then the transit time is $\tau_t = L/v_{sat}$ where $L$ is the channel length. If $v_{sat} = 10^7$ cm/sec and $L = 1$ $\mu$m, then $\tau_t = 10$ psec, which translates into a maximum frequency of 100 GHz. This frequency is much larger than the typical maximum frequency response of a MOSFET. The transit time of carriers through the channel is usually not the limiting factor in the frequency response of MOSFETs.

The second limiting factor is the gate or capacitance charging time. If we neglect $r_s$, $r_d$, $r_{ds}$, and $C_{ds}$, the resulting equivalent small-signal circuit is shown in Figure 12–55 where $R_L$ is a load resistance.

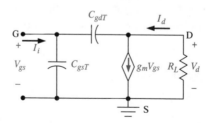

**Figure 12–55** High-frequency small-signal equivalent circuit of common-source n-channel MOSFET.

The input gate impedance in this equivalent circuit is no longer infinite. Summing currents at the input gate node, we have

$$I_i = j\omega C_{gsT} V_{gs} + j\omega C_{gdT}(V_{gs} - V_d) \tag{12-76}$$

where $I_i$ is the input current. Likewise, summing currents at the output drain node, we have

$$\frac{V_d}{R_L} + g_m V_{gs} + j\omega C_{gdT}(V_d - V_{gs}) = 0 \tag{12-77}$$

Combining Equations (12–76) and (12–77) to eliminate the voltage variable $V_d$, we can determine the input current as

$$I_i = j\omega \left\{ C_{gsT} + C_{gdT} \left[ \frac{1 + g_m R_L}{1 + j\omega R_L C_{gdT}} \right] \right\} V_{gs} \tag{12-78}$$

Normally, $\omega R_L C_{gdT}$ is much less than unity; therefore we may neglect the $(j\omega R_L C_{gdT})$ term in the denominator. Equation (12–78) then simplifies to

$$I_i = j\omega [C_{gsT} + C_{gdT}(1 + g_m R_L)] V_{gs} \tag{12-79}$$

Figure 12–56 shows the equivalent circuit with the equivalent input impedance described by Equation (12–79). The parameter $C_M$ is the Miller capacitance and is given by

$$C_M = C_{gdT}(1 + g_m R_L) \tag{12-80}$$

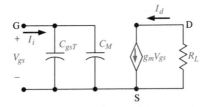

**Figure 12–56**  Small-signal equivalent circuit including Miller capacitance.

The serious effect of the drain overlap capacitance now becomes apparent. When the transistor is operating in the saturation region, $C_{gd}$ essentially becomes zero, but $C_{gdp}$ is a constant. This parasitic capacitance is multiplied by the gain of the transistor and can become a significant factor in the input impedance.

The cutoff frequency $f_T$ is defined to be the frequency at which the magnitude of the current gain of the device is unity, or when the magnitude of the input current $I_i$ is equal to the ideal load current $I_d$. From Figure 12–56, we can see that

$$I_i = j\omega (C_{gsT} + C_M) V_{gs} \tag{12-81}$$

and the ideal load current is

$$I_d = g_m V_{gs} \tag{12-82}$$

The magnitude of the current gain is then

$$\left| \frac{I_d}{I_i} \right| = \frac{g_m}{2\pi f(C_{gsT} + C_M)} \tag{12-83}$$

Setting the magnitude of the current gain equal to unity at the cutoff frequency, we find

$$f_T = \frac{g_m}{2\pi(C_{gsT} + C_M)} = \frac{g_m}{2\pi C_G} \tag{12-84}$$

where $C_G$ is the equivalent input gate capacitance.

In the ideal MOSFET, the overlap or parasitic capacitances, $C_{gsp}$ and $C_{gdp}$, are zero. Also, when the transistor is biased in the saturation region, $C_{gd}$ approaches zero and $C_{gs}$ is approximately $C_{ox}WL$. The transconductance of the ideal MOSFET biased in the saturation region and assuming a constant mobility was given by Equation (12–68) as

$$g_{ms} = \frac{W\mu_n C_{ox}}{L}(V_{GS} - V_T)$$

Then, for this ideal case, the cutoff frequency is

$$f_T = \frac{g_m}{2\pi C_G} = \frac{\dfrac{W\mu_n C_{ox}}{L}(V_{GS} - V_T)}{2\pi(C_{ox}WL)} = \frac{\mu_n(V_{GS} - V_T)}{2\pi L^2} \tag{12-85}$$

---

**Example 12–11**

**Objective:** To calculate the cutoff frequency of an ideal MOSFET with a constant mobility.

Assume that the electron mobility in an n-channel device is $\mu_n = 400$ cm²/V-sec and that the channel length is $L = 4\ \mu$m. Also assume that $V_T = 1$ volt and let $V_{GS} = 3$ volts.

**Solution:** From Equation (12–85), the cutoff frequency is

$$f_T = \frac{\mu_n(V_{GS} - V_T)}{2\pi L^2} = \frac{400(3 - 1)}{2\pi(4 \times 10^{-4})^2} = 796 \text{ MHz}$$

**Comment:** In an actual MOSFET, the effect of the parasitic capacitance will substantially reduce the cutoff frequency from that calculated in this example.

# *12.5 MOS TECHNOLOGY

The primary objective of this text is to present the basic physics of semiconductor materials and devices without considering in detail the various fabrication processes: this important subject is left to other texts. However, there are some MOS technologies that are used extensively, for which the basic fabrication techniques must be considered in order to understand essential characteristics of these devices and circuits. The two MOS technologies we will consider briefly are the complementary MOS, or CMOS process, and power MOSFETs.

## 12.5.1 The CMOS Technology

We have considered the physics of both n-channel and p-channel enhancement mode MOSFETs. Both devices are used in a CMOS inverter, which is the basis of CMOS digital logic circuits. The dc power dissipation in a digital circuit can be reduced to very low levels by using a complementary p-channel and n-channel pair.

It is necessary to form electrically isolated p- and n-substrate regions in an integrated circuit to accommodate the n- and p-channel transistors. The p-well process has been a commonly used technique for CMOS circuits. The process starts with a fairly low doped n-type silicon substrate in which the p-channel MOSFET will be fabricated. A diffused p-region, called a p-well, is formed in which the n-channel MOSFET will be fabricated. In most cases, the p-type substrate doping level must be larger than the n-type substrate doping level to obtain the desired threshold voltages. The larger p-doping can easily compensate the initial n-doping to form the p-well. A simplified cross section of the p-well CMOS structure is shown in Figure 12–57a. The notation FOX stands for field oxide, which is a relatively thick oxide separating the devices. The field oxide prevents either the n- or p-substrates from becoming inverted and helps maintain isolation between the two devices. In practice, additional processing steps must be included; for example, providing connections so that the p-well and n-substrate can be electrically connected to the appropriate voltages. The n-substrate must always be at a higher potential than the p-well; therefore, this pn junction will always be reverse biased.

With ion implantation now being extensively used for threshold voltage control, both the n-well CMOS process and twin-well CMOS process can be used. The n-well CMOS process, shown in Figure 12–57b, starts with an optimized p-type substrate that is used to form the n-channel MOSFETs. (The n-channel MOSFETs, in general, have superior characteristics so this starting point should yield excellent n-channel devices.) The n-well is then added in which the p-channel devices are fabricated. The n-well doping can be controlled by ion implantation.

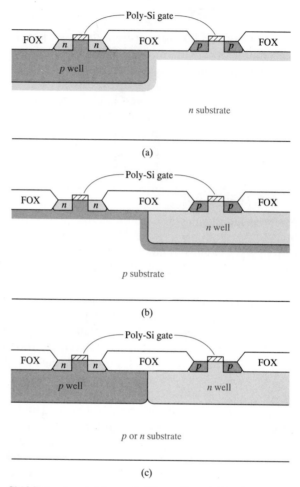

**Figure 12–57**   CMOS structures: (a) p-well, (b) n-well, and (c) twin well. (From Yang [14].)

The twin-well CMOS process, shown in Figure 12–57c, allows both the p-well and n-well regions to be optimally doped to control the threshold voltage and transconductance of each transistor. The twin-well process allows a higher packing density because of self-aligned channel stops.

One major problem in CMOS circuits has been latch-up. Latch-up refers to a high-current, low-voltage condition that may occur in a four-layer pnpn structure. Figure 12–58a shows the circuit of a CMOS inverter and Figure 12–58b shows a simplified integrated circuit layout of the inverter circuit. In the CMOS layout, the $p^+$-source to $n$-substrate to $p$-well to $n^+$-source forms such a four-layer structure. This four-layer, or SCR, device is discussed in more detail in Chapter 15, but we will consider the basic effect here.

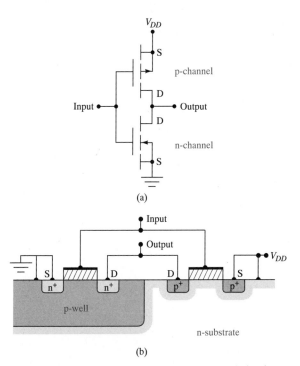

$V_{DD}$

S

p-channel

D

Input •

• Output

D

n-channel

S

(a)

• Input

• Output

$V_{DD}$

S | D | D | S

$n^+$ | $n^+$ | $p^+$ | $p^+$

p-well

n-substrate

(b)

Figure 12–58  (a) CMOS inverter circuit. (b) Simplified integrated circuit cross section of CMOS inverter.

The equivalent circuit of this four-layer structure is shown in Figure 12–59. The SCR action involves the interaction of the parasitic pnp and npn transistors. The npn transistor corresponds to the vertical $n^+$-source to $p$-well to $n$-substrate structure and the pnp transistor corresponds to the lateral $p$-well to $n$-substrate to $p^+$-source structure. Under normal CMOS operation, both parasitic bipolar transistors are cut off. However, under certain conditions, avalanche breakdown may occur in the $p$-well to $n$-substrate junction, driving both bipolar transistors into saturation. This high-current, low-voltage condition—latch-up—can sustain itself by positive feedback. The condition can prevent the CMOS circuit from operating and can also cause permanent damage and burn-out of the circuit.

Latch-up can be prevented if the product $\beta_n\beta_p$ is less than unity at all times, where $\beta_n$ and $\beta_p$ are the common emitter current gains of the npn and pnp parasitic bipolar transistors, respectively. One method of preventing latch-up is to "kill" the minority carrier lifetime. Minority carrier lifetime degradation can be accomplished by gold doping or neutron irradiation, which introduce deep traps within the semiconductor. The deep traps increase the excess minority carrier recombination rate and reduce current

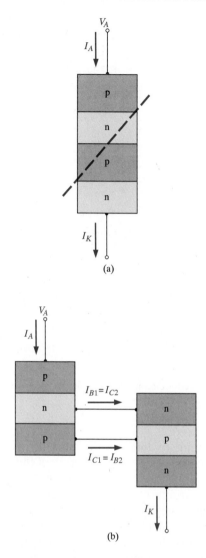

**Figure 12–59**   (a) The splitting of the basic pnpn structure. (b) The two-transistor equivalent circuit of the four-layered pnpn device.

gain. A second method of preventing latch-up is by using proper circuit layout techniques. If the two bipolar transistors can be effectively decoupled, then latch-up can be minimized or prevented. The two parasitic bipolar transistors can also be decoupled by using a different fabrication technology. The silicon-on-insulator technology, for example, allows the n-channel and the p-channel MOSFETs to be isolated from each other by an insulator. This isolation decouples the parasitic bipolar transistors.

### 12.5.2 Power MOSFETs

The basic operation of the power MOSFET is the same as that of any MOSFET. However, the current handling capability of these devices is usually in the ampere range and the drain-to-source blocking voltage may be in the range of 50 to 100 volts or even higher. One big advantage that a power MOSFET has over a bipolar power device is that the control signal is applied to the gate whose input impedance is extremely large. Even during switching between "on" and "off" states, the gate current is small, so that relatively large currents can be switched with very small control currents.

Large currents can be obtained in a MOSFET with a very large channel width. To achieve a large channel width device with good characteristics, power MOSFETs are fabricated with a repetitive pattern of small cells operating in parallel. An important parameter of a power MOSFET is the "on-resistance," which can be written as

$$R_{on} = R_S + R_{CH} + R_D \qquad (12-86)$$

where $R_S$ is the resistance associated with the source contact, $R_{CH}$ is the channel resistance, and $R_D$ is the resistance associated with the drain contact. The $R_S$ and $R_D$ resistance values are not necessarily negligible in power MOSFETs since small resistances and high currents can produce considerable power dissipation.

In the linear region of operation, we may write the channel resistance as

$$R_{CH} = \frac{L}{Z\mu_n C_{ox}(V_{GS} - V_T)} \qquad (12-87)$$

We have noted in previous chapters that mobility decreases with increasing temperature. The threshold voltage varies only slightly with temperature so that, as current in a device increases and produces additional power dissipation, the temperature of the device increases, the carrier mobility decreases, and $R_{CH}$ increases, which inherently limits the channel current. The resistances $R_S$ and $R_D$ are proportional to semiconductor resistivity and so are also inversely proportional to mobility and have the same temperature characteristics as $R_{CH}$.

The increase in resistance with temperature provides stability for the power MOSFET. If the current in any particular cell begins to increase, the resulting temperature rise will increase the on-resistance, thus limiting the current. With this particular characteristic, the total current in a power MOSFET tends to be evenly distributed among the parallel cells, not concentrated in any single cell, a condition that can cause burn-out.

There are two basic power MOSFET structures. The first is called a DMOS device and is shown in Figure 12–60. The DMOS device uses a double diffusion process: the p-base or the p-substrate region and the $n^+$ source contact are diffused through a common window defined by the edge of the gate. The p-base region is diffused deeper than the $n^+$ source and the

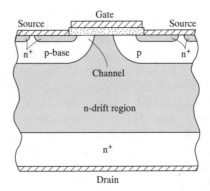

**Figure 12–60**   Cross section of the basic DMOS structure. (From Baliga [1].)

difference in the lateral diffusion distance between the p-base and the $n^+$ source defines the surface channel length.

Electrons enter the source terminal and flow laterally through the inversion layer under the gate to the n-drift region. The electrons then flow vertically through the n-drift region to the drain terminal. The conventional current direction is from the drain to the source. The n-drift region must be moderately doped so that the drain breakdown voltage is sufficiently large. However, the thickness of the n-drift region should also be as thin as possible to minimize drain resistance.

A parasitic $n^+pnn^+$ vertical bipolar transistor structure exists in this device which can enhance the breakdown process. The breakdown effect will be considered in the next chapter. It is important to electrically short the p-base to the $n^+$ source as shown by the source metalization so that the bipolar transistor is maintained inactive or cut off during the operation of the MOSFET.

The second power MOSFET structure, shown in Figure 12–61, is a VMOS structure. The vertical channel or VMOS power device is a nonplanar structure that requires a different type of fabrication process. In this

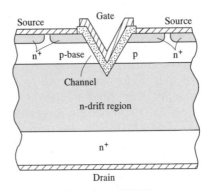

**Figure 12–61**   Cross section of the basic VMOS structure. (From Baliga [1].)

case, a p-base diffusion is performed over the entire surface followed by the $n^+$ source diffusions. A V-shaped groove is then formed, extending through to the n-drift region. It has been found that certain chemical solutions etch the (111) planes in silicon at a much slower rate than the other planes. If (100) oriented silicon is etched through a window at the surface, these chemical etches will create the V-shaped groove. A gate oxide is then grown in the V-groove and the metal gate material is deposited. An electron inversion layer is formed in the base so that current is again essentially a vertical current between the source and drain. Note again that the source and base are short-circuited by the source contact. This connection tends to keep the parasitic bipolar transistor in cutoff. The relatively low doped n-drift region supports the drain voltage since the depletion region extends mainly into this low doped region.

We mentioned that many individual MOSFET cells are connected in parallel to fabricate a power MOSFET with the proper width-to-length ratio. Figure 12–62 shows a HEXFET structure. Each cell is a DMOS device with

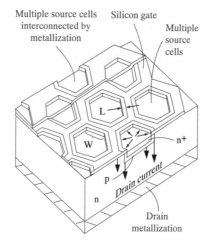

**Figure 12–62**   The HEXFET structure. (From Sze [10].)

an $n^+$ polysilicon gate. The HEXFET has a very high packing density—it may be on the order of $10^5$ cells per cm². In the VMOS structure, the anisotropic etching of the grooves must be along the ⟨110⟩ direction on the (100) surface. This constraint limits the design options available for this type of device.

## 12.6  SUMMARY AND REVIEW

The fundamental physics and characteristics of the metal-oxide-semiconductor field-effect transistor have been considered in this chapter. The two basic types of MOSFETs are the n-channel, in which current is due to the flow of electrons, and the p-channel, in which current is due to the flow of holes. Each of these can be either enhancement mode, in which the device is normally "off" and is turned on by applying a gate voltage, or depletion mode, in which the device is normally "on" and is turned off by applying a voltage to the control gate terminal. Each type of MOSFET, in conjunction with a circuit, is capable of voltage gain and signal-power gain.

The heart of the MOSFET is the metal-oxide-semiconductor structure, or MOS capacitor. The energy bands in the semiconductor adjacent to the oxide–semiconductor interface bend depending upon the voltage applied across the MOS capacitor so that the position of the conduction and valence bands relative to the Fermi level is a function of the MOS capacitor voltage. The change in energy bands means that the semiconductor surface can be inverted from p-type to n-type by applying a positive gate voltage, or from n-type to p-type by applying a negative gate voltage. An important characteristic of the MOS structure is the creation of the inversion layer charge in the semiconductor adjacent to the oxide. The basic MOS field-effect transistor action is the modulation of the inversion charge density, or channel conductance, by the gate voltage.

Two important parameters of the MOSFET are the flat-band voltage and threshold voltage. The flat-band voltage is the gate voltage that must be applied to achieve the flat-band condition, in which the conduction and valence bands in the semiconductor do not bend and there is no space charge region in the semiconductor. The flat-band voltage is a function of the metal-oxide barrier height, the semiconductor-oxide barrier height, and the amount of fixed trapped oxide charge. The threshold voltage is the applied gate voltage required to reach the inversion transition point, which is the condition at which the inversion charge density is equal in magnitude to the semiconductor doping concentration. The threshold voltage is a function of the flat-band voltage, the semiconductor doping concentration, and oxide thickness.

A great deal of information about the MOS system can be obtained from the capacitance-voltage characteristics of the MOS capacitor. The amount of equivalent fixed oxide charge and the density of interface states, for example, can be determined from the C-V measurements. These measurements are used extensively as a diagnostic tool to determine the quality of the gate oxide.

The current in a MOSFET is due to the flow of carriers in the inversion layer between the source and drain terminals. The inversion layer charge density and channel conductance are controlled by the gate voltage, which means that the channel current is also controlled by the gate voltage. The ideal current-voltage relationship was derived by starting with Ohm's law applied to the channel conductance. The drain current is a function of both the gate-to-source and drain-to-source voltages in the nonsaturation region, but is independent of the drain-to-source voltage in the saturation region for the ideal MOSFET. The transconductance, or transistor gain, is the rate of change of drain current with respect to the corresponding change in gate voltage. This parameter is a function of the electrical and geometrical properties of the device. The MOSFET is actually a four-terminal device with the substrate, or body, being the fourth connection. The threshold voltage is a function of the voltage across the source-to-substrate pn junction, which must be zero- or reverse biased. As the magnitude of the reverse-bias source-to-substrate voltage increases, the magnitude of the threshold voltage increases. This body effect applies to both n- and p-channel MOSFETs. The substrate bias effect is significant for MOSFETs in integrated circuits in which the source and substrate are not electrically tied together. The drain current becomes a function of the source-to-substrate voltage as well as being a function of the gate voltage.

A small-signal equivalent circuit of the MOSFET was developed. The equivalent circuit includes capacitances; these introduce frequency effects into the transistor model. The various physical factors in the MOSFET that affect the frequency limitations were considered. The principal limiting factor is the gate capacitance charging time. To increase the frequency capability, device size and parasitic capacitances must be minimized. In particular, the drain overlap capacitance may be a limiting factor in the frequency response of the MOSFET because of the Miller effect. The transistor cutoff frequency, a figure of merit for the frequency response of the device, is inversely proportional to channel length; thus a reduction in channel length results in an increased frequency capability of the MOSFET.

We considered two particular MOS technologies in this chapter: the CMOS process and high-power MOSFETs. Both n- and p-channel devices are fabricated on the same semiconductor chip in CMOS digital electronic circuits, so it is necessary to have electrically isolated p- and n-substrate regions to accommodate the two types of transistors. Various processes are used to fabricate this structure. One potentially serious problem encountered in the CMOS structure is latch-up—the high-current, low-voltage condition that may occur in a four-layer pnpn structure. One possible method of preventing latch-up is to use the silicon-on-insulator (SOI) technology, which involves forming a thin layer of silicon on an insulator material. Power MOSFETs operate in the same manner as the basic MOSFETs considered previously, but special care must be taken in the design so that the required stand-off voltage and high current density can be tolerated. The DMOS and the VMOS designs were described.

We have considered the basic physics and characteristics of the MOSFET in this chapter. In the next chapter we continue the discussion of the MOSFET and explore nonideal effects, small geometry effects, and radiation and hot electron effects. The tendency in device design is to make the device smaller and smaller; therefore, the topics considered in the next chapter are real problems that must be addressed in a VLSI circuit design.

## GLOSSARY OF IMPORTANT TERMS

**Accumulation layer charge:** The induced charge directly under an oxide that is in excess of the thermal-equilibrium majority carrier concentration.

**Bulk charge effect:** The deviation in drain current from the ideal due to the space charge width variation along the channel length caused by a drain-to-source voltage.

**Channel conductance:** The ratio of drain current to drain-to-source voltage in the limit as $V_{DS} \to 0$.

**Channel conductance modulation:** The process whereby the channel conductance varies with gate-to-source voltage.

**CMOS:** Complementary MOS; the technology which uses both p- and n-channel devices in an electronic circuit fabricated in a single semiconductor chip.

**Cutoff frequency:** The signal frequency at which the input ac gate current is equal to the output ac drain current.

**Depletion mode MOSFET:** The type of MOSFET in which a gate voltage must be applied to turn the device off.

**Enhancement mode MOSFET:** The type of MOSFET in which a gate voltage must be applied to turn the device on.

**Equivalent fixed oxide charge:** The effective fixed charge in the oxide, $Q'_{ss}$, directly adjacent to the oxide–semiconductor interface.

**Flat-band voltage:** The gate voltage that must be applied to create the flat-band condition in which there is no space charge region in the semiconductor under the oxide.

**Gate capacitance charging time:** The time during which the input gate capacitance is being charged or discharged due to a step change in the gate signal.

**Interface states:** The allowed electronic energy states within the bandgap energy at the oxide–semiconductor interface.

**Inversion layer charge:** The induced charge directly under the oxide, which is the opposite type compared with the semiconductor doping.

**Inversion layer mobility:** The mobility of carriers in the inversion layer.

**Latch-up:** The high-current, low-voltage condition that may occur in a four-layer pnpn structure such as in CMOS.

**Maximum induced space charge width:** The width of the induced space charge region under the oxide at the threshold inversion condition.

**Metal–semiconductor work function difference:** The parameter $\phi_{ms}$, a function of the difference between the metal work function and semiconductor electron affinity.

**Moderate inversion:** The condition in which the induced space charge width is changing slightly when the gate voltage is at or near the threshold voltage and the inversion charge density is of the same order of magnitude as the semiconductor doping concentration.

**Oxide capacitance:** The ratio of oxide permittivity to oxide thickness which is the capacitance per unit area, $C_{ox}$.

**Saturation:** The condition in which the inversion charge density is zero at the drain and the drain current is no longer a function of the drain-to-source voltage.

**Strong inversion:** The condition in which the inversion charge density is larger than the magnitude of the semiconductor doping concentration.

**Threshold inversion point:** The condition in which the inversion charge density is equal in magnitude to the semiconductor doping concentration.

**Threshold voltage:** The gate voltage that must be applied to achieve the threshold inversion point.

**Transconductance:** The ratio of an incremental change in drain current to the corresponding incremental change in gate voltage.

**Weak inversion:** The condition in which the inversion charge density is less than the magnitude of the semiconductor doping concentration.


## PROBLEMS

(Note: In the following problems, assume the semiconductor and oxide in the MOS system are silicon and silicon dioxide, respectively, and assume the temperature is $T = 300°K$ unless otherwise stated.)


### Section 12.1

1. (a) Calculate the maximum space charge width $x_{dT}$ and the maximum space charge density $|Q'_{SD}(max)|$ in p-type silicon, gallium arsenide, and germanium semiconductors of a MOS structure. Let $T = 300°K$ and assume $N_a = 10^{16}$ cm$^{-3}$. (b) Repeat part (a) if $T = 200°K$.

2. Consider n-type silicon in a MOS structure. Let $T = 300°K$. Determine the semiconductor doping so that $|Q'_{SD}(max)| = 7.5 \times 10^{-9}$ coul/cm$^2$.

3. Determine the metal–semiconductor work function difference $\phi_{ms}$ in a MOS structure with p-type silicon for the case when the gate is (a) aluminum, (b) $n^+$ polysilicon, and (c) $p^+$ polysilicon. Let $N_a = 3 \times 10^{15}$ cm$^{-3}$.

4. Consider a MOS structure with n-type silicon. A metal–semiconductor work function difference of $\phi_{ms} = -0.35$ volt is required. Determine the silicon doping required to meet this specification when the gate is (a) $n^+$ polysilicon, (b) $p^+$ polysilicon, and (c) aluminum. If a particular gate cannot meet this requirement, explain why.

5.  Consider an $n^+$ polysilicon-silicon dioxide-n-type silicon MOS capacitor. Let $N_d = 10^{15}$ cm$^{-3}$. Calculate the flat-band voltage for (a) $t_{ox} = 500$Å when $Q'_{ss}$ is (i) $10^{10}$ cm$^{-2}$, (ii) $10^{11}$ cm$^{-2}$, and (iii) $5 \times 10^{11}$ cm$^{-2}$. (b) Repeat part (a) when $t_{ox} = 1000$Å.

6.  Consider an aluminum gate-silicon dioxide-p-type silicon MOS structure with $t_{ox} = 450$Å. The silicon doping is $N_a = 2 \times 10^{16}$ cm$^{-3}$ and the flat-band voltage is $V_{FB} = -1.0$ volt. Determine the fixed oxide charge $Q'_{ss}$.

7.  A MOS transistor is fabricated on a p-type silicon substrate with $N_a = 2 \times 10^{15}$ cm$^{-3}$. The oxide thickness is $t_{ox} = 650$Å and the equivalent fixed oxide charge is $Q'_{ss} = 2 \times 10^{11}$ cm$^{-2}$. Calculate the threshold voltage for (a) an aluminum gate, (b) $n^+$ polysilicon gate, and (c) $p^+$ polysilicon gate.

8.  Repeat problem 7 for an n-type silicon substrate with $N_d = 10^{15}$ cm$^{-3}$.

*9. A MOS transistor with an aluminum gate is fabricated on a p-type silicon substrate. The oxide thickness is $t_{ox} = 750$Å and the equivalent fixed oxide charge is $Q'_{ss} = 10^{11}$ cm$^{-2}$. The measured threshold voltage is $V_T = +0.80$ volt. Determine the p-type doping.

*10. Repeat problem 9 for an n-type silicon substrate if the measured threshold voltage is $V_T = -1.50$ volts. Determine the n-type doping.

11. Consider the MOS transistor in problem 9. If the measured threshold voltage is $V_T = +0.80$ volt and the p-type doping is $N_a = 8 \times 10^{16}$ cm$^{-3}$, determine the equivalent fixed oxide charge $Q'_{ss}$.

12. An n-channel depletion mode MOSFET with an $n^+$ polysilicon gate is shown in Figure 12–40. The n-channel doping is $N_d = 10^{15}$ cm$^{-3}$ and the oxide thickness is $t_{ox} = 500$Å. The equivalent fixed oxide charge is $Q'_{ss} = 10^{10}$ cm$^{-2}$. The n-channel thickness $t_c$ is equal to the maximum induced space charge width. (Disregard the space charge region at the n-channel–p-substrate junction.) (a) Determine the channel thickness $t_c$, and (b) calculate the threshold voltage.

13. Consider a MOS capacitor with an $n^+$ polysilicon gate and a p-type silicon substrate. Assume $N_a = 10^{16}$ cm$^{-3}$ and let $E_F - E_c = 0.2$ eV in the $n^+$ polysilicon. Assume the oxide has a thickness of $t_{ox} = 700$Å. Also assume that $\chi'$ (polysilicon) $= \chi'$ (single-crystal silicon). (a) Sketch the energy-band diagrams (i) for $V_G = 0$ and (ii) at flat-band. (b) Calculate the metal–semiconductor work function difference. (c) Calculate the threshold voltage for the ideal case of zero fixed oxide charge and zero interface states.

**Section 12.2**

14. An ideal MOS capacitor with an aluminum gate has a silicon dioxide thickness of $t_{ox} = 1000$Å on a p-type silicon substrate doped

with an acceptor concentration of $N_a = 10^{16}$ cm$^{-3}$. Determine the capacitances $C_{ox}$, $C'_{FB}$, $C'_{min}$, and $C'(inv)$ at (a) $f = 1$ Hz and (b) $f = 1$ MHz. (c) Determine $V_{FB}$ and $V_T$. Sketch $C'/C_{ox}$ versus $V_G$ for parts (a) and (b).

15. Repeat problem 14 for an n-type silicon substrate doped with a donor concentration of $N_d = 5 \times 10^{14}$ cm$^{-3}$.

*16. Using superposition, show that the shift in the flat-band voltage due to a fixed charge distribution $\rho(x)$ in the oxide is given by

$$\Delta V_{FB} = -\frac{1}{C_{ox}} \int_0^{t_{ox}} \frac{x\rho(x)}{t_{ox}} \, dx$$

17. Using the results of problem 16, calculate the shift in the flat-band voltage for the following oxide charge distributions: (a) $Q'_{ss} = 5 \times 10^{11}$ cm$^{-2}$ is entirely located at the oxide–semiconductor interface. Let $t_{ox} = 750$Å. (b) $Q'_{ss} = 5 \times 10^{11}$ cm$^{-2}$ is uniformly distributed throughout the oxide which has a thickness of $t_{ox} = 750$Å. (c) $Q'_{ss} = 5 \times 10^{11}$ cm$^{-2}$ forms a triangular distribution with the peak at $x = t_{ox} = 750$Å (the oxide–semiconductor interface) and which goes to zero at $x = 0$ (the metal–oxide interface).

18. Consider a MOS capacitor with a p-type substrate. Assume that donor type interface traps exist only at midgap, i.e., at $E_{Fi}$. Sketch the high-frequency C-V curve from accumulation to inversion. Compare this sketch to the ideal C-V plot.

## Section 12.3

19. An expression that includes the inversion charge density was given by Equation (12–55). Consider the definition of threshold voltage and show that the inversion charge density goes to zero at the drain terminal at saturation. (Hint: Let $V_x = V_{DS} = V_{DS}(sat)$.)

20. An ideal n-channel MOSFET has the following parameters:

   $W = 30$ $\mu$m        $\mu_n = 450$ cm$^2$/V-sec
   $L = 2$ $\mu$m         $t_{ox} = 350$Å
   $V_T = +0.80$ volt

   (a) Plot $I_D$ versus $V_{DS}$ for $0 \leq V_{DS} \leq 5$ volts and for $V_{GS} = 0, 1, 2, 3, 4,$ and 5 volts. Indicate on each curve the $V_{DS}(sat)$ point. (b) Plot $\sqrt{I_D(sat)}$ versus $V_{GS}$ for $0 \leq V_{GS} \leq 5$ volts. (c) Plot $I_D$ versus $V_{GS}$ for $V_{DS} = 0.1$ V and for $0 \leq V_{GS} \leq 5$ volts.

21. An ideal p-channel MOSFET has the following parameters:

   $W = 15$ $\mu$m        $\mu_p = 300$ cm$^2$/V-sec
   $L = 1.5$ $\mu$m       $t_{ox} = 350$Å
   $V_T = -0.80$ volt

(a) Plot $I_D$ versus $V_{SD}$ for $0 \leq V_{SD} \leq 5$ volts and for $V_{SG} = 0, 1, 2,$ 3, 4, and 5 volts. Indicate on each curve the $V_{SD}(\text{sat})$ point. (b) Plot $I_D$ versus $V_{SG}$ for $V_{SD} = 0.1$V and for $0 \leq V_{SG} \leq 5$ volts.

22.   The channel conductance for a p-channel MOSFET is defined as

$$g_d = \left.\frac{\partial I_D}{\partial V_{SD}}\right|_{V_{SD} \to 0}$$

Plot the channel conductance for the p-channel MOSFET in problem 21 for $0 \leq V_{SG} \leq 5$ volts.

23.   The experimental characteristics of an ideal n-channel MOSFET biased in the saturation region are shown in Figure 12–63. If $W/L = 10$ and $t_{ox} = 425$Å, determine $V_T$ and $\mu_n$.

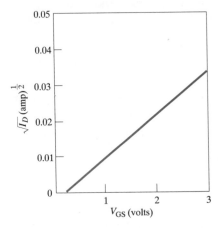

**Figure 12–63**    Figure for problem 23.

24.   (a) An ideal n-channel MOSFET has an inversion carrier mobility $\mu_n = 525$ cm$^2$/V-sec, a threshold voltage $V_T = +0.75$ volt, and an oxide thickness $t_{ox} = 400$Å. When biased in the saturation region, the required rated current is $I_D(\text{sat}) = 6$ mA when $V_{GS} = 5$ volts. Determine the required $W/L$ ratio. (b) A p-channel MOSFET has the same requirements when $V_{SG} = 5$ volts and has the same parameters as part (a) except $\mu_p = 300$ cm$^2$/V-sec and $V_T = -0.75$ volt. Determine the $W/L$ ratio.

25.   Consider the transistor described in problem 20. (a) Calculate $g_{mL}$ for $V_{DS} = 0.5$ volt. (b) Calculate $g_{ms}$ for $V_{GS} = 4$ volts.

26.   Consider the transistor described in problem 21. (a) Calculate $g_{mL}$ for $V_{SD} = 0.5$ volt. (b) Calculate $g_{ms}$ for $V_{SG} = 4$ volts.

27. An n-channel MOSFET has the following parameters:

$$t_{ox} = 400\text{Å} \qquad N_a = 5 \times 10^{16} \text{ cm}^{-3}$$
$$V_{FB} = -0.5 \text{ volt} \qquad L = 2 \ \mu\text{m}$$
$$W = 10 \ \mu\text{m} \qquad \mu_n = 450 \text{ cm}^2/\text{V-sec}$$

Plot $\sqrt{I_D}$ versus $V_{GS}$ over the range $0 \le I_D \le 1$ mA when the transistor is biased in the saturation region for the following values of source-to-body voltage: $V_{SB} = 0, 1, 2,$ and 4 volts.

28. Consider a p-channel MOSFET with $t_{ox} = 600\text{Å}$ and $N_d = 5 \times 10^{15}$ $\text{cm}^{-3}$. Determine the body-to-source voltage, $V_{BS}$, such that the shift in threshold voltage, $\Delta V_T$, from the $V_{BS} = 0$ curve is $\Delta V_T = -1.5$ volts.

## Section 12.4

29. Consider an ideal n-channel MOSFET with a width-to-length ratio of $(W/L) = 10$, an electron mobility of $\mu_n = 400 \text{ cm}^2/\text{V-sec}$, an oxide thickness of $t_{ox} = 475\text{Å}$, and a threshold voltage of $V_T = +0.65$ volt. (a) Determine the maximum value of source resistance so that the saturation transconductance $g_{ms}$ is reduced by no more than 20 percent from its ideal value when $V_{GS} = 5$ volts. (b) Using the value of $r_s$ calculated in part (a), how much is $g_{ms}$ reduced from its ideal value when $V_{GS} = 3$ volts?

30. An n-channel MOSFET has the following parameters:

$$\mu_n = 400 \text{ cm}^2/\text{V-sec} \qquad t_{ox} = 500\text{Å}$$
$$L = 2 \ \mu\text{m} \qquad W = 20 \ \mu\text{m}$$
$$V_T = +0.75 \text{ volt}$$

Assume the transistor is biased in the saturation region at $V_{GS} = 4$ volts. (a) Calculate the ideal cutoff frequency. (b) Assume that the gate oxide overlaps both the source and drain contacts by 0.75 $\mu$m. If a load resistance of $R_L = 10 \ k\Omega$ is connected to the output, calculate the cutoff frequency.

31. Repeat problem 30 for the case when the electrons are traveling at a saturation velocity of $v_{sat} = 4 \times 10^6$ cm/sec.

## Summary and Review

*32. Design an ideal silicon n-channel MOSFET with a polysilicon gate to have a threshold voltage of $V_T = 0.65$V. Assume an oxide thickness of $t_{ox} = 300$ Å, a channel length of $L = 1.25 \ \mu$m, and a nominal value of $Q'_{ss} = 1.5 \times 10^{11}$ cm$^{-2}$. It is desired to have a drain current of $I_D = 50 \ \mu$A at $V_{GS} = 2.5$V and $V_{DS} = 0.1$V. Determine the substrate doping concentration, channel width, and type of gate required.

*33. Design an ideal silicon n-channel depletion mode MOSFET with a polysilicon gate to have a threshold voltage of $V_T = -0.65$V. Assume an oxide thickness of $t_{ox} = 300$ Å, a channel length of $L = 1.25$ $\mu$m, and a nominal value of $Q'_{ss} = 1.5 \times 10^{11}$ cm$^{-2}$. It is desired to have a drain current of $I_D(\text{sat}) = 50$ $\mu$A at $V_{GS} = 0$. Determine the type of gate, substrate doping concentration, and channel width required.

*34. Consider the CMOS inverter circuit shown in Figure 12–58a. Ideal n- and p-channel devices are to be designed with channel lengths of $L = 2.5$ $\mu$m and oxide thicknesses of $t_{ox} = 450$ Å. Assume the inversion channel mobilities are one half the bulk values. The threshold voltages of the n- and p-channel transistors are to be $+0.5$V and $-0.5$V, respectively. The drain current is to be $I_D = 0.256$ $mA$ when the input voltage to the inverter is 1.5V and 3.5V with $V_{DD} = 5$V. The gate material is to be the same in each device. Determine the type of gate, substrate doping concentrations, and channel widths.

## READING LIST

1. Baliga, B. J. *Modern Power Devices*. New York: Wiley, 1987.

2. Muller, R. S., and T. I. Kamins. *Device Electronics for Integrated Circuits*. 2nd ed. New York: Wiley, 1986.

3. Nicollian, E. H., and J. R. Brews. *MOS Physics and Technology*. New York: Wiley, 1982.

4. Ong, D. G. *Modern MOS Technology: Processes, Devices, and Design*. New York: McGraw-Hill, 1984.

5. Pierret, R. F. *Field Effect Devices*. Vol. 4 of the *Modular Series on Solid State Devices*. 2nd ed. Reading, Mass.: Addison-Wesley, 1990.

6. Pulfrey, D. L., and N. G. Tarr. *Introduction to Microelectronic Devices*. Englewood Cliffs, N.J.: Prentice Hall, 1989.

7. Schroder, D. K. *Advanced MOS Devices, Modular Series on Solid State Devices*. Reading, Mass.: Addison-Wesley, 1987.

*8. Shur, M. *Physics of Semiconductor Devices*. Englewood Cliffs, N.J.: Prentice Hall, 1990.

9. Streetman, B. G. *Solid State Electronic Devices*. 3rd ed. Englewood Cliffs, N.J.: Prentice Hall, 1990.

10. Sze, S. M. *Physics of Semiconductor Devices*. 2nd ed. New York: Wiley, 1981.

11. Tsividis, Y. P. *Operation and Modeling of the MOS Transistor*. New York: McGraw-Hill, 1987.

12. Werner, W. M. "The Work Function Difference of the MOS System with Aluminum Field Plates and Polycrystalline Silicon Field Plates." *Solid State Electronics* 17 (1974), pp. 769–75.

13. Yamaguchi, T.; S. Morimoto; G. H. Kawamoto; and J. C. DeLacy. "Process and Device Performance of $1\mu$m-Channel n-Well CMOS Technology." *IEEE Transactions on Electron Devices* ED-31 (February 1984), pp. 205–14.

14. Yang, E. S. *Microelectronic Devices*. New York: McGraw-Hill, 1988.

15. Zambuto, M. *Semiconductor Devices*. New York: McGraw-Hill, 1989.

# METAL-OXIDE-SEMICONDUCTOR FIELD-EFFECT TRANSISTOR: ADDITIONAL CONCEPTS

## CHAPTER 13

## PREVIEW

In this chapter we will present additional concepts that are commonly encountered in metal-oxide-semiconductor field-effect transistors. These concepts include nonideal effects, small device geometry, breakdown, threshold voltage adjustment by ion implantation, and radiation effects.

Nonideal effects in MOSFETs alter the ideal characteristics we have been developing. A few nonideal effects we will consider include subthreshold conduction, channel length modulation, variations in channel mobility, and carrier velocity saturation. The tendency in MOSFET design has been to make the devices smaller and smaller. Transconductance and cutoff frequency increase as channel length is reduced and the packing density of MOSFETs increases as the channel width becomes narrower. Short channel lengths and narrow channel widths introduce deviations from the ideal threshold voltage developed in the previous chapter.

Breakdown voltage in MOSFETs also becomes a more important consideration as device size is reduced. We will analyze the gate oxide breakdown voltage and breakdown mechanisms in the semiconductor junctions. We will describe a particular MOSFET design that reduces breakdown effects: the lightly doped drain transistor. A problem often encountered in MOS fabrication is that a predetermined design or fabrication sequence may produce a device with an unacceptable threshold voltage. Ion implantation can be used to .change and adjust the substrate doping near the oxide-semiconductor surface in order to produce the desired threshold voltage. We will discuss this method of threshold voltage adjustment.

We have previously considered the effect of fixed charge trapped in the oxide and the effect of charge trapped in interface states on MOSFET characteristics. Ionizing radiation and hot electron effects are two processes that generate these types of charges; we will discuss both. Excess electron-hole pairs are generated in the oxide by ionizing radiation and hot electrons are generated by impact ionization in the space charge region near the drain terminal. A fraction of these generated charges may become trapped in the oxide and may create additional interface states.

## 13.1 NONIDEAL EFFECTS

As with any semiconductor device, the experimental characteristics of MOSFETs deviate to some degree from the ideal relations that have been theoretically derived using the various assumptions and approximations. In this section, we will consider four effects that cause deviations from the assumptions used in the ideal derivations. These effects are subthreshold conduction, channel length modulation, mobility variations, and velocity saturation.

### 13.1.1 Subthreshold Conduction

The ideal current-voltage relationship predicts zero drain current when the gate-to-source voltage is less than or equal to the threshold voltage. Experimentally, $I_D$ is not zero when $V_{GS} \leq V_T$. Figure 13–1 shows a comparison

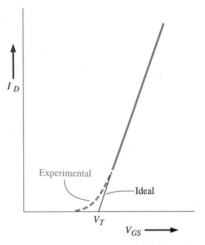

**Figure 13–1**    Comparison of ideal and experimental plots of $\sqrt{I_D}$ versus $V_{GS}$.

between the ideal characteristic that was derived, and the experimental results. The drain current, which exists for $V_{GS} \leq V_T$, is known as the subthreshold current.

The energy-band diagram of a MOS structure with a p-type substrate biased so that $\phi_s < 2\phi_{fp}$ is shown in Figure 13–2. At the same time, the Fermi level is closer to the conduction band than the valence band, so the semiconductor surface develops the characteristics of a lightly doped n-type mate-

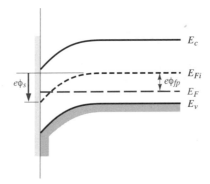

**Figure 13–2** Energy-band diagram when $\phi_{fp} < \phi_s < 2\phi_{fp}$.

rial. We would expect, then, to observe some conduction between the $n^+$ source and drain contacts through this weakly inverted channel. The condition for $\phi_{fp} < \phi_s < 2\phi_{fp}$ is known as *weak inversion*.

Figure 13–3 shows the surface potential along the length of the channel at accumulation, weak inversion, and threshold for the case when a small drain voltage is applied. The bulk p-substrate is assumed to be at zero potential. Figures 13–3b and 13–3c are the accumulation and weak inversion cases. There is a potential barrier between the $n^+$ source and channel region which the electrons must overcome in order to generate a channel current. A comparison of these barriers with those in pn junctions would suggest that the channel current is an exponential function of $V_{GS}$. In the inversion mode, shown in Figure 13–3d, the barrier is so small that we lose the exponential dependence since the junction is more like an ohmic contact.

The actual derivation of the subthreshold current is beyond the scope of this text. We can write that

$$I_D(\text{sub}) \propto \left[ \exp\left(\frac{eV_{GS}}{kT}\right) \right] \cdot \left[ 1 - \exp\left(\frac{-eV_{DS}}{kT}\right) \right] \qquad (13\text{–}1)$$

If $V_{DS}$ is larger than a few $(kT/e)$ volts, then the subthreshold current is independent of $V_{DS}$.

Figure 13–4 shows the exponential behavior of the subthreshold current for several body-to-source voltages. Also shown on the curves are the threshold voltage values. Ideally, a change in gate voltage of approximately 60 mV produces an order of magnitude change in the subthreshold current. A detailed analysis of the subthreshold condition shows that the slope of the $\ln I_D$ versus $V_{GS}$ curve is a function of the semiconductor doping and is also a function of the interface state density. The measurement of the slope of these curves has been used to experimentally determine the oxide-semiconductor interface state density.

If a MOSFET is biased at or even slightly below the threshold voltage, the drain current is not zero. The subthreshold current may add significantly to

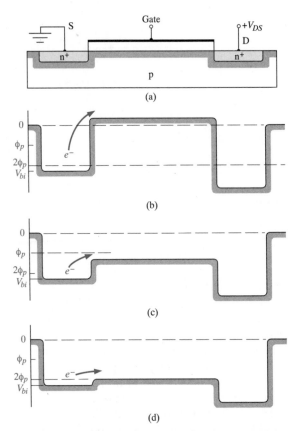

**Figure 13–3** (a) Cross section along channel length of n-channel MOSFET. Energy-band diagrams along channel length at (b) accumulation, (c) weak inversion, and (d) inversion.

power dissipation in a large-scale integrated circuit in which hundreds or thousands of MOSFETs are used. The circuit design must include the subthreshold current or ensure that the MOSFET is biased sufficiently below the threshold voltage in the "off" state.

### 13.1.2 Channel Length Modulation

We assumed in the derivation of the ideal current-voltage relationship that the channel length $L$ was a constant. However, when the MOSFET is biased in the saturation region, the depletion region at the drain terminal extends laterally into the channel, reducing the effective channel length. Since the depletion region width is bias dependent, the effective channel length is also bias dependent and is modulated by the drain-to-source voltage. This channel length modulation effect is shown in Figure 13–5 for an n-channel MOSFET.

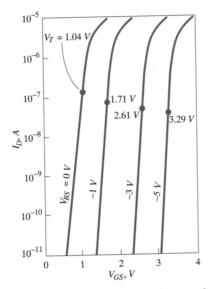

**Figure 13-4** Subthreshold current-voltage characteristics for several values of substrate voltage (the threshold voltage is indicated on each curve). (From Schroder [14].)

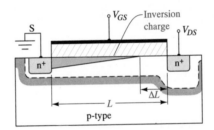

**Figure 13-5** Cross section of n-channel MOSFET showing channel length modulation effect.

As a first approximation, the incremental change in the depletion layer at the drain is given by

$$\Delta L = \sqrt{\frac{2\varepsilon_s}{eN_a}} \left[ \sqrt{\phi_{fp} + V_{DS}(\text{sat}) + \Delta V_{DS}} - \sqrt{\phi_{fp} + V_{DS}(\text{sat})} \right] \quad (13\text{-}2)$$

where 

$$\Delta V_{DS} = V_{DS} - V_{DS}(\text{sat}) \quad (13\text{-}3)$$

The applied drain-to-source voltage is $V_{DS}$ and we are assuming that $V_{DS} > V_{DS}(\text{sat})$. The drain terminal is very heavily doped so the depletion region extends only into the channel region. Since the drain current is inversely proportional to the channel length, we may write

$$I'_D = \left( \frac{L}{L - \Delta L} \right) I_D \quad (13\text{-}4)$$

where $I_D'$ is the actual drain current and $I_D$ is the ideal drain current. Since $\Delta L$ is a function of $V_{DS}$, $I_D'$ is now also a function of $V_{DS}$ even though the transistor is biased in the saturation region. Figure 13–6 shows some typical $I_D'$ versus $V_{DS}$ curves with positive slopes in the saturation region due to channel length modulation. As the MOSFET dimensions become smaller, the change in the channel length $\Delta L$ becomes a larger fraction of the original length $L$ and the channel length modulation becomes more severe.

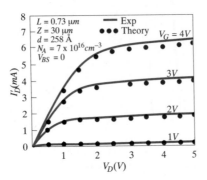

**Figure 13–6**   Current-voltage characteristics of a MOSFET showing short-channel effects. (From Sze [17].)

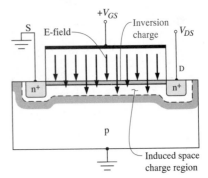

**Figure 13–7**   Vertical electric field in an n-channel MOSFET.

### 13.1.3  Mobility Variation

In the derivation of the ideal I-V relationship, we explicitly assumed that the mobility was a constant. However, this assumption must be modified for two reasons. The first effect to be considered is the variation in mobility with gate voltage. The second reason for a mobility variation is that the effective carrier mobility decreases as the carrier approaches the velocity saturation limit. This effect will be discussed in the next section.

The inversion layer charge is induced by a vertical electric field, which is shown in Figure 13–7 for an n-channel device. A positive gate voltage produces a force on the electrons in the inversion layer toward the surface. As the electrons travel through the channel toward the drain, they are attracted to the surface, but then are repelled by localized coulombic forces. This effect, schematically shown in Figure 13–8, is called *surface scattering*. The surface scattering effect reduces mobility. If there is a positive fixed oxide charge near the oxide-semiconductor interface, the mobility will be further reduced due to the additional coulomb interaction.

The relationship between the inversion charge mobility and the transverse electric field is usually measured experimentally. An effective transverse

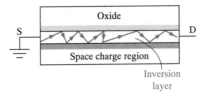

Figure 13-8    Schematic of carrier surface scattering effects.

electric field can be defined as

$$E_{eff} = \frac{1}{\varepsilon_s}\left(|Q'_{SD}(max)| + \frac{1}{2}Q'_n\right) \qquad (13\text{--}5)$$

The effective inversion charge mobility can be determined from the channel conductance as a function of gate voltage. Figure 13–9 shows the effective

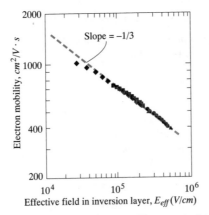

Figure 13-9    Measured inversion layer electron mobility versus electric field at the inversion layer. (From Yang [19].)

electron mobility at $T = 300°K$ for different doping levels and different oxide thicknesses. The effective mobility is only a function of the electric field at the inversion layer and is independent of oxide thickness. The effective mobility may be represented by

$$\mu_{eff} = \mu_0\left(\frac{E_{eff}}{E_0}\right)^{-1/3} \qquad (13\text{--}6)$$

where $\mu_0$ and $E_0$ are constants determined from experimental results.

The effective inversion charge mobility is a strong function of temperature because of lattice scattering. As the temperature is reduced, the mobility increases.

---

**Example 13–1**

**Objective:** To calculate the effective electric field at threshold for a given semiconductor doping.

Consider a p-type silicon substrate at $T = 300°K$ doped to $N_a = 3 \times 10^{16}$ cm$^{-3}$.

**Solution:** From the results of Chapter 12, we can calculate

$$\phi_{fp} = V_t \ln \left(\frac{N_a}{n_i}\right) = (0.0259) \ln \left(\frac{3 \times 10^{16}}{1.5 \times 10^{10}}\right) = 0.376 \text{ volt}$$

and

$$x_{dT} = \left\{\frac{4\varepsilon_s \phi_{fp}}{eN_a}\right\}^{1/2} = \left\{\frac{4(11.7)(8.85 \times 10^{-14})(0.376)}{(1.6 \times 10^{-19})(3 \times 10^{16})}\right\}^{1/2}$$

which is $x_{dT} = 0.18 \ \mu$m. Then

$$|Q'_{SD}(\text{max})| = eN_a x_{dT} = 8.64 \times 10^{-8} \text{ coul/cm}^2$$

At the threshold inversion point, we may assume that $Q'_n = 0$, so the effective electric field from Equation (13–5) is found as

$$E_{\text{eff}} = \frac{1}{\varepsilon_s} |Q'_{SD}(\text{max})| = \frac{8.64 \times 10^{-8}}{(11.7)(8.85 \times 10^{-14})} = 8.34 \times 10^4 \text{ V/cm}$$

**Comment:** We can see, from Figure 13–9, that this value of effective transverse electric field at the surface is sufficient for the effective inversion charge mobility to be significantly less than the bulk semiconductor value.

---

The effective mobility is a function of gate voltage through the inversion charge density in Equation (13–5). As the gate voltage increases, the carrier mobility decreases even further.

### 13.1.4 Velocity Saturation

The mobility variation due to the horizontal drain-to-source electric field is caused by the velocity saturation effect. Velocity saturation will become more prominent in shorter-channel devices since the corresponding horizontal electric field is generally larger.

In the ideal I-V relationship, current saturation occurred when the inversion charge density became zero at the drain terminal, or when

$$V_{DS} = V_{GS} - V_T \tag{13–7}$$

for the n-channel MOSFET. However, velocity saturation can change this saturation condition. Velocity saturation will occur when the horizontal electric field is approximately $10^4$ V/cm. If $V_{DS} = 5$ volts in a device with a channel length of $L = 1$ $\mu$m, the average electric field is $5 \times 10^4$ V/cm. Velocity saturation, then, is very likely to occur in short-channel devices.

The modified $I_D(\text{sat})$ characteristics are described approximately by

$$I_D(\text{sat}) = WC_{ox}(V_{GS} - V_T)v_{sat} \tag{13-8}$$

where $v_{sat}$ is the saturation velocity (approximately $10^7$ cm/sec for electrons in bulk silicon) and $C_{ox}$ is the gate oxide capacitance per cm². The saturation velocity will decrease somewhat with applied gate voltage because of the vertical electric field and surface scattering. Velocity saturation will yield an $I_D(\text{sat})$ value which is smaller than that predicted by the ideal relation and will yield a smaller $V_{DS}(\text{sat})$ value than predicted. The $I_D(\text{sat})$ current is also approximately linear with $V_{GS}$ rather than having the ideal square law dependence predicted previously.

There are several models of mobility versus electric field. One particular relation that is commonly used is

$$\mu = \frac{\mu_{eff}}{\left[ 1 + \left( \dfrac{\mu_{eff}E}{v_{sat}} \right)^2 \right]^{1/2}} \tag{13-9}$$

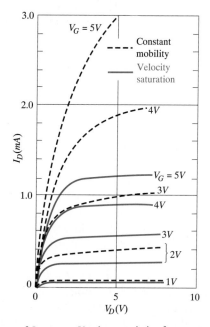

**Figure 13-10** Comparison of $I_D$ versus $V_D$ characteristics for constant mobility (dashed curves) and for field-dependent mobility and velocity saturation effects (solid curves). (From Sze [17].)

Figure 13–10 shows a comparison of drain current versus drain-to-source voltage characteristics for constant mobility and for field-dependent mobility. The smaller values of $I_D$(sat) and the approximate linear dependence on $V_{GS}$ may be noted for the field-dependent mobility curves.

The transconductance is found from

$$g_{ms} = \frac{\partial I_D(\text{sat})}{\partial V_{GS}} = WC_{\text{ox}}v_{\text{sat}} \tag{13–10}$$

which is now independent of $V_{GS}$ and $V_{DS}$ when velocity saturation occurs. The drain current is saturated by the velocity saturation effect, which leads to a constant transconductance.

When velocity saturation occurs, the cutoff frequency is given by

$$f_T = \frac{g_m}{2\pi C_G} = \frac{WC_{\text{ox}}v_{\text{sat}}}{2\pi(C_{\text{ox}}WL)} = \frac{v_{\text{sat}}}{2\pi L} \tag{13–11}$$

where the parasitic capacitances are assumed to be negligible.

## 13.2 SMALL DEVICE GEOMETRIES

We derived the ideal MOSFET relations in the previous chapter including expressions for threshold voltage and for the current-voltage characteristics. We have now considered some of the nonideal effects including channel length modulation. Additional effects on threshold voltage occur as the devices shrink in size. A reduction in channel length will increase the transconductance and frequency response of the MOSFET and a reduction in channel width will increase the packing density in an integrated circuit. A reduction in either or both the channel length and channel width can affect the threshold voltage.

### 13.2.1 Short-Channel Effects

In the ideal MOSFET, the threshold voltage was derived using the concept of charge neutrality in which the sum of charges in the metal, oxide, inversion layer, and semiconductor space charge region is zero. We also assumed that the gate area was the same as the active area in the semiconductor. Using this assumption, we considered only equivalent surface charge densities and neglected any effects on threshold voltage that might occur due to source and drain space charge regions that extend into the active channel region.

Figure 13–11a shows the cross section of a long n-channel MOSFET at flat-band, with zero source and drain voltage applied. The space charge regions at the source and drain extend into the channel region, but occupy only a small fraction of the entire channel region. The gate voltage, then, will control essentially all of the space charge induced in the channel region at inversion as shown in Figure 13–11b.

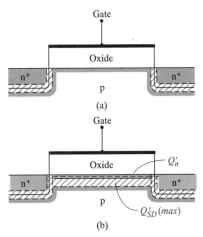

**Figure 13–11**  Cross section of a long n-channel MOSFET (a) at flat-band and (b) at inversion.

As the channel length decreases, the fraction of charge in the channel region controlled by the gate decreases. This effect can be seen in Figure 13–12 for the flat-band condition. As the drain voltage increases, the re-

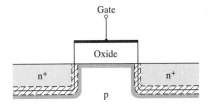

**Figure 13–12**  Cross section of a short n-channel MOSFET at flat-band.

verse-biased space charge region at the drain extends further into the channel area and the gate will control even less bulk charge. The amount of charge in the channel region, $Q'_{SD}(\text{max})$, controlled by the gate, affects the threshold voltage as can be seen from Equation (13–12).

$$V_{TN} = (|Q'_{SD}(\text{max})| - Q'_{ss}) \left(\frac{t_{ox}}{\varepsilon_{ox}}\right) + \phi_{ms} + 2\phi_{fp} \qquad (13\text{–}12)$$

We can quantitatively determine the short-channel effects on the threshold voltage by considering the parameters shown in Figure 13–13. The source and drain junctions are characterized by a diffused junction depth $r_j$. We will assume that the lateral diffusion distance under the gate is the same as the vertical diffusion distance. This assumption is a reasonably good approximation for diffused junctions, but becomes less accurate for ion im-

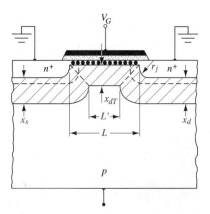

**Figure 13–13**    Charge sharing in the short-channel threshold voltage model. (From Yau [20].)

planted junctions. We will initially consider the case when the source, drain, and body contacts are all at ground potential.

The basic assumption in this analysis is that the bulk charge in the trapezoidal region under the gate is controlled by the gate. The potential difference across the bulk space charge region is $2\phi_{fp}$ at the threshold inversion point and the built-in potential barrier height of the source and drain junctions is also approximately $2\phi_{fp}$, implying that the three space charge widths are essentially equal. We can then write

$$x_s \approx x_d \approx x_{dT} \equiv x_{dT} \tag{13–13}$$

Using the geometrical approximation, the average bulk charge per unit area $Q_B'$ in the trapezoid is

$$|Q_B'| \cdot L = eN_a x_{dT}\left(\frac{L + L'}{2}\right) \tag{13–14}$$

From the geometry, we can show that

$$\frac{L + L'}{2L} = \left\{1 - \frac{r_j}{L}\left[\sqrt{1 + \frac{2x_{dT}}{r_j}} - 1\right]\right\} \tag{13–15}$$

Then    $$|Q_B'| = eN_a x_{dT}\left\{1 - \frac{r_j}{L}\left[\sqrt{1 + \frac{2x_{dT}}{r_j}} - 1\right]\right\} \tag{13–16}$$

Equation (13–16) is now used in place of $|Q_{SD}'(\text{max})|$ in the expression for the threshold voltage.

Since $|Q_{SD}'(\text{max})| = eN_a x_{dT}$, we can find $\Delta V_T$ as

$$\Delta V_T = -\frac{eN_a x_{dT}}{C_{\text{ox}}}\left\{\frac{r_j}{L}\left[\sqrt{1 + \frac{2x_{dT}}{r_j}} - 1\right]\right\} \tag{13–17}$$

where    $$\Delta V_T = V_{T(\text{short channel})} - V_{T(\text{long channel})} \tag{13–18}$$

As the channel length decreases, the threshold voltage shifts in the negative direction so that an n-channel MOSFET shifts toward depletion mode.

---

**Example 13–2**

**Objective:** To calculate the threshold voltage shift due to short-channel effects.

Consider an n-channel MOSFET with $N_a = 3 \times 10^{16}$ cm$^{-3}$ and $t_{ox} = 450$ Å. Let $L = 1.25$ $\mu$m and assume that $r_j = 0.5$ $\mu$m.

**Solution:** We can determine the oxide capacitance as

$$C_{ox} = \frac{\varepsilon_{ox}}{t_{ox}} = \frac{(3.9)(8.85 \times 10^{-14})}{450 \times 10^{-8}} = 7.67 \times 10^{-8} \text{ F/cm}^2$$

and calculate the potential as

$$\phi_{fp} = V_t \ln\left(\frac{N_a}{n_i}\right) = (0.0259) \ln\left(\frac{3 \times 10^{16}}{1.5 \times 10^{10}}\right) = 0.376 \text{ volt}$$

Then the maximum space charge width can be found as

$$x_{dT} = \left\{\frac{4\varepsilon_s \phi_{fp}}{eN_a}\right\}^{1/2} = \left\{\frac{4(11.7)(8.85 \times 10^{-14})(0.376)}{(1.6 \times 10^{-19})(3 \times 10^{16})}\right\}^{1/2} = 0.18 \text{ } \mu\text{m}$$

Finally, the threshold voltage shift, from Equation (13–17), is

$$\Delta V_T = -\frac{(1.6 \times 10^{-19})(3 \times 10^{16})(0.18 \times 10^{-4})}{7.67 \times 10^{-8}} \left\{\frac{0.5}{1.25}\left[\sqrt{1 + \frac{2(0.18)}{0.5}} - 1\right]\right\}$$

which gives

$$\Delta V_T = -0.140 \text{ volt}$$

**Comment:** If the threshold voltage value of an n-channel MOSFET is to be $V_T = 0.5$ volt, for example, a shift of $\Delta V_T = -0.140$ volt due to short-channel effects is significant and needs to be taken into account in the design of the device.

---

The effect of short channels becomes more pronounced as the channel length is reduced further.

The shift in threshold voltage with channel length for an n-channel MOSFET is shown in Figure 13–14. As the substrate doping increases, the initial threshold voltage increases, as we saw in the last chapter, and the short-channel threshold shift also becomes larger. The short-channel effects on threshold voltage do not become significant until the channel length becomes less than approximately 2 $\mu$m. The threshold voltage shift also becomes smaller as the diffusion depth $r_j$ becomes smaller so that very shallow junctions reduce the threshold voltage dependence on channel length.

Equation (13–17) was derived using the assumption that the source, channel, and drain space charge widths were all equal. If we now apply a drain

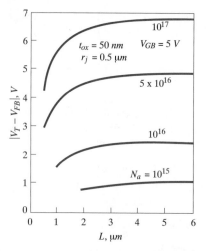

**Figure 13–14**    Threshold voltage versus channel length for various substrate dopings. (From Yau [20].)

voltage, the space charge width at the drain terminal widens, which makes $L'$ smaller, and the amount of bulk charge controlled by the gate voltage decreases. This effect makes the threshold voltage a function of drain voltage. As the drain voltage increases, the threshold voltage of an n-channel MOSFET decreases. The threshold voltage versus channel length is plotted in Figure 13–15 for two values of drain-to-source voltage and two values of body-to-source voltage.

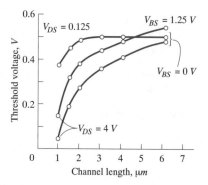

**Figure 13–15**    Threshold voltage versus channel length for two values of drain-to-source and body-to-source voltage. (From Yang [19].)

### 13.2.2  Narrow-Channel Effects

Figure 13–16 shows the cross section along the channel width of an n-channel MOSFET biased at inversion. The current is perpendicular to the

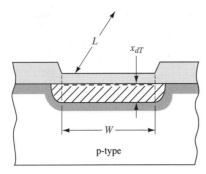

**Figure 13–16** Cross section of an n-channel MOSFET showing the depletion region along the width of the device.

channel width through the inversion charge. We may note in the figure that there is an additional space charge region at each end of the channel width. This additional charge is controlled by the gate voltage but was not included in the derivation of the ideal threshold voltage relation. The threshold voltage expression must be modified to include this additional charge.

If we neglect short-channel effects, the gate controlled bulk charge can be written as

$$Q_B = Q_{B0} + \Delta Q_B \tag{13–19}$$

where $Q_B$ is the total bulk charge, $Q_{B0}$ is the ideal bulk charge, and $\Delta Q_B$ is the additional bulk charge at the ends of the channel width. For a uniformly doped p-type semiconductor biased at the threshold inversion point, we may write

$$|Q_{B0}| = eN_a W L x_{dT} \tag{13–20}$$

and

$$\Delta Q_B = eN_a L x_{dT}(\xi x_{dT}) \tag{13–21}$$

where $\xi$ is a fitting parameter that accounts for the lateral space charge width. The lateral space charge width may not be the same as the vertical width $x_{dT}$ due to the thicker field oxide at the ends, and/or due to the nonuniform semiconductor doping created by an ion implantation. If the ends were a semicircle, then $\xi = \pi/2$.

We may now write

$$|Q_B| = |Q_{B0}| + |\Delta Q_B| = eN_a W L x_{dT} + eN_a L x_{dT}(\xi x_{dT})$$
$$= eN_a W L x_{dT} \left(1 + \frac{\xi x_{dT}}{W}\right) \tag{13–22}$$

The effect of the end space charge regions becomes significant as the width $W$ decreases and the factor $(\xi x_{dT})$ becomes a significant fraction of the width $W$.

The change in threshold voltage due to the additional space charge is

$$\Delta V_T = \frac{e N_a x_{dT}}{C_{ox}} \left( \frac{\xi x_{dT}}{W} \right) \tag{13-23}$$

The shift in threshold voltage due to a narrow channel is in the positive direction for the n-channel MOSFET. As the width $W$ becomes smaller, the shift in threshold voltage becomes larger.

---

**Example 13–3**

**Objective:** To calculate the channel width that will produce a particular threshold voltage shift due to narrow channel effects.

Consider an n-channel MOSFET with $N_a = 3 \times 10^{16}$ cm$^{-3}$ and $t_{ox} = 450$ Å. Let $\xi = \pi/2$. Assume that we want to limit the threshold shift to $\Delta V_T = 0.2$ volt.

**Solution:** From Example 13–2, we have

$$C_{ox} = 7.67 \times 10^{-8} \text{ F/cm}^2 \quad \text{and} \quad x_{dT} = 0.18 \ \mu\text{m}$$

From Equation (13–23), we can express the channel width as

$$W = \frac{e N_a (\xi x_{dT}^2)}{C_{ox}(\Delta V_T)} = \frac{(1.6 \times 10^{-19})(3 \times 10^{16})\left(\frac{\pi}{2}\right)(0.18 \times 10^{-4})^2}{(7.67 \times 10^{-8})(0.2)} = 1.59 \ \mu\text{m}$$

**Comment:** We can note that the threshold shift of $\Delta V_T = 0.2$ volt occurs at a channel width of $W = 1.59 \ \mu$m, which is approximately 10 times larger than the induced space charge width $x_{dT}$.

---

Figure 13–17 shows the threshold voltage as a function of channel width. We can again note that the threshold voltage shift begins to become apparent for channel widths that are large compared to the induced space charge width.

Figures 13–18a and 13–18b show qualitatively the threshold voltage shifts due to short-channel and narrow-channel effects, respectively, in n-channel MOSFETs. The narrow-channel device produces a larger threshold voltage; the short-channel device produces a smaller threshold voltage. For devices exhibiting both short-channel and narrow-channel effects, the two models need to be combined into a three-dimensional volume approximation of the space charge region controlled by the gate.

## 13.3 ADDITIONAL ELECTRICAL CHARACTERISTICS

There is a tremendous volume of information on MOSFETs that cannot be included in an introductory text on semiconductor physics and devices.

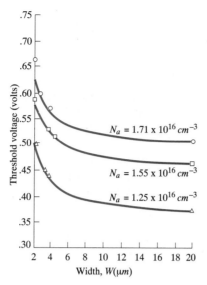

**Figure 13-17**   Threshold voltage versus channel width (solid curves, theoretical; points, experimental). (From Akers [1].)

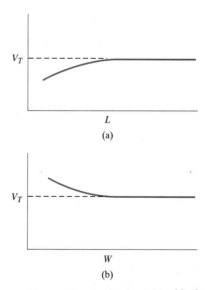

**Figure 13-18**   Qualitative variation of threshold voltage (a) with channel length and (b) with channel width.

However, two additional topics should be included here: breakdown voltage and threshold adjustment by ion implantation.

### 13.3.1 Breakdown Voltage

Several voltage breakdown mechanisms in the MOSFET must be considered, including voltage breakdown across the oxide, as well as the various voltage breakdown mechanisms in the semiconductor junctions.

**Oxide Breakdown**    We have assumed that the oxide is a perfect insulator. However, if the electric field in the oxide becomes large enough, breakdown can occur, which can lead to a catastrophic failure. In silicon dioxide, the electric field at breakdown is on the order of $6 \times 10^6$ V/cm. This breakdown field is larger than that in silicon, but the gate oxides are also quite thin. A gate voltage of approximately 30 volts would produce breakdown in an oxide with a thickness of 500 Å. However, a safety margin of a factor of 3 is common, so that the maximum safe gate voltage with $t_{ox} = 500$ Å would be 10 volts. A safety margin is necessary since there may be defects in the oxide that lower the breakdown field. Oxide breakdown is normally not a serious problem except in power devices and ultrathin oxide devices. Other oxide degradation problems will be discussed later in this chapter.

**Avalanche Breakdown**    Avalanche breakdown may occur by impact ionization in the space charge region near the drain terminal. We considered avalanche breakdown in pn junctions in Chapter 8. In an ideal planar one-sided pn junction, breakdown is a function primarily of the doping concentration in the low-doped region of the junction. For the MOSFET, the low-doped region corresponds to the semiconductor substrate. If a p-type substrate doping is $N_a = 3 \times 10^{16}$ cm$^{-3}$, for example, the pn junction breakdown voltage would be approximately 25 volts for a planar junction. However, the $n^+$ drain contact may be a fairly shallow diffused region with a large curvature. The electric field in the depletion region tends to be concentrated at the curvature, which lowers the breakdown voltage. This curvature effect is shown in Figure 13–19.

**Near Avalanche and Snapback Breakdown**    Another breakdown mechanism results in the S-shaped breakdown curve shown in Figure 13–20. This breakdown process is due to second order effects and can be explained with the aid of Figure 13–21. The n-channel enhancement mode MOSFET geometry in Figure 13–21a shows the n-type source and drain contacts along with the p-type substrate. The source and body are at ground potential. The n(source)-p(substrate)-n(drain) structure also forms a parasitic bipolar transistor. The equivalent circuit is shown in Figure 13–21b.

Figure 13–22a shows the device when avalanche breakdown is just beginning in the space charge region near the drain. We tend to think of the

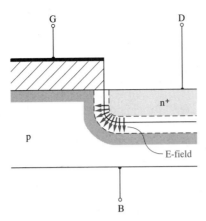

**Figure 13–19**   Curvature effect on the electric field in the drain junction.

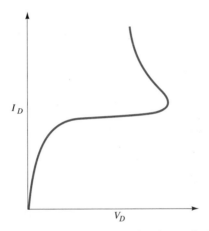

**Figure 13–20**   Current-voltage characteristic showing the snapback breakdown effect.

avalanche breakdown suddenly occurring at a particular voltage. However, avalanche breakdown is a gradual process that starts at low current levels and for electric fields somewhat below the breakdown field. The electrons generated by the avalanche process flow into the drain and contribute to the drain current. The avalanche-generated holes generally flow through the substrate to the body terminal. Since the substrate has a nonzero resistance, a voltage drop is produced as shown. This potential difference drives the source-to-substrate pn junction into forward bias near the source terminal. The source is heavily doped n-type; thus, a large number of electrons can be injected from the source contact into the substrate under forward bias. This process will become severe as the voltage drop in the substrate approaches 0.6 to 0.7 volt. A fraction of the injected electrons will diffuse across the parasitic base region into the reverse-biased drain space charge region where they also add to the drain current.

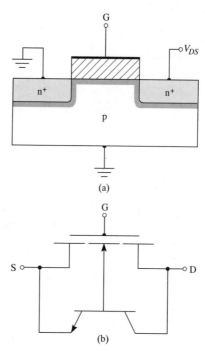

**Figure 13–21**   (a) Cross section of the n-channel MOSFET. (b) Equivalent circuit including the parasitic bipolar transistor.

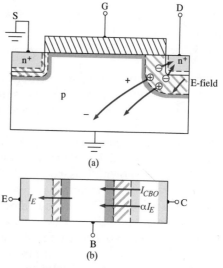

**Figure 13–22**   (a) Substrate current-induced voltage drop caused by avalanche multiplication at the drain. (b) Currents in the parasitic bipolar transistor.

The avalanche breakdown process is a function of not only the electric field but the number of carriers involved. The rate of avalanche breakdown increases as the number of carriers in the drain space charge region increases. We now have a regenerative or positive feedback mechanism. Avalanche breakdown near the drain terminal produces the substrate current, which produces the forward-biased source-substrate pn junction voltage. The forward-biased junction injects carriers that can diffuse back to the drain and increase the avalanche process. The positive feedback produces an unstable system.

The snapback or negative resistance portion of the curve shown in Figure 13–20 can now be explained using the parasitic bipolar transistor. The potential of the base of the bipolar transistor near the emitter (source) is almost floating since this voltage is determined primarily by the avalanche-generated substrate current rather than an externally applied voltage.

For the open-base bipolar transistor shown in Figure 13–22, we can write

$$I_C = \alpha I_E + I_{CB0} \tag{13–24}$$

where $\alpha$ is the common base current gain and $I_{CB0}$ is the base-collector leakage current. For an open-base, $I_C = I_E$, so Equation (13–24) becomes

$$I_C = \alpha I_C + I_{CB0} \tag{13–25}$$

At breakdown, the current in the B-C junction is multiplied by the multiplication factor $M$, so we have

$$I_C = M(\alpha I_C + I_{CB0}) \tag{13–26}$$

Solving for $I_C$ we obtain

$$I_C = \frac{M I_{CB0}}{1 - \alpha M} \tag{13–27}$$

Breakdown is defined as the condition which produces $I_C \rightarrow \infty$. For a single reverse-biased pn junction, $M \rightarrow \infty$ at breakdown. However, from equation (13–27), breakdown is now defined to be the condition when $\alpha M \rightarrow 1$ or, for the open-base condition, breakdown occurs when $M \rightarrow 1/\alpha$, which is a much lower multiplication factor than for the simple pn junction.

An empirical relation for the multiplication factor is usually written as

$$M = \frac{1}{1 - (V_{CE}/V_{BD})^m} \tag{13–28}$$

where $m$ is an empirical constant in the range of 3 to 6 and $V_{BD}$ is the junction breakdown voltage.

The common base current gain factor $\alpha$ is a strong function of collector current for small values of collector current. This effect was discussed in Chapter 10 on bipolar transistors. At low currents, the recombination current in the B-E junction is a significant fraction of the total current so that the common base current gain is small. As the collector current increases, the value of $\alpha$ increases. As avalanche breakdown begins and $I_C$ is small, particular values of $M$ and $V_{CE}$ are required to produce the condition of $\alpha M = 1$. As the collector current increases, $\alpha$ increases; therefore, smaller values of $M$ and $V_{CE}$ are required to produce the avalanche breakdown condition. The snapback, or negative resistance, breakdown characteristic is then produced.

Only a fraction of the injected electrons from the forward-biased source-substrate junction are collected by the drain terminal. A more exact calculation of the snapback characteristic would necessarily involve taking into account this fraction; thus the simple model would need to be modified. However, the above discussion qualitatively describes the snapback effect. The snapback effect can be minimized by using a heavily doped substrate that will prevent any significant voltage drop from being developed. A thin epitaxial p-type layer with the proper doping concentration to produce the required threshold voltage can be grown on a heavily doped substrate.

**Near Punch-Through Effects** Punch-through is the condition at which the drain-to-substrate space charge region extends completely across the channel region to the source-to-substrate space charge region. In this situation, the barrier between the source and drain is completely eliminated and a very large drain current would exist.

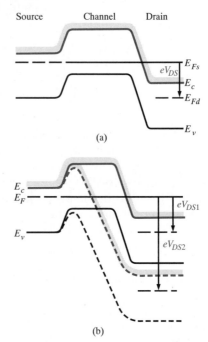

**Figure 13–23** (a) Equipotential plot along the surface of a long-channel MOSFET. (b) Equipotential plot along the surface of a short-channel MOSFET before and after punch-through.

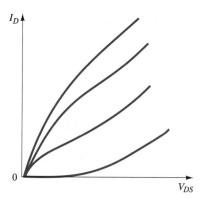

**Figure 13–24**   Typical I-V characteristics of a MOSFET exhibiting punch-through effects.

However, the drain current will begin to increase rapidly before the actual punch-through condition is reached. This characteristic is referred to as the near punch-through condition. Figure 13–23a shows the ideal energy-band diagram from source to drain for a long n-channel MOSFET for the case when $V_{GS} < V_T$ and when the drain-to-source voltage is relatively small. The large potential barriers prevent significant current between the drain and source. Figure 13–23b shows the energy-band diagram when a relatively large drain voltage $V_{DS2}$ is applied. The space charge region near the drain terminal is beginning to interact with the source space charge region and the potential barrier is being lowered. Since the current is an exponential function of barrier height, the current will increase very rapidly with drain voltage once this near punch-through condition has been reached. Figure 13–24 shows some typical characteristics of a short-channel device with a near punch-through condition.

---

**Example 13–4**

**Objective:** To calculate the theoretical punch-through voltage assuming the abrupt junction approximation.

Consider an n-channel MOSFET with source and drain doping concentrations of $N_d = 10^{19}$ cm$^{-3}$ and a channel region doping of $N_a = 10^{16}$ cm$^{-3}$. Assume a channel length of $L = 1.2$ $\mu$m and assume the source and body are at ground potential.

**Solution:** The pn junction built-in potential barrier is given by

$$V_{bi} = V_t \ln\left(\frac{N_a N_d}{n_i^2}\right) = (0.0259) \ln\left[\frac{(10^{16})(10^{19})}{(1.5 \times 10^{10})^2}\right] = 0.874 \text{ volt}$$

The zero-biased source-substrate pn junction width is

$$x_{d0} = \left\{\frac{2\varepsilon_s V_{bi}}{eN_a}\right\}^{1/2} = \left\{\frac{2(11.7)(8.85 \times 10^{-14})(0.874)}{(1.6 \times 10^{-19})(10^{16})}\right\}^{1/2} = 0.336 \text{ } \mu\text{m}$$

The reverse-biased drain-substrate pn junction width is given by

$$x_d = \left\{ \frac{2\varepsilon_s(V_{bi} + V_{DS})}{eN_a} \right\}^{1/2}$$

At punch-through, we will have

$$x_{d0} + x_d = L \qquad \text{or} \qquad 0.336 + x_d = 1.2$$

which gives $x_d = 0.864$ $\mu$m at the punch-through condition. We can then find

$$V_{bi} + V_{DS} = \frac{x_d^2 e N_a}{2\varepsilon_s} = \frac{(0.864 \times 10^{-4})^2(1.6 \times 10^{-19})(10^{16})}{2(11.7)(8.85 \times 10^{-14})}$$

$$= 5.77 \text{ volts}$$

The punch-through voltage is then found as

$$V_{DS} = 5.77 - 0.874 = 4.9 \text{ volts}$$

**Comment:** As the two space charge regions approach punch-through, the abrupt junction approximation is no longer a good assumption.

---

For a doping of $10^{16}$ cm$^{-3}$, the two space charge regions will begin to interact when the abrupt depletion layers are approximately 0.25 $\mu$m apart. The drain voltage at which this near punch-through condition, also known as drain-induced barrier lowering, occurs is significantly less than the ideal punch-through voltage such as calculated in Example 13–4 (see problem 13).

### * 13.3.2 The Lightly Doped Drain Transistor

The junction breakdown voltage is a function of the maximum electric field. As the channel length becomes smaller, the bias voltages may not be scaled down accordingly so the junction electric fields become larger. As the electric field increases, near avalanche breakdown and near punch-through effects become more serious. In addition, as device geometries are scaled down, the parasitic bipolar device becomes more dominant and breakdown effects are enhanced.

One approach that reduces these breakdown effects is to alter the doping profile of the drain contact. The *Lightly Doped Drain* (LDD) design and doping profiles are shown in Figure 13–25a; the conventional MOSFET and doping profiles are shown in Figure 13–25b for comparison. By introducing the lightly doped region, the peak electric field in the space charge region is reduced and the breakdown effects are minimized. The peak electric field at the drain junction is a function of the semiconductor doping as well as the curvature of the $n^+$ drain region. Figure 13–26 shows the physical geometries of a conventional $n^+$ drain contact and an LDD structure superimposed on the same plot. The magnitude of the electric field at the oxide–semiconductor interface in the LDD structure is less than in the conventional structure. The electric field in the conventional device peaks approximately at the

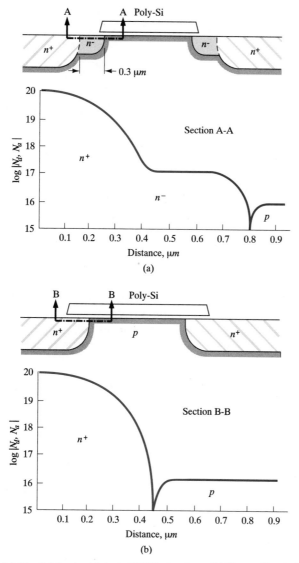

**Figure 13-25**  (a) The lightly doped drain (LDD) structure. (b) Conventional structure. (From Ogura et al. [10].)

metallurgical junction and drops quickly to zero in the drain because no field can exist in the highly conductive $n^+$ region. On the other hand, the electric field in the LDD device extends across the n-region before dropping to zero at the drain. This effect will minimize breakdown and the hot electron effects which we will discuss in Section 13.4.3.

Two disadvantages of the LDD device are an increase in both fabrication complexity and drain resistance. The added processing steps, however, produce a device with significant improvements in performance. The cross

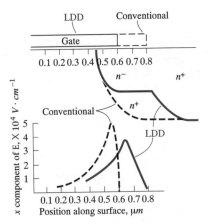

**Figure 13-26** Magnitude of the electric field at the Si-SiO$_2$ interface as a function of distance; $V_{DS} = 10V$, $V_{SB} = 2V$, $V_{GS} = V_T$. (From Ogura et al. [10].)

section of the LDD device shown in Figure 13–25 indicates a lightly doped n-region at the source terminal also. The inclusion of this region does not improve device performance, but does reduce the fabrication complexity as much as possible. The added series resistances will increase power dissipation in the device; this must be taken into account in high-power devices.

### 13.3.3 Threshold Adjustment by Ion Implantation

Several factors, such as fixed oxide charge, metal-semiconductor work function difference, oxide thickness, and semiconductor doping, influence the threshold voltage. All of the above parameters may be fixed in a particular design and fabrication process although the resulting threshold voltage may not be acceptable for all applications. Ion implantation can be used to change and adjust the substrate doping near the oxide-semiconductor surface to provide the desired threshold voltage.

To change the doping and thereby change the threshold voltage, a precise, controlled number of either donor or acceptor ions are implanted into the semiconductor near the oxide surface. When a MOS device is biased in either depletion or inversion and when the implanted dopant atoms are within the induced space charge region, then the ionized dopant charge adds to (or subtracts from) the maximum space charge density, which controls the threshold voltage. An implant of acceptor ions into either a p- or n-type substrate will shift the threshold voltage to more positive values while an implant of donor ions will shift the threshold voltage to more negative values.

As a first approximation, assume that $D_I$ acceptor atoms per cm$^2$ are implanted into a p-type substrate directly adjacent to the oxide–semiconduc-

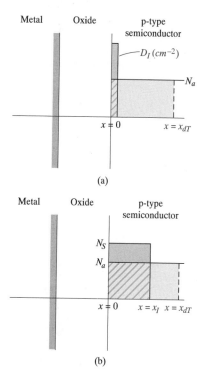

**Figure 13–27** (a) Ion-implanted profile approximated by a delta function. (b) Ion-implanted profile approximated by a step function, in which the depth $x_I$ is less than the space charge width $x_{dT}$.

tor interface as shown in Figure 13–27a. The shift in threshold voltage due to the implant is

$$\Delta V_T = + \frac{e D_I}{C_{\text{ox}}} \qquad (13\text{–}29)$$

If donor atoms were implanted into the p-type substrate, the space charge density would be reduced; thus, the threshold voltage would shift in the negative voltage direction.

A second type of implant approximation is the step junction, shown in Figure 13–27b. If the induced space charge width at the threshold inversion point is less than $x_I$, then the threshold voltage is determined on the basis of a semiconductor with a uniform doping concentration of $N_s$ atoms per cm³. On the other hand, if the induced space charge width is greater than $x_I$ at the threshold inversion point, then a new expression for $x_{dT}$ must be derived. We can apply Poisson's equation and show that the maximum induced space charge width after the step implant is

$$x_{dT} = \sqrt{\frac{2\varepsilon_s}{eN_a} \left[ 2\phi_{fp} - \frac{ex_I^2}{2\varepsilon_s} (N_s - N_a) \right]^{1/2}} \qquad (13\text{-}30)$$

The threshold voltage after a step implant for the case when $x_{dT} > x_I$ can be written as

$$V_T = V_{T0} + \frac{eD_I}{C_{ox}} \qquad (13\text{-}31)$$

where $V_{T0}$ is the pre-implant threshold voltage. The parameter $D_I$ is given by

$$D_I = (N_s - N_a)x_I \qquad (13\text{-}32)$$

which is the number per $cm^2$ of implanted ions. The pre-implant threshold voltage is

$$V_{T0} = V_{FB0} + 2\phi_{fp0} + \frac{eN_a x_{dT0}}{C_{ox}} \qquad (13\text{-}33)$$

where the subscript "0" indicates the pre-implant values.

---

### Example 13–5

**Objective:** To determine the ion implant density required to adjust the threshold voltage if a step implant is used.

Consider an n-channel MOSFET with a doping of $N_a = 5 \times 10^{15}$ $cm^{-3}$, an oxide thickness of $t_{ox} = 500$ Å, and an initial flat-band voltage of $V_{FB0} = -1.25$ volts. A threshold voltage of $V_T = +0.70$ volt is desired.

**Solution:** We may calculate the necessary parameters as

$$\phi_{fp0} = V_t \ln \left( \frac{N_a}{n_i} \right) = (0.0259) \ln \left( \frac{5 \times 10^{15}}{1.5 \times 10^{10}} \right) = 0.329 \text{ volt}$$

$$x_{dT0} = \left\{ \frac{4\varepsilon_s \phi_{fp0}}{eN_a} \right\}^{1/2} = \left\{ \frac{4(11.7)(8.85 \times 10^{-14})(0.329)}{(1.6 \times 10^{-19})(5 \times 10^{15})} \right\}^{1/2} = 0.413 \ \mu m$$

$$C_{ox} = \frac{\varepsilon_{ox}}{t_{ox}} = \frac{(3.9)(8.85 \times 10^{-14})}{500 \times 10^{-8}} = 6.9 \times 10^{-8} \ F/cm^2$$

The initial pre-implant threshold voltage is

$$V_{T0} = V_{FB0} + 2\phi_{fp0} + \frac{eN_a x_{dT0}}{C_{ox}}$$

$$= -1.25 + 2(0.329) + \frac{(1.6 \times 10^{-19})(5 \times 10^{15})(0.413 \times 10^{-4})}{6.9 \times 10^{-8}}$$

$$= -0.113 \text{ volt}$$

The threshold voltage after implant, from Equation (13–31), is

$$V_T = V_{T0} + \frac{eD_I}{C_{ox}}$$

so that

$$+0.70 = -0.113 + \frac{(1.6 \times 10^{-19}) D_I}{6.9 \times 10^{-8}}$$

which gives

$$D_I = 3.51 \times 10^{11} \text{ cm}^{-2}$$

If the uniform step implant extends to a depth of $x_I = 0.15 \ \mu$m, for example, then the equivalent acceptor concentration at the surface is

$$N_s - N_a = \frac{D_I}{x_I} = \frac{3.51 \times 10^{11}}{0.15 \times 10^{-4}} = 2.34 \times 10^{16} \text{ cm}^{-3}$$

or

$$N_s = 2.84 \times 10^{16} \text{ cm}^{-3}$$

**Comment:** The required implant dose to achieve the desired threshold voltage is $D_I = 3.51 \times 10^{11} \text{ cm}^{-2}$. This calculation has assumed that the induced space charge width in the channel region is greater than the ion implant depth $x_I$. We can show that this requirement is indeed satisfied in this example.

---

The actual implant dose versus distance is neither a delta function nor a step function; it tends to be a Gaussian type distribution. The threshold shift due to a nonuniform ion implant density may be defined as the shift in curves of $N_{\text{inv}}$ versus $V_G$ where $N_{\text{inv}}$ is the inversion carrier density per cm$^2$. This shift corresponds to an experimental shift of drain current versus $V_G$ when the transistor is biased in the linear mode. The criteria of the threshold inversion point as $\phi_s = 2\phi_{fp}$ in the implanted devices have an uncertain meaning because of the nonuniform doping in the substrate. The determination of the threshold voltage becomes more complicated and will not be done here.

## * 13.4  RADIATION AND HOT ELECTRON EFFECTS

We have considered the effects of fixed trapped oxide charge and interface state charge on the capacitance-voltage characteristics of MOS capacitors and on the MOSFET characteristics. These charges can exist because the oxide is essentially a perfect dielectric and a net charge density can exist in a dielectric material. Two processes that generate these charges are ionizing radiation and impact ionization in the drain region of a MOSFET operating near avalanche breakdown.

MOS devices are exposed to ionizing radiation, for example, in communication satellites orbiting through the Van Allen radiation belts. The ionizing radiation can produce additional fixed oxide charge and also additional interface states. In this short discussion of radiation effects in MOSFETs, we will be concerned only with the permanent effects that occur in the device characteristics.

Another source can generate oxide charge and interface states: the hot electron effect. Electrons near the drain terminal of a MOSFET operating near avalanche breakdown can have energies that are much larger than the thermal-equilibrium value. These are called *hot electrons* and have energies sufficient to penetrate the oxide-semiconductor barrier.

### 13.4.1 Radiation-Induced Oxide Charge

Gamma-rays or X-rays incident on semiconductor or oxide materials can interact with valence band electrons. The incident radiation photons can impart enough energy to a valence electron to elevate the electron into the conduction band; an empty state or hole is also produced in the valence band. This process generates electron-hole pairs. These newly generated electrons and holes can move through a material under the influence of an electric field.

Figure 13–28 shows the energy-band diagram of a MOS device with a p-type substrate and a positive gate voltage. The bandgap energy of silicon

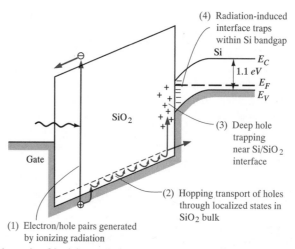

**Figure 13–28**    Schematic of ionizing radiation-induced processes in a MOS capacitor with a positive gate bias. (From Ma et al. [5].)

dioxide is approximately 9 eV. The figure schematically shows the creation of an electron-hole pair in the oxide by ionizing radiation. The force on the radiation-induced electron is toward the gate and the force on the radiation-induced hole is toward the semiconductor. It has been found that generated electrons in the oxide are fairly mobile with a mobility value on the order of 20 $cm^2/V$-sec. At high electric fields, the electron velocity in the oxide also saturates at approximately $10^7$ cm/sec so that the electron transit time for typical gate oxide thicknesses is on the order of a few picoseconds. For positive gate voltages, the vast majority of radiation-induced electrons flow out through the gate terminal; for this reason, in general these electrons do not play a significant role in the radiation response of MOS devices.

The generated holes, on the other hand, undergo a stochastic hopping transport process through the oxide (shown schematically in Figure 13–28). The hole transport process is dispersive in time and is a function of the electric field, temperature, and oxide thickness. The effective hole mobility in silicon dioxide is typically in the range of $10^{-4}$ to $10^{-11}$ cm$^2$/$V$-sec; thus, holes are relatively immobile when compared with electrons.

When holes reach the silicon–silicon dioxide interface, a fraction are captured in trapping sites while the remainder flow into the silicon. A net positive radiation-induced charge is now trapped in the oxide due to these captured holes. This trapped charge can last from hours to years. As we have seen, a positive oxide charge causes a negative shift in threshold voltage.

The measured areal hole trap densities are in the range of $10^{12}$ to $10^{13}$ cm$^{-2}$ depending upon oxide and device processing. In general, these traps are located within approximately 50 Å of the Si–SiO$_2$ interface. The hole trap is usually associated with a trivalent silicon defect which has an oxygen vacancy in the SiO$_2$ structure. The oxygen vacancies are located in a silicon-rich region near the Si–SiO$_2$ interface.

Since the threshold or flat-band voltage shift is a function of the amount of trapped charge, the voltage shift is a function of applied voltage across the oxide. Figure 13–29 shows the flat-band voltage shift of a MOS capacitor as

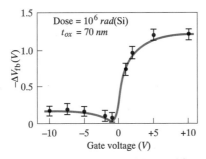

**Figure 13–29**   Radiation-induced flat-band voltage shift in a MOS capacitor as a function of applied gate bias during irradiation. (From Ma et al. [5].)

a function of gate voltage applied during irradiation. For small values of gate voltage, some radiation-generated holes and electrons recombine in the oxide. Hence the amount of charge reaching the Si–SiO$_2$ interface and being trapped is less than for a large positive gate voltage, where essentially all radiation-generated holes reach the interface without recombining with electrons. If the fraction of generated holes that become trapped is relatively constant, then the voltage shift becomes independent of positive gate bias, as shown in the figure. For negative applied gate voltages, the radiation-induced holes move toward the gate terminal. There can be positive charge trapping in the oxide near the gate, but the effect of this trapped charge on the threshold voltage is small.

---

**Example 13–6**

**Objective:** To calculate the threshold voltage shift due to radiation-induced oxide charge trapping.

Consider a MOS device with a 1000 Å oxide. Assume that a pulse of ionizing radiation creates $10^{17}$ electron-holes pairs per $cm^3$ in the oxide. Also assume that the electrons are swept out through the gate terminal with zero recombination, and that 20 percent of the generated holes are trapped at the oxide–semiconductor interface.

**Solution:** The areal density of holes generated in the oxide is

$$N_h = (10^{17})(1000 \times 10^{-8}) = 10^{12} \text{ cm}^{-2}$$

The equivalent surface charge density trapped is then

$$Q'_{ss} = (10^{12})(0.2) = 2 \times 10^{11} \text{ cm}^{-2}$$

The threshold voltage shift is

$$\Delta V_T = -\frac{Q'_{ss}}{C_{ox}}$$

where

$$C_{ox} = \frac{\varepsilon_{ox}}{t_{ox}} = \frac{(3.9)(8.85 \times 10^{-14})}{1000 \times 10^{-8}} = 3.45 \times 10^{-8} \text{ F/cm}^2$$

We then have

$$\Delta V_T = -\frac{(2 \times 10^{11})(1.6 \times 10^{-19})}{3.45 \times 10^{-8}} = -0.93 \text{ volt}$$

**Comment:** As we have seen previously, a positive fixed oxide charge shifts the threshold voltage in the negative voltage direction.

---

One failure mechanism caused by the radiation-induced oxide charge in an n-channel MOSFET in an integrated circuit is a shift from enhancement mode to depletion mode. The device will be turned on rather than off at zero gate voltage; consequently, the circuit function may be disrupted or an excessive power supply current may be generated in the circuit.

The gate voltage in a p-channel MOSFET is normally negative with respect to the substrate. Radiation-generated holes in the oxide are forced to the gate-oxide interface. The trapped charge in this region has less effect on the threshold voltage, so threshold shifts in p-channel MOSFETs are normally smaller if the trap concentrations at the gate-oxide and oxide–semiconductor interfaces are of the same order of magnitude.

### 13.4.2 Radiation-Induced Interface States

We have considered the effect of interface states on the C-V characteristics of a MOS capacitor and on the MOSFET characteristics. The net charge in the interface states of an n-channel MOS device at the threshold inversion point is negative. This negative charge will cause a shift in threshold voltage in the positive voltage direction, which is opposite to the shift due to the positive oxide charge. In addition, since the interface states can be charged, they are another source of coulomb interaction with the inversion charge carrier, which means that the inversion carrier mobility is a function of the interface state density through surface-scattering effects. Interface states, then, affect both threshold voltage and carrier mobility.

When MOS devices are exposed to ionizing radiation, additional interface states are generated at the Si–SiO$_2$ interface. The radiation-induced interface states tend to be donor states in the lower half of the bandgap and acceptor states in the upper half. Figure 13–30 shows the threshold voltage in an n-channel and p-channel MOSFET as a function of ionizing radiation dose. We initially see the negative threshold voltage shift in both devices due to the radiation-induced positive oxide charge. The reversal in threshold shift at the higher dose levels is attributable to the creation of radiation-induced interface states that tend to compensate the radiation-induced positive oxide charge.

In our discussion of subthreshold conduction, we mentioned that the slope of the ln $I_D$ versus $V_{GS}$ curves in the subthreshold region is a function of the density of interface states. Figure 13–31 shows the subthreshold cur-

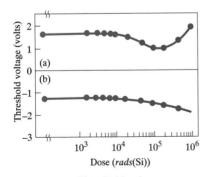

**Figure 13–30** Threshold voltage versus total ionizing radiation dose of (a) an n-channel MOSFET, and (b) a p-channel MOSFET. (From Ma et al. [5].)

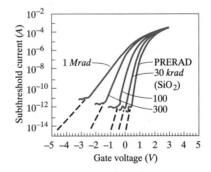

**Figure 13–31** Subthreshold current versus gate voltage of a MOSFET prior to irradiation and at four total radiation dose levels. (From Ma et al. [5].)

rent at several total ionizing dose levels. The change in slope indicates that the density of interface states is increasing with total dose.

The buildup of radiation-induced interface states occurs over a relatively long time period and is a very strong function of the applied electric field in the oxide. Figure 13–32 shows the radiation-induced interface state density

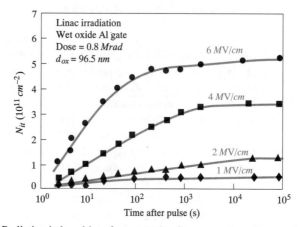

**Figure 13–32** Radiation-induced interface state density versus time after a pulse of ionizing radiation for several values of oxide electric field. (From Ma et al. [5].)

versus time for several applied fields. The final interface state density is reached between 100 to 10,000 seconds after a pulse of ionizing radiation. Almost all models for the generation of radiation-induced interface states depend on the transport or trapping of radiation-generated holes near the Si–SiO$_2$ interface. This transport and trapping process is time and field dependent, supporting the time and field dependence of the interface state buildup.

The sensitivity of the Si–SiO$_2$ interface to the buildup of radiation-induced interface states is a strong function of device processing. The interface state buildup in aluminum-gate MOSFETs tends to be smaller than in polysilicon-gate devices. This difference is probably more a result of variations between the two processing technologies than an inherent difference. Hydrogen appears to be important in the radiation-induced interface state buildup—hydrogen tends to passivate dangling silicon bonds at the interface, reducing the preradiation density of interface states. However, devices passivated with hydrogen appear to be more susceptible to the buildup of radiation-induced interface states. The silicon–hydrogen bond at the interface may be broken by the radiation process, which leaves a dangling silicon bond that acts like an interface state trap. These traps at the interface have been identified through electron spin resonance experiments.

Interface states may seriously affect the MOSFET characteristics, which in turn can affect MOSFET circuit performance. Radiation-induced interface states can cause shifts in threshold voltage, affecting circuit performance as we have discussed. A reduction in mobility can affect the speed and output drive capability of a circuit.

### 13.4.3 Hot Electron Charging Effects

We have considered breakdown voltage effects in a MOSFET. In particular, as the electric field in the drain junction space charge region increases, electron-hole pairs can be generated by impact ionization. The generated electrons tend to be swept to the drain and generated holes swept into the substrate in an n-channel MOSFET.

Some of the electrons generated in the space charge region are attracted to the oxide due to the electric field induced by a positive gate voltage; this effect is shown in Figure 13–33. These generated electrons have energies far

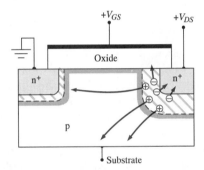

**Figure 13–33**   Hot carrier generation, current components, and electron injection into the oxide.

greater than the thermal-equilibrium value and are called hot electrons. If the electrons have energies on the order of 1.5 eV, they may be able to tunnel into the oxide; or in some cases they may be able to overcome the silicon-oxide potential barrier and produce a gate current, which may be in the range of $fA(10^{-15} A)$ or perhaps $pA(10^{-12} A)$. A fraction of the electrons traveling through the oxide may be trapped, producing a net negative charge density in the oxide. The probable cause of electron trapping is usually less than that of hole trapping; but a hot electron induced gate current may exist over a long period of time, therefore the negative charging effect may build up. The negative oxide charge trapping will cause a local positive shift in the threshold voltage.

The energetic electrons, as they cross the $Si–SiO_2$ interface, can generate additional interface states. The probable cause of interface state generation is due to the breaking of silicon-hydrogen bonds—a dangling silicon bond is produced, which acts as an interface state. The charge trapping in interface states causes a shift in threshold voltage, additional surface scattering, and reduced mobility. The hot electron charging effects are continuous processes so the device degrades over a period of time. This degradation is obviously an undesirable effect and may tend to limit the useful life of the device. The lightly doped drain (LDD) structure was discussed in Section 13.3.2. The maximum electric field is reduced in this device, decreasing the probability of impact ionization and hot electron effects.

## 13.5 SUMMARY AND REVIEW

In the previous chapter we covered the fundamental physics of the MOSFET; in this chapter we considered some additional concepts commonly encountered in these devices.

We discussed four nonideal MOSFET characteristics. One of these is subthreshold conduction. In the idealized MOSFET model, the drain current is zero when the gate voltage is equal to the threshold voltage. However, the channel can be biased in weak inversion and small drain currents will be present. These subthreshold currents may contribute significantly to power dissipation in a VLSI circuit because of the large number of devices. The second nonideal effect considered was channel length modulation. The effective electrical length of the channel decreases with drain voltage as the depletion region at the drain extends into the channel when the MOSFET is biased in the saturation region. The channel length becomes bias dependent; the drain current is then a function of drain-to-source voltage, even in the saturation region. Channel length modulation results in a finite output impedance at the drain terminal. The channel carrier mobility is a function of surface scattering, which leads to the third nonideal effect considered. As the gate voltage increases, the transverse electric field at the oxide–semiconductor interface increases, causing additional surface scattering. Since the inversion carrier mobility is a strong function of surface scattering, mobility is not constant as we had assumed in the ideal I-V derivation, but is a function of gate voltage. In addition, carriers may reach their saturation velocity in the channel; therefore, the drain current will saturate at a lower value.

The tendency in MOSFET design is to make the device smaller and smaller. The transconductance and cutoff frequency, for example, increase as channel length decreases. However, the threshold voltage begins to deviate from its predicted value as the channel length shrinks. The threshold voltage is a function of the amount of bulk charge in the semiconductor which is controlled by the gate voltage. As channel length decreases, the amount of bulk charge controlled by the source and drain regions increases, so that the fraction controlled by the gate decreases. The threshold voltage of an n-channel MOSFET, then, shifts in the negative-voltage direction as the channel length is reduced. As the channel width decreases, the packing density of MOSFETs in an integrated circuit increases. However, as channel width decreases, the amount of bulk charge at the ends of the channel can become a significant fraction of the total charge controlled by the gate. This charge must be included in the threshold voltage calculation. The additional bulk charge causes the threshold voltage of an n-channel MOSFET to shift in the positive voltage direction as the channel width is reduced.

We considered breakdown and threshold adjustment by ion implantation. There are several voltage breakdown mechanisms in the MOSFET. Oxide breakdown occurs when the electric field in the oxide reaches approximately $6 \times 10^6$ V/cm. Oxide breakdown is usually not a problem except in high-voltage devices or in ultrathin oxide devices. Avalanche breakdown can occur in the space charge region near the drain terminal and, because of the curvature of the drain region, increasing the concentration of electric field lines, the breakdown voltage is less than that of a planar pn junction. Avalanche breakdown can also be enhanced by the parasitic bipolar transistor associated with the MOSFET. Positive feedback provided by the parasitic bipolar device can produce the snapback breakdown characteristic. Breakdown also occurs when the space charge region at the drain contact extends completely across the channel region to the source space charge region, causing punch-through. However, the drain current will begin to increase rapidly before the actual punch-through condition is reached, as a result of the near punch-through condition. The various breakdown mechanisms are all functions of the maximum electric field in the drain space charge region. This field can be reduced by using a lightly doped drain (LDD) MOSFET.

The semiconductor doping concentration in the channel region is one of the variables that determines the threshold voltage of a MOSFET. The threshold voltage can be modified, or adjusted, by changing the doping concentration with ion implantation. A precise, controlled number of either donor or acceptor ions can be implanted in the semiconductor near the oxide surface. An implant of acceptor ions into either a p- or n-type substrate will shift the threshold voltage to more positive values while an implant of donor ions will shift the threshold voltage to more negative values. All other parameters involved in the threshold voltage may be fixed in a particular design or fabrication process; thus, ion implantation provides a flexible way of adjusting the threshold voltage to a desired value.

Ionizing radiation and hot electron effects are two processes that generate fixed oxide charge and additional interface states. Ionizing radiation produces electron-hole pairs in the gate oxide. Since electrons are quite mobile, they are swept out of the oxide through the gate terminal under a positive gate bias. Holes, on the other hand, undergo a stochastic hopping transport process through the oxide toward the oxide–semiconductor interface. A fraction of the holes are captured in trapping sites near the oxide-semiconductor interface; this produces a negative threshold voltage shift in the device. Additional interface states can also be generated during this process. The buildup of interface states occurs over a relatively long time period and is a very strong function of the applied electric field in the oxide, the presence of hydrogen, and various processing characteristics of the oxide. Radiation-generated interface states can cause additional shifts in threshold voltage as well as reducing carrier mobility.

Other charging effects can occur due to hot electron effects. Electrons created in the space charge region near the drain terminal by impact ionization have energies far greater than the thermal-equilibrium value and may tunnel or be swept into the oxide. A fraction of these hot electrons trapped in the oxide may produce a net trapped charge that causes local shifts in threshold voltage. Additional interface states can also be created by the energetic electrons crossing the oxide–semiconductor interface, producing a shift in the threshold voltage as well as a degradation in the carrier mobility. Since hot electron charging effects occur over a long time period, the MOSFET characteristics may degrade with time and the useful lifetime of the device may be limited.

## GLOSSARY OF IMPORTANT TERMS

**Channel length modulation:** The change in effective channel length with drain-to-source voltage when the MOSFET is biased in saturation.

**Hot electrons:** Electrons with energies far greater than the thermal-equilibrium value caused by acceleration in high electric fields.

**Lightly Doped Drain (LDD):** A MOSFET with a lightly doped drain region adjacent to the channel to reduce voltage breakdown effects.

**Narrow-channel effects:** The shift in threshold voltage as the channel width narrows.

**Near punch-through:** The reduction in the potential barrier between source and substrate by the drain-to-substrate voltage, resulting in a rapid increase in drain current.

**Short-channel effects:** The shift in threshold voltage as the channel length becomes smaller.

**Snapback:** The negative resistance effect during breakdown in a MOSFET caused by the variable current gain in a parasitic bipolar transistor.

**Subthreshold conduction:** The process of current conduction in a MOSFET when the transistor is biased below the threshold inversion point.

**Surface scattering:** The process of electric field attraction and coulomb repulsion of carriers at the oxide–semiconductor interface as the carriers drift between source and drain.

**Threshold adjustment:** The process of altering the threshold voltage by changing the semiconductor doping concentration through ion implantation.

# PROBLEMS

(Note: In the following problems, assume the semiconductor and oxide in the MOS system are silicon and silicon dioxide, respectively, and assume the temperature is $T = 300°K$ unless otherwise stated.)

## Section 13.1

1. A silicon n-channel MOSFET has an acceptor doping of $N_a = 10^{16}$ cm$^{-3}$ and a threshold voltage of $V_T = +0.75$ volt. (a) Determine the minimum channel length so that the incremental change $\Delta L$ is no more than 10 percent of the original length $L$ when $V_{DS} = 5$ volts and $V_{GS} = 5$ volts. (b) Repeat part (a) for $V_{GS} = 2$ volts.

2. Consider the n-channel MOSFET described in problem 20 of Chapter 12. Let $N_a = 3 \times 10^{16}$ cm$^{-3}$. (a) Using Equations (13–4) and (13–2), calculate the output conductance defined as $g_0 = \partial I_D'/\partial V_{DS}$ for $V_{GS} = 2$ volts and $\Delta V_{DS} = 1$ volt. (b) Repeat part (a) if the channel length is reduced to $L = 1$ $\mu$m.

3. (a) Consider an n-channel enhancement mode MOSFET with $(W/L) = 10$, $C_{ox} = 6.9 \times 10^{-8}$ F/cm$^2$, and $V_T = +1$ volt. Assume a constant mobility of $\mu_n = 500$ cm$^2$/V-sec. Plot $\sqrt{I_D}$ versus $V_{GS}$ for $0 \le V_{GS} \le 5$ volts when the transistor is biased in the saturation region. (b) Now assume that the effective mobility in the channel is given by

$$\mu_{\text{eff}} = \mu_0 \left(\frac{E_{\text{eff}}}{E_c}\right)^{-1/3}$$

where $\mu_0 = 1000$ cm$^2$/V-sec and $E_c = 2.5 \times 10^4$ V/cm. As a first approximation, let $E_{\text{eff}} = V_{GS}/t_{ox}$. Using $\mu_{\text{eff}}$ in place of $\mu_n$ in the $\sqrt{I_D}$ versus $V_{GS}$ relation, plot $\sqrt{I_D}$ versus $V_{GS}$ over the same $V_{GS}$ range as in part (a). (c) Plot the curves from parts (a) and (b) on the same graph. What can be said about the slopes of the two curves?

4. An n-channel enhancement mode MOSFET has the following parameters:

$t_{ox} = 400$ Å          $N_a = 5 \times 10^{16}$ cm$^{-3}$

$V_{FB} = -1.2$ volt       $L = 2$ $\mu$m

$W = 20$ $\mu$m

(a) Assuming a constant mobility of $\mu_n = 400$ cm$^2$/V-sec, plot $I_D$ versus $V_{DS}$ for $V_{GS} - V_T = 1$ volt and $V_{GS} - V_T = 2$ volts over the range $0 \le V_{DS} \le 5$ volts. (b) Consider the piecewise linear

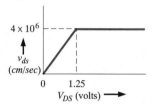

**Figure 13–34**    Figure for problem 4.

model of the carrier velocity versus $V_{DS}$ shown in Figure 13–34. Again plot $I_D$ versus $V_{DS}$ for the same voltage values given in part (a). Compare the $V_{DS}$(sat) values for the curves in parts (a) and (b).

## Section 13.2

5.  Consider an n-channel MOSFET with $N_a = 10^{16}$ cm$^{-3}$ and $t_{ox} = 450$ Å. If $r_j = 0.3$ $\mu$m and $L = 1$ $\mu$m, determine the threshold shift due to the short channel effect.

6.  An n-channel MOSFET is doped to $N_a = 3 \times 10^{16}$ cm$^{-3}$ and has an oxide thickness of $t_{ox} = 800$ Å. The diffused junction radius is $r_j = 0.60$ $\mu$m. If the threshold shift due to short-channel effects is to be no more than $\Delta V_T = -0.20$ volt, determine the minimum channel length $L$.

*7.  The shift in threshold voltage due to short-channel effects given by Equation (13–17) assumed all space charge regions were of equal width. If a drain voltage is applied, this condition is no longer valid. Using the same trapezoidal approximation, show that the threshold voltage shift is given by

$$\Delta V_T = -\frac{eN_a x_{dT}}{C_{ox}} \cdot \frac{r_j}{2L} \left\{ \left[ \sqrt{1 + \frac{2x_{ds}}{r_j} + \alpha^2} - 1 \right] \right.$$
$$\left. + \left[ \sqrt{1 + \frac{2x_{dD}}{r_j} + \beta^2} - 1 \right] \right\}$$

where

$$\alpha^2 = \frac{x_{ds}^2 - x_{dT}^2}{r_j^2} \qquad \beta^2 = \frac{x_{dD}^2 - x_{dT}^2}{r_j^2}$$

and where $x_{ds}$ and $x_{dD}$ are the source and drain space charge widths, respectively.

8.  An n-channel MOSFET has a substrate doping of $N_a = 10^{16}$ cm$^{-3}$ and an oxide thickness of $t_{ox} = 450$ Å. The channel width is $W = 2.5$ $\mu$m. Neglecting short-channel effects, calculate the threshold

voltage shift due to narrow-channel effects. (Assume the fitting parameter is $\xi = \pi/2$.)

9.  Consider an n-channel MOSFET with $N_a = 3 \times 10^{16}$ cm$^{-3}$ and $t_{ox} = 800$ Å. The depletion regions at the end of the channel width can be approximated by triangular regions shown in Figure 13–35. Assume both the lateral and vertical depletion widths are equal to $x_{dT}$. If the threshold shift due to narrow-channel effects is $\Delta V_T = +0.25$ volt, calculate the channel width $W$.

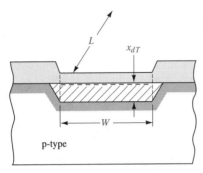

**Figure 13–35**   Figure for problem 9.

## Section 13.3

10.  A MOS device has a silicon dioxide gate insulator with a thickness of $t_{ox} = 250$ Å. (a) Calculate the ideal oxide breakdown voltage. (b) If a safety factor of three is required, determine the maximum safe gate voltage that may be applied.

11.  In a power MOS device, a maximum gate voltage of 20 volts is to be applied. If a safety factor of three is specified, determine the minimum thickness necessary for the silicon dioxide gate insulator.

*12.  The snapback breakdown condition is defined to be when $\alpha M = 1$, where $\alpha$ is the common base current gain and $M$ is the multiplication constant given by Equation (13–28). Let $m = 3$ and let $V_{BD} = 15$ volts. The common base current gain is a very strong function of junction current $I_D$. Assume $\alpha$ is described by the relation

$$\alpha = (0.18) \log_{10}\left(\frac{I_D}{3 \times 10^{-9}}\right)$$

where $I_D$ is given in amperes. Plot the curve of $I_D$ versus $V_{CE}$ which satisfies the snapback condition over the range $10^{-8} \le I_D \le 10^{-3}$ amps. (Use a log scale for the current.)

13. Near punch-through occurs when the two depletion regions are within approximately six Debye lengths of each other. The extrinsic Debye length $L_D$ is defined as

$$L_D = \left\{ \frac{\varepsilon_s(kT/e)}{eN_a} \right\}^{1/2}$$

Consider the n-channel MOSFET in Example 13–4. Calculate the near punch-through voltage. How does this voltage compare to the ideal punch-through voltage determined in the example?

14. The near punch-through voltage (see problem 13) of an n-channel MOSFET is to be no less than $V_{DS} = 5$ volts. The source and drain regions are doped $N_d = 10^{19}$ cm$^{-3}$ and the channel region is doped $N_a = 3 \times 10^{16}$ cm$^{-3}$. The source and body are at ground potential. Determine the minimum channel length.

15. Repeat problem 14 if a source-substrate voltage $V_{SB} = 2$ volts is applied.

16. Consider an n-channel MOSFET with an $n^+$ polysilicon gate and with $N_a = 2 \times 10^{15}$ cm$^{-3}$, $t_{ox} = 650$ Å, and $Q'_{ss} = 2 \times 10^{11}$ cm$^{-2}$. (a) Calculate the threshold voltage. (b) The desired threshold voltage is $V_T = +0.80$ volt. Determine the type and ion implant density required to achieve this specification. Assume the implant is directly adjacent to the oxide–semiconductor interface.

17. A MOS transistor with an aluminum gate is fabricated on an n-type semiconductor substrate. The doping is $N_d = 10^{16}$ cm$^{-3}$, the oxide thickness is $t_{ox} = 750$ Å, and the equivalent fixed oxide charge is $Q'_{ss} = 5 \times 10^{11}$ $cm^{-2}$. (a) Calculate the threshold voltage. (b) The desired threshold voltage is $V_T = -0.50$ volt. Determine the type and ion implant density required to achieve this specification. Assume the implant is directly adjacent to the oxide–semiconductor interface.

18. Consider an n-channel MOSFET with a doping of $N_a = 10^{15}$ cm$^{-3}$, an oxide thickness of $t_{ox} = 750$ Å, and an initial flat-band voltage of $V_{FB} = -1.50$ volts. (a) Calculate the threshold voltage. (b) Determine the type and the ion implant density $D_I$ required to achieve a threshold voltage of $V_T = +0.90$ volt with zero volts applied to the substrate. (c) Use the results of part (b) and determine the threshold voltage if a source-to-body voltage $V_{SB} = 2$ volts is applied.

### Section 13.4

19. One rad(Si)[1] produces on the average $8 \times 10^{12}$ electron-hole pairs/cm$^3$ in silicon dioxide. Assume that a pulse of ionizing radiation

---

[1] One rad(Si) is equivalent to 100 ergs of energy deposited per cm$^3$ in silicon. We normally use this same total dose notation for the total dose effects in silicon dioxide.

with a total dose of $10^5$ rads(Si) is incident on a MOS device with a 750 Å oxide. Assume that there is no electron-hole recombination and that the electrons are swept out through the gate terminal. If 10 percent of the generated holes are trapped at the oxide–semiconductor interface, calculate the threshold voltage shift.

20. Reconsider problem 19. If the threshold voltage shift is to be no more than $\Delta V_T = -0.50$ volt, calculate the maximum percentage of holes that can be trapped.

21. Show that, for the simple model of radiation-induced hole trapping we have considered, the threshold voltage shift is proportional to $\Delta V_T \propto -t_{ox}^2$. Thin oxides are one requirement for radiation-tolerant MOS devices.

## Summary and Review

*22. Reconsider problem 32 in Chapter 12 taking into account short-channel effects.

*23. A particular process produces an n-channel MOSFET with the following properties:

$$t_{ox} = 325 \text{ Å} \qquad L = 0.8 \ \mu\text{m}$$
$$N_a = 10^{16} \text{ cm}^{-3} \qquad W = 20 \ \mu\text{m}$$
$$n^+ \quad \text{polysilicon gate} \quad r_j = 0.35 \ \mu\text{m}$$
$$Q'_{ss} = 10^{11} \text{ cm}^{-2}$$

The desired threshold voltage is $V_T = 0.35$ V at $T = 300°$K. Design an additional process to achieve this objective by using ion implantation, which produces a step function profile that is 0.35 $\mu$m deep.

## READING LIST

1. Akers, L. A., and J. J. Sanchez. "Threshold Voltage Models of Short, Narrow, and Small Geometry MOSFETs: A Review." *Solid State Electronics* 25 (July 1982), pp. 621–41.

2. Baliga, B. J. *Modern Power Devices.* New York: Wiley, 1987.

3. Brews, J. R. "Threshold Shifts Due to Nonuniform Doping Profiles in Surface Channel MOSFETs." *IEEE Transactions on Electron Devices* ED-26 (November 1979), pp. 1696–1710.

4. Klaassen, F. M., and W. Hes. "On the Temperature Coefficient of the MOSFET Threshold Voltage." *Solid State Electronics* 29 (August 1986), pp. 787–89.

5. Ma, T. P., and P. V. Dressendorfer. *Ionizing Radiation Effects in MOS Devices and Circuits.* New York: Wiley, 1989.

6. Muller, R. S., and T. I. Kamins. *Device Electronics for Integrated Circuits.* 2nd ed. New York: Wiley, 1986.

7.  Neamen, D. A.; B. Buchanan; and W. Shedd. "Ionizing Radiation Effects in SOS Structures." *IEEE Transactions on Nuclear Science* NS-22 (December 1975), pp. 2197–2202.

8.  Nicollian, E. H., and J. R. Brews. *MOS Physics and Technology*. New York: Wiley, 1982.

9.  Ning, T. H.; P. W. Cook; R. H. Dennard; C. M. Osburn; S. E. Schuster; and H. N. Yu. "1 $\mu$m MOSFET VLSI Technology: Part IV—Hot Electron Design Constraints." *IEEE Transactions on Electron Devices* ED-26 (April 1979), pp. 346–53.

10. Ogura, S.; P. J. Tsang; W. W. Walker; D. L. Critchlow; and J. F. Shepard. "Design and Characteristics of the Lightly Doped Drain-Source (LDD) Insulated Gate Field-Effect Transistor." *IEEE Transactions on Electron Devices* ED-27 (August 1980), pp. 1359–67.

11. Ong, D. G. *Modern MOS Technology: Processes, Devices, and Design*. New York: McGraw-Hill, 1984.

12. Pierret, R. F. *Field Effect Devices*. Vol. 4 of the *Modular Series on Solid State Devices*. 2nd ed. Reading, Mass.: Addison-Wesley, 1990.

13. Sanchez, J. J.; K. K. Hsueh; and T. A. DeMassa. "Drain-Engineered Hot-Electron-Resistant Device Structures: A Review." *IEEE Transactions on Electron Devices* ED-36 (June 1989), pp. 1125–32.

14. Schroder, D. K. *Advanced MOS Devices*. *Modular Series on Solid State Devices*. Reading, Mass.: Addison-Wesley, 1987.

*15. Shur, M. *Physics of Semiconductor Devices*. Englewood Cliffs, N.J.: Prentice Hall, 1990.

16. Streetman, B. G. *Solid State Electronic Devices*. 3rd ed. Englewood Cliffs, N.J.: Prentice Hall, 1990.

17. Sze, S. M. *Physics of Semiconductor Devices*. 2nd ed. New York: Wiley, 1981.

18. Tsividis, Y. P. *Operation and Modeling of the MOS Transistor*. New York: McGraw-Hill, 1987.

19. Yang, E. S. *Microelectronic Devices*. New York: McGraw-Hill, 1988.

20. Yau, L. D. "A Simple Theory to Predict the Threshold Voltage of Short-Channel IGFETs." *Solid-State Electronics* 17 (October 1974), pp. 1059–63.

# OPTICAL DEVICES

In previous chapters, we considered the basic physics of transistors used to amplify or switch electrical signals. Semiconductor devices can also be designed and fabricated to detect and generate optical signals. In this chapter, we discuss the basic principles of solar cells, photodetectors, light emitting diodes, and laser diodes. Solar cells and photodetectors convert optical power into electrical power; light emitting diodes and laser diodes convert electrical power into optical power.

The characteristics of solar cells and photodetectors are a function of optical energy, which is absorbed in a semiconductor and generates excess electron-hole pairs producing photocurrents. The output terminals of the solar cell are connected to a resistive load so that the input optical power is converted to electrical power. The simple pn junction solar cell is considered first. The characteristics of solar cells we will discuss include short circuit current, open circuit voltage, maximum power, and conversion efficiency. We will consider the effect of nonuniform generation of excess carriers in the solar cell, and then discuss the heterojunction and amorphous silicon solar cells.

Photodetectors are semiconductor devices that convert optical signals into electrical signals. The photoconductor, in which photon-generated excess electron-hole pairs change the conductivity of a semiconductor, is perhaps the simplest type of photodetector. A photodiode is a pn junction diode operated with a reverse-bias voltage applied. The excess electrons and holes created in the space charge region are separated very quickly by the electric field so that a photocurrent is generated. The PIN and avalanche photodiodes are variations of the simple pn photodiode. The photocurrent generated in the B-C junction of a bipolar transistor is multiplied by the gain of the transistor. This photodetector is referred to as a phototransistor.

The inverse mechanism of a photodetector is electroluminescence. Excess carriers are generated and then recombine, which may result in the

609

emission of photons in a forward-biased pn junction—such a device is referred to as a light emitting diode, or LED. The LED photon emission is due
to a spontaneous transition of an electron from the conduction to the valence
band, resulting in the fairly wide spectral output bandwidth. The spectral
output of an LED has a relatively wide bandwidth of wavelengths but the
emission spectrum is narrow enough so that a particular color will be visible
to the eye, provided that the output is in the visible range. If the LED device
structure and operating mode are modified, the device can operate in a new
mode in which the bandwidth is very narrow and the photon output is coherent. This new device is called a laser diode. The laser diode relies on the
concepts of stimulated emission and population inversion, and on the use of
an optical resonant cavity.

## 14.1  OPTICAL ABSORPTION

In Chapter 2, we discussed the wave–particle duality principle and indicated
that light waves could be treated as particles, which are referred to as photons. The energy of a photon is $E = h\nu$ where $h$ is Plank's constant and $\nu$ is
the frequency. We can also relate the wavelength and energy by

$$\lambda = \frac{c}{\nu} = \frac{hc}{E} = \frac{1.24}{E}\ \mu m \qquad (14\text{--}1)$$

where $E$ is the photon energy in eV and $c$ is the speed of light.

   There are several possible photon-semiconductor interaction mechanisms. For example, photons can interact with the semiconductor lattice
whereby the photon energy is converted into heat. Photons can also interact
with impurity atoms, either donors or acceptors, or they can interact with
defects within the semiconductor. However, the basic photon interaction
process of greatest interest is the interaction with valence electrons. When a
photon collides with a valence electron, enough energy may be imparted to
elevate the electron into the conduction band. Such a process generates
electron-hole pairs and creates excess carrier concentrations. The behavior
of excess carriers in a semiconductor was considered in Chapter 6.

### 14.1.1  Photon Absorption Coefficient

When a semiconductor is illuminated with light, the photons may be absorbed or they may propagate through the semiconductor, depending on the
photon energy and on the bandgap energy $E_g$. If the photon energy is less
than $E_g$, the photons are not readily absorbed. In this case, the light is
transmitted through the material and the semiconductor appears to be transparent.

   If $E = h\nu > E_g$, the photon can interact with a valence electron and
elevate the electron into the conduction band. The valence band contains
many electrons and the conduction band contains many empty states, so the

probability of this interaction is high when $h\nu > E_g$. This interaction creates an electron in the conduction band and a hole in the valence band—an electron-hole pair. The basic absorption processes for different values of $h\nu$ are shown in Figure 14–1. When $h\nu > E_g$, an electron-hole pair is created and the excess energy may give the electron or hole additional kinetic energy, which will be dissipated as heat in the semiconductor.

The intensity of the photon flux is denoted by $I_\nu(x)$ and is expressed in terms of energy/cm²-sec. Figure 14–2 shows an incident photon intensity

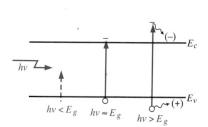

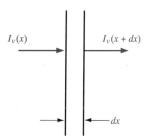

**Figure 14–1**   Optically generated electron-hole pair formation in a semiconductor.

**Figure 14–2**   Optical absorption in a differential length.

at a position $x$ and the photon flux emerging at a distance $x + dx$. The energy absorbed per unit time in the distance $dx$ is given by

$$\alpha I_\nu(x) \, dx \qquad (14\text{–}2)$$

where $\alpha$ is the absorption coefficient. The absorption coefficient is the relative number of photons absorbed per unit distance, given in units of cm⁻¹.

From Figure 14–2, we can write

$$I_\nu(x + dx) - I_\nu(x) = \frac{dI_\nu(x)}{dx} \cdot dx = -\alpha I_\nu(x) \, dx \qquad (14\text{–}3)$$

or

$$\frac{dI_\nu(x)}{dx} = -\alpha I_\nu(x) \qquad (14\text{–}4)$$

If the initial condition is given as $I_\nu(0) = I_{\nu 0}$, then the solution to the differential equation, Equation (14–4), is

$$I_\nu(x) = I_{\nu 0} e^{-\alpha x} \qquad (14\text{–}5)$$

The intensity of the photon flux decreases exponentially with distance through the semiconductor material. The photon intensity as a function of $x$ for two general values of absorption coefficient is shown in Figure 14–3. If

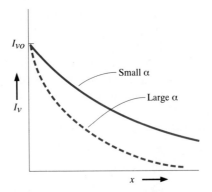

**Figure 14–3** Photon intensity versus distance for two absorption coefficients.

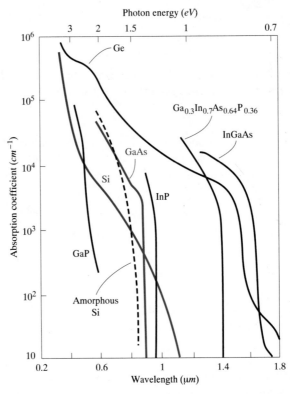

**Figure 14–4** Absorption coefficient as a function of wavelength for several semiconductors. (From Shur [8].)

the absorption coefficient is large, the photons are absorbed over a relatively short distance.

The absorption coefficient in the semiconductor is a very strong function of photon energy and bandgap energy. Figure 14–4 shows the absorption coefficient $\alpha$ plotted as a function of wavelength for several semiconductor materials. The absorption coefficient increases very rapidly for $hv > E_g$, or for $\lambda < 1.24/E_g$. The absorption coefficients are very small for $hv < E_g$, so the semiconductor appears transparent to photons in this energy range.

---

**Example 14–1**

**Objective:** To calculate the thickness of a semiconductor that will absorb 90 percent of the incident photon energy.

Consider silicon and assume that in the first case the incident wavelength is $\lambda = 1.0$ $\mu$m and in the second case, the incident wavelength is $\lambda = 0.5$ $\mu$m.

**Solution:** From Figure 14–4, the absorption coefficient is $\alpha \approx 10^2$ cm$^{-1}$ for $\lambda = 1.0$ $\mu$m. If 90 percent of the incident flux is to be absorbed in a distance $d$, then the flux emerging at $x = d$ will be 10 percent of the incident flux. We can write

$$\frac{I_v(d)}{I_{v0}} = 0.1 = e^{-\alpha d}$$

Solving for the distance $d$, we have

$$d = \frac{1}{\alpha} \ln \left(\frac{1}{0.1}\right) = \frac{1}{10^2} \ln (10) = 0.0230 \text{ cm}$$

In the second case, the absorption coefficient is $\alpha \approx 10^4$ cm$^{-1}$ for $\lambda = 0.5$ $\mu$m. The distance $d$, then, in which 90 percent of the incident flux is absorbed, is

$$d = \frac{1}{10^4} \ln \left(\frac{1}{0.1}\right) = 2.30 \times 10^{-4} \text{ cm} = 2.30 \text{ } \mu\text{m}$$

**Comment:** As the incident photon energy increases, the absorption coefficient increases rapidly, so that the photon energy can be totally absorbed in a very narrow region at the surface of the semiconductor.

---

The relation between the bandgap energies of some of the common semiconductor materials and the light spectrum is shown in Figure 14–5. We may note that silicon and gallium arsenide will absorb all of the visible spectrum, whereas gallium phosphide, for example, will be transparent to the red spectrum.

## 14.1.2 Electron-Hole Pair Generation Rate

We have shown that photons with energy greater than $E_g$ can be absorbed in a semiconductor, thereby creating electron-hole pairs. The intensity $I_v(x)$ is

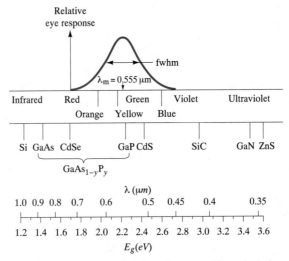

**Figure 14–5** Light spectrum versus wavelength and energy. Figure includes relative response of the human eye. (From Sze [11].)

in units of energy/cm²-sec and $\alpha I_\nu(x)$ is the rate at which energy is absorbed per unit volume. If we assume that one absorbed photon at an energy $h\nu$ creates one electron-hole pair, then the generation rate of electron-hole pairs is

$$g' = \frac{\alpha I_\nu(x)}{h\nu} \tag{14–6}$$

which is in units of #/cm³-sec. We may note that the ratio $I_\nu(x)/h\nu$ is the photon flux. If, on the average, one absorbed photon produces less than one electron-hole pair, then Equation (14–6) must be multiplied by an efficiency factor.

---

**Example 14–2**

**Objective:** To calculate the generation rate of electron-hole pairs given an incident intensity of photons.

Consider gallium arsenide at $T = 300°K$. Assume the photon intensity at a particular point is $I_\nu(x) = 0.05$ Watts/cm² at a wavelength of $\lambda = 0.75 \ \mu$m. This intensity is typical of sunlight, for example.

**Solution:** The absorption coefficient for gallium arsenide at this wavelength is $\alpha \approx 0.7 \times 10^4$ cm⁻¹. The photon energy, using Equation (14–1), is

$$E = h\nu = \frac{1.24}{0.75} = 1.65 \text{ eV}$$

Then, from Equation (14–6) and including the conversion factor between joules and eV, we have, for a unity efficiency factor,

$$g' = \frac{\alpha I_\nu(x)}{h\nu} = \frac{(0.7 \times 10^4)(0.05)}{(1.6 \times 10^{-19})(1.65)} = 1.33 \times 10^{21} \text{ cm}^{-3} \text{ sec}^{-1}$$

If the incident photon intensity is a steady-state intensity, then, from Chapter 6, the steady-state excess carrier concentration is $\delta n = g'\tau$, where $\tau$ is the excess minority carrier lifetime. If $\tau = 10^{-7}$ sec, for example, then

$$\delta n = (1.33 \times 10^{21})(10^{-7}) = 1.33 \times 10^{14} \text{ cm}^{-3}$$

**Comment:** This example gives an indication of the magnitude of the electron-hole generation rate and the magnitude of the excess carrier concentration. Obviously, as the photon intensity decreases with distance in the semiconductor, the generation rate also decreases.

## 14.2 SOLAR CELLS

A solar cell is a pn junction device with no voltage directly applied across the junction. The solar cell converts photon power into electrical power and delivers this power to a load. These devices have long been used for the power supply of satellites and space vehicles, and also as the power supply to some calculators. We will first consider the simple pn junction solar cell with uniform generation of excess carriers. We will also discuss briefly the heterojunction and amorphous silicon solar cells.

### 14.2.1 The pn Junction Solar Cell

Consider the pn junction shown in Figure 14–6 with a resistive load. Even with zero bias applied to the junction, an electric field exists in the space

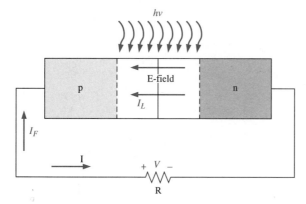

**Figure 14–6** A pn junction solar cell with resistive load.

charge region as shown in the figure. Incident photon illumination can create electron-hole pairs in the space charge region that will be swept out producing the photocurrent $I_L$ in the reverse-bias direction as shown.

The photocurrent $I_L$ produces a voltage drop across the resistive load which forward biases the pn junction. The forward-bias voltage produces a forward-bias current $I_F$ as indicated in the figure. The net pn junction current, in the reverse-bias direction, is

$$I = I_L - I_F = I_L - I_S \left[ \exp \left( \frac{eV}{kT} \right) - 1 \right] \tag{14-7}$$

where the ideal diode equation has been used. As the diode becomes forward biased, the magnitude of the electric field in the space charge region decreases, but does not go to zero or change direction. The photocurrent is always in the reverse-bias direction and the net solar cell current is also always in the reverse-bias direction.

There are two limiting cases of interest. The short circuit condition occurs when $R = 0$ so that $V = 0$. The current in this case is referred to as the *short-circuit current*, or

$$I = I_{SC} = I_L \tag{14-8}$$

The second limiting case is the open-circuit condition and occurs when $R \rightarrow \infty$. The net current is zero and the voltage produced is the *open-circuit voltage*. The photocurrent is just balanced by the forward-biased junction current so we have

$$I = 0 = I_L - I_S \left[ \exp \left( \frac{eV_{oc}}{kT} \right) - 1 \right] \tag{14-9}$$

We can find the open circuit voltage $V_{oc}$ as

$$V_{oc} = V_t \ln \left( 1 + \frac{I_L}{I_S} \right) \tag{14-10}$$

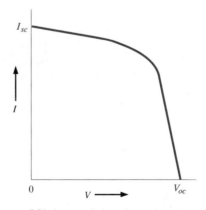

**Figure 14–7**   I-V characteristics of a pn junction solar cell.

A plot of the diode current $I$ as a function of the diode voltage $V$ from Equation (14–7) is shown in Figure 14–7. We may note the short-circuit current and open-circuit voltage points on the figure.

---

**Example 14–3**

**Objective:** To calculate the open-circuit voltage of a silicon pn junction solar cell.
Consider a silicon pn junction at $T = 300°K$ with the following parameters:

$N_a = 5 \times 10^{18}$ cm$^{-3}$     $N_d = 10^{16}$ cm$^{-3}$

$D_n = 25$ cm$^2$/sec     $D_p = 10$ cm$^2$/sec

$\tau_{n0} = 5 \times 10^{-7}$ sec     $\tau_{p0} = 10^{-7}$ sec

Let the photocurrent density be $J_L = I_L/A = 15$ mA/cm$^2$.

**Solution:** We have that

$$J_S = \frac{I_S}{A} = \left(\frac{eD_n n_{p0}}{L_n} + \frac{eD_p p_{n0}}{L_p}\right) = en_i^2\left(\frac{D_n}{L_n N_a} + \frac{D_p}{L_p N_d}\right)$$

We may calculate

$$L_n = \sqrt{D_n \tau_{n0}} = \sqrt{(25)(5 \times 10^{-7})} = 35.4 \ \mu m$$

and

$$L_p = \sqrt{D_p \tau_{p0}} = \sqrt{(10)(10^{-7})} = 10.0 \ \mu m$$

Then

$$J_S = (1.6 \times 10^{-19})(1.5 \times 10^{10})^2 \times \left\{\frac{25}{(35.4 \times 10^{-4})(5 \times 10^{18})} + \frac{10}{(10 \times 10^{-4})(10^{16})}\right\}$$

$$= 3.6 \times 10^{-11} \ A/cm^2$$

Then from Equation (14–10), we can find

$$V_{oc} = V_t \ln\left(1 + \frac{I_L}{I_S}\right) = V_t \ln\left(1 + \frac{J_L}{J_S}\right)$$

$$= (0.0259) \ln\left(1 + \frac{15 \times 10^{-3}}{3.6 \times 10^{-11}}\right) = 0.514 \ volt$$

**Comment:** We may note that $J_S$ is a function of the semiconductor doping concentrations. As the doping concentrations increase, $J_S$ decreases, which increases the open-circuit voltage. However, since $V_{oc}$ is a function of the log of $I_L$ and $I_S$, the open-circuit voltage is not a strong function of these parameters.

---

The power delivered to the load is

$$P = I \cdot V = I_L \cdot V - I_S\left[\exp\left(\frac{eV}{kT}\right) - 1\right] \cdot V \qquad (14\text{–}11)$$

We may find the current and voltage which will deliver the maximum power to the load by setting the derivative equal to zero, or $dP/dV = 0$. Using Equation (14–11), we find

$$\frac{dP}{dV} = 0 = I_L - I_S \left[ \exp\left(\frac{eV_m}{kT}\right) - 1 \right] - I_S V_m \left(\frac{e}{kT}\right) \exp\left(\frac{eV_m}{kT}\right) \quad (14\text{–}12)$$

where $V_m$ is the voltage which produces the maximum power. We may rewrite Equation (14–12) in the form

$$\left(1 + \frac{V_m}{V_t}\right) \exp\left(\frac{eV_m}{kT}\right) = 1 + \frac{I_L}{I_S} \quad (14\text{–}13)$$

The value of $V_m$ may be determined by trial and error. Figure 14–8 shows the maximum power rectangle where $I_m$ is the current when $V = V_m$.

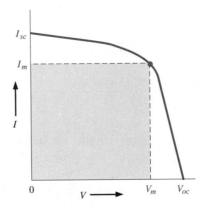

**Figure 14–8**   Maximum power rectangle of the solar cell I-V characteristics.

### 14.2.2 Conversion Efficiency and Solar Concentration

The conversion efficiency of a solar cell is defined as the ratio of output electrical power to incident optical power. For the maximum power output, we can write

$$\eta = \frac{P_m}{P_{\text{in}}} \times 100 \text{ percent} = \frac{I_m V_m}{P_{\text{in}}} \times 100 \text{ percent} \quad (14\text{–}14)$$

The maximum possible current and the maximum possible voltage in the solar cell are $I_{SC}$ and $V_{oc}$, respectively. The ratio $I_m V_m / I_{SC} V_{oc}$ is called the fill factor and is a measure of the realizable power from a solar cell. Typically, the fill factor is between 0.7 and 0.8.

The conventional pn junction solar cell has a single semiconductor bandgap energy. When the cell is exposed to the solar spectrum, a photon with energy less than $E_g$ will have no effect on the electrical output power of the solar cell. A photon with energy greater than $E_g$ will contribute to the

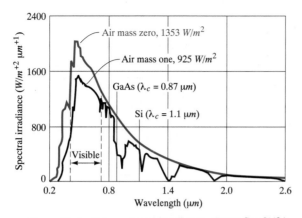

**Figure 14-9** Solar spectral irradiance. (From Sze [11].)

solar cell output power, but the fraction of photon energy that is greater than $E_g$ will eventually only be dissipated as heat. Figure 14-9 shows the solar spectral irradiance (power per unit area per unit wavelength) where air mass zero represents the solar spectrum outside the earth's atmosphere and air mass one is the solar spectrum at the earth's surface at noon. The maximum efficiency of a silicon pn junction solar cell is approximately 28 percent. Nonideal factors, such as series resistance and reflection from the semiconductor surface, will lower the conversion efficiency typically to the range of 10 to 15 percent.

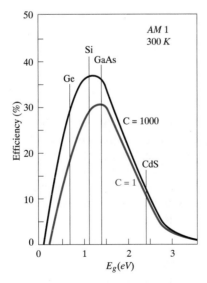

**Figure 14-10** Ideal solar cell efficiency at $T = 300°K$ for $C = 1$ sun and for a $C = 1000$ sun concentration as a function of bandgap energy. (From Sze [11].)

A large optical lens can be used to concentrate sunlight onto a solar cell so that the light intensity can be increased up to several hundred times. The short-circuit current increases linearly with light concentration while the open-circuit voltage increases only slightly with concentration. Figure 14–10 shows the ideal solar cell efficiency at 300°K for two values of solar concentration. We can see that the conversion efficiency increases only slightly with optical concentration. The primary advantage of using concentration techniques is to reduce the overall system cost since an optical lens is less expensive than an equivalent area of solar cells.

---

**Example 14–4**

**Objective:** To calculate the open-circuit voltage when solar concentration is used.

Consider the silicon pn junction solar cell described in Example 14–3. Let the solar intensity increase by a factor of 10.

**Solution:** The photocurrent density in Example 14–3 for one sun was $J_L = 15$ mA/cm$^2$. If the intensity of sunlight increases by a factor of 10, then the photocurrent density for ten suns is $J_L = 150$ mA/cm$^2$. The reverse-saturation current density $J_S$ remains unchanged at $J_S = 3.6 \times 10^{-11}$ A/cm$^2$ (assuming the temperature remains constant). The open-circuit voltage from Equation (14–10) is

$$V_{oc} = V_t \ln \left(1 + \frac{J_L}{J_S}\right) = (0.0259) \ln \left(1 + \frac{150 \times 10^{-3}}{3.6 \times 10^{-11}}\right) = 0.574 \text{ volt}$$

**Comment:** The open-circuit voltage increases slightly as the solar concentration increases, which means that the efficiency will increase slightly with solar concentration.

---

### 14.2.3 Nonuniform Absorption Effects

We have seen from the previous section that the photon absorption coefficient in a semiconductor is a very strong function of the incident photon energy or wavelength. Figure 14–4 showed the absorption coefficient as a function of wavelength for several semiconductor materials. As the absorption coefficient increases, more photon energy will be absorbed near the surface than deeper into the semiconductor. In this case, then, we will not have uniform excess-carrier generation in a solar cell.

The number of photons absorbed per cm$^3$ per sec as a function of distance $x$ from the surface can be written as

$$\alpha \Phi_0 e^{-\alpha x} \tag{14–15}$$

where $\Phi_0$ is the incident photon flux (cm$^{-2}$ sec$^{-1}$) on the surface of the semiconductor. We can also take into account the reflection of photons from the surface. Let $R(\lambda)$ be the fraction of photons which are reflected. (For bare silicon, $R \approx 35$ percent). If we assume that each photon absorbed creates one electron-hole pair, then the generation rate of electron-hole pairs as a function of distance $x$ from the surface is

$$G_L = \alpha(\lambda)\Phi_0(\lambda)[1 - R(\lambda)]e^{-\alpha(\lambda)x} \qquad (14\text{--}16)$$

where each parameter may be a function of the incident wavelength. Figure 14–11 shows the excess minority carrier concentrations in this pn solar cell for two values of wavelength and for the case when $s = 0$ at the surface.

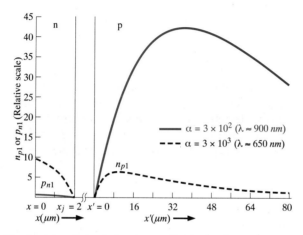

**Figure 14–11** Steady-state, photon-induced normalized minority carrier concentration in the pn junction solar cell for two values of incident photon wavelength ($x_j = 2$ $\mu$m, $W = 1$ $\mu$m, $L_p = L_n = 40$ $\mu$m).

## 14.2.4 The Heterojunction Solar Cell

As we have mentioned in previous chapters, a heterojunction is formed between two semiconductors with different bandgap energies. A typical pN heterojunction energy-band diagram in thermal equilibrium is shown in Figure 14–12. Assume that photons are incident on the wide-bandgap material. Photons with energy less than $E_{gN}$ will pass through the wide-bandgap material, which acts as an optical window, and photons with energies greater than $E_{gp}$ will be absorbed in the narrow bandgap material. On the average, excess carriers created in the depletion region and within a diffusion length of the junction will be collected and will contribute to the photocurrent. Photons with an energy greater than $E_{gN}$ will be absorbed in the wide-bandgap material, and excess carriers generated within one diffusion length of the junction will be collected. If $E_{gN}$ is large enough, then the high-energy photons will be absorbed in the space charge region of the narrow-bandgap material. This heterojunction solar cell should have better characteristics than a homojunction cell, especially at the shorter wavelengths.

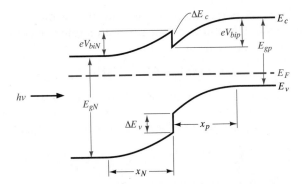

**Figure 14–12**    The energy-band diagram of a pN heterojunction in thermal equilibrium.

A variation of the heterojunction is shown in Figure 14–13. A pn homo-junction is formed and then a wide-bandgap material is grown on top. Again, the wide-bandgap material acts as an optical window for photon energies $h\nu < E_{g1}$. Photons with energies $E_{g2} < h\nu < E_{g1}$ will create excess carriers in the homojunction and photons with energies $h\nu > E_{g1}$ will create excess carriers in the window type material. If the absorption coefficient in the

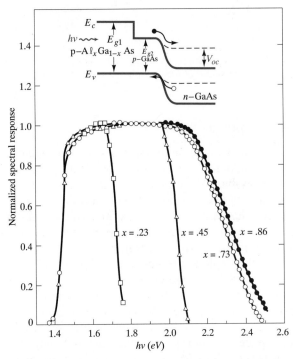

**Figure 14–13**    The normalized spectral response of several AlGaAs/GaAs solar cells with different compositions. (From Sze [10].)

narrow bandgap material is high, then essentially all of the excess carriers will be generated within a diffusion length of the junction, so the collection efficiency will be very high. Figure 14–13 also shows the normalized spectral response for various mole fractions $x$ in the $Al_xGa_{1-x}As$.

### 14.2.5 Amorphous Silicon Solar Cells

Single-crystal silicon solar cells tend to be expensive and are limited to approximately six inches in diameter. A system powered by solar cells requires, in general, a very large area solar cell array to generate the required power. Amorphous silicon solar cells provide the possibility of fabricating large area and relatively inexpensive solar cell systems.

When silicon is deposited by CVD techniques at temperatures below 600°C, an amorphous film is formed regardless of the type of substrate. In amorphous silicon, there is only very short range order, and no crystalline regions are observed. Hydrogen may be incorporated in the silicon to reduce the number of dangling bonds, creating a material called hydrogenated amorphous silicon.

The density of states versus energy for amorphous silicon is shown in Figure 14–14. Amorphous silicon contains large numbers of electronic energy states within the normal bandgap of single-crystal silicon. However, because of the short range order, the effective mobility is quite small, typi-

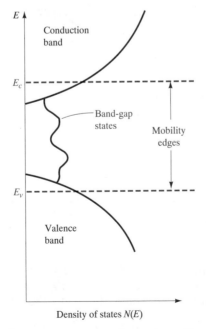

Figure 14–14  Density of states versus energy of amorphous silicon. (From Yang [15].)

cally in the range between $10^{-6}$ and $10^{-3}$ cm²/V-sec. The mobilities in the states above $E_c$ and below $E_v$ are between 1 and 10 cm²/V-sec. Consequently, conduction through the energy states between $E_c$ and $E_v$ is negligible because of the low mobility. Because of the difference in mobility values, $E_c$ and $E_v$ are referred to as the mobility edges and the energy between $E_c$ and $E_v$ is referred to as the mobility gap. The mobility gap can be modified by adding specific types of impurities. Typically, the mobility gap is on the order of 1.7 eV.

Amorphous silicon has a very high optical absorption coefficient so most sunlight is absorbed within approximately 1 $\mu$m of the surface. Consequently, only a very thin layer of amorphous silicon is required for a solar cell. A typical amorphous silicon solar cell is a PIN device shown in Figure 14–15. The amorphous silicon is deposited on an optically transparent

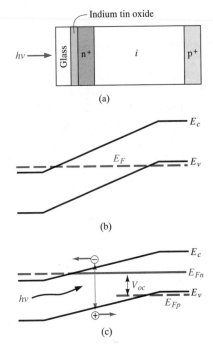

**Figure 14–15**   The (a) cross section, (b) energy-band diagram at thermal equilibrium, and (c) energy-band diagram under photon illumination of an amorphous silicon PIN solar cell. (From Yang [15].)

indium-tin-oxide coated glass substrate. If aluminum is used as the back contact it will reflect any transmitted photons back through the PIN device. The $n^+$ and $p^+$ regions can be quite thin while the intrinsic region may be in the range of 0.5 to 1.0 $\mu$m thick. The energy-band diagram for the thermal equilibrium case is shown in the figure. Excess carriers generated in the intrinsic region are separated by the electric field and produce the photocurrent, as we have discussed. Conversion efficiencies are smaller than in single-crystal silicon, but the reduced cost makes this technology attractive. Amorphous silicon solar cells approximately 40 cm wide and many meters long have been fabricated.

## 14.3 PHOTODETECTORS

There are several semiconductor devices that can be used to detect the presence of photons. These devices are known as photodetectors; they convert optical signals into electrical signals. When excess electrons and holes are generated in a semiconductor, there is an increase in the conductivity of the material. This change in conductivity is the basis of the photoconductor, perhaps the simplest type of photodetector. If electrons and holes are generated within the space charge region of a pn junction, then they will be separated by the electric field and a current will be produced. The pn junction is the basis of several photodetector devices including the photodiode and the phototransistor.

### 14.3.1 Photoconductor

Figure 14–16 shows a bar of semiconductor material with ohmic contacts at each end and a voltage applied between the terminals. The initial thermal-

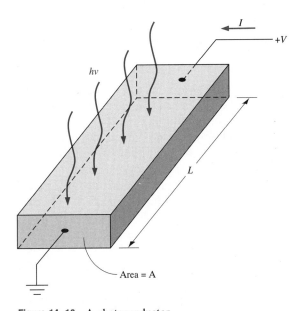

**Figure 14–16**  A photoconductor.

equilibrium conductivity is

$$\sigma_0 = e(\mu_n n_0 + \mu_p p_0) \tag{14-17}$$

If excess carriers are generated in the semiconductor, the conductivity becomes

$$\sigma = e[\mu_n(n_0 + \delta n) + \mu_p(p_0 + \delta p)] \tag{14-18}$$

where $\delta n$ and $\delta p$ are the excess electron and hole concentrations, respectively. If we consider an n-type semiconductor, then, from charge neutrality, we can assume that $\delta n = \delta p \equiv \delta p$. We will use $\delta p$ as the concentration of excess carriers. In steady state, the excess carrier concentration is given by $\delta p = G_L \tau_p$ where $G_L$ is the generation rate of excess carriers ($cm^{-3}$ $sec^{-1}$) and $\tau_p$ is the excess minority carrier lifetime.

The conductivity from Equation (14-18) can be rewritten as

$$\sigma = e(\mu_n n_0 + \mu_p p_0) + e(\delta p)(\mu_n + \mu_p) \tag{14-19}$$

The change in conductivity due to the optical excitation, known as the *photoconductivity*, is then

$$\Delta\sigma = e(\delta p)(\mu_n + \mu_p) \tag{14-20}$$

An electric field is induced in the semiconductor by the applied voltage, which produces a current. The current density can be written as

$$J = (J_0 + J_L) = (\sigma_0 + \Delta\sigma)E \tag{14-21}$$

where $J_0$ is the current density in the semiconductor prior to optical excitation and $J_L$ is the photocurrent density. The photocurrent density is $J_L = \Delta\sigma \cdot E$. If the excess electrons and holes are generated uniformly throughout the semiconductor, then the photocurrent is given by

$$I_L = J_L \cdot A = \Delta\sigma \cdot AE = eG_L\tau_p(\mu_n + \mu_p)AE \tag{14-22}$$

where $A$ is the cross-sectional area of the device. The photocurrent is directly proportional to the excess carrier generation rate, which in turn is proportional to the incident photon flux.

If excess electrons and holes are not generated uniformly throughout the semiconductor material, then the total photocurrent is found by integrating the photoconductivity over the cross-sectional area.

Since $\mu_n E$ is the electron drift velocity, the electron transit time, that is, the time required for an electron to flow through the photoconductor, is

$$t_n = \frac{L}{\mu_n E} \tag{14-23}$$

The photocurrent, from Equation (14-22), can be rewritten as

$$I_L = eG_L \left(\frac{\tau_p}{t_n}\right)\left(1 + \frac{\mu_p}{\mu_n}\right) AL \tag{14-24}$$

We may define a photoconductor gain, $\Gamma_{ph}$, as the ratio of the rate at which charge is collected by the contacts to the rate at which charge is generated within the photoconductor. We can write the gain as

$$\Gamma_{ph} = \frac{I_L}{eG_L AL} \tag{14-25}$$

which, using Equation (14–24), can be written

$$\Gamma_{ph} = \frac{\tau_p}{t_n}\left(1 + \frac{\mu_p}{\mu_n}\right) \tag{14-26}$$

---

**Example 14–5**

**Objective:** To calculate the gain of a silicon photoconductor.

Consider an n-type silicon photoconductor with a length $L = 100$ $\mu$m, cross-sectional area $A = 10^{-7}$ cm$^2$, and minority carrier lifetime $\tau_p = 10^{-6}$ sec. Let the applied voltage be $V = 10$ volts.

**Solution:** The electron transit time is determined as

$$t_n = \frac{L}{\mu_n E} = \frac{L^2}{\mu_n V} = \frac{(100 \times 10^{-4})^2}{(1350)(10)} = 7.41 \times 10^{-9} \text{ sec}$$

The photoconductor gain is then

$$\Gamma_{ph} = \frac{\tau_p}{t_n}\left(1 + \frac{\mu_p}{\mu_n}\right) = \frac{10^{-6}}{7.41 \times 10^{-9}}\left(1 + \frac{480}{1350}\right) = 1.83 \times 10^2$$

**Comment:** The fact that a photoconductor—a bar of semiconductor material—has a gain may be surprising.

---

Let's consider physically what happens to a photon-generated electron, for example. After the excess electron is generated, it drifts very quickly out of the photoconductor at the anode terminal. In order to maintain charge neutrality throughout the entire photoconductor, another electron immediately enters the photoconductor at the cathode and drifts toward the anode. This process will continue during a time period equal to the mean carrier lifetime. At the end of this period, on the average, the photoelectron will recombine with a hole.

The electron transit time, using the parameters from Example 14–5, is $t_n = 7.41 \times 10^{-9}$ sec. In a simplistic sense, the photoelectron will circulate around the photoconductor circuit 135 times during the $10^{-6}$ sec time duration, which is the mean carrier lifetime. If we take into account the photon-generated hole, the total number of charges collected at the photoconductor contacts for every electron generated is 183.

When the optical signal ends, the photocurrent will decay exponentially with a time constant equal to the minority carrier lifetime. The switching speed or frequency response is inversely proportional to the lifetime. From the photoconductor gain expression, we would like a large minority carrier lifetime, but the switching speed is enhanced by a small minority carrier lifetime. There is obviously a tradeoff between gain and speed. In general, the performance of a photodiode, which we will discuss next, is superior to that of a photoconductor.

### 14.3.2 Photodiode

A photodiode is a pn junction diode operated with an applied reverse-bias voltage. We will initially consider a long diode in which excess carriers are generated uniformly throughout the entire semiconductor device. Figure 14–17a shows the reverse-biased diode and Figure 14–17b shows the minority carrier distribution in the reverse-biased junction prior to photon illumination.

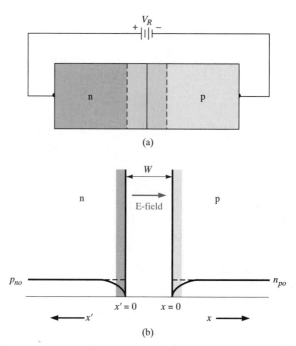

(a)

(b)

**Figure 14–17**  (a) A reverse-biased pn junction. (b) Minority carrier concentration in the reverse-biased pn junction.

Let $G_L$ be the generation rate of excess carriers. The excess carriers generated within the space charge region are swept out of the depletion region very quickly by the electric field; the electrons are swept into the n-region and the holes into the p-region. The photon-generated current density from the space charge region is given by

$$J_{L1} = e \int G_L \, dx \qquad (14\text{--}27)$$

where the integral is over the space charge region width. If $G_L$ is constant throughout the space charge volume, then

$$J_{L1} = eG_L W \qquad (14\text{--}28)$$

where $W$ is the space charge width. We may note that $J_{L1}$ is in the reverse-bias direction through the pn junction. This component of photocurrent responds very quickly to the photon illumination and is known as the prompt photocurrent.

We may note, by comparing Equations (14–28) and (14–25), that the photodiode gain is unity. The speed of the photodiode is limited by the carrier transport through the space charge region. If we assume that the saturation drift velocity is $10^7$ cm/sec and the depletion width is 2 $\mu$m, the transit time is $\tau_t = 20 ps$. The ideal modulating frequency has a period of $2\tau_t$, so the frequency is $f = 25$ GHz. This frequency response is substantially higher than that of photoconductors.

Excess carriers are also generated within the neutral n- and p-regions of the diode. The excess minority carrier electron distribution in the p-region is found from the ambipolar transport equation, which is

$$D_n \frac{\partial^2(\delta n_p)}{\partial x^2} + G_L - \frac{\delta n_p}{\tau_{n0}} = \frac{\partial(\delta n_p)}{\partial t} \qquad (14\text{--}29)$$

We will assume that the E-field is zero in the neutral regions. In steady state, $\partial(\delta n_p)/\partial t = 0$, so that Equation (14–29) can be written as

$$\frac{d^2(\delta n_p)}{dx^2} - \frac{\delta n_p}{L_n^2} = -\frac{G_L}{D_n} \qquad (14\text{--}30)$$

where $L_n^2 = D_n\tau_{n0}$.

The solution to Equation (14–30) can be found as the sum of the homogeneous and particular solutions. The homogeneous solution is found from the equation

$$\frac{d^2(\delta n_{ph})}{dx^2} - \frac{\delta n_{ph}}{L_n^2} = 0 \qquad (14\text{--}31)$$

where $\delta n_{ph}$ is the homogeneous solution and is given by

$$\delta n_{ph} = Ae^{-x/L_n} + Be^{+x/L_n} \qquad (x \geq 0) \qquad (14\text{--}32)$$

One boundary condition is that $\delta n_{ph}$ must remain finite, which implies that $B \equiv 0$ for the "long" diode.

The particular solution is found from

$$-\frac{\delta n_{pp}}{L_n^2} = -\frac{G_L}{D_n} \qquad (14\text{--}33)$$

which yields

$$\delta n_{pp} = \frac{G_L L_n^2}{D_n} = \frac{G_L(D_n \tau_{no})}{D_n} = G_L \tau_{no} \qquad (14\text{--}34)$$

The total steady-state solution for the excess minority carrier electron concentration in the p-region is then

$$\delta n_p = A e^{-x/L_n} + G_L \tau_{no} \qquad (14\text{--}35)$$

The total electron concentration is zero at $x = 0$ for the reverse-biased junction. The excess electron concentration $x = 0$ is then

$$\delta n_p(x = 0) = -n_{po} \qquad (14\text{--}36)$$

Using the boundary condition from Equation (14–36), the electron concentration given by Equation (14–35) becomes

$$\delta n_p = G_L \tau_{no} - (G_L \tau_{no} + n_{po}) e^{-x/L_n} \qquad (14\text{--}37)$$

We can find the excess minority carrier hole concentration in the n-region using the same type of analysis. Using the $x'$ notation which was shown in Figure 14–17, we can write

$$\delta p_n = G_L \tau_{po} - (G_L \tau_{po} + p_{no}) e^{-x'/L_p} \qquad (14\text{--}38)$$

Equations (14–37) and (14–38) are plotted in Figure 14–18. We may note that the steady-state values far from the space charge region are the same as were given previously.

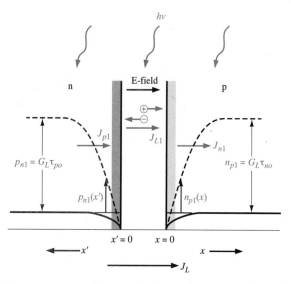

**Figure 14–18**   Steady-state, photo-induced minority carrier concentrations and photocurrents in a "long" reverse-biased pn junction.

The gradient in the minority carrier concentrations will produce diffusion currents in the pn junction. The diffusion current density at $x = 0$ due to minority carrier electrons is

$$J_{n1} = eD_n \frac{d(\delta n_p)}{dx}\bigg|_{x=0} = eD_n \frac{d}{dx}[G_L \tau_{no} - (G_L \tau_{no} + n_{po})e^{-x/L_n}]\bigg|_{x=0} \tag{14-39}$$

$$= \frac{eD_n}{L_n}(G_L \tau_{no} + n_{po})$$

Equation (14–39) can be written as

$$J_{n1} = eG_L L_n + \frac{eD_n n_{po}}{L_n} \tag{14-40}$$

The first term in Equation (14–40) is the steady-state photocurrent density while the second term is the ideal reverse saturation current density due to the minority carrier electrons.

The diffusion current density (in the $x$-direction) at $x' = 0$ due to the minority carrier holes is

$$J_{p1} = eG_L L_p + \frac{eD_p p_{no}}{L_p} \tag{14-41}$$

Similarly, the first term is the steady-state photocurrent density and the second term is the ideal reverse-saturation current density.

The total steady-state diode photocurrent density for the long diode is now

$$J_L = eG_L W + eG_L L_n + eG_L L_p = e(W + L_n + L_p)G_L \tag{14-42}$$

Again note that the photocurrent is in the reverse-bias direction through the diode. The photocurrent given by Equation (14–42) is the result of assuming: uniform generation of excess carriers throughout the entire structure, a long diode, and steady state.

The time response of the diffusion components of the photocurrent is relatively slow, since these currents are the result of the diffusion of minority carriers toward the depletion region. The diffusion components of photocurrent are referred to as the delayed photocurrent.

---

**Example 14–6**

**Objective:** To calculate the steady-state photocurrent density in a reverse-biased, long pn diode.

Consider a silicon pn diode at $T = 300°K$ with the following parameters:

$N_a = 10^{16}$ cm$^{-3}$      $N_d = 10^{16}$ cm$^{-3}$

$D_n = 25$ cm$^2$/sec      $D_p = 10$ cm$^2$/sec

$\tau_{no} = 5 \times 10^{-7}$ sec      $\tau_{po} = 10^{-7}$ sec

Assume that a reverse-bias voltage of $V_R = 5$ volts is applied and let $G_L = 10^{21}$ cm$^{-3}$ sec$^{-1}$.

**Solution:** We may calculate various parameters as follows:

$$L_n = \sqrt{D_n \tau_{no}} = \sqrt{(25)(5 \times 10^{-7})} = 35.4 \; \mu\text{m}$$
$$L_p = \sqrt{D_p \tau_{po}} = \sqrt{(10)(10^{-7})} = 10.0 \; \mu\text{m}$$
$$V_{bi} = V_t \ln\left(\frac{N_a N_d}{n_i^2}\right) = (0.0259) \ln\left[\frac{(10^{16})(10^{16})}{(1.5 \times 10^{10})^2}\right] = 0.695 \text{ volt}$$
$$W = \left\{\frac{2\varepsilon_s}{e}\left(\frac{N_a + N_d}{N_a N_d}\right)(V_{bi} + V_R)\right\}^{1/2}$$
$$= \left\{\frac{2(11.7)(8.85 \times 10^{-14})}{1.6 \times 10^{-19}} \cdot \frac{(2 \times 10^{16})}{(10^{16})(10^{16})} \cdot (0.695 + 5)\right\}^{1/2} = 1.21 \; \mu\text{m}$$

Finally, the steady-state photocurrent density is

$$J_L = e(W + L_n + L_p)G_L$$
$$= (1.6 \times 10^{-19})(1.21 + 35.4 + 10.0) \times 10^{-4}(10^{21}) = 0.75 \text{ Amp/cm}^2$$

**Comment:** Again, keep in mind that this photocurrent is in the reverse-bias direction through the diode and is many orders of magnitude larger than the reverse-bias saturation current density in the pn junction diode.

---

In this example calculation, $L_n \gg W$ and $L_p \gg W$. In many pn junction structures, the assumption of a long diode will not be valid so that the photocurrent expression will have to be modified. In addition, the photon energy absorption may not be uniform throughout the pn structure. The effect of nonuniform absorption will be considered in the next section.

### 14.3.3 PIN Photodiode

In many photodetector applications, the speed of response is important; therefore the prompt photocurrent generated in the space charge region is the only photocurrent of interest. To increase the photodetector sensitivity, the depletion region width should be made as large as possible. This can be achieved in a PIN photodiode.

The PIN diode consists of a p-region and an n-region separated by an intrinsic region. A sketch of a PIN diode is shown in Figure 14–19a. The intrinsic region width $W$ is much larger than the space charge width of a normal pn junction. If a reverse bias is applied to the PIN diode, the space charge region extends completely through the intrinsic region.

Assume that a photon flux $\Phi_0$ is incident on the $p^+$ region. If we assume that the $p^+$ region width $W_p$ is very thin, then the photon flux, as a function of distance, in the intrinsic region is $\Phi(x) = \Phi_0 e^{-\alpha x}$, where $\alpha$ is the photon absorption coefficient. This nonlinear photon absorption is shown in Figure

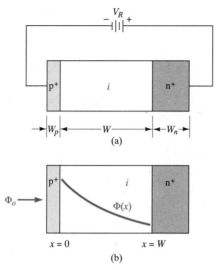

**Figure 14–19** (a) A reverse-biased PIN photodiode. (b) Geometry showing nonuniform photon absorption.

**14–19b.** The photocurrent density generated in the intrinsic region can be found as

$$J_L = e \int_0^W G_L \, dx = e \int_0^W \Phi_0 \alpha e^{-\alpha x} \, dx = e \Phi_0 (1 - e^{-\alpha W}) \qquad (14\text{–}43)$$

This equation assumes that there is no electron-hole recombination within the space charge region and also that each photon absorbed creates one electron-hole pair.

**Example 14–7**

**Objective:** To calculate the photocurrent density in a PIN photodiode.

Consider a silicon PIN diode with an intrinsic region width of $W = 20 \ \mu m$. Assume that the photon flux is $10^{17}$ cm$^{-2}$ sec$^{-1}$ and the absorption coefficient is $\alpha = 10^3$ cm$^{-1}$.

**Solution:** The generation rate of electron-hole pairs at the front edge of the intrinsic region is

$$G_{L1} = \alpha \Phi_0 = (10^3)(10^{17}) = 10^{20} \text{ cm}^{-3} \text{ sec}^{-1}$$

and the generation rate at the back edge of the intrinsic region is

$$G_{L2} = \alpha \Phi_0 e^{-\alpha W} = (10^3)(10^{17}) \exp\left[-(10^3)(20 \times 10^{-4})\right]$$
$$= 0.135 \times 10^{20} \text{ cm}^{-3} \text{ sec}^{-1}$$

The generation rate is obviously not uniform throughout the intrinsic region. The photocurrent density is then

$$J_L = e\Phi_0(1 - e^{-\alpha W})$$
$$= (1.6 \times 10^{-19})(10^{17})\{1 - \exp[-(10^3)(20 \times 10^{-4})]\}$$
$$= 13.8 \text{ mA/cm}^2$$

**Comment:** The prompt photocurrent density of a PIN photodiode will be larger than that of a regular photodiode since the space charge region is larger in a PIN photodiode.

In most situations, we will not have a long diode; thus, the steady-state photocurrent described by Equation (14–42) will not apply for most photodiodes.

### 14.3.4 Avalanche Photodiode

The avalanche photodiode is similar to the pn or PIN photodiode except that the bias applied to the avalanche photodiode is sufficiently large to cause impact ionization. Electron-hole pairs are generated in the space charge region by photon absorption as we have discussed previously. The photon-generated electrons and holes now generate additional electron-hole pairs through impact ionization. The avalanche photodiode now has a current gain introduced by the avalanche multiplication factor.

The electron-hole pairs generated by photon absorption and by impact ionization are swept out of the space charge region very quickly. If the saturation velocity is $10^7$ cm/sec in a depletion region which is 10 $\mu$m wide, then the transit time is

$$\tau_t = \frac{10^7}{10 \times 10^{-4}} = 100 \text{ psec}$$

The period of a modulation signal would be $2\tau_t$, so that the frequency would be

$$f = \frac{1}{2\tau_t} = \frac{1}{200 \times 10^{-12}} = 5 \text{ GHz}$$

If the avalanche photodiode current gain is 20, then the gain-bandwidth product is 100 GHz. The avalanche photodiode could respond to light waves modulated at microwave frequencies.

### 14.3.5 Phototransistor

A bipolar transistor can also be used as a photodetector. The phototransistor can have high gain through the transistor action. An npn bipolar phototransistor is shown in Figure 14–20a. This device has a large base-collector

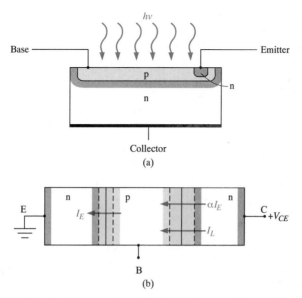

**Figure 14–20**  (a) A bipolar phototransistor. (b) Block diagram of the open-base phototransistor.

junction area and is usually operated with the base open-circuited. Figure 14–20b shows the block diagram of the phototransistor. Electrons and holes generated in the reverse-biased B-C junction are swept out of the space charge region, producing a photocurrent $I_L$. Holes are swept into the p-type base making the base positive with respect to the emitter. Since the B-E becomes forward-biased, electrons will be injected from the emitter back into the base, leading to the normal transistor action.

From Figure 14–20b, we see that

$$I_E = \alpha I_E + I_L \qquad (14\text{--}44)$$

where $I_L$ is the photon-generated current and $\alpha$ is the common base current gain. Since the base is an open circuit, we have $I_C = I_E$ so Equation (14–44) can be written as

$$I_C = \alpha I_C + I_L \qquad (14\text{--}45)$$

Solving for $I_C$, we find

$$I_C = \frac{I_L}{1 - \alpha} \qquad (14\text{--}46)$$

Relating $\alpha$ to $\beta$, the dc common emitter current gain, Equation (14–46) becomes

$$I_C = (1 + \beta)I_L \qquad (14\text{--}47)$$

Equation (14–47) shows that the basic B-C photocurrent is multiplied by the factor $(1 + \beta)$. The phototransistor, then, amplifies the basic photocurrent.

With the relatively large B-C junction area, the frequency response of the phototransistor is limited by the B-C junction capacitance. Since the base is essentially the input to the device, the large B-C capacitance is multiplied by the Miller effect, so the frequency response of the phototransistor is further reduced. The phototransistor, however, is a lower-noise device than the avalanche photodiode.

Phototransistors can also be fabricated in heterostructures. The injection efficiency is increased due to the bandgap differences, as we discussed in Chapter 10. With the bandgap difference, the lightly doped base restriction no longer applies. A fairly heavily doped, narrow-base device can be fabricated with a high blocking voltage and a high gain.

## 14.4 PHOTOLUMINESCENCE AND ELECTROLUMINESCENCE

In the first section of this chapter, we discussed the creation of excess electron-hole pairs by photon absorption. Eventually, excess electrons and holes recombine, and in direct bandgap materials the recombination process may result in the emission of a photon. The general property of light emission is referred to as luminescence. When excess electrons and holes are created by photon absorption, then photon emission from the recombination process is called photoluminescence.

Electroluminescence is the process of generating photon emission when the excitation of excess carriers is a result of an electric current caused by an applied electric field. We will be mainly concerned with injection electroluminescence, the result of injecting carriers across a pn junction. The light emitting diode and the pn junction laser diode are examples of this phenomenon. In these devices electric energy, in the form of a current, is converted directly into photon energy.

### 14.4.1 Basic Transitions

Once electron-hole pairs are formed, there are several possible processes by which the electrons and holes can recombine. Some recombination processes may result in photon emission from direct bandgap materials, whereas other recombination processes in the same material may not.

Figure 14–21a shows the basic interband transitions. Curve *(i)* corresponds to an intrinsic emission very close to the bandgap energy of the material. Curves *(ii)* and *(iii)* correspond to energetic electrons or holes. If either of these recombinations result in the emission of a photon, the energy of the emitted photon will be slightly larger than the bandgap energy. There will then be an emission spectrum and a bandwidth associated with the emission.

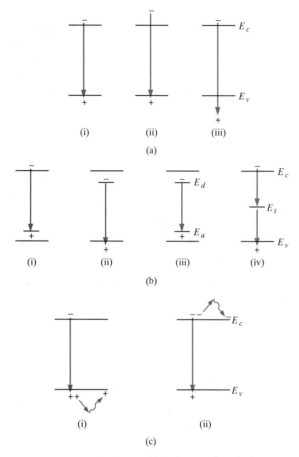

**Figure 14–21**   Basic transitions in a semiconductor.

The possible recombination processes involving impurity or defect states is shown in Figure 14–21b. Curve *(i)* is the conduction band to acceptor transition, curve *(ii)* is the donor to valence band transition, curve *(iii)* is the donor to acceptor transition, and curve *(iv)* is the recombination due to a deep trap. Curve *(iv)* is a nonradiative process corresponding to the Shockley-Read-Hall recombination process discussed in Chapter 6. The other recombination processes may or may not result in the emission of a photon.

Figure 14–21c shows the Auger recombination process, which can become important in direct bandgap materials with high doping concentrations. The Auger recombination process is a nonradiative process. The Auger recombination, in one case, shown in curve *(i),* is a recombination between an electron and hole, accompanied by the transfer of energy to another free hole. Similarly, in the second case, the recombination between an electron and hole can result in the transfer of energy to a free electron as shown in curve *(ii)*. The third particle involved in this process will eventually lose its energy to the lattice in the form of heat. The process involving two holes and an electron would occur predominantly in heavily doped p-type materials, and the process involving two electrons and a hole would occur primarily in a heavily doped n-type material.

The recombination processes shown in Figure 14–21a indicate that the emission of a photon is not necessarily at a single, discrete energy, but can occur over a range of energies. The spontaneous emission rate generally has the form

$$I(\nu) \propto \nu^2 (h\nu - E_g)^{1/2} \exp\left[\frac{-(h\nu - E_g)}{kT}\right] \qquad (14\text{–}48)$$

where $E_g$ is the bandgap energy. Figure 14–22 shows the emission spectra from gallium arsenide. The peak photon energy decreases with temperature because the bandgap energy decreases with temperature. We will show that the bandwidth of the emission spectra can be greatly reduced in a LASER diode by using an optical resonator.

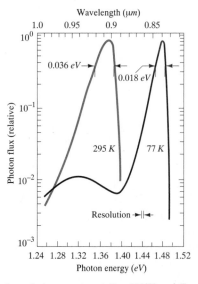

**Figure 14–22**   GaAs diode emission spectra at $T = 300°$K and $T = 77°$K. (From Sze [10].)

## 14.4.2 Luminescent Efficiency

We have shown that not all recombination processes are radiative. An efficient luminescent material is one in which radiative transitions predominate. The quantum efficiency is defined as the ratio of the radiative recombination rate to the total recombination rate for all processes. We can write

$$\eta_q = \frac{R_r}{R} \qquad (14\text{–}49)$$

where $\eta_q$ is the quantum efficiency, $R_r$ is the radiative recombination rate, and $R$ is the total recombination rate of the excess carriers. Since the recombination rate is inversely proportional to lifetime, we can write the quantum efficiency in terms of lifetimes as

$$\eta_q = \frac{\tau_{nr}}{\tau_{nr} + \tau_r} \qquad (14\text{–}50)$$

where $\tau_{nr}$ is the nonradiative lifetime and $\tau_r$ is the radiative lifetime. For a high luminescent efficiency, the nonradiative lifetimes must be large; thus, the probability of a nonradiative recombination is small compared to the radiative recombination.

The interband recombination rate of electrons and holes will be directly proportional to the number of electrons available and directly proportional to the number of available empty states (holes). We can write

$$R_r = Bnp \qquad (14\text{–}51)$$

where $R_r$ is the band-to-band radiative recombination rate and $B$ is the constant of proportionality. The values of $B$ for direct-bandgap materials are on the order of $10^6$ larger than for indirect bandgap materials. The probability of a direct band-to-band radiative recombination transition in an indirect bandgap material is very unlikely.

One problem encountered with the emission of photons from a direct bandgap material is the reabsorption of the emitted photons. In general, the emitted photon will have energies $h\nu > E_g$, which means that the absorption coefficient is not zero for this energy. In order to generate a light output from a light emitting device, the process must take place near the surface. One possible solution to the reabsorption problem is to use heterojunction devices. These will be discussed in later sections.

### 14.4.3 Materials

An important direct bandgap semiconductor material for optical devices is gallium arsenide. Another compound material that is of great interest is $Al_xGa_{1-x}As$. This material is a compound semiconductor in which the ratio of aluminum atoms to gallium atoms can be varied to achieve specific characteristics. Figure 14–23 shows the bandgap energy as a function of the mole fraction between aluminum and gallium. We can note from the figure that for $0 < x < 0.45$, the alloy material is a direct bandgap material. For $x > 0.45$, the material becomes an indirect bandgap material, not suitable for optical devices. For $0 < x < 0.35$, the bandgap energy can be expressed as

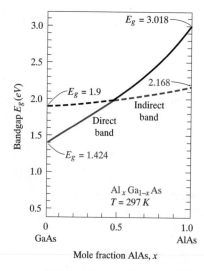

Figure 14–23    Bandgap energy of Al$_x$Ga$_{1-x}$As as a function of the mole fraction $x$. (From Sze [11].)

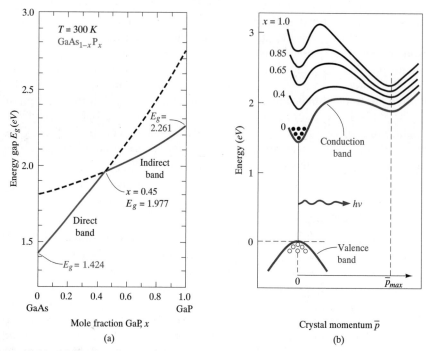

(a)                                                    (b)

Figure 14–24    (a) Bandgap energy of GaAs$_{1-x}$P$_x$ as a function of mole fraction $x$. (b) $E$ versus $k$ diagram of GaAs$_{1-x}$P$_x$ for various values of $x$. (From Sze [11].)

$$E_g = 1.424 + 1.247x \quad \text{eV} \qquad (14-52)$$

Another compound semiconductor used for optical devices is the $GaAs_{1-x}P_x$ system. Figure 14–24a shows the bandgap energy as a function of the mole fraction $x$. For $0 \le x \le 0.45$, this material is also a direct bandgap material, and for $x > 0.45$, the bandgap becomes indirect. Figure 14–24b is the $E$ versus $k$ diagram, showing how the bandgap changes from direct to indirect as the mole fraction changes.

---

**Example 14–8**

**Objective:** To determine the output wavelength of a $GaAs_{1-x}P_x$ material for two different mole fractions.

Consider first GaAs and then $GaAs_{1-x}P_x$.

**Solution:** GaAs has a bandgap energy of $E_g = 1.42$ eV. This material would produce a photon output at a wavelength of

$$\lambda = \frac{1.24}{E} = \frac{1.24}{1.42} = 0.873 \ \mu m$$

This wavelength is in the infrared range and not in the visible range. If we desire a visible output with a wavelength of $\lambda = 0.653 \ \mu m$, for example, the bandgap energy would have to be

$$E = \frac{1.24}{\lambda} = \frac{1.24}{0.653} = 1.90 \text{ eV}$$

This bandgap energy would correspond to a mole fraction of approximately $x = 0.4$.

**Comment:** By changing the mole fraction in the $GaAs_{1-x}P_x$ system, the output can change from the infrared to the red spectrum.

---

## 14.5 LIGHT EMITTING DIODES

Photodetectors and solar cells convert optical energy into electrical energy—the photons generate excess electrons and holes, which produce an electric current. We might also apply a voltage across a pn junction resulting in a diode current, which in turn can produce photons and a light output. This inverse mechanism is called injection electroluminescence. This device is known as a *Light Emitting Diode* (LED). The spectral output of an LED may have a relatively wide wavelength bandwidth of between 30 and 40 nm. However, this emission spectrum is narrow enough so that a particular color will be observed, provided the output is in the visible range.

### 14.5.1 Generation of Light

As we discussed previously, photons may be emitted if an electron and hole recombine by a direct band-to-band recombination process in a direct bandgap material. The emission wavelength, from Equation (14–1), is

$$\lambda = \frac{hc}{E_g} = \frac{1.24}{E_g} \ \mu\text{m} \qquad (14\text{–}53)$$

where $E_g$ is the bandgap energy measured in electron-volts.

When a voltage is applied across a pn junction, electrons and holes are injected across the space charge region where they become excess minority carriers. These excess minority carriers diffuse into the neutral semiconductor regions where they recombine with majority carriers. If this recombination process is a direct band-to-band process, photons are emitted. The diode diffusion current is directly proportional to the recombination rate, so the output photon intensity will also be proportional to the ideal diode diffusion current. In gallium arsenide, electroluminescence originates primarily on the p-side of the junction because the efficiency for electron injection is higher than that for hole injection.

### 14.5.2 Internal Quantum Efficiency

The *internal quantum efficiency* of an LED is the fraction of diode current that will produce luminescence. The internal quantum efficiency is a function of the injection efficiency and a function of the percentage of radiative recombination events compared with the total number of recombination events.

The three current components in a forward-biased diode are the minority carrier electron diffusion current, the minority carrier hole diffusion current, and the space charge recombination current. These current densities can be written, respectively, as

$$J_n = \frac{eD_n n_{po}}{L_n} \left[ \exp\left(\frac{eV}{kT}\right) - 1 \right] \qquad (14\text{–}54a)$$

$$J_p = \frac{eD_p p_{no}}{L_p} \left[ \exp\left(\frac{eV}{kT}\right) - 1 \right] \qquad (14\text{–}54b)$$

and

$$J_R = \frac{en_i W}{2\tau_0} \left[ \exp\left(\frac{eV}{2kT}\right) - 1 \right] \qquad (14\text{–}54c)$$

The recombination of electrons and holes within the space charge region is, in general, through traps near midgap and is a nonradiative process. Since luminescence is due primarily to the recombination of minority carrier electrons in GaAs, we can define an injection efficiency as the fraction of electron current to total current. Then

$$\gamma = \frac{J_n}{J_n + J_p + J_R} \tag{14–55}$$

where $\gamma$ is the injection efficiency. We can make $\gamma$ approach unity by using an $n^+ p$ diode so that $J_p$ is a small fraction of the diode current and by forward biasing the diode sufficiently so that $J_R$ is a small fraction of the total diode current.

Once the electrons are injected into the p-region, not all electrons will recombine radiatively. We can define the radiative and nonradiative recombination rates as

$$R_r = \frac{\delta n}{\tau_r} \tag{14–56a}$$

and

$$R_{nr} = \frac{\delta n}{\tau_{nr}} \tag{14–56b}$$

where $\tau_r$ and $\tau_{nr}$ are the radiative and nonradiative recombination lifetimes, respectively, and $\delta n$ is the excess carrier concentration. The total recombination rate is

$$R = R_r + R_{nr} = \frac{\delta n}{\tau} = \frac{\delta n}{\tau_r} + \frac{\delta n}{\tau_{nr}} \tag{14–57}$$

where $\tau$ is the net excess carrier lifetime.

The radiative efficiency is defined as the fraction of recombinations that are radiative. We can write

$$\eta = \frac{R_r}{R_r + R_{nr}} = \frac{\dfrac{1}{\tau_r}}{\dfrac{1}{\tau_r} + \dfrac{1}{\tau_{nr}}} = \frac{\tau}{\tau_r} \tag{14–58}$$

where $\eta$ is the radiative efficiency. The nonradiative recombination rate is proportional to $N_t$ which is the density of nonradiative trapping sites within the forbidden bandgap. Obviously, the radiative efficiency increases as $N_t$ is reduced.

The internal quantum efficiency is now written as

$$\eta_i = \gamma\eta \tag{14–59}$$

The radiative recombination rate is proportional to the p-type doping. As the p-type doping increases, the radiative recombination rate increases. However, the injection efficiency decreases as the p-type doping increases; therefore, there is an optimum doping that will maximize the internal quantum efficiency.

### 14.5.3 External Quantum Efficiency

One very important parameter of the LED is the *external quantum efficiency:* the fraction of generated photons that are actually emitted from the semiconductor. The external quantum efficiency is normally a much smaller number than the internal quantum efficiency. Once a photon has been produced in the semiconductor, there are three loss mechanisms the photon may encounter: photon absorption within the semiconductor, Fresnel loss, and critical angle loss.

Figure 14–25 shows a pn junction LED. Photons can be emitted in any direction. Since the emitted photon energy must be $h\nu \geq E_g$, these emitted

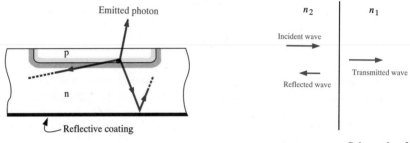

**Figure 14–25**  Schematic of photon emission at the pn junction of an LED.

**Figure 14–26**  Schematic of incident, reflected, and transmitted photons at a dielectric interface.

photons can be reabsorbed within the semiconductor material. The majority of photons will actually be emitted away from the surface and reabsorbed in the semiconductor.

Photons must be emitted from the semiconductor into air; thus, the photons must be transmitted across a dielectric interface. Figure 14–26 shows the incident, reflected, and transmitted waves. The parameter $\bar{n}_2$ is the index of refraction for the semiconductor and $\bar{n}_1$ is the index of refraction for air. The reflection coefficient is

$$\Gamma = \left(\frac{\bar{n}_2 - \bar{n}_1}{\bar{n}_2 + \bar{n}_1}\right)^2 \tag{14–60}$$

This effect is called Fresnel loss. The reflection coefficient $\Gamma$ is the fraction of incident photons that are reflected back into the semiconductor.

---

### Example 14–9

**Objective:** To calculate the reflection coefficient at a semiconductor–air interface.
   Consider the interface between a GaAs semiconductor and air.

**Solution:** The index of refraction for GaAs is $\bar{n}_2 = 3.66$ and for air is $\bar{n}_1 = 1.0$. The reflection coefficient is

$$\Gamma = \left(\frac{\bar{n}_2 - \bar{n}_1}{\bar{n}_2 + \bar{n}_1}\right)^2 = \left(\frac{3.66 - 1.0}{3.66 + 1.0}\right)^2 = 0.33$$

**Comment:** A reflection coefficient of $\Gamma = 0.33$ means that 33 percent of the photons incident from the gallium arsenide on the gallium arsenide–air interface are reflected back into the semiconductor.

---

Photons incident on the semiconductor–air interface at an angle are refracted as shown in Figure 14–27. If the photons are incident on the interface

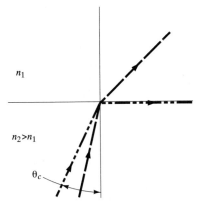

**Figure 14–27** Schematic showing refraction and total internal reflection at the critical angle at a dielectric interface.

at an angle greater than the critical angle $\theta_c$, the photons experience total internal reflection. The critical angle is determined from Snell's law and is given by

$$\theta_c = \sin^{-1}\left(\frac{\bar{n}_1}{\bar{n}_2}\right) \tag{14–61}$$

---

**Example 14–10**

**Objective:** To calculate the critical angle at a semiconductor–air interface.
   Consider the interface between GaAs and air.

**Solution:** For GaAs, $\bar{n}_2 = 3.66$ and for air, $\bar{n}_1 = 1.0$. The critical angle is

$$\theta_c = \sin^{-1}\left(\frac{\bar{n}_1}{\bar{n}_2}\right) = \sin^{-1}\left(\frac{1.0}{3.66}\right) = 15.9°$$

**Comment:** Any photon that is incident at an angle greater than 15.9° will be reflected back into the semiconductor.

---

Figure 14–28a shows the external quantum efficiency plotted as a function of the p-type doping concentration and Figure 14–28b is a plot of the external

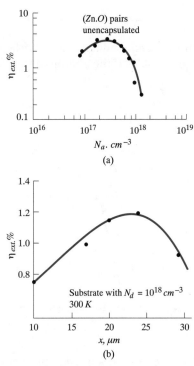

Figure 14–28    (a) External quantum efficiency of a GaP LED versus acceptor doping. (b) External quantum efficiency of a GaAs LED versus junction depth. (From Yang [15].)

efficiency as a function of junction depth below the surface. Both figures show that the external quantum efficiency is in the range of 1 to 3 percent.

### 14.5.4 LED Devices

The wavelength of the output signal of an LED is determined by the bandgap energy of the semiconductor. Gallium arsenide, a direct bandgap material, has a bandgap energy of $E_g = 1.42$ eV, which yields a wavelength of $\lambda = 0.873$ $\mu$m. Comparing this wavelength to the visible spectrum which was shown in Figure 14–5, the output of a GaAs LED is not in the visible range.

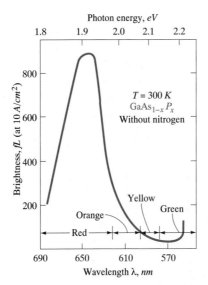

**Figure 14-29** Brightness of GaAsP diodes versus wavelength (or versus bandgap energy). (From Yang [15].)

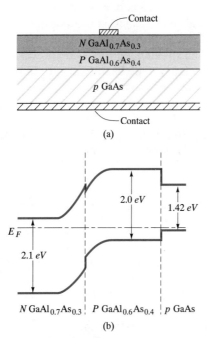

**Figure 14-30** The (a) cross section and (b) thermal-equilibrium energy-band diagram of a GaAlAs heterojunction LED. (From Yang [15].)

For a visible output, the wavelength of the signal should be in the range of 0.4 to 0.72 $\mu$m. This range of wavelengths corresponds to bandgap energies between approximately 1.7 and 3.1 eV.

GaAs$_{1-x}$P$_x$ is a direct bandgap material for $0 \leq x \leq 0.45$ as was shown in Figure 14–24. At $x = 0.40$, the bandgap energy is approximately $E_g = 1.9$ eV, which would produce an optical output in the red range. Figure 14–29 shows the brightness of GaAs$_{1-x}$P$_x$ diodes for different values of $x$. The peak also occurs in the red range. By using planar technology, GaAs$_{0.6}$P$_{0.4}$ monolithic arrays have been fabricated for numeric and alphanumeric displays. When the mole fraction $x$ is greater than 0.45, the material changes to an indirect bandgap semiconductor so that the quantum efficiency is greatly reduced.

GaAl$_x$As$_{1-x}$ can be used in a heterojunction structure to form an LED. A device structure is shown in Figure 14–30. Electrons are injected from the wide-bandgap $N-$GaAl$_{0.7}$As$_{0.3}$ into the narrow-bandgap $p-$GaAl$_{0.6}$As$_{0.4}$. The minority carrier electrons in the p-material can recombine radiatively. Since $E_{gp} < E_{gN}$, the photons are emitted through the wide-bandgap N-material with essentially no absorption. The wide bandgap N-material acts as an optical window and the external quantum efficiency increases.

## 14.6 LASER DIODES

The photon output of the LED is due to an electron giving up energy as it makes a transition from the conduction band to the valence band. The LED photon emission is spontaneous in that each band-to-band transition is an independent event. The spontaneous emission process yields a spectral output of the LED with a fairly wide bandwidth. If the structure and operating condition of the LED are modified, the device can operate in a new mode, producing a coherent spectral output with a bandwidth of wavelengths less than 0.1 nm. This new device is a laser diode, where LASER stands for *Light Amplification by Stimulated Emission of Radiation*. Although there are many different types of lasers, we will only be concerned with the pn junction laser diode.

### 14.6.1 Stimulated Emission and Population Inversion

Figure 14–31a shows the case when an incident photon is absorbed and an electron is elevated from an energy state $E_1$ to an energy state $E_2$. This process is known as induced absorption. If the electron spontaneously makes the transition back to the lower energy level with a photon being emitted, we have a spontaneous emission process as indicated in Figure 14–31b. On the other hand, if there is an incident photon at a time when an electron is in the higher energy state as shown in Figure 14–31c, the incident photon can interact with the electron, causing the electron to make a transition downward. The downward transition produces a photon. Since this process was initiated by the incident photon, the process is called *stimulated*

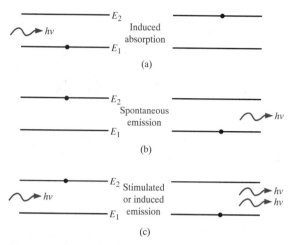

**Figure 14–31**  Schematic diagram showing (a) induced absorption, (b) spontaneous emission, and (c) stimulated emission processes.

or *induced emission*. Note that this stimulated emission process has produced two photons; thus, we can have optical gain or amplification. The two emitted photons are in phase so that the spectral output will be coherent.

In thermal equilibrium, the electron distribution in a semiconductor is determined by the Fermi-Dirac statistics. If the Boltzmann approximation applies, then we can write

$$\frac{N_2}{N_1} = \exp\left[\frac{-(E_2 - E_1)}{kT}\right] \tag{14–62}$$

where $N_1$ and $N_2$ are the electron concentrations in the energy levels $E_1$ and $E_2$, respectively, and where $E_2 > E_1$. In thermal equilibrium, $N_2 < N_1$. The probability of an induced absorption event is exactly the same as that of an induced emission event. The number of photons absorbed is proportional to $N_1$ and the number of additional photons emitted is proportional to $N_2$. In order to achieve optical amplification or for lasing action to occur, we must have $N_2 > N_1$; this is called population inversion. We cannot achieve lasing action at thermal equilibrium.

Figure 14–32 shows the two energy levels with a light wave at an intensity $I_\nu$ propagating in the $z$-direction. The change in intensity as a function of $z$ can be written as

$$\frac{dI_\nu}{dz} \propto \frac{\text{\# photons emitted}}{\text{cm}^3} - \frac{\text{\# photons absorbed}}{\text{cm}^3}$$

or

$$\frac{dI_\nu}{dz} = N_2 W_i \cdot h\nu - N_1 W_i \cdot h\nu \tag{14–63}$$

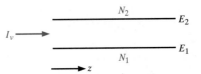

$$N_2 \qquad\qquad E_2$$

$I_\nu \longrightarrow$

$$\qquad\qquad E_1$$
$$N_1$$

$\longrightarrow z$

**Figure 14–32**   Light propagating in z-direction through a material with two energy levels.

where $W_i$ is the induced transition probability. Equation (14–63) assumes no loss mechanisms and neglects the spontaneous transitions.

Equation (14–63) can be written as

$$\frac{dI_\nu}{dz} = \gamma(\nu)I_\nu \qquad\qquad (14\text{–}64)$$

where $\gamma(\nu) \propto (N_2 - N_1)$ and is the amplification factor. From Equation (14–64), the intensity is

$$I_\nu = I_\nu(0)e^{\gamma(\nu)z} \qquad\qquad (14\text{–}65)$$

Amplification occurs when $\gamma(\nu) > 0$ and absorption occurs when $\gamma(\nu) < 0$.

We can achieve population inversion and lasing in a forward-biased pn homojunction diode, if both sides of the junction are degenerately doped. Figure 14–33a shows the energy-band diagram of a degenerately doped pn junction in thermal equilibrium. The Fermi level is in the conduction band in the n-region and the Fermi level is in the valence band in the p-region. Figure

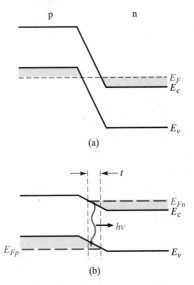

**Figure 14–33**   (a) Degenerately doped pn junction at zero bias. (b) Degenerately doped pn junction under forward bias with photon emission.

14–33b shows the energy bands of the pn junction when a forward bias is applied. The gain factor in a pn homojunction diode is given by

$$\gamma(\nu) \propto \left\{1 - \exp\left[\frac{h\nu - (E_{Fn} - E_{Fp})}{kT}\right]\right\} \qquad (14\text{–}66)$$

In order for $\gamma(\nu) > 1$, we must have $h\nu < (E_{Fn} - E_{Fp})$, which implies that the junction must be degenerately doped since we also have the requirement that $h\nu \geq E_g$. In the vicinity of the junction, there is a region in which population inversion occurs. There are large numbers of electrons in the conduction band directly above a large number of empty states. If band-to-band recombination occurs, photons will be emitted with energies in the range $E_g < h\nu < (E_{Fn} - E_{Fp})$.

### 14.6.2 Optical Cavity

Population inversion is one requirement for lasing action to occur. Coherent emission output is achieved by using an optical cavity. The cavity will cause a buildup of the optical intensity from positive feedback. A resonant cavity consisting of two parallel mirrors is known as a Fabry-Perot resonator. The resonant cavity can be fabricated, for example, by cleaving a gallium arsenide crystal along the (110) planes as shown in Figure 14–34. The optical wave propagates through the junction in the z-direction, bouncing back and forth between the end mirrors. The mirrors are actually only partially reflecting so that a portion of the optical wave will be transmitted out of the junction.

For resonance, the length of the cavity $L$ must be an integral number of half wavelengths, or

$$N\left(\frac{\lambda}{2}\right) = L \qquad (14\text{–}67)$$

where $N$ is an integer. Since $\lambda$ is small and $L$ is relatively large, there can be many resonant modes in the cavity. Figure 14–35a shows the resonant modes as a function of wavelength.

When a forward-bias current is applied to the pn junction, spontaneous emission will initially occur. The spontaneous emission spectrum is relatively broadband and is superimposed on the possible lasing modes as shown in Figure 14–35b. In order for lasing to be initiated, the spontaneous emission gain must be larger than the optical losses. By positive feedback in the cavity, lasing can occur at several specific wavelengths as indicated in Figure 14–35c.

### 14.6.3 Threshold Current

The optical intensity in the device can be written from Equation (14–65) as $I_\nu \propto e^{\gamma(\nu)z}$ where $\gamma(\nu)$ is the amplification factor. We have two basic loss

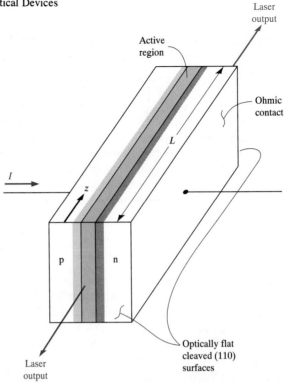

**Figure 14-34**  A pn junction laser diode with cleaved (110) planes forming the Fabry-Perot cavity. (After Yang [15].)

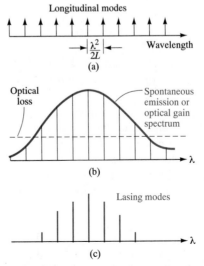

**Figure 14-35**  Schematic diagram showing (a) resonant modes of a cavity with length $L$, (b) spontaneous emission curve, and (c) actual emission modes of a laser diode. (After Yang [15].)

mechanisms. The first is the photon absorption in the semiconductor material. We can write

$$I_\nu \propto e^{-\alpha(\nu)z} \qquad (14\text{--}68)$$

where $\alpha(\nu)$ is the absorption coefficient. The second loss mechanism is due to the partial transmission of the optical signal through the ends, or through the partially reflecting mirrors.

At the onset of lasing, which is known as threshold, the optical loss of one round trip through the cavity is just offset by the optical gain. The threshold condition is then expressed as

$$\Gamma_1\Gamma_2 \exp[(2\gamma_t(\nu) - 2\alpha(\nu))L] = 1 \qquad (14\text{--}69)$$

where $\Gamma_1$ and $\Gamma_2$ are the reflectivity coefficients of the two end mirrors. For the case when the optical mirrors are cleaved (110) surfaces of gallium arsenide, the reflectivity coefficients are given approximately by

$$\Gamma_1 = \Gamma_2 = \left(\frac{\bar{n}_2 - \bar{n}_1}{\bar{n}_2 + \bar{n}_1}\right)^2 \qquad (14\text{--}70)$$

where $\bar{n}_2$ and $\bar{n}_1$ are the index of refraction parameters for the semiconductor and air, respectively. The parameter $\gamma_t(\nu)$ is the optical gain at threshold.

The optical gain at threshold, $\gamma_t(\nu)$, may be determined from Equation (14–69) as

$$\gamma_t(\nu) = \alpha + \frac{1}{2L} \ln\left(\frac{1}{\Gamma_1\Gamma_2}\right) \qquad (14\text{--}71)$$

Since the optical gain is a function of the pn junction current, we can define a threshold current density as

$$J_{th} = \frac{1}{\beta}\left[\alpha + \frac{1}{2L} \ln\left(\frac{1}{\Gamma_1\Gamma_2}\right)\right] \qquad (14\text{--}72)$$

where $\beta$ can be determined theoretically or experimentally. Figure 14–36 shows the threshold current density as a function of the mirror losses. We may note the relatively high threshold current density for a pn junction laser diode.

### 14.6.4 Device Structures and Characteristics

We have seen that in a homojunction LED, the photons may be emitted in any direction, which lowers the external quantum efficiency. Significant improvement in device characteristics can be made if the emitted photons are confined to a region near the junction. This confinement can be achieved by using an optical dielectric waveguide. The basic device is a three-layered, double heterojunction structure known as a double heterojunction laser. A requirement for a dielectric waveguide is that the index of refraction of the

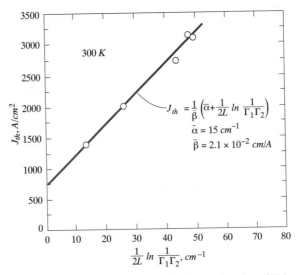

**Figure 14–36**   Threshold current density of a laser diode as a function of Fabry-Perot cavity end losses. (After Yang [15].)

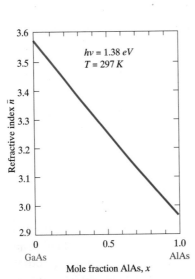

**Figure 14–37**   Index of refraction of $Al_xGa_{1-x}As$ as a function of mole fraction $x$. (From Sze [11].)

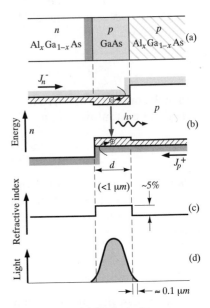

**Figure 14–38**   (a) Basic double heterojunction structure. (b) Energy-band diagram under forward bias. (c) Refractive index change through the structure. (d) Confinement of light in the dielectric waveguide. (From Yang [15].)

center material be larger than that of the other two dielectrics. Figure 14–37 shows the index of refraction for the AlGaAs system. We may note that GaAs has the highest index of refraction.

An example of a double heterojunction laser is shown in Figure 14–38a. A thin p-GaAs layer is between P-AlGaAs and N-AlGaAs layers. A simplified energy-band diagram is shown in Figure 14–38b for the forward-biased diode. Electrons are injected from the N-AlGaAs into the p-GaAs. Population inversion is easily obtained since the conduction band potential barrier prevents the electrons from diffusing into the P-AlGaAs region. Radiative recombination is then confined to the p-GaAs region. Since the index of refraction of GaAs is larger than AlGaAs, the light wave is also confined to the GaAs region. An optical cavity can be formed by cleaving the semiconductor perpendicular to the N-AlGaAs–p-GaAs junction.

Typical optical output versus diode current characteristics are shown in Figure 14–39. The threshold current is defined to be the current at the breakpoint. At low currents, the output spectrum is very wide and is the

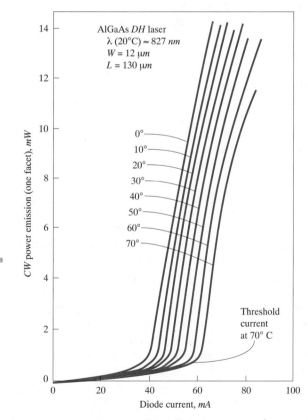

**Figure 14–39**  Typical output power versus laser diode current at various temperatures. (From Yang [15].)

result of the spontaneous transitions. When the diode current is slightly above the threshold value, the various resonant frequencies are observed. When the diode current becomes large, a single dominant mode with a narrow bandwidth is produced.

The performance of the laser diode can be further improved if a very narrow recombination region is used with a somewhat wider optical waveguide. Very complex structures using multilayers of compound semiconductor materials have been fabricated in a continuing effort to improve semiconductor laser performance.

## 14.7 SUMMARY AND REVIEW

In this chapter, we have considered the basic principles of solar cells, photodetectors, light emitting diodes, and laser diodes. The operation of these devices depends upon the creation of excess carriers by the absorbed energy of incident photons, or by the emission of photons due to recombination of excess carriers.

Solar cells convert optical power into electrical power. The short-circuit current, open-circuit voltage, and conversion efficiency were considered. Conversion efficiency takes into account incident photons with energies less than $E_g$ that are not absorbed, and also incident photons with energies greater than $E_g$ whose excess energy produces heat. For these reasons, ideal solar cell conversion efficiencies are typically less than 30 percent. The pn junction solar cell as well as heterojunction and amorphous silicon solar cells were considered. Complicated cascade heterojunction cells can be fabricated that tend to increase the conversion efficiency and produce relatively large open-circuit voltages. Amorphous silicon offers the possibility of low cost, large area solar cell arrays.

Photodetectors are semiconductor devices that convert optical signals into electrical signals. We considered the photoconductor, photodiode, PIN photodiode, avalanche photodiode, and the phototransistor. The change in conductivity of a semiconductor due to the creation of excess electrons and holes is the basis of the photoconductor. The photoconductivity is directly proportional to the incident photon flux and can be measured as a change in current, the photocurrent, in the semiconductor. The other photodetector devices considered have pn junctions that are reverse biased. Excess carriers generated within the reverse-biased space charge regions are separated very quickly, producing a photocurrent that is directly proportional to the incident photon flux. The space charge region in a reverse-biased PIN photodiode extends completely through the intrinsic region, so that the space charge width and in turn the magnitude of the photocurrent can be designed or controlled. The induced photocurrent in the avalanche photodiode is multiplied by the avalanche multiplication factor and, in addition, the time response of this device is very fast. The photocurrent in a phototransistor is multiplied by the transistor gain; however, the time response of the phototransistor tends to be slower because of the Miller effect and Miller capacitance.

The inverse mechanism of photon absorption in a pn junction is injection electroluminescence. The recombination of excess electrons and holes in a direct bandgap material can result in the emission of photons. Because the bandgap energy of the semiconductor determines the wavelength of the optical output signal, the output wavelength can be engineered within certain limits by using compound semiconductors such as $GaAs_{1-x}P_x$.

The light emitting diodes (LEDs) are the class of pn junction diodes whose photon output is the result of spontaneous recombinations of excess electrons and holes. The excess carriers are created in a forward-biased pn junction diode. A fairly wide bandwidth in the output signal, on the order of 30 nm, is a result of the spontaneous emission process. One characteristic of interest in the LED is the external quantum efficiency: the ratio of photons emitted from the semiconductor to the total number of recombination processes. Photons can be reabsorbed within the semiconductor, be reflected at the semiconductor–air interface, or experience total internal reflection if the incident angle is greater than the critical angle. These three loss mechanisms result in quantum efficiencies of from 1 to 3 percent for typical LEDs.

The output of a laser diode is the result of stimulated emission. An optical cavity, or Fabry-Perot resonator, is also used in conjunction with the diode so that the photon output is in phase, or coherent. Bandwidths on the order of 0.1 nm can be obtained in pn junction laser diodes. The efficiency of the laser diode is greatly improved by confining the emitted photons to a region near the junction; this confinement is achieved with an optical dielectric waveguide. One such laser diode is a three-layered, double heterojunction device. Other more complex multilayered structures can be fabricated to improve the laser diode characteristics.

## GLOSSARY OF IMPORTANT TERMS

**Absorption coefficient:** The relative number of photons absorbed per unit distance in a semiconductor and denoted by the parameter $\alpha$.

**Auger recombination:** The recombination of an electron and hole accompanied by the absorption of the released energy by a third particle; a nonradiative recombination process.

**Conversion efficiency:** The ratio of output electrical power to incident optical power in a solar cell.

**Delayed photocurrent:** The component of photocurrent in a semiconductor device due to diffusion currents.

**External quantum efficiency:** The ratio of emitted photons to generated photons in a semiconductor device.

**Fill factor:** The ratio $I_m V_m$ to $I_{SC} V_{oc}$, which is a measure of the realizable power from a solar cell. The parameters $I_m$ and $V_m$ are the current and voltage at the maximum power point, respectively, and $I_{SC}$ and $V_{oc}$ are the short-circuit current and open-circuit voltage.

**Fresnel loss:** The ratio of reflected to incident photons at an interface due to a change in the index of refraction.

**Internal quantum efficiency:** The fraction of diode current that produces luminescence.

**LASER diode:** An acronym for *Light Amplification by Stimulated Emission of Radiation;* the stimulated emission of photons produced in a forward-biased pn junction in conjunction with an optical cavity.

**LED:** An acronym for *Light Emitting Diode;* the spontaneous photon emission due to electron-hole recombination in a forward-biased pn junction.

**Luminescence:** The general property of light emission.

**Nonradiative recombination:** A recombination process of electrons and holes that does not produce a photon, such as the indirect transition between conduction and valence bands in silicon.

**Open-circuit voltage:** The voltage generated across the open-circuited terminals of a solar cell.

**Photocurrent:** The current generated in a semiconductor device due to the flow of excess carriers generated by the absorption of photons.

**Population inversion:** The condition whereby the concentration of electrons in one energy state is greater than that in a lower energy state; a nonequilibrium condition.

**Prompt photocurrent:** The component of photocurrent generated within the space charge region of a semiconductor device.

**Radiative recombination:** The recombination process of electrons and holes that produces a photon, such as the direct band-to-band transition in gallium arsenide.

**Shockley-Read-Hall recombination:** The recombination of electron-hole pairs via a deep-level trap; a nonradiative recombination process.

**Short-circuit current:** The current produced in a solar cell when the two terminals are shorted together.

**Stimulated emission:** The process whereby an electron is induced by an incident photon to make a transition to a lower energy state, emitting a second photon.

## PROBLEMS

### Section 14.1

1. (*a*) Calculate the maximum wavelength, $\lambda$, of a light source that can generate electron-hole pairs in Ge, Si, and GaAs. (*b*) Two sources

generate light at wavelengths of $\lambda = 570$ nm and $\lambda = 700$ nm. What are the corresponding photon energies?

2.  (a) A sample of GaAs is 0.35 $\mu$m thick. The sample is illuminated with a light source with $h\nu = 2$ eV. Determine the absorption coefficient and determine the percentage of light that is absorbed in the sample. (b) Repeat part (a) for silicon.

3.  A light source with $h\nu = 1.3$ eV and at a power density of $10^{-2}$ watts/cm$^2$ is incident on a thin slab of silicon. The excess minority carrier lifetime is $10^{-6}$ sec. Determine the electron-hole generation rate and the steady-state excess carrier concentration. Neglect surface effects.

4.  Consider an n-type GaAs sample with $\tau_p = 10^{-7}$ sec. (a) It is desired to generate a steady-state excess carrier concentration of $\delta p = 10^{15}$ cm$^{-3}$ at the surface. The incident photon energy is $h\nu = 1.9$ eV. Determine the incident power density required. (Neglect surface effects.) (b) At what distance in the semiconductor does the generation rate drop to 20 percent of that at the surface.

5.  (a) Consider a GaAs semiconductor illuminated with photons at an energy of $h\nu = 1.65$ eV. Determine the thickness of the material so that 75 percent of the energy is absorbed. (b) Determine the thickness so that 75 percent of the energy is transmitted.

6.  If the thickness of a GaAs semiconductor is 1 $\mu$m and 50 percent of the incident monochromic photon energy is absorbed, determine the incident photon energy and wavelength.

*7. Consider monochromatic light at an intensity $I_{\nu0}$ incident on the surface at $x = 0$ of an n-type semiconductor that extends to $x = \infty$. Assume the electric field is zero in the semiconductor and assume a surface recombination velocity, $s$. Taking into account the absorption coefficient, determine the steady-state excess hole concentration as a function of $x$.

*8. Monochromatic light with intensity $I_{\nu0}$ is incident on a p-type semiconductor as shown in Figure 14–40. Assume the surface recombination velocity at $x = 0$ is $s = \infty$ and assume the surface recombination

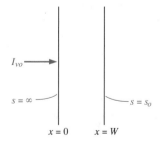

**Figure 14–40**  Figure for problem 8.

velocity at $x = W$ is $s = s_0$. Derive the expression for the steady state excess electron concentration as a function of $x$.

## Section 14.2

9. Consider an ideal long $n^+ p$ junction GaAs solar cell at $T = 300°K$ in which excess carriers are uniformly generated. The parameters of the diode are as follows:

$$N_d = 10^{19} \text{ cm}^{-3} \qquad\qquad D_n = 225 \text{ cm}^2/\text{sec}$$
$$\tau_{n0} = \tau_{p0} = 5 \times 10^{-8} \text{ sec} \qquad D_p = 7 \text{ cm}^2/\text{sec}$$

The generated photocurrent density is $J_L = 30$ mA/cm$^2$. Plot the open-circuit voltage as a function of the acceptor doping concentration for $10^{15} \le N_a \le 10^{18}$ cm$^{-3}$.

10. A long silicon pn junction solar cell with an area of 2 cm$^2$ has the following parameters:

$$N_d = 10^{19} \text{ cm}^{-3} \qquad N_a = 3 \times 10^{16} \text{ cm}^{-3}$$
$$D_p = 6 \text{ cm}^2/\text{sec} \qquad D_n = 18 \text{ cm}^2/\text{sec}$$
$$\tau_{p0} = 5 \times 10^{-7} \text{ sec} \qquad \tau_{n0} = 5 \times 10^{-6} \text{ sec}$$

Assume that excess carriers are uniformly generated in the solar cell and that $J_L = 25$ mA/cm$^2$. Let $T = 300°K$. (a) Plot the I-V characteristics of the diode, (b) determine the maximum power output of the solar cell, and (c) calculate the external load resistance that will produce the maximum power.

11. Consider the solar cell in problem 10. If the solar intensity is increased by a factor of 10, determine the maximum power output of the solar cell. By what factor has the power increased from that in problem 10?

*12. Consider the pn junction solar cell with nonuniform absorption. Derive the expression for the excess minority carrier electron concentration for the short-circuit condition and for the case when the p-region is very long and the n-region is short.

13. The absorption coefficient in amorphous silicon is approximately $10^4$ cm$^{-1}$ at $h\nu = 1.7$ eV and $10^5$ cm$^{-1}$ at $h\nu = 2.0$ eV. Determine the amorphous silicon thickness for each case so that 90 percent of the photons are absorbed.

## Section 14.3

14. Consider a silicon photoconductor at $T = 300°K$ with the following parameters:

$$N_d = 10^{16} \text{ cm}^{-3} \qquad N_a = 10^{15} \text{ cm}^{-3}$$
$$\mu_n = 1000 \text{ cm}^2/\text{V-sec} \qquad \mu_p = 430 \text{ cm}^2/\text{V-sec}$$
$$\tau_{n0} = 10^{-6} \text{ sec} \qquad \tau_{p0} = 10^{-7} \text{ sec}$$
$$A = 10^{-3} \text{ cm}^2 \qquad L = 100 \ \mu\text{m}$$

Assume that a voltage of 5 volts is applied and assume that excess electrons and holes are uniformly generated at a rate of $G_L = 10^{20}$ cm$^{-3}$sec$^{-1}$. Calculate (a) the steady-state excess carrier concentration, (b) the photoconductivity, (c) the steady-state photocurrent, and (d) the photoconductor gain.

15. Excess carriers are uniformly generated in a GaAs photoconductor at a rate of $G_L = 10^{21}$ cm$^{-3}$sec$^{-1}$. The area is $A = 10^{-4}$ cm$^2$ and the length is $L = 100 \ \mu$m. The other parameters are:

$$N_d = 5 \times 10^{16} \text{ cm}^{-3} \qquad N_a = 0$$
$$\mu_n = 8000 \text{ cm}^2/\text{V-sec} \qquad \mu_p = 250 \text{ cm}^2/\text{V-sec}$$
$$\tau_{n0} = 10^{-7} \text{ sec} \qquad \tau_{p0} = 10^{-8} \text{ sec}$$

If a voltage of 5 volts is applied, calculate (a) the steady-state excess carrier concentration, (b) the photoconductivity, (c) the steady-state photocurrent, and (d) the photoconductor gain.

*16. Consider an n-type silicon photoconductor that is 1 $\mu$m thick, 50 $\mu$m wide, and has an applied electric field in the longitudinal dimension of 50V/cm. If the incident photon flux is $\Phi_0 = 10^{16}$ cm$^{-2}$sec$^{-1}$ and the absorption coefficient is $\alpha = 5 \times 10^4$ cm$^{-1}$, calculate the steady-state photocurrent if $\mu_n = 1200$ cm$^2$/V-sec, $\mu_p = 450$ cm$^2$/V-sec, and $\tau_{p0} = 2 \times 10^{-7}$ sec.

17. Consider a long silicon pn junction photodiode at $T = 300°$K with the following parameters:

$$N_a = 2 \times 10^{16} \text{ cm}^{-3} \qquad N_d = 10^{18} \text{ cm}^{-3}$$
$$D_n = 25 \text{ cm}^2/\text{sec} \qquad D_p = 10 \text{ cm}^2/\text{sec}$$
$$\tau_{n0} = 2 \times 10^{-7} \text{ sec} \qquad \tau_{p0} = 10^{-7} \text{ sec}$$

Assume a reverse-bias voltage of $V_R = 5$ volts is applied and assume a uniform generation rate of $G_L = 10^{21}$ cm$^{-3}$sec$^{-1}$ exists throughout the entire photodiode. Calculate (a) the prompt photocurrent density and (b) the total steady-state photocurrent density.

*18. Starting with the ambipolar transport equation for minority carrier holes, derive Equation (14–41) using the geometry shown in Figure 14–17.

19. Consider a silicon PIN photodiode at $T = 300°$K. Consider intrinsic layer widths of 1, 10, and 100 $\mu$m. If the incident photon flux is

$\Phi_0 = 10^{17}$ cm$^{-2}$sec$^{-1}$ and the absorption coefficient is $\alpha = 3 \times 10^3$ cm$^{-1}$, calculate the prompt photocurrent density for each diode.

20. Consider a silicon PIN photodiode exposed to sunlight. Calculate the intrinsic region width so that at least 90 percent of all photons with wavelengths $\lambda \leq 1$ $\mu$m are absorbed in the intrinsic region. Neglect any absorption in the $p^+$ or $n^+$ regions.

## Section 14.4

21. Consider the Al$_x$Ga$_{1-x}$As system. Determine the range of the direct bandgap energies possible and the corresponding range of wavelengths.

22. Consider the GaAs$_{1-x}$P$_x$ system. Assume that $x = 0.35$. Determine (a) the bandgap energy and (b) the photon wavelength.

## Section 14.5

23. Consider a pn junction GaAs LED. Assume that photons are generated uniformly in all directions in a plane perpendicular to the junction at a distance of 0.50 $\mu$m from the surface. (a) Taking into account total internal reflection, calculate the fraction of photons that have the potential of being emitted from the semiconductor. (b) Using the results of part (a) and including Fresnel loss, determine the fraction of generated photons that will be emitted from the semiconductor into air (neglect absorption losses).

*24. In a pn junction LED, consider a point source in the semiconductor at the junction and assume that photons are emitted uniformly in all directions. Show that (neglecting photon absorption) the external quantum efficiency of the LED is given by

$$\eta_{ext} = \frac{2\bar{n}_1 \bar{n}_2}{(\bar{n}_1 + \bar{n}_2)^2} (1 - \cos \theta_c)$$

where $\bar{n}_1$ and $\bar{n}_2$ are the index of refraction parameters for the air and semiconductor, respectively, and $\theta_c$ is the critical angle.

## Section 14.6

25. Consider an optical cavity. If $N \gg 1$, show that the wavelength separation between two adjacent resonant modes is $\Delta\lambda = \lambda^2/2L$.

26. If the photon output of a laser diode is equal to the bandgap energy, find the wavelength separation between adjacent resonant modes in a GaAs laser with $L = 75$ $\mu$m.

# READING LIST

1.  Carlson, D. E. "Amorphous Silicon Solar Cells." *IEEE Transactions on Electron Devices* ED-24 (April 1977), pp. 449–53.

2.  Fonash, S. J. *Solar Cell Device Physics.* New York: Academic Press, 1981.

3.  Kressel, H. *Semiconductor Devices for Optical Communications: Topics in Applied Physics.* Vol. 39. New York: Springer-Verlag, 1987.

4.  MacMillan, H. F.; H. C. Hamaker; G. F. Virshup; and J. G. Werthen. "Multijunction III-V Solar Cells: Recent and Projected Results." *Twentieth IEEE Photovoltaic Specialists Conference.* (1988), pp. 48–54.

5.  Madan, A. "Amorphous Silicon: From Promise to Practice." *IEEE Spectrum* 23 (September 1986), pp. 38–43.

6.  Pankove, J. I. *Optical Processes in Semiconductors.* New York: Dover Publications, 1971.

7.  Roulston, D. J. *Bipolar Semiconductor Devices.* New York: McGraw-Hill, 1990.

*8.  Shur, M. *Physics of Semiconductor Devices.* Englewood Cliffs, N.J.: Prentice Hall, 1990.

9.  Streetman, B. G. *Solid State Electronic Devices.* 3rd ed. Englewood Cliffs, N.J.: Prentice Hall, 1990.

10.  Sze, S. M. *Physics of Semiconductor Devices.* 2nd ed. New York: Wiley, 1981.

11.  ———. *Semiconductor Devices: Physics and Technology.* New York: Wiley, 1985.

*12.  Wang, S. *Fundamentals of Semiconductor Theory and Device Physics.* Englewood Cliffs, N.J.: Prentice Hall, 1989.

13.  Wilson, J., and J. F. B. Hawkes. *Optoelectronics: An Introduction.* Englewood Cliffs, N.J.: Prentice Hall, 1983.

*14.  Wolfe, C. M.; N. Holonyak, Jr.; and G. E. Stillman. *Physical Properties of Semiconductors.* Englewood Cliffs, N.J.: Prentice Hall, 1989.

15.  Yang, E. S. *Microelectronic Devices.* New York: McGraw-Hill, 1988.

## CHAPTER 15

# OTHER SEMICONDUCTOR DEVICES

## PREVIEW

Although we have considered a wide range of semiconductor devices up to this point in the text, there are still many others that exist, with specialized applications. In this chapter, we will briefly consider four other types of semiconductor devices; these include the charge-coupled device, thyristor, GUNN effect diode, and avalanche transit time diode.

The charge-coupled device consists of a series of closely spaced MOS capacitors. Charge can be stored in a capacitor and transferred along the chain by a series of clock pulses. The operation of the device depends on the dynamic pulsed characteristics of the MOS capacitor, which we will consider first. The charge-coupled device has applications in signal processing and optical imaging. The charge in the MOS capacitor can be electronically injected, for signal processing, or optically injected, for optical imaging.

One application for electronic devices is in the switching of electrical currents or power. A semiconductor device used in switching applications is the thyristor. The thyristor refers to a general class of semiconductor pnpn switching devices that can be switched between a high impedance, low current state and a low impedance, high current state. The thyristor is particularly useful and efficient in switching large currents at low frequencies.

Another general class of solid state devices are those that can be used in the generation of microwave frequency signals. Two devices used in this application are the GUNN effect diode and the avalanche transit time diode. The GUNN effect device makes use of the negative resistance property of the drift velocity versus electric field characteristic of gallium arsenide or other III-V compound semiconductors. The avalanche transit time diode also relies on a negative resistance effect, which is obtained from the time delays associated with avalanche injection and with charge drift through a semiconductor region. These devices can be used to generate signals in the frequency range of 1 to 100 GHz.

Entire texts and many articles have been written on each type of solid state device that we will consider in this chapter. The objective of this chapter is to give the reader an introduction to these devices in sufficient detail so that further in-depth study is facilitated.

## 15.1 THE CHARGE-COUPLED DEVICE

A general application of solid state electronics is in the field of optical imaging such as would be used in a lightweight, low-power television camera. The *Charge-Coupled Device* (CCD) forms the basis of most solid state imagers. The CCD consists of a series of closely spaced MOS capacitors. Charge packets can be stored in each MOS capacitor and transferred along the chain by a series of clock pulses. We will first consider the dynamic effects of the MOS capacitor, and then the basic CCD structure. The CCD can also be useful in signal-processing applications. These two applications will be briefly discussed.

### 15.1.1 Dynamic Effects in the MOS Capacitor

The basic element in a CCD is the MOS capacitor which is operated in a pulsed condition, so that it is not in thermal equilibrium in most conditions. A MOS capacitor with a p-type substrate and an applied positive gate voltage is shown in Figure 15–1. We will assume that the gate voltage is applied

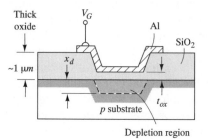

**Figure 15–1**   A MOS capacitor with a p-type substrate and a positive applied gate voltage.

at $t = 0$ and is large enough to create an inversion layer of electrons. At $t = 0^+$ the inversion layer has not been created, since a relatively long time is required before sufficient generation occurs to charge the channel. The energy band and depletion region under the gate for this condition, called *deep depletion*, is shown in Figure 15–2a. A substantial portion of the applied gate voltage is across the depletion layer; therefore, the surface potential is large and the depletion region depth is larger than the thermal equilibrium value. For $t > 0$, an inversion layer begins to form due to electron-hole pair generation within the space charge region. This condition is shown in Figure 15–2b.

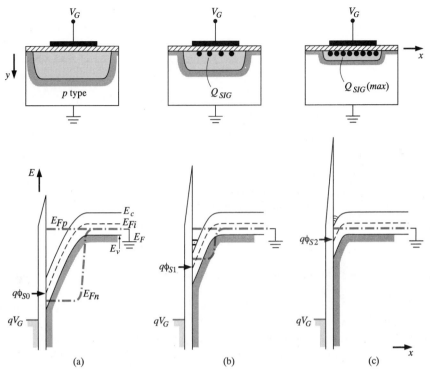

**Figure 15-2** The space charge region and energy-band diagram of a MOS capacitor (a) in deep depletion with no inversion charge, (b) in deep depletion with some inversion charge, and (c) in thermal equilibrium. (After Schroder [12].)

As the inversion charge begins to form, the surface potential decreases and the depletion region width decreases. After a relatively long time, thermal equilibrium is reached as shown in Figure 15–2c. In this case, the surface potential is $\phi_{s2} \cong 2\phi_{fp}$ and most of the gate voltage is now dropped across the oxide. The actual time required to reach this thermal-equilibrium condition is known as the thermal relaxation time and is usually on the order of seconds.

The energy-band diagrams shown in Figure 15–2 are plotted as a function of $y$, or as a function of distance from the surface into the semiconductor. We can also plot the energy-band diagram in the $x$-direction at $y = 0$, or along the surface. These diagrams, corresponding to the conditions given in Figure 15–2, are shown in Figure 15–3. These figures now begin to show the concept of a potential well under the gate. The electrons in the inversion layer at the oxide–semiconductor interface are constrained in the vertical or $y$-direction by the semiconductor potential barrier and the oxide potential barrier. This effect is observed in Figures 15–2b and c. The inversion layer electrons are also confined in the $x$-direction by the potential barriers as seen

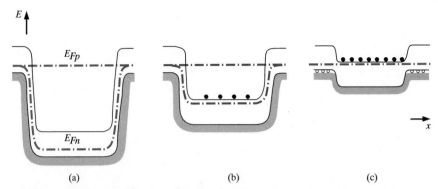

**Figure 15–3** The energy-band diagram along the surface of a MOS capacitor (a) in deep depletion with no inversion charge, (b) in deep depletion with some inversion charge, and (c) in thermal equilibrium. (After Schroder [12].)

in Figures 15–3b and c. The electronic charge, or *charge packet,* is confined to the *potential well.*

In plotting the energy-band diagrams along the surface, we can note that the conduction band, valence band, and intrinsic Fermi level are always parallel. We do not, therefore, need to plot each energy level to convey the idea of the potential well. Figure 15–4 shows plots of the intrinsic Fermi

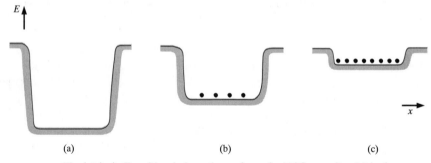

**Figure 15–4** The intrinsic Fermi level along the surface of a MOS capacitor (a) in deep depletion with no inversion charge, (b) in deep depletion with some inversion charge, and (c) in thermal equilibrium.

level for the same conditions as in Figure 15–3. These diagrams also convey the idea of the potential well being empty or filled with electrons. Figure 15–4a is a dry or empty well and Figure 15–4c is a full well. We must be a little careful in applying the analogy of a well since potential barriers still exist in the case of a full well. However, the concept of a potential well is useful in later discussions of the CCD.

## 15.1.2 The Basic CCD Shift Register

The transfer of charge can be implemented with different MOS structures and electrode arrangements. A basic three-phase CCD is a linear array of closely spaced MOS capacitors with three electrodes per stage as shown in Figure 15–5a. Every third electrode is connected to the same clock voltage

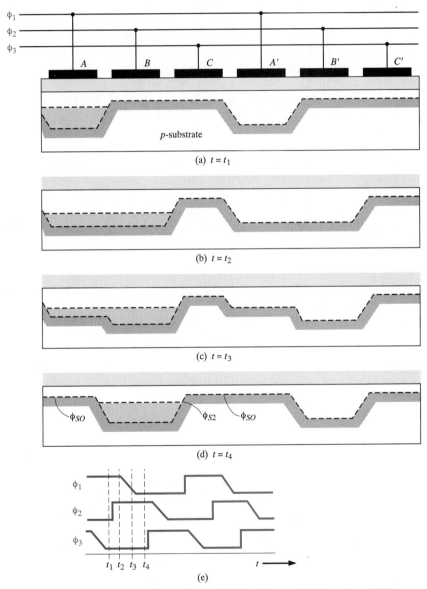

**Figure 15–5** Potential wells, charge transfer, and timing diagram of a three-phase CCD. (After Yang [19].)

so that three separate clock generators are required. The shape of the clock pulses and the relative phases are shown in Figure 15–5e. We will see that these waveforms allow better efficiency of the transfer of charge.

Consider first the case at $t = t_1$ when $\phi_1$ is high and both $\phi_2$ and $\phi_3$ are low. The potential wells are formed under the $\phi_1$ electrodes. Assume that a charge packet exists under the $A$ electrode but that no charge exists under the $A'$ electrode. At $t = t_2$, the $\phi_2$ pulse goes high and a potential well is also formed under the $B$ and $B'$ electrodes. The charge in the $A$ well flows into the $B$ well and both wells will be filled to the same depth. Potential wells exist under the $A'$ and $B'$ electrodes, but still no charge is present. The potential $\phi_1$ now decreases linearly with time and the potential well under $\phi_1$ rises linearly with time. The charge packet in the $A$ well then spills into the $B$ well. The partial emptying of the $A$ well at $t = t_3$ is shown. The relatively slow increase in the well depth of the $\phi_1$ wells ensures maximum charge transfer. At $t = t_4$, the complete transfer of charge from the $\phi_1$ well into the $\phi_2$ well has been completed. The charge packet under electrode $A$ has been shifted to the well under electrode $B$ and the zero charge under $A'$ has been shifted to electrode $B'$. The charge packet is shifted into the $\phi_3$ well during the time $\phi_3$ is high, and finally the charge packet from $A$ is shifted into the $A'$ well during the next phase when $\phi_1$ goes high. During one complete clock cycle, the charge packet shifts one stage to the right, where a stage contains three gates. Note that the packet is prevented from shifting to the left by the potential barrier.

One parameter of interest in CCDs is the charge storage capability, or the maximum amount of charge that can be stored under the gate of a CCD. Consider the situation shown in Figure 15–5d at time $t = t_4$. Let the gate voltage on the $\phi_1$ and $\phi_3$ lines be $\phi_1 = \phi_3 = V_{G1}$, which produces a surface potential $\phi_{s0}$. The gate voltage, depletion width, and surface potential for zero inversion charge are related by

$$V_{G1} - V_{FB} = -\frac{eN_a x_{d0}}{C_{ox}} + \phi_{s0} \qquad (15\text{–}1)$$

where the depletion width is

$$x_{d0} = \sqrt{\frac{2\varepsilon_s \phi_{s0}}{eN_a}} \qquad (15\text{–}2)$$

The depletion width $x_{d0}$ will, in general, be larger than the value $x_{dT}$ at thermal equilibrium.

When a charge packet is present under the $\phi_2$ gate with a voltage $V_{G2}$, we have

$$V_{G2} - V_{FB} = -\frac{Q'_{sig}}{C_{ox}} - \frac{eN_a x_{d2}}{C_{ox}} + \phi_{s2} \qquad (15\text{–}3)$$

where $Q'_{sig}$ is the signal charge per unit area and $\phi_{s2}$ is the corresponding surface potential. The depletion width is

$$x_{d2} = \sqrt{\frac{2\varepsilon_s \phi_{s2}}{eN_a}} \qquad (15-4)$$

The maximum signal charge under the $\phi_2$ gate will exist when the surface potentials are equal, or $\phi_{s0} = \phi_{s2}$. Combining Equations (15–1) and (15–3), we find that

$$Q'_{sig}(max) = -C_{ox}(V_{G2} - V_{G1}) \qquad (15-5)$$

If the signal charge packet were larger than this value, the charge would spill over into the adjacent potential well and the CCD shift register would not function properly.

---

**Example 15–1**

**Objective:** To calculate the maximum charge that can be stored under the gate in a CCD.

Consider a silicon MOS capacitor with a gate area of 5 $\mu$m $\times$ 10 $\mu$m and an oxide thickness $t_{ox} = 500$ Å. Let $V_{G2} - V_{G1} = 5$ volts.

**Solution:** From Equation (15–5) we can write

$$Q'_{sig}(max) = -C_{ox}(V_{G2} - V_{G1}) = -\frac{\varepsilon_{ox}}{t_{ox}}(V_{G2} - V_{G1})$$

$$= -\frac{(3.9)(8.85 \times 10^{-14})(5)}{500 \times 10^{-8}} = -3.45 \times 10^{-7} \text{ coul/cm}^2$$

Considering the electronic charge and the gate area, the total number of electrons in the charge packet is

$$\frac{Q_{sig}(max)}{(-e)} = \frac{(-3.45 \times 10^{-7})(5 \times 10^{-4})(10 \times 10^{-4})}{(-1.6 \times 10^{-19})} = 1.08 \times 10^6 \text{ electrons}$$

**Comment:** The charge packet size is obviously a function of the geometry, but typical values are in the range of 1 to 5 $\times$ 10^6 electrons.

---

In the particular CCD design shown in Figure 15–5, three gates and three clock pulses per stage are required to transfer the charge. Adjacent gates must be spaced sufficiently close together so that, when the same gate voltage is applied, the space charge regions overlap and there is no potential barrier between the two gates. If the spacing becomes too large, potential barriers will exist and total charge transfer will not occur. The actual fabrication of CCDs becomes quite complex so as to ensure the maximum possible charge transfer.

We have considered the transfer or shifting of charge packets along a CCD. The charge packets must be injected into the CCD by some means and then must be detected at the end of the CCD register. Charge packets may be optically injected in an image sensor or electrically injected in a shift register or delay line.

Figure 15–6 shows one such charge injection technique. The n-type region, or source, is connected to the p-type substrate. When a positive pulse

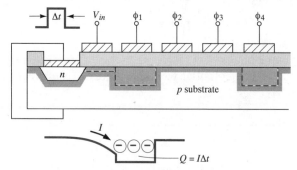

**Figure 15–6**  One technique for charge injection into a CCD delay line. (From Yang [19].)

is applied to $V_{in}$, electrons flow from the n-region filling the potential well under the $V_{in}$ electrode. This process is exactly the same as in a MOSFET when the gate voltage is applied and the inversion layer is created. When $\phi_1$ goes high, the charge shifts to the $\phi_1$ electrode and $V_{in}$ can now go low. The normal transfer or shifting process can begin.

Figure 15–7 shows the situation when light is incident on the back surface of a CCD. Electron-hole pairs are generated within the space charge regions and are collected under the gates with the deepest wells. The light intensity may not be uniform across the surface. The collected charge will be proportional to the photon intensity and to the time duration of the illumination. The charges can now be shifted through the CCD to the output in the usual manner.

One detection scheme is known as a charge-sensing floating-gate amplifier, shown in Figure 15–8a. The output circuit consists of an output gate $\phi_G$, a surface MOSFET with a floating $n^+$-source, a reset gate $\phi_R$, and a drain

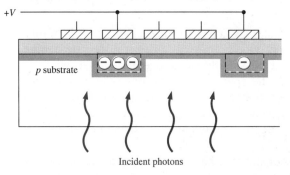

**Figure 15–7**  Charge collection under a CCD gate due to photon illumination on the back surface. (From Yang [19].)

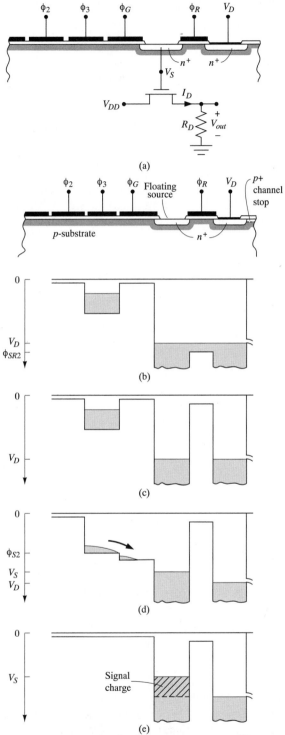

**Figure 15–8** (a) A CCD charge-sensing floating-gate output stage. The potentials in the output stage (b) during reset, (c) immediately after reset, (d) during the time the output gate is pulsed, and (e) after the charge is transferred to the output. (From Schroder [12].)

biased at $V_D$. The floating $n^+$-source is also connected to the gate of an output MOSFET.

Initially, a reset voltage $\phi_{R2} > V_D$ is applied. The surface potential under the reset gate and the potential of the floating $n^+$-region become equal to $V_D$, since any charge above $V_D$ will flow into the drain. This effect is shown in Figure 15–8b. At the end of the reset pulse, the potential of the $n^+$-region remains at $V_D$ as shown in Figure 15–8c. If a signal charge is in the $\phi_3$ well as shown in Figure 15–8c, then when the output gate is pulsed, the signal charge packet can flow into the floating $n^+$-well, changing the surface potential from $V_D$ to $V_S$ as shown in Figures 15–8d and e. The potential $V_S$ of the floating $n^+$-region is directly proportional to the signal charge which is transferred. The change in potential $\Delta V$ is given by

$$\Delta V = + \frac{Q_{\text{sig}}}{C} \qquad (15\text{–}6)$$

where $C$ is the sum of the $n^+$-$p$ junction capacitance and the output MOSFET gate capacitance.

---

### Example 15–2

**Objective:** To determine the change in voltage $\Delta V$ for a given charge packet and a given capacitance.

Consider the charge packet from Example 15–1 and let the capacitance be $C = 0.2$ pF.

**Solution:** The charge packet density from Example 15–1 was $Q'_{\text{sig}} = -3.45 \times 10^{-7}$ coul/cm$^2$. The gate dimension was 5 $\mu$m $\times$ 10 $\mu$m so

$$Q_{\text{sig}} = -(3.45 \times 10^{-7})(5 \times 10^{-4})(10 \times 10^{-4}) = -1.73 \times 10^{-13} \text{ coul}$$

Then the change in voltage is

$$\Delta V = \frac{Q_{\text{sig}}}{C} = \frac{-1.73 \times 10^{-13}}{0.2 \times 10^{-12}} = -0.86 \text{ volt}$$

**Comment:** The 0.86 volt potential change appears at the gate of the output MOSFET. This change in gate voltage will produce a change in the drain current that is easily detected.

---

An important parameter of the CCD is the charge transfer efficiency. Charge transfer from one cell to the next may be imperfect for a variety of reasons, including the existence of potential barriers between wells. In the CCD we have discussed, the charge packet is located at the oxide–semiconductor interface. This CCD is then known as a surface-channel CCD. An important factor in the surface-channel CCD which may limit the transfer efficiency is the presence of interface states. Some charge in the charge packet may be trapped in the interface states; if the emission time from an interface state is longer than the transfer time, this charge will not be transferred. The transfer efficiency can be relatively low in this particular device. One solution to this problem is to build bulk channel CCDs. The charge packet in these devices is located within the semiconductor away from the surface. The basic operation is the same as we have described so we will not specifically discuss these devices.

### 15.1.3 Applications of CCDs

We have considered a CCD shift register in which charge is transferred along a one-dimensional array of MOS capacitors. A useful property of CCDs is that the charge transfer process can be extended to two dimensions. The charge can be transferred in one direction and then shifted orthogonally to that direction. This property is useful in imaging applications.

One extremely important application of CCDs is optical imaging, or the television camera. The process of electronic imaging is the detection, read-out, and display of, in general, a two-dimensional array of photons from a source that is being imaged. We considered the basic concept of the CCD imaging process in Figure 15-7. The amount of charge collected in a potential well is directly proportional to the incident photon intensity and to the integration time.

After the charge has been collected in the potential wells, it must be transferred out and read. One technique for this is shown in Figure 15-9, and referred to as an interline-transfer imager. After the optical integration time,

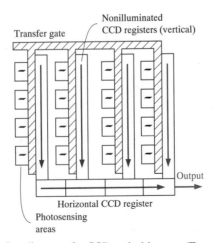

**Figure 15-9**   Interline-transfer CCD optical imager. (From Yang [19].)

the charge packets are transferred into parallel shift register lines, and then shifted down to the horizontal output register. A high-quality solid state imager consists of several hundred thousand picture elements, or pixels, along with the associated shift registers.

Another application of CCDs has been in signal processing. A variable time delay in the shift register can be implemented by simply changing the clock frequency. These delay lines have many applications in electronic systems. One such application is a comb filter, shown in Figure 15–10a. The

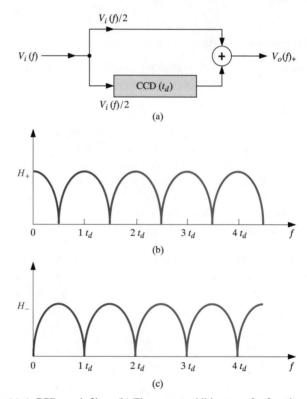

Figure 15–10   (a) A CCD comb filter. (b) The output addition transfer function. (c) The output subtraction transfer function. (From Schroder [12].)

input signal $V_i(f)$ is divided equally into two components. One component goes through a CCD delay line whose transfer function, assuming unity transfer efficiency, is

$$G(f) = \exp\left(-j2\pi f t_d\right) \tag{15–7}$$

where $t_d$ is the delay time. The other component of the input signal is not delayed. After summation at the output, we have

$$V_0(f)_+ = \frac{V_i(f)}{2} [1 + \exp(-j2\pi f t_d)] \tag{15-8}$$

The magnitude of the transfer function is

$$H_+ = \left| \frac{V_0(f)_+}{V_i(f)} \right| = \left| \frac{1 + \exp(-j2\pi f t_d)}{2} \right| \tag{15-9}$$

Equation (15-9) can be simplified to

$$H_+ = |\cos(\pi f t_d)| \tag{15-10}$$

which is plotted in Figure 15-10b. We may note that $H_+ = 0$ at $f = (n + \frac{1}{2})/t_d$ where $n = 0, 1, 2, \ldots$. The transfer function has zeros at well-defined and periodic frequencies. Obviously, no information passes through this filter at these frequencies.

If the two component signals in the circuit in Figure 15-10a are subtracted instead of added at the output, the magnitude of the transfer function becomes

$$H_- = \left| \frac{V_0(f)_-}{V_i(f)} \right| = |\sin(\pi f t_d)| \tag{15-11}$$

This transfer function is plotted in Figure 15-10c and has zeros at $f = n/t_d$ where $n = 0, 1, 2, \ldots$. We may note that the transfer function zeros of $H_+$ and $H_-$ are shifted in frequency. This type of filter is referred to as a comb filter and is ideally suited for color television systems, for example, in which both luminance and chrominance control signals are contained in a common communications channel.

## 15.2 THE THYRISTOR

One of the important applications of electronic devices is in switching between an OFF or blocking state to an ON or low impedance state. Thyristor is the name given to a general class of semiconductor pnpn switching devices that exhibit bistable regenerative switching characteristics. We have considered the transistor which may be switched ON with the application of a base drive or a gate voltage. The base drive or gate voltage must be applied as long as the transistor is to remain ON. There are a number of applications in which it is useful to have a device remain in a blocking state until switched to the low impedance state by a control signal which then does not necessarily have to remain ON. These devices are efficient in switching large currents at low frequencies, such as industrial control circuits operating at 60 Hz.

A *Semiconductor Controlled Rectifier* (SCR) is the common name given to a three-terminal thyristor. The SCR (sometimes referred to as a silicon controlled rectifier) is a four-layer pnpn structure with a gate control terminal. As with most semiconductor devices, there are several variations of the device structure. We will consider the basic SCR operation and limitations, and then discuss some variations of the basic four-layer device.

### 15.2.1 The Basic Characteristics

The four-layer pnpn structure is shown in Figure 15–11a. The upper p-region is called the anode and the lower n-region is called the cathode. If a positive voltage is applied to the anode, the device is said to be forward biased. However, the junction $J_2$ is reverse biased so that only a very small current exists. If a negative voltage is applied to the anode, then junctions $J_1$ and $J_3$ are reverse biased—again only a very small current will exist. Figure 15–11b

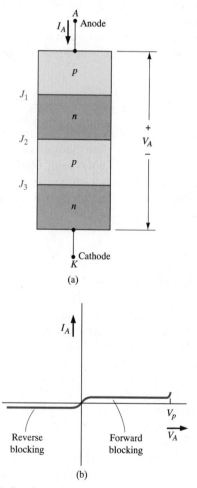

(a)

(b)

**Figure 15–11**   (a) The basic four-layer pnpn structure. (b) The initial current-voltage characteristic of the pnpn device.

shows the I-V characteristics for these conditions. The voltage $V_p$ is the breakdown voltage of the $J_2$ junction. For properly designed devices, the blocking voltage can be several thousand volts.

To consider the characteristics of the device as it goes into its conducting state, we can model the structure as coupled npn and pnp bipolar transistors. Figure 15–12a shows how we can split the four-layer structure and Figure 15–12b shows the two-transistor equivalent circuit with the associated currents. Since the base of the pnp device is the same as the collector of the npn transistor, the base current $I_{B1}$ must in fact be the same as the collector

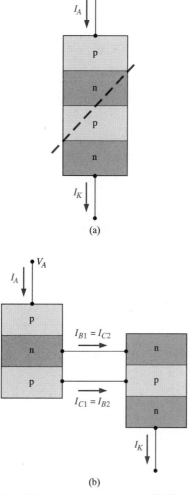

(a)

(b)

**Figure 15–12** (a) The splitting of the basic pnpn structure. (b) The two-transistor equivalent circuit of the four-layer pnpn device.

current $I_{C2}$. Similarly, since the collector of the pnp transistor is the same as the base of the npn device, the collector current $I_{C1}$ must be the same as the base current $I_{B2}$. In this bias configuration, the B-C of the pnp and the B-C of the npn devices are reverse biased while the B-E junctions are both forward biased. The parameters $\alpha_1$ and $\alpha_2$ are the common base current gains of the pnp and npn transistors, respectively.

We can write

$$I_{C1} = \alpha_1 I_A + I_{C01} = I_{B2} \tag{15-12a}$$

and

$$I_{C2} = \alpha_2 I_K + I_{C02} = I_{B1} \tag{15-12b}$$

where $I_{C01}$ and $I_{C02}$ are the reverse B-C junction saturation currents in the two devices. In this particular configuration, $I_A = I_K$ and $I_{C1} + I_{C2} = I_A$. If we add Equations (15–12a) and (15–12b), we obtain

$$I_{C1} + I_{C2} = I_A = (\alpha_1 + \alpha_2)I_A + I_{C01} + I_{C02} \tag{15-13}$$

The anode current $I_A$, from Equation (15–13), can be found as

$$I_A = \frac{I_{C01} + I_{C02}}{1 - (\alpha_1 + \alpha_2)} \tag{15-14}$$

As long as $(\alpha_1 + \alpha_2)$ is much smaller than unity, the anode current is small, as we indicated in Figure 15–11b.

The common base current gains, $\alpha_1$ and $\alpha_2$, are very strong functions of collector current as we discussed in Chapter 10. For small values of $V_A$, the collector current in each device is just the reverse saturation current, which is very small. The small collector current implies that both $\alpha_1$ and $\alpha_2$ are much smaller than unity. The four-layer structure maintains this blocking condition until the junction $J_2$ starts into breakdown or until a current is induced in the $J_2$ junction by some external means.

Consider, initially, the condition when the applied anode voltage is sufficiently large to cause the $J_2$ junction to start into avalanche breakdown. This effect is shown in Figure 15–13a. The electrons generated by impact ionization are swept into the $n_1$ region, making the $n_1$ region more negative, and the holes generated by impact ionization are swept into the $p_2$ region, making the $p_2$ region more positive. The more negative voltage of the $n_1$ region and the more positive voltage of the $p_2$ region means that the forward-bias junction voltages $V_1$ and $V_3$ both increase. The increase in the respective B-E junction voltages causes an increase in current, which results in an increase in the common base current gains $\alpha_1$ and $\alpha_2$, causing a further increase in $I_A$ as seen in Equation (15–14). We now have a regenerative positive feedback situation, so the current $I_A$ will increase very rapidly.

As the anode current $I_A$ increases and $\alpha_1 + \alpha_2$ increases, the two equivalent bipolar transistors are driven into saturation and the junction $J_2$ becomes forward biased. The total voltage across the device decreases and is approxi-

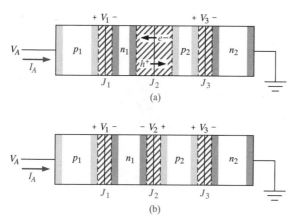

**Figure 15-13** (a) The pnpn device when the $J_2$ junction starts into avalanche breakdown. (b) The junction voltages in the pnpn structure when the device is in the high-current, low-impedance state.

mately equal to one diode drop as shown in Figure 15–13b. The current in the device is limited by the external circuit. If the current is allowed to increase, ohmic losses may become important so that the voltage drop across the device may increase slightly with current. The $I_A$ versus $V_A$ characteristic is shown in Figure 15–14.

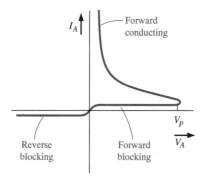

**Figure 15-14** The current-voltage characteristics of the pnpn device.

## 15.2.2 Triggering the SCR

In the last section, we considered the case when the four-layer pnpn device is turned on by the avalanche breakdown process in the center junction. The turn-on condition can also be initiated by other means. Figure 15–15a shows a three-terminal SCR in which the third terminal is the gate control. We can determine the effect of the gate current by reconsidering Equations (15–12a) and (15–12b).

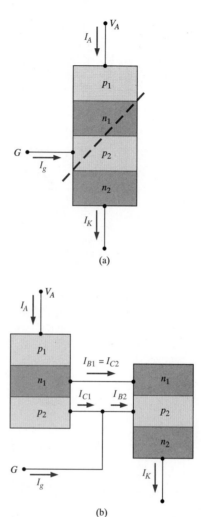

**Figure 15–15**  (a) The three-terminal SCR. (b) The two-transistor equivalent circuit of the three-terminal SCR.

Figure 15–15b again shows the two-transistor equivalent circuit including the gate current. We can write

$$I_{C1} = \alpha_1 I_A + I_{C01} \tag{15–15a}$$

and

$$I_{C2} = \alpha_2 I_K + I_{C02} \tag{15–15b}$$

We now have $I_K = I_A + I_g$ and we can still write $I_{C1} + I_{C2} = I_A$. Adding Equations (15–15a) and (15–15b), we find that

$$I_{C1} + I_{C2} = I_A = (\alpha_1 + \alpha_2)I_A + \alpha_2 I_g + I_{C01} + I_{C02} \tag{15–16}$$

Solving for $I_A$, we find

$$I_A = \frac{\alpha_2 I_g + (I_{C01} + I_{C02})}{1 - (\alpha_1 + \alpha_2)} \qquad (15\text{–}17)$$

We can think of the gate current as the flow of holes into the $p_2$ region. The additional holes increase the potential of this region, which increases the forward-biased B-E voltage of the npn bipolar transistor, and the transistor action. The transistor action of the npn increases the collector current $I_{C2}$, which starts the transistor action of the pnp bipolar transistor, and the entire pnpn device can be turned on into its low impedance state. The gate current required to switch the SCR into its ON condition is typically in the milliamp range. The SCR can be turned on with a small gate current, which can control hundreds of amperes of anode current. The gate current can be turned off and the SCR will remain in its conducting state. The gate loses control of the device once the SCR is triggered into its conducting state. The current-voltage characteristics of the SCR as a function of gate current is shown in Figure 15–16.

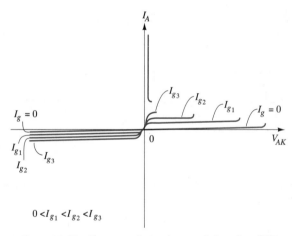

**Figure 15–16**   Current-voltage characteristics of an SCR.

A simple application of an SCR in a half-wave control circuit is shown in Figure 15–17a. The input signal is an ac voltage and a trigger pulse will control the turn-on of the SCR. We will assume that the trigger pulse occurs at time $t_1$ during the ac voltage cycle. Prior to $t_1$, the SCR is off so that the current in the load is zero, thus there is a zero output voltage. At $t = t_1$, the SCR is triggered on and the input voltage appears across the load (neglecting the voltage drop across the SCR). The SCR turns off when the anode-to-cathode voltage becomes zero even though the trigger pulse has been turned off prior to this time. The time at which the SCR is triggered during the

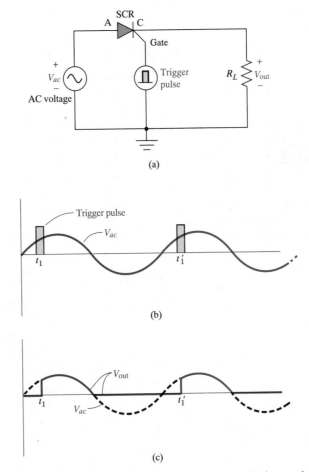

**Figure 15–17**   (a) Simple SCR circuit. (b) Input ac voltage signal and trigger pulse. (c) Output voltage versus time.

voltage cycle can be varied, changing the amount of power delivered to the load. Full-wave control circuits can be designed to increase efficiency and degree of control.

The gate allows control of the turn-on of the SCR. However, the four-layer pnpn structure can also be triggered ON by other means. In many integrated circuits, parasitic pnpn structures exist. One such example is the CMOS structure that we considered in Chapter 12. A transient ionizing radiation pulse can trigger the parasitic four-layer device by generating electron-hole pairs, particularly in the $J_2$ junction, producing a photocurrent. The photocurrent is equivalent to a gate current in an SCR so the parasitic device can be switched into its conducting state. Again, once the device is switched on, it will remain in its conducting state even when the radiation ceases. An optical signal can also trigger the device in the same manner by generating electron-hole pairs.

Another triggering mechanism in the pnpn device is by $dV/dt$ triggering. If the forward-bias anode voltage is applied rapidly, the voltage across the $J_2$ junction will also change quickly. This changing reverse-biased $J_2$ junction voltage means that the space charge region width is increasing; thus, electrons are being removed from the $n_1$ side of the junction and holes are being removed from the $p_2$ side of the junction. If $dV/dt$ is large, the rate of removal of these carriers is rapid, which leads to a large transient current that is equivalent to a gate current and can trigger the device into a low impedance conducting state. In SCR devices, a $dV/dt$ rating is usually specified. However, in parasitic pnpn structures, the $dV/dt$ triggering mechanism is a potential problem.

### 15.2.3 SCR Turn-Off

Switching the four-layer pnpn structure from its conducting state to its blocking state can be accomplished if the current $I_A$ is reduced below the value creating the $\alpha_1 + \alpha_2 = 1$ condition. This critical $I_A$ current is called the holding current. If a parasitic four-layer structure is triggered into the conducting state, the effective anode current in the device must be reduced below the corresponding holding current in order to turn off the device. This requirement essentially implies that all power supplies must be turned off in order to bring the parasitic device back into its blocking state.

The SCR can be triggered on by supplying holes to the $p_2$ region of the device. The SCR can perhaps be turned off by removing holes from this same region. If the reverse gate current is large enough to bring the npn bipolar transistor out of saturation, then the SCR can be switched from the conducting state into the blocking state. However, the lateral dimensions of the device may be large enough so that nonuniform biasing in the $J_2$ and $J_3$ junctions occurs during a negative gate current and the device will remain in the low impedance conducting state. The four-layer pnpn device must be specifically designed for a turn-off capability.

### 15.2.4 Device Structures

Many thyristor structures have been fabricated with specific characteristics for specific applications. We will consider a few of these types of device to gain an appreciation for the variety of structures.

**Basic SCR** There are many variations of diffusion, implantation, and epitaxial growth that can be used in the fabrication of the SCR device. The basic structure is shown in Figure 15–18. The $p_1$ and $p_2$ regions are diffused into a fairly high resistivity $n_1$ material. The $n^+$ cathode is formed and the $p^+$ gate contact is made. High thermal conductivity materials can be used for the anode and cathode ohmic contacts to aid in heat dissipation for high-power

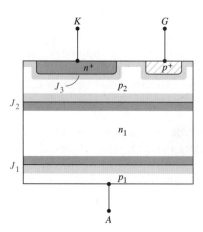

**Figure 15-18**    The basic SCR device structure.

devices. The $n_1$ region width may be on the order of 250 $\mu$m in order to support very large reverse-bias voltages across the $J_2$ junction. The $p_1$ and $p_2$ regions may be on the order of 75 $\mu$m wide while the $n^+$ and $p^+$ regions are normally quite thin.

**Bilateral Thyristor**    Since thyristors are often used in ac power applications, it may be useful to have a device that switches symmetrically in the positive and negative cycles of the ac voltage. There are a number of such devices, but the basic concept is to connect two conventional thyristors in antiparallel as shown in Figure 15–19a. The integration of this concept into a single device is shown in Figure 15–19b. Symmetrical n-regions can be diffused into a p-n-p structure. Figure 15–19c shows the current-voltage characteristics in which the triggering into the conduction mode would be due to breakdown triggering. The two terminals alternately share the role of anode and cathode during successive half cycles of the ac voltage.

Triggering by a gate control is more complex for this device since a single gate region must serve for both of the antiparallel thyristors. One such device is known as a *triac*. Figure 15–20a shows the cross section of such a device. This device can be triggered into conduction by gate signals of either polarity and with anode-to-cathode voltages of either polarity.

One particular gate control situation is shown in Figure 15–20b. Terminal 1 is positive with respect to terminal 2, and a negative gate voltage is applied with respect to terminal 1, so the gate current is negative. This polarity arrangement induces the current $I_1$ and the junction $J_4$ becomes forward biased. Electrons are injected from $n_3$, diffuse across $p_2$, and are collected in the $n_1$ region. In this case $n_3 p_2 n_1$ behaves like a saturated transistor. The collected electrons in $n_1$ lower the potential of $n_1$ with respect to $p_2$. The current across the $p_2 n_1$ junction increases, which can trigger the $p_2 n_1 p_1 n_4$ thyristor into its conducting mode.

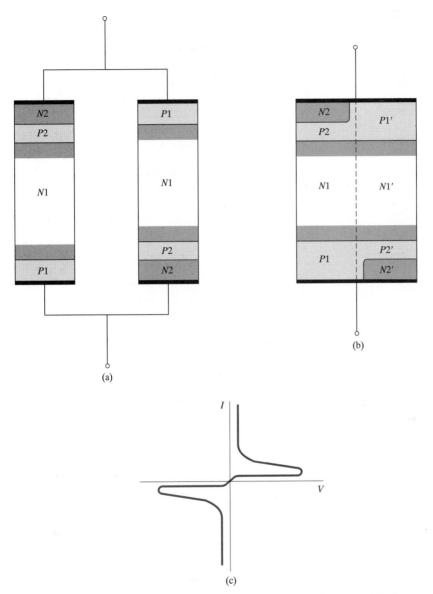

**Figure 15–19** (a) The antiparallel connection of two thyristors to form a bilateral device. (b) The bilateral thyristor as an integrated device. (c) The current-voltage characteristics of the bilateral thyristor. (From Ghandi [5].)

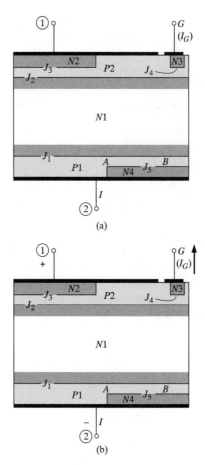

**Figure 15–20**  (a) The triac device. (b) The triac with a specific bias configuration. (From Ghandhi [5].)

We can show that the other combinations of gate, anode, and cathode voltages will also trigger the triac into conduction. Figure 15–21 shows the terminal characteristics.

**MOS Gated Thyristor**  The operation of a MOS gated thyristor is based upon controlling the gain of the npn bipolar transistor. Figure 15–22 shows a V-groove MOS gated thyristor. The MOS gate structure must extend into the n-drift region. If the gate voltage is zero, the depletion edge in the p-base remains essentially flat and parallel to the junction $J_2$; the gain of the npn transistor is low. This effect is shown in the figure by the dashed line. When a positive gate voltage is applied, the surface of the p-base becomes depleted—the depletion region in the p-base adjacent to the gate is shown by the dotted line. The undepleted base width $W_\mu$ of the npn bipolar device narrows and the gain of the device increases.

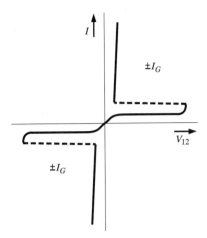

**Figure 15–21**   The current-voltage characteristics of the triac.

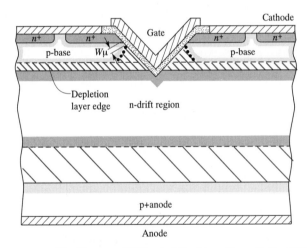

**Figure 15–22**   The V-groove MOS gated thyristor. (From Baliga [1].)

At a gate voltage approximately equal to the threshold voltage, electrons from the $n^+$ emitter are injected through the depletion region into the n-drift region. The potential of the n-drift region is lowered, which further forward biases the $p^+$-anode to n-drift junction voltage, and the regenerative process is initiated. The gate voltage required to initiate turn-on is approximately the threshold voltage of the MOS device. One advantage of this device is that the input impedance to the control terminal is very high; relatively large currents can be switched with very small capacity coupled gate currents.

**MOS Turn-Off Thyristor**    The MOS turn-off thyristor can both turn on and turn off the anode current by applying a signal to a MOS gate terminal. The basic device structure is shown in Figure 15–23. By applying a positive gate

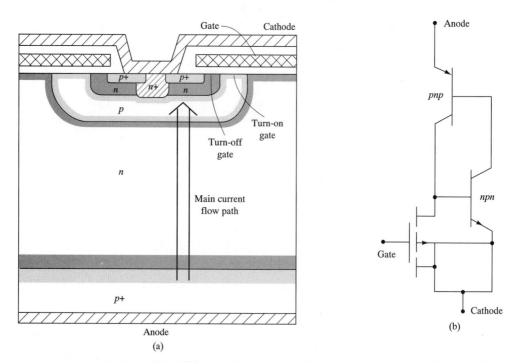

**Figure 15–23**    (a) The MOS turn-off thyristor. (b) Equivalent circuit for the MOS turn-off thyristor. (From Baliga [1].)

voltage, the $n^+pn$ bipolar transistor can be turned on as just discussed. Once the thyristor is turned on, the device can be turned off by applying a negative gate voltage: the negative gate voltage turns on the p-channel MOS transistor which effectively short circuits the B-E junction of the $n^+pn$ bipolar transistor. Holes that now enter the p-base have an alternative path to the cathode. If the resistance of the p-channel MOSFET becomes low enough, all current will be diverted away from the $n^+p$ emitter and the $n^+pn$ device will effectively be turned off.

## 15.3  THE GUNN EFFECT DEVICES

An important class of solid state devices are those that can be used to generate microwave signals. One such class involves the III-V compound semiconductors that exhibit the negative resistance effect in the drift velocity versus electric field characteristic. These types of device are referred to as *GUNN effect,* or transferred electron, devices. There are several possible modes of operation for these devices, but each mode depends on the transfer of electrons from a high-mobility state to a higher-energy, low-mobility state.

### 15.3.1 The GUNN Effect Diode

The form of oscillation that depends on the transferred electron mechanism was first demonstrated by J. B. Gunn in 1963. This device is called the GUNN diode.[1]

Figure 15–24 shows the simplified $E$ versus $k$ diagram of gallium arsenide. In n-type gallium arsenide, the valence band is nearly full and the central or

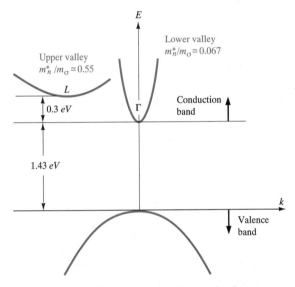

**Figure 15–24**  A simplified $E$ versus $k$ diagram for GaAs.

lower valley of the conduction band contains essentially all of the conduction electrons when the semiconductor is in thermal equilibrium. There is also a satellite or upper valley whose minimum energy is approximately 0.3 eV, or many kT electron-volts, above the minimum energy of the lower valley. The satellite valley normally contains very few electrons at thermal equilibrium.

If an electric field larger than a critical or threshold value ($\approx 3 \times 10^3$ V/cm) is applied to the semiconductor, electrons in the lower valley can gain

---

[1] This device is usually referred to as a diode since it has two terminals, although there is no junction involved as in the diodes we have considered.

enough energy and be scattered into the upper valley. The effective density of states of the upper valley is much larger than the lower valley, so the probability of electrons being scattered into the upper valley is high as long as the electric field is larger than the critical field. The effective mass of the electrons in the upper valley is much larger than the effective mass in the lower valley, which means that the electron mobility in the upper valley is much smaller than that in the lower valley. As the number of electrons transferred to the upper valley increases, the summation of all electron drift velocities decreases. The drift velocity versus electric field characteristic of gallium arsenide is shown in Figure 15–25. The decrease in drift velocity

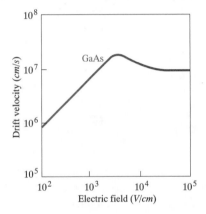

**Figure 15–25**    Drift velocity versus electric field in GaAs. (From Wang [18].)

with an increase in electric field is the differential negative mobility or differential negative resistance characteristic and is a property of the bulk semiconductor material rather than being a function of a pn junction, for example.

Figure 15–26a shows the drift current versus electric field characteristic of gallium arsenide and Figure 15–26b shows the two terminal n-type semiconductor device with an applied electric field $E_A$. Suppose that a small fluctuation in the electron concentration occurs as indicated in the diagram near the cathode terminal. A small dipole can form as shown in Figure 15–26c. This dipole layer is called a domain. The electric field within the dipole increases and the electric field outside the dipole decreases if the applied voltage across the device remains constant. If the initial electric field $E_A$ is on the negative slope of the $J$ versus E curve, then the change in electric field inside and outside of the domain results in a lower value of electron drift velocity inside the dipole compared with outside. Electrons to the right of the domain drift away while electrons pile up on the left side of the domain. This difference in electron velocity causes the accumulation and

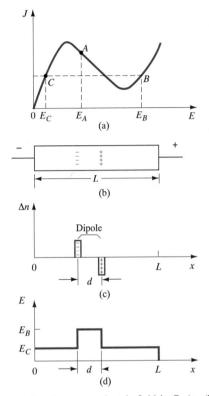

**Figure 15–26** (a) Drift current density versus electric field in GaAs. (b) Formation of a dipole when the GaAs is biased at $E = E_A$. (c) Change in electron concentration with the formation of a dipole. (d) Electric field in the device with a dipole. (After Sze [16].)

depletion layers of the domain to grow, further increasing the size of the domain.

A steady-state or stable condition can be reached when the field inside the domain is at point B and the field outside the domain is at point C as indicated in Figure 15–26a. The electric field through the device is shown in Figure 15–26d. In this condition, all electrons drift at the same velocity $v_s$, thus the domain drifts toward the anode with no further growth. As the domain passes through the anode a fluctuation in current can be measured. Also as the domain passes through the anode, the electric field in the semiconductor again increases above the critical value to $E_A$, so the process can begin all over again. We want domains to nucleate or form near the cathode and drift to the anode. The domains tend to nucleate at a high resistance region; this can be designed into the device at the cathode.

We have assumed that the length of the semiconductor is large enough for a stable domain to be formed. The formation of a stable domain before it reaches the anode places limits on the length and carrier concentration of the semiconductor. The transit time through the device must be larger than the magnitude of the negative dielectric relaxation time. We can write this condition as

$$\frac{L}{v_s} > \tau_d = \frac{\varepsilon_s}{\sigma} = \frac{\varepsilon_s}{e|\mu^*|n_o} \tag{15-18}$$

where $\mu^*$ is the average value of the differential negative mobility.

We can rewrite Equation (15–18) in the form

$$Ln_0 > \frac{\varepsilon_s v_s}{e|\mu^*|} \tag{15-19}$$

If $|\mu^*| = 100$ cm²/V-sec and $v_s = 10^7$ cm/sec, then for gallium arsenide

$$Ln_0 > \frac{\varepsilon_s v_s}{e\mu^*} = \frac{(13.1)(8.85 \times 10^{-14})(10^7)}{(1.6 \times 10^{-19})(100)} \approx 10^{12} \text{ cm}^{-2} \tag{15-20}$$

The general condition for the transit time domain mode of operation is $10^{12} < Ln_0 < 10^{14}$ cm$^{-2}$.

---

**Example 15–3**

**Objective:** To design the length and doping concentration of a gallium arsenide GUNN diode.

Consider a GUNN diode which is to oscillate at $f = 10$ GHz. Determine the length $L$ and the electron concentration $n_0$.

**Solution:** The transit time of a domain through the device corresponds to one period. Then

$$\tau_t = \frac{L}{v_s} = \frac{1}{f}$$

where $f$ is the frequency of oscillation. The device length is found as

$$L = \frac{v_s}{f} = \frac{10^7}{10 \times 10^9} = 10 \ \mu\text{m}$$

The electron concentration can be determined from

$$Ln_0 > 10^{12} \text{ cm}^{-2}$$

so that

$$n_0 > \frac{10^{12}}{L} = \frac{10^{12}}{10 \times 10^{-4}} = 10^{15} \text{ cm}^{-3}$$

**Comment:** Theoretical studies show that the maximum efficiency of the transit time domain mode device occurs when the $n_0 L$ product is one to several times $10^{12}$ cm$^{-2}$, so that the domain fills about one half of the sample and the current waveform more closely approximates a sine wave.

---

The maximum dc-to-rf conversion efficiency for this mode is only approximately 10 percent. Because of the low efficiency, the rf output power of this device is limited to approximately 1 watt or less because of the heating effects in the semiconductor.

## 15.3.2 Modes of Operation

The dipole layer or transit time domain mode of operation was the first mode of oscillation observed. The dipole mode was also discussed because it is fairly straightforward and simple to visualize. The formation of stable domains, however, is not the only mode of operation for transferred electron devices. There are several other possible modes of operation; we will consider only two.

The high-frequency transferred electron devices are usually operated in a microwave cavity, so the applied voltage across the device will vary with time. In the transit time dipole layer mode of operation, a large fraction of the voltage across the device is dropped across the high field domain region. During the negative portion of the rf voltage in the cavity, the total voltage applied across the device is reduced, thus decreasing the domain width. If the minimum voltage across the device is small enough, the domain width will continue to decrease until the domain accumulation layer and depletion layer neutralize each other and the domain is quenched. During the positive rf voltage cycle, another domain will form near the cathode. In this quenched domain mode of operation, the operating frequency is higher than the transit time domain mode frequency, so that the domain is quenched before it is collected at the anode. The frequency of operation is now determined by the resonant cavity rather than by the transit time or length of the device.

The transit time domain mode begins with the formation of a dipole near the cathode. We may, instead, simply start with an accumulation layer of electrons near the cathode as shown in Figure 15–27. The electric field through the device changes as shown in the figure. The electric field to the left of the accumulation layer is reduced so that the electron velocity increases. The electric field to the right of the accumulation layer increases so that the electron velocity decreases in this region. The change in these velocities means that the accumulation charge will increase. If the device is long enough, a stable accumulation layer will be formed which drifts to the anode.

When the frequency is very high, the dipoles or domains do not have sufficient time to form during the time the electric field is above the threshold value. An accumulation layer of electrons can form at the cathode and most of the semiconductor remains homogeneous and biased in the negative resistance state during most of the rf voltage cycle. If the frequency is high enough, the accumulation layer will be completely damped before it reaches the anode. This mode of operation is the limited space charge accumulation

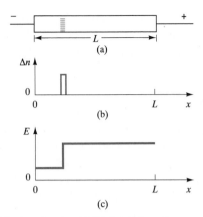

**Figure 15–27**   A GUNN device showing (a) the formation of an accumulation layer near the cathode, (b) the change in electron concentration with the formation of an accumulation layer, and (c) the electric field in the device with an accumulation layer. (After Sze [16].)

(LSA) mode. The frequency of operation is again determined by the resonant cavity and is again much larger than the domain transit time frequency. The efficiency of the LSA mode can be as high as 20 percent. When LSA diodes operate in a pulsed mode, peak powers of a few hundred watts can be obtained. The LSA diodes can be used in moderate pulsed power transmitters for radar systems, for example.

## 15.4  AVALANCHE TRANSIT TIME DEVICES

Solid state microwave oscillators can also be fabricated from pn junctions which inject carriers by avalanche multiplication into a drift region in the semiconductor. The combined time delays of the avalanche multiplication process and drift region transit time result in a negative resistance effect. We will first consider the operation of the basic IMPATT diode and then discuss a more efficient and higher-power TRAPATT mode device.

### 15.4.1  The Read Diode

The name IMPATT is an acronym for *IMP*act ionization *A*valanche *T*ransit *T*ime. The basic principle of operation of the IMPATT diode can most easily be understood by referring to the first proposed avalanche diode called the *Read diode*. One Read diode is the $n^+pip^+$ structure shown in Figure 15–28a with the doping profile shown in Figure 15–28b. A reverse-bias voltage is applied so that the space charge region extends completely through the i-region. The electric field profile is shown in Figure 15–28c. Avalanche multiplication will occur in the narrow p-region. Similar devices can be fabricated in a $p^+nin^+$ structure.

The Read diode is mounted in a microwave cavity and an ac voltage at the cavity resonant frequency will be superimposed on the applied dc voltage

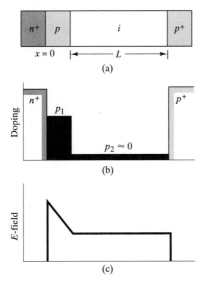

**Figure 15–28** (a) An $n^+pip^+$ Read diode. (b) Doping concentration in the Read diode. (c) Electric field profile in the Read diode.

across the diode as shown in Figure 15–29a. The total reverse-bias voltage is

$$V_R = V_{dc} + V_{ac} \qquad (15\text{–}21)$$

The dc voltage, $V_{dc}$, biases the diode right at the breakdown point so that the total field will cause impact ionization at the $n^+p$ junction during the positive half-cycle of the ac voltage. The hole current, $I_0(t)$, generated at the $n^+p$ junction ($x = 0$) by avalanche multiplication, increases exponentially with time while the field is above the breakdown value. During the negative half cycle, the field is below the breakdown value and the generated hole current decreases exponentially. The generated hole current at $x = 0$ is shown in Figure 15–29b. The peak value of current occurs at $\omega t = \pi$ since this current

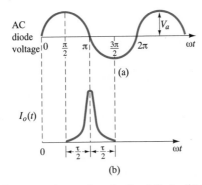

**Figure 15–29** (a) AC voltage superimposed on the Read diode. (b) Hole current generated at the $n^+p$ junction.

is the result of integrating the total number of holes generated by impact ionization during the positive voltage cycle. The generated holes will now drift through the intrinsic region inducing a current in the external circuit.

The induced current in the external circuit can be determined by considering the analogy of positive charge drifting between parallel plates as shown in Figure 15–30. Figure 15–30a shows the situation when a positive charge is

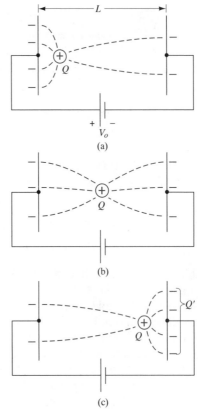

**Figure 15–30**   A drift region between parallel plates. A positive charge is (a) near the positive conductor, (b) in the center, and (c) near the negative conductor.

near the positive terminal. An electric field is created by the positive charge and induces a negative charge on the two conductors as shown. Figure 15–30b shows the situation when the positive charge has drifted halfway between the two conductors. The relative amount of charge induced on each conductor has changed. Figure 15–30c shows the situation when the positive charge has drifted close to the negative terminal. The relative amount of negative charge induced on each conductor has again changed. The change in the induced charge on each conductor occurs by the charge flowing through the external circuit. We see that, as the positive charge drifts between the two conductors, a current is induced in the external circuit.

We can derive an expression for the induced external circuit current. The work done on the positive charge is

$$W_h = F \cdot x \tag{15-22}$$

where the force is

$$F = (+Q)E_0 \tag{15-23}$$

and the electric field is

$$E_0 = \frac{V_0}{L} \tag{15-24}$$

The dc power supply does work in transferring the charge $Q'$ through the external circuit which is given by

$$W_e = V_0 \cdot Q' \tag{15-25}$$

where $Q'$ is the magnitude of the transferred charge. The induced charge sets up the fields which do the same amount of work on the positive charge, so equating Equations (15–22) and (15–25), we have

$$\frac{(+Q)V_0 x}{L} = (Q')V_0 \tag{15-26}$$

or

$$Q' = (+Q)\frac{x}{L} \tag{15-27}$$

The magnitude of the induced charge on the negative conductor increases linearly with distance as the positive charge drifts between the conductors. The current in the external circuit associated with the changing induced charge is

$$I_e(t) = \frac{dQ'}{dt} = \frac{(+Q)}{L} \cdot \frac{dx}{dt} = \frac{(+Q)v_s}{L} \tag{15-28}$$

where $v_s$ is the drift velocity of the positive charge. Equation (15–28) shows that, if the drift velocity is constant, the induced external circuit current is also a constant during the drift time.

The length of the drift region in the Read diode should be designed to be equal to one-half period of the ac voltage. The ac voltage, hole current $I_0(t)$, and induced external current $I_e(t)$ waveforms are shown in Figure 15–31. The induced external circuit current is constant assuming the carriers are drifting at their saturation velocity. The ac voltage and external current are out of phase by 180°, which yields the negative resistance characteristic required in oscillators. The resonant frequency of the cavity should be designed so that

$$f = \frac{v_s}{2L} \tag{15-29}$$

where $L$ is the length of the intrinsic drift region.

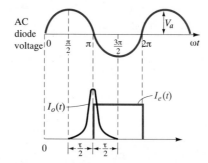

**Figure 15–31**   The ac voltage, the generated hole current at $x = 0$, and the induced external circuit current in a Read diode.

---

**Example 15–4**

**Objective:** To determine the external current and frequency of oscillation of a Read diode.

Consider a diode with an intrinsic drift region length $L = 5.0 \ \mu$m, a hole saturation drift velocity $v_s = 10^7$ cm/sec, and an avalanche-generated number of holes equal to $6 \times 10^7$ holes.

**Solution:** The total positive charge generated by the avalanche multiplication process is

$$+Q = (6 \times 10^7)(1.6 \times 10^{-19}) = 9.6 \times 10^{-12} \ \text{coul}$$

The induced external circuit current from Equation (15–28) is

$$I_e(t) = \frac{(+Q)v_s}{L} = \frac{(9.6 \times 10^{-12})(10^7)}{5 \times 10^{-4}} = 192 \ \text{mA}$$

The resonant, or oscillation, frequency is

$$f = \frac{v_s}{2L} = \frac{10^7}{2(5 \times 10^{-4})} = 10 \ \text{GHz}$$

**Comment:** The current in an IMPATT diode can be quite large and the frequency is easily in the microwave range for reasonable dimensions of the device.

---

### 15.4.2 The IMPATT Diode

The Read diode is one of the simplest IMPATT diodes that can be fabricated. The basic physical mechanisms of all IMPATT diodes involve the creation of carriers by impact ionization and the drift of these carriers through a drift region to create the negative resistance characteristic.

An IMPATT diode may be fabricated using a one-sided abrupt $p^+n$ diode as shown in Figure 15–32a. The space charge region should extend through the n-region to minimize any series resistance so that the electric field profile will be as shown in the figure. Figure 15–32b shows a two-sided abrupt pn junction, which means that the avalanche region is located at the center of the depletion region. Both electrons and holes will drift, contributing to the induced external circuit current. The efficiency of this double-drift device is greater than that of a single-drift region device. Figure 15–32c shows a modified Read diode, also called a hi-lo structure. The electric field is shown and the avalanche process occurs near the $p^+n_1$ junction. The space charge region should extend completely through to the $n^+$ region. Another IMPATT diode structure, referred to as a lo-hi-lo structure, is shown in Figure 15–32d. The electric field is nearly a constant in the $n_1$ region; thus, the avalanche region width is equal to the width of the lo $n_1$ region.

The efficiency of the IMPATT diode is, in general, larger than that of the GUNN diode. The expected maximum efficiency is about 15 percent for a single-drift silicon diode, about 21 percent for a double-drift silicon diode, and about 38 percent for a single-drift gallium arsenide diode. At lower frequencies, the maximum cw output power is primarily limited by thermal considerations. At frequencies in the range of 1 to 10 GHz, cw power outputs of 10 to 20 watts can be obtained. If the diodes are operated in a pulsed mode, higher peak output power can be obtained.

### 15.4.3 The TRAPATT Diode

The TRAPATT diode stands for *TRA*pped *P*lasma *A*valanche *T*riggered *T*ransit mode device. The TRAPATT diode is a high-power, high-efficiency device, usually operated in a pulsed mode. The peak power can be on the order of a kilowatt and the device operates in a frequency range of approximately 0.5 to 5 GHz. The efficiency may be as high as 75 percent.

Consider an $n^+pp^+$ diode. Assume that a current step is applied at time $t = 0$ when the maximum electric field in the diode is smaller than the breakdown value. The current and electric field characteristics are shown in Figures 15–33a and b. The displacement current causes the electric field to increase as a function of time as indicated in Figure 15–33b, so that the breakdown field will be reached at a certain point in time. As the field continues to increase, the point in the diode at which the breakdown field, $E_m$, is reached moves across the p-region at a particular velocity, as shown in Figure 15–33c. The velocity at which the leading edge of the avalanche region moves through the diode is a function of the diode current and doping concentration. This velocity is much larger than the carrier saturation velocity; thus, the avalanche zone will quickly sweep across the diode, creating a region filled with a highly conductive plasma of electrons and holes. The velocity at which the edge of the avalanche zone moves is not the drift

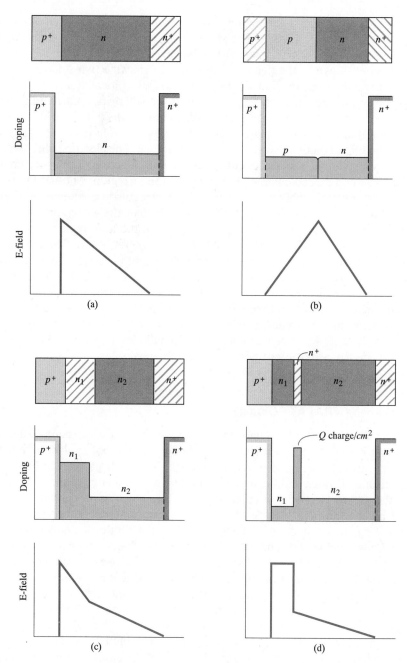

**Figure 15–32**   The structure, doping profile, and electric field in a (a) single-drift $p^+nn^+$ IMPATT diode, (b) double-drift IMPATT diode, (c) hi-lo IMPATT diode, and (d) lo-hi-lo IMPATT diode.

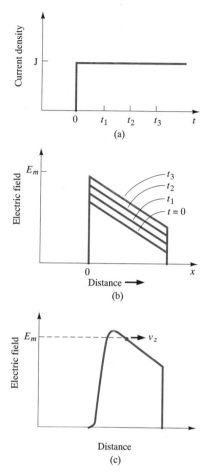

**Figure 15–33** (a) A step current pulse, (b) the electric field versus time, and (c) the electric field versus distance in a TRAPATT diode. (From Sze [16].)

velocity of the generated carriers. This space charge causes the voltage across the diode to drop to a low value.

The field and carrier density calculated at a particular point in time is shown in Figure 15–34a. The corresponding current and voltage waveforms are shown in Figures 15–34b and c, respectively. The dot in Figures 15–34b and c correspond to the condition shown in Figure 15–34a. We can note that the voltage at the beginning of the cycle can be considerably larger than the steady-state breakdown voltage. As the plasma is formed, the voltage across the device decreases substantially. Since the field in the device becomes small, the electrons and holes in the device drift at their low field mobility values. The transit time of the carriers across the diode, then, is usually much longer than the carrier saturation transit time, $L/v_s$, which implies that

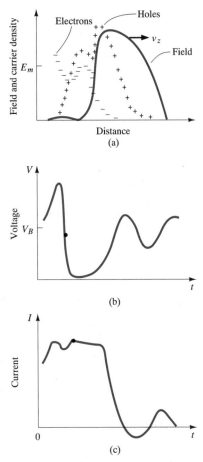

**Figure 15–34**   (a) The field and carrier concentrations at a particular time, (b) the voltage versus time, and (c) the current versus time in a TRAPATT diode. (From Sze [16].)

the TRAPATT diode operates at relatively low frequencies compared with a corresponding IMPATT device.

The TRAPATT diode requires a circuit that can support harmonics of the fundamental frequency. For this reason, the design of the TRAPATT diode and the associated circuit is more complicated than the IMPATT diode. However, the large peak output power makes these devices attractive in pulsed radar systems, for example.

## 15.5  SUMMARY AND REVIEW

We have considered, in this chapter, a few of the many semiconductor devices with specialized applications. The charge-coupled device, used in signal processing and imaging, was first considered. We then discussed the thyristor, which is used in switching applications. Finally, we discussed the generation of microwave signals using GUNN effect devices and avalanche transit time devices.

The charge-coupled device consists of a series of closely spaced MOS capacitors operated in a pulsed, deep-depletion transient mode. Charge can be stored under the gate of the MOS capacitor and transferred along the chain by a series of clock pulses. We developed an expression for the amount of charge that can be stored under a gate. The charge in the MOS capacitor can be electronically injected and shifted through the CCD delay line for signal-processing applications, such as in a comb filter. Charge can also be optically injected into the space charge region under a MOS gate. Since the amount of charge collected is directly proportional to the optical intensity, the CCD can be used as an optical imager. The optical imager is one important application of CCDs.

The thyristor refers to a general class of pnpn switching devices that can be switched between a high-impedance, low-current state and a low-impedance, high-current state. These devices exhibit a bistable regenerative positive-feedback switching characteristic. We modeled the pnpn device as coupled npn and pnp bipolar transistors. In the ON state, both bipolar transistors are driven into saturation, creating the high-current, low-voltage condition. In the OFF or blocking state, large voltages can be applied to the device and the current is essentially zero. The turn-on characteristics of the four-layer pnpn device can be controlled through a gate control terminal. The three-terminal thyristors are referred to as semiconductor controlled rectifiers (SCRs). A bilateral SCR, the triac, was discussed. The thyristor is particularly useful and more efficient than a transistor in switching large currents at low frequencies.

Another general class of solid state devices are those used to generate microwave frequency signals. One such device is the GUNN or transferred electron device, which relies on the negative resistance property of the drift velocity versus electric field characteristic in III-V compound semiconductor materials. These devices have two terminals and depend on the transfer of electrons from a high-mobility conduction state to a higher energy, low-mobility conduction state. The differential negative resistance is a property of the bulk semiconductor material. Several modes of operation are possible, including the transit time domain mode whose frequency of operation depends on the transit time of the domain through the semiconductor material. Other modes, including the limited space charge accumulation (LSA) mode, are more dependent upon the resonant frequency of a microwave cavity rather than directly on the size of the semiconductor device.

Another group of semiconductor devices used to generate microwave signals is the pn junction devices that rely on avalanche injection and a transit time delay to achieve a negative resistance characteristic. The IMPATT diode is operated in a microwave cavity so that a sinusoidal voltage is superimposed upon a dc reverse-biased junction voltage. During the positive rf voltage cycle, carriers are created by impact ionization. These carriers then move through a drift region so that the resulting ac current and ac voltage are 180° out of phase, which is the basis for the design of an oscillator. The IMPATT device may also be operated in a TRAPATT mode, in which a plasma of electrons and holes drifts through the device. The TRAPATT diode is a high-power, high-efficiency device, but operates at lower frequencies than the IMPATT diode.

## GLOSSARY OF IMPORTANT TERMS

**Charge packet:** The inversion charge stored in a potential well.

**Deep depletion:** The nonequilibrium condition in a MOS capacitor in which the induced depletion region is larger than the value at thermal equilibrium.

**Domain:** The dipole layer of charge that can form in a GUNN effect diode.

**GUNN effect diode:** A two-terminal gallium arsenide device that is biased on the negative differential mobility portion of the drift velocity versus electric field curve and exhibits high frequency oscillations.

**IMPATT diode:** *IMP*act ionization *A*valanche *T*ransit *T*ime diode; a pn junction that injects carriers by avalanche multiplication into a drift region and, because of time delays, produces a negative differential resistance effect.

**Potential well:** The space charge region under the oxide in a MOS capacitor in which an inversion charge may be held by potential barriers.

**Read diode:** The name given to the first proposed IMPATT diode.

**SCR:** *S*emiconductor *C*ontrolled *R*ectifier; the common name given to a three-terminal thyristor.

**Thyristor:** The name given to a general class of semiconductor pnpn switching devices exhibiting bistable regenerative switching characteristics.

**TRAPATT diode:** *TRA*pped *P*lasma *A*valanche *T*riggered *T*ransit mode diode; a pn junction in which breakdown produces a high-density plasma of electrons and holes that drifts through the device and, because of the time delays, produces a negative differential resistance effect.

**Triac:** The name of a bilateral three-terminal thyristor.

## PROBLEMS

### Section 15.1

1. Consider a MOS capacitor with the following parameters:

$$N_a = 10^{15} \text{ cm}^{-3} \qquad V_{FB} = 1.5 \text{ volts}$$
$$t_{ox} = 1000 \text{ Å} \qquad Area = 10 \ \mu\text{m} \times 20 \ \mu\text{m}$$

Calculate: *(a)* The oxide capacitance. *(b)* The surface potential when $V_G = 10$ volts and there is no inversion charge. *(c)* The deep-depletion region width. *(d)* The depletion-layer charge.

2. A CCD has silicon MOS capacitors with an area of 5 $\mu$m $\times$ 5 $\mu$m, a p-type substrate with $N_a = 2 \times 10^{15}$ cm$^{-3}$, and a gate oxide with $t_{ox} = 500$ Å. *(a)* Determine the flat-band voltage so that the threshold inversion point is at $V_G = 0$. Calculate $x_d$. *(b)* Using the results of part *(a)*, determine the deep depletion space charge width and surface potential if $V_G = 5$ volts. *(c)* What is the maximum signal charge per packet when $V_G = 5$ volts?

3. A CCD is fabricated on a p-type silicon substrate with $N_a = 5 \times 10^{14}$ cm$^{-3}$. The oxide thickness is $t_{ox} = 1200$ Å and the gate area is 10 $\mu$m $\times$ 20 $\mu$m. *(a)* Calculate the surface potential and depletion-layer depth for $V_G = 10$ volts and for $V_G = 20$ volts. Assume $V_{FB} = 0$ and $Q_{sig} = 0$. *(b)* Repeat part *(a)* if $10^6$ electrons are introduced in a charge packet under each gate.

4. Consider the MOS capacitor in problem 2. If the signal charge present at $V_G = 5$ volts is 10 percent of the maximum value possible, and if electron-hole pairs are being generated in the space charge region at the rate of $G_L = 10^{15}$ cm$^{-3}$ sec$^{-1}$, estimate the time required for the signal charge to increase by 5 percent.

5. A three-phase CCD delay line has 1000 stages and is operated at a clock frequency of 1 MHz. The transfer inefficiency per stage is $10^{-4}$. If a charge packet of $10^6$ electrons is injected at the input at time $t = 0$, *(a)* determine the time at which the charge packet reaches the output and *(b)* determine the number of electrons in the charge packet at the output.

6. Show that the transfer function of a subtractive comb filter is $H_- = |\sin(\pi f t_d)|$, which was given in Equation (15–11).

## Section 15.2

7. One condition for switching a thyristor is that $\alpha_1 + \alpha_2 = 1$. Show that this condition corresponds to $\beta_1 \beta_2 = 1$, where $\beta_1$ and $\beta_2$ are the common emitter current gains of the pnp and npn bipolar transistors in the equivalent circuit of the thyristor.

8. Explain how a pulse of ionizing radiation could trigger a basic CMOS structure into a high-current, low-impedance state.

9. Show that the triac can be triggered into its ON state by gate signals of either polarity and with anode-to-cathode voltages of either polarity. Consider each voltage polarity combination.

## Section 15.3

10. If $N_L$ and $N_U$ are the density of states in the lower and upper valleys, respectively, of the GaAs conduction band, show that, when the Boltzmann approximation holds,

$$\frac{n_U}{n_L} = \frac{N_U}{N_L} \exp\left(\frac{-\Delta E}{kT}\right)$$

where $n_U$ and $n_L$ are the electron concentrations in the two valleys.

11. (a) Using the results of problem 10, calculate $n_U/n_L$ at $T = 300°K$ if $N_U = 28\,N_L$. (b) If an electric field is applied so that $n_U = n_L$, calculate the effective electron temperature $T$.

12. A GaAs GUNN diode is to operate in the stable domain mode. The device length is $L = 5\ \mu m$ and $n_0 L = 5 \times 10^{12}\ cm^{-2}$. (a) What is the electron concentration required and what will be the approximate oscillating frequency. (b) If the device is biased at the critical electric field, calculate the power dissipation per unit volume.

13. An LSA diode is operating at $f = 10$ GHz in a pulsed mode. The peak output power is 250 watts and the duty cycle is 0.001. The efficiency is 12 percent. (a) Calculate the average output power and (b) the dc input power.

## Section 15.4

14. A Read diode has the following parameters: drift region length $L = 8\ \mu m$; total hole charge $+Q = 15\ pC$; hole drift velocity $v_s = 7 \times 10^6$ cm/sec. Calculate (a) the external circuit current $I_e(t)$, and (b) the resonant frequency.

15. A silicon $p^+ n i n^+$ Read diode is to be designed to operate at $f = 40$ GHz with an external circuit current of $I_e(t) = 20$ mA. Determine (a) the i-region length and (b) the electron drift charge required.

16. Consider a silicon Read diode which has an intrinsic region length of $L = 10\ \mu m$ and an external circuit current of $I_e(t) = 40$ mA. (a) Estimate the carrier drift time. (b) Estimate the drift charge (coul). (c) Determine the operating frequency.

## READING LIST

1. Baliga, B. J. *Modern Power Devices*. New York: Wiley, 1987.
2. Bulman, P. J.; G. S. Hobson; and B. C. Taylor. *Transferred Electron Devices*. New York: Academic Press, 1972.
3. DeLoach, B. C., Jr. and D. L. Scharfetter. "Device Physics of TRAPATT Oscillators." *IEEE Transactions on Electron Devices* ED-17 (January 1970), pp. 9–21.

4. Gentry, F. E.; F. W. Gutzwiller; N. Holonyak, Jr.; and E. E. Von Zastrow. *Semiconductor Controlled Rectifiers: Principles and Applications of pnpn Devices.* Englewood Cliffs, N.J.: Prentice Hall, 1964.

5. Ghandhi, S. K. *Semiconductor Power Devices: Physics of Operation and Fabrication Technology.* New York: Wiley, 1977.

6. Gibbons, G. *Avalanche-Diode Microwave Oscillators.* Oxford: Clarendon Press, 1973.

7. Liao, S. Y. *Microwave Solid-State Devices.* Englewood Cliffs, N.J.: Prentice Hall, 1985.

8. Mauro, R. *Engineering Electronics: A Practical Approach.* Englewood Cliffs, N.J.: Prentice Hall, 1989.

9. Milnes, A. G. *Semiconductor Devices and Integrated Electronics.* New York: Van Nostrand Reinhold, 1980.

10. Muller, R. S., and T. I. Kamins. *Device Electronics for Integrated Circuits.* 2nd ed. New York: Wiley, 1986.

11. Roulston, D. J. *Bipolar Semiconductor Devices.* New York: McGraw-Hill, 1990.

12. Schroder, D. K. *Advanced MOS Devices: Modular Series on Solid State Devices.* Reading, Mass.: Addison-Wesley, 1987.

*13. Shur, M. *GaAs Devices and Circuits.* New York: Plenum Press, 1987.

*14. _____. *Physics of Semiconductor Devices.* Englewood Cliffs, N.J.: Prentice Hall, 1990.

15. Streetman, B. G. *Solid State Electronic Devices.* 3rd ed. Englewood Cliffs, N.J.: Prentice Hall, 1990.

16. Sze, S. M. *Physics of Semiconductor Devices.* 2nd ed. New York: Wiley, 1981.

17. _____. *Semiconductor Devices: Physics and Technology.* New York: Wiley, 1985.

*18. Wang, S. *Fundamentals of Semiconductor Theory and Device Physics.* Englewood Cliffs, N.J.: Prentice Hall, 1989.

19. Yang, E. S. *Microelectronic Devices.* New York: McGraw-Hill, 1988.

# SELECTED LIST OF SYMBOLS

This list does not include some symbols that are defined and used specifically in only one section. Some symbols have more than one meaning; however, the context in which the symbol is used should make the meaning unambiguous. The usual unit associated with each symbol is given.

| | |
|---|---|
| $a$ | Unit cell dimension (Å), potential well width, acceleration, gradient of impurity concentration, channel thickness of a one-sided JFET (cm) |
| $a_0$ | Bohr radius (Å) |
| $c$ | Speed of light (cm/sec) |
| $d$ | Distance (cm) |
| $e$ | Electronic charge (magnitude) (coul), Napierian base |
| $f$ | Frequency (Hz) |
| $f_F(E)$ | Fermi-Dirac probability function |
| $f_T$ | Cutoff frequency (Hz) |
| $g$ | Generation rate (cm$^{-3}$sec$^{-1}$) |
| $g'$ | Generation rate of excess carriers (cm$^{-3}$sec$^{-1}$) |
| $g(E)$ | Density of states function (cm$^{-3}$) |
| $g_c, g_v$ | Density of states function in the conduction band and valence band (cm$^{-3}$) |
| $g_d$ | Channel conductance ($S$), small signal diffusion conductance ($S$) |
| $g_m$ | Transconductance (A/V) |
| $g_n, g_p$ | Generation rate for electrons and holes (cm$^{-3}$sec$^{-1}$) |
| $h$ | Planck's constant (J-sec), induced space charge width in a JFET (cm) |
| $\hbar$ | Modified Plank's constant ($h/2\pi$) |
| $h_f$ | Small-signal common emitter current gain |
| $j$ | Imaginary constant, $\sqrt{-1}$ |
| $k$ | Boltzmann's constant (J/K), wavenumber (cm$^{-1}$) |
| $k_n$ | Conduction parameter (A/V$^2$) |

| $m$ | Mass (kg) |
|---|---|
| $m_0$ | Rest mass of the electron (kg) |
| $m^*$ | Effective mass (kg) |
| $m_n^*, m_p^*$ | Effective mass of an electron and hole (kg) |
| $n$ | Integer |
| $n, l, m, s$ | Quantum numbers |
| $n, p$ | Electron and hole concentration (cm$^{-3}$) |
| $\bar{n}$ | Index of refraction |
| $n', p'$ | Constants related to the trap energy (cm$^{-3}$) |
| $n_{B0}, p_{E0}, p_{C0}$ | Thermal-equilibrium minority carrier electron concentration in the base and minority carrier hole concentration in the emitter and collector (cm$^{-3}$) |
| $n_d$ | Density of electrons in the donor energy level (cm$^{-3}$) |
| $n_i$ | Intrinsic concentration of electrons (cm$^{-3}$) |
| $n_0, p_0$ | Thermal-equilibrium concentration of electrons and holes (cm$^{-3}$) |
| $n_p, p_n$ | Minority carrier electron and minority carrier hole concentration (cm$^{-3}$) |
| $n_{p0}, p_{n0}$ | Thermal-equilibrium minority carrier electron and minority carrier hole concentration (cm$^{-3}$) |
| $n_s$ | Density of a two-dimensional electron gas (cm$^{-2}$) |
| $p$ | Momentum |
| $p_a$ | Density of holes in the acceptor energy level (cm$^{-3}$) |
| $p_i$ | Intrinsic hole concentration ($= n_i$)(cm$^{-3}$) |
| $q$ | Charge (coul) |
| $r, \theta, \phi$ | Spherical coordinates |
| $r_d, r_\pi$ | Small-signal diffusion resistance (ohm) |
| $r_{ds}$ | Small-signal drain-to-source resistance (ohm) |
| $s$ | Surface recombination velocity (cm/sec) |
| $t$ | Time (sec) |
| $t_d$ | Delay time (sec) |
| $t_{ox}$ | Gate oxide thickness (cm or Å) |
| $t_s$ | Storage time (sec) |
| $u(x)$ | Periodic wave function |
| $v$ | Velocity (cm/sec) |
| $v_d$ | Carrier drift velocity (cm/sec) |
| $v_{ds}, v_s, v_{sat}$ | Carrier saturation drift velocity (cm/sec) |
| $x, y, z$ | Cartesian coordinates |
| $x$ | Mole fraction in compound semiconductors |
| $x_B, x_E, x_C$ | Neutral base, emitter, and collector region widths (cm) |

| | |
|---|---|
| $x_d$ | Induced space charge width (cm) |
| $x_{dT}$ | Maximum space charge width (cm) |
| $x_n$, $x_p$ | Depletion width from the metallurgical junction into n-type and p-type semiconductor regions (cm) |
| $A$ | Area (cm$^2$) |
| $A^*$ | Effective Richardson constant (A/$^\circ$K$^2$/cm$^2$) |
| $B$ | Magnetic flux density (Wb/m$^2$) |
| $B$, $E$, $C$ | Base, emitter, and collector |
| $BV_{CBO}$ | Breakdown voltage of collector-base junction with emitter open (volt) |
| $BV_{CEO}$ | Breakdown voltage between collector-emitter with base open (volt) |
| $C$ | Capacitance (F) |
| $C'$ | Capacitance per unit area (F/cm$^2$) |
| $C_d$, $C_\pi$ | Diffusion capacitance (F) |
| $C_{FB}$ | Flat-band capacitance (F) |
| $C_{gs}$, $C_{gd}$, $C_{ds}$ | Gate-source, gate-drain, and drain-source capacitance (F) |
| $C_j'$ | Junction capacitance per unit area (F/cm$^2$) |
| $C_M$ | Miller capacitance (F) |
| $C_n$, $C_p$ | Constants related to capture rate of electrons and holes |
| $C_{ox}$ | Gate oxide capacitance per unit area (F/cm$^2$) |
| $C_\mu$ | Reverse-biased B-C junction capacitance (F) |
| $D$, $S$, $G$ | Drain, source, and gate of an FET |
| $D'$ | Ambipolar diffusion coefficient (cm$^2$/sec) |
| $D_B$, $D_E$, $D_C$ | Base, emitter, and collector minority carrier diffusion coefficients (cm$^2$/sec) |
| $D_{it}$ | Density of interface states (#/eV-cm$^3$) |
| $D_n$, $D_p$ | Minority carrier electron and minority carrier hole diffusion coefficient (cm$^2$/sec) |
| $E$ | Energy (joule or eV) |
| $E_a$ | Acceptor energy level (eV) |
| $E_c$, $E_v$ | Energy at the bottom edge of the conduction band and top edge of the valence band (eV) |
| $\Delta E_c$, $\Delta E_v$ | Difference in conduction band energies and valence band energies at a heterojunction (eV) |
| $E_d$ | Donor energy level (eV) |
| $E_F$ | Fermi energy (eV) |
| $E_{Fi}$ | Intrinsic Fermi energy (eV) |
| $E_{Fn}$, $E_{Fp}$ | Quasi-Fermi energy levels for electrons and holes (eV) |
| $E_g$ | Bandgap energy (eV) |

| | |
|---|---|
| $\Delta E_g$ | Bandgap narrowing factor (eV), difference in bandgap energies at a heterojunction (eV) |
| $E_t$ | Trap energy level (eV) |
| $F$ | Force $(N)$ |
| $F_n^-, F_p^+$ | Electron and hole particle flux (cm$^{-2}$sec$^{-1}$) |
| $F_{1/2}(\eta)$ | Fermi-Dirac integral function |
| $G$ | Generation rate of electron-hole pairs (cm$^{-3}$sec$^{-1}$) |
| $G_L$ | Excess carrier generation rate (cm$^{-3}$sec$^{-1}$) |
| $G_{n0}, G_{p0}$ | Thermal equilibrium generation rate for electrons and holes (cm$^{-3}$sec$^{-1}$) |
| $G_{01}$ | Conductance (S) |
| $I$ | Current (A) |
| $I_A$ | Anode current (A) |
| $I_B, I_E, I_C$ | Base, emitter, and collector current (A) |
| $I_{CBO}$ | Reverse-bias collector-base junction current with emitter open (A) |
| $I_{CEO}$ | Reverse-bias collector-emitter current with base open (A) |
| $I_D$ | Diode current (A), drain current (A) |
| $I_D(\text{sat})$ | Saturation drain current (A) |
| $I_g$ | Thyristor gate current (A) |
| $I_K$ | Cathode current (A) |
| $I_L$ | Photocurrent (A) |
| $I_{P1}$ | Pinchoff current (A) |
| $I_S$ | Ideal reverse-bias saturation current (A) |
| $I_{SC}$ | Short-circuit current (A) |
| $I_\nu$ | Photon intensity (energy/cm$^2$/sec) |
| $J$ | Electric current density (A/cm$^2$) |
| $J_{\text{gen}}$ | Generation current density (A/cm$^2$) |
| $J_L$ | Photocurrent density (A/cm$^2$) |
| $J_n, J_p$ | Electron and hole electric current density (A/cm$^2$) |
| $J_n^-, J_p^+$ | Electron and hole particle current density (cm$^{-2}$sec$^{-1}$) |
| $J_{\text{rec}}$ | Recombination current density (A/cm$^2$) |
| $J_{r0}$ | Zero-bias recombination current density (A/cm$^2$) |
| $J_R$ | Reverse-bias current density (A/cm$^2$) |
| $J_S$ | Ideal reverse-bias saturation current density (A/cm$^2$) |
| $J_{sT}$ | Ideal reverse saturation current density in a Schottky diode (A/cm$^2$) |
| $L$ | Length (cm), inductance (H), channel length (cm) |
| $\Delta L$ | Channel length modulation factor (cm) |
| $L_B, L_E, L_C$ | Minority carrier diffusion length in the base, emitter, and collector (cm) |

| | |
|---|---|
| $L_D$ | Debye length (cm) |
| $L_n, L_p$ | Minority carrier electron and hole diffusion length (cm) |
| $M, M_n$ | Multiplication constant |
| $N$ | Number density (cm$^{-3}$) |
| $N_a$ | Density of acceptor impurity atoms (cm$^{-3}$) |
| $N_B, N_E, N_C$ | Base, emitter, and collector doping concentrations (cm$^{-3}$) |
| $N_c, N_v$ | Effective density of states function in the conduction band and valence band (cm$^{-3}$) |
| $N_d$ | Density of donor impurity atoms (cm$^{-3}$) |
| $N_{it}$ | Interface state density (cm$^{-2}$) |
| $N_t$ | Trap density (cm$^{-3}$) |
| $P$ | Power (watt) |
| $P(r)$ | Probability density function |
| $Q$ | Charge (coul) |
| $Q'$ | Charge per unit area (coul/cm$^2$) |
| $Q_B$ | Gate controlled bulk charge (coul) |
| $Q'_n$ | Inversion channel charge density per unit area (coul/cm$^2$) |
| $Q'_{sig}$ | Signal charge density per unit area (coul/cm$^2$) |
| $Q'_{SD}(\text{max})$ | Maximum space charge density per unit area (coul/cm$^2$) |
| $Q'_{ss}$ | Equivalent trapped oxide charge per unit area (coul/cm$^2$) |
| $R$ | Reflection coefficient, recombination rate (cm$^{-3}$sec$^{-1}$), resistance (ohms) |
| $R(r)$ | Radial wave function |
| $R_c$ | Specific contact resistance (ohm-cm$^2$) |
| $R_{cn}, R_{cp}$ | Capture rate for electrons and holes (cm$^{-3}$sec$^{-1}$) |
| $R_{en}, R_{ep}$ | Emission rate for electrons and holes (cm$^{-3}$sec$^{-1}$) |
| $R_n, R_p$ | Recombination rate of electrons and holes (cm$^{-3}$sec$^{-1}$) |
| $R_{n0}, R_{p0}$ | Thermal equilibrium recombination rate of electrons and holes (cm$^{-3}$sec$^{-1}$) |
| $T$ | Temperature (kelvin), kinetic energy (joule or eV), transmission coefficient |
| $V$ | Potential (volt), potential energy (joule or eV) |
| $V_a$ | Applied forward-bias voltage (volt) |
| $V_A$ | Early voltage (volt), anode voltage (volt) |
| $V_{bi}$ | Built-in potential barrier (volt) |
| $V_B$ | Breakdown voltage (volt) |
| $V_{BD}$ | Breakdown voltage at the drain (volt) |
| $V_{BE}, V_{CB}, V_{CE}$ | Base-emitter, collector-base, and collector-emitter voltage (volt) |
| $V_{DS}, V_{GS}$ | Drain-source and gate-source voltage (volt) |

| | |
|---|---|
| $V_{DS}(\text{sat})$ | Drain-source saturation voltage (volt) |
| $V_{FB}$ | Flat-band voltage (volt) |
| $V_G$ | Gate voltage (volt) |
| $V_H$ | Hall voltage (volt) |
| $V_{oc}$ | Open-circuit voltage (volt) |
| $V_{ox}$ | Potential difference across an oxide (volt) |
| $V_{p0}$ | Pinchoff voltage (volt) |
| $V_{pt}$ | Punch-through voltage (volt) |
| $V_R$ | Applied reverse-bias voltage (volt) |
| $V_{SB}$ | Source-body voltage (volt) |
| $V_t$ | Thermal voltage ($kT/e$) |
| $V_T$ | Threshold voltage (volt) |
| $\Delta V_T$ | Threshold voltage shift (volt) |
| $W$ | Total space charge width (cm), channel width (cm) |
| $W_B$ | Metallurgical base width (cm) |
| $Y$ | Admittance |
| $\alpha$ | Photon absorption coefficient ($\text{cm}^{-1}$), ac common base current gain |
| $\alpha_n$, $\alpha_p$ | Electron and hole ionization rates ($\text{cm}^{-1}$) |
| $\alpha_0$ | dc common base current gain |
| $\alpha_T$ | Base transport factor |
| $\beta$ | common emitter current gain |
| $\gamma$ | Emitter injection efficiency factor |
| $\delta$ | Recombination factor |
| $\delta n$, $\delta p$ | Excess electron and hole concentration ($\text{cm}^{-3}$) |
| $\delta n_p$, $\delta p_n$ | Excess minority carrier electron and excess minority carrier hole concentration ($\text{cm}^{-3}$) |
| $\varepsilon$ | Permittivity ($\text{F/cm}^2$) |
| $\varepsilon_0$ | Permittivity of free space ($\text{F/cm}^2$) |
| $\varepsilon_{ox}$ | Permittivity of an oxide ($\text{F/cm}^2$) |
| $\varepsilon_r$ | Relative permittivity or dielectric constant |
| $\varepsilon_s$ | Permittivity of a semiconductor ($\text{F/cm}^2$) |
| $\theta_c$ | Critical angle for total internal reflection |
| $\lambda$ | Wavelength (cm or $\mu$m) |
| $\mu$ | Permeability (H/cm) |
| $\mu'$ | Ambipolar mobility ($\text{cm}^2/\text{V-sec}$) |
| $\mu_n$, $\mu_p$ | Electron and hole mobility ($\text{cm}^2/\text{V-sec}$) |
| $\mu_0$ | Permeability of free space (H/cm) |
| $\nu$ | Frequency (Hz) |

| | |
|---|---|
| $\rho$ | Resistivity (ohm-cm), volume charge density (coul/cm$^3$) |
| $\sigma$ | Conductivity (ohm$^{-1}$cm$^{-1}$) |
| $\Delta\sigma$ | Photoconductivity (ohm$^{-1}$cm$^{-1}$) |
| $\sigma_i$ | Intrinsic conductivity (ohm$^{-1}$cm$^{-1}$) |
| $\sigma_n$, $\sigma_p$ | Conductivity of n-type and p-type semiconductor (ohm$^{-1}$cm$^{-1}$) |
| $\tau$ | Lifetime (sec) |
| $\tau_n$, $\tau_p$ | Electron and hole lifetime (sec) |
| $\tau_{n0}$, $\tau_{p0}$ | Excess minority carrier electron and hole lifetime (sec) |
| $\tau_0$ | Lifetime in space charge region (sec) |
| $\phi$ | Potential (volt) |
| $\phi(t)$ | Time-dependent wave function |
| $\Delta\phi$ | Schottky barrier lowering potential (volt) |
| $\phi_{Bn}$ | Schottky barrier height (volt) |
| $\phi_{B0}$ | Ideal Schottky barrier height (volt) |
| $\phi_{fn}$, $\phi_{fp}$ | Potential difference (magnitude) between $E_{Fi}$ and $E_F$ in n-type and p-type semiconductor (volt) |
| $\phi_{Fn}$, $\phi_{Fp}$ | Potential difference (with sign) between $E_{Fi}$ and $E_F$ in n-type and p-type semiconductor (volt) |
| $\phi_m$ | Metal work function (volt) |
| $\phi_m'$ | Modified metal work function (volt) |
| $\phi_{ms}$ | Metal-semiconductor work function difference (volt) |
| $\phi_n$, $\phi_p$ | Potential difference (magnitude) between $E_c$ and $E_F$ in n-type and between $E_v$ and $E_F$ in p-type semiconductor (volt) |
| $\phi_s$ | Semiconductor work function (volt), surface potential (volt) |
| $\chi$ | Electron affinity (volt) |
| $\chi'$ | Modified electron affinity (volt) |
| $\psi(x)$ | Time-independent wave function |
| $\omega$ | Radian frequency (sec$^{-1}$) |
| $\Gamma$ | Reflection coefficient |
| $\Gamma_{ph}$ | Photoconductor gain |
| E | Electric field (V/cm) |
| $E_H$ | Hall electric field (V/cm) |
| $E_{crit}$ | Critical electric field at breakdown (V/cm) |
| $\Theta(\theta)$ | Angular wave function |
| $\Phi$ | Photon flux (cm$^{-2}$sec$^{-1}$) |
| $\Phi(\phi)$ | Angular wave function |
| $\Psi(x,t)$ | Total wave function |

# SYSTEM OF UNITS, CONVERSION FACTORS, AND GENERAL CONSTANTS

**B.1**   International system of units*

| Quantity | Unit | Symbol | Dimension |
|---|---|---|---|
| Length | meter | m | |
| Mass | kilogram | kg | |
| Time | second | s or sec | |
| Temperature | kelvin | K | |
| Current | ampere | A | |
| Frequency | hertz | Hz | $1/s$ |
| Force | newton | N | $kg\text{-}m/s^2$ |
| Pressure | pascal | Pa | $N/m^2$ |
| Energy | joule | J | N-m |
| Power | watt | W | J/s |
| Electric charge | coulomb | C | A-s |
| Potential | volt | V | J/C |
| Conductance | siemens | S | A/V |
| Resistance | ohm | $\Omega$ | V/A |
| Capacitance | farad | F | C/V |
| Magnetic flux | weber | Wb | V-s |
| Magnetic flux density | tesla | T | $Wb/m^2$ |
| Inductance | henry | H | Wb/A |

\* The cm is the common unit of length and the electron-volt is the common unit of energy (see Appendix F) used in the study of semiconductors. However, the joule and in some cases the meter should be used in most formulas.

**B.2**   Conversion factors

| | Prefixes | | |
|---|---|---|---|
| 1 Å (angstrom) $= 10^{-8}$ cm $= 10^{-10}$ m | $10^{-15}$ | femto- | $= f$ |
| 1 $\mu$m (micron) $= 10^{-4}$ cm | $10^{-12}$ | pico- | $= p$ |
| 1 mil $= 10^{-3}$ in. $= 25.4$ $\mu$m | $10^{-9}$ | nano- | $= n$ |
| 2.54 cm $= 1$ in. | $10^{-6}$ | micro- | $= \mu$ |
| 1 eV $= 1.6 \times 10^{-19}$ J | $10^{-3}$ | milli- | $= m$ |
| 1 J $= 10^7$ erg | $10^{+3}$ | kilo- | $= k$ |
| | $10^{+6}$ | mega- | $= M$ |
| | $10^{+9}$ | giga- | $= G$ |
| | $10^{+12}$ | tera- | $= T$ |

**B.3**  Physical constants

| | |
|---|---|
| Avogadro's number | $N_A = 6.02 \times 10^{+23}$ atoms per gram molecular weight |
| Boltzmann's constant | $k = 1.38 \times 10^{-23}$ J/K = $8.62 \times 10^{-5}$ eV/K |
| Electronic charge (magnitude) | $e = 1.60 \times 10^{-19}$ C |
| Free electron rest mass | $m_o = 9.11 \times 10^{-31}$ kg |
| Permeability of free space | $\mu_o = 4\pi \times 10^{-7}$ H/m |
| Permittivity of free space | $\varepsilon_o = 8.85 \times 10^{-14}$ F/cm = $8.85 \times 10^{-12}$ F/m |
| Planck's constant | $h = 6.625 \times 10^{-34}$ J-s = $4.135 \times 10^{-15}$ eV-s |
| $\dfrac{h}{2\pi} =$ | $\hbar = 1.054 \times 10^{-34}$ J-s |
| Proton rest mass | $M = 1.67 \times 10^{-27}$ kg |
| Speed of light in vacuum | $c = 2.998 \times 10^{10}$ cm/s |
| Thermal voltage ($T = 300°$K) | $V_t = \dfrac{kT}{e} = 0.0259$ volt |
| | $kT = 0.0259$ eV |

**B.4**  Silicon, gallium arsenide, and germanium properties ($T = 300°$K)

| Property | Si | GaAs | Ge |
|---|---|---|---|
| Atoms (cm$^{-3}$) | $5.0 \times 10^{22}$ | $4.42 \times 10^{22}$ | $4.42 \times 10^{22}$ |
| Atomic weight | 28.09 | 144.63 | 72.60 |
| Crystal structure | Diamond | Zincblende | Diamond |
| Density (g/cm$^{-3}$) | 2.33 | 5.32 | 5.33 |
| Lattice constant (Å) | 5.43 | 5.65 | 5.65 |
| Melting point (°C) | 1415 | 1238 | 937 |
| Dielectric constant | 11.7 | 13.1 | 16.0 |
| Bandgap energy (eV) | 1.12 | 1.42 | 0.66 |
| Electron affinity, $\chi$, (volts) | 4.01 | 4.07 | 4.13 |
| Effective density of states in conduction band, $N_c$, (cm$^{-3}$) | $2.8 \times 10^{19}$ | $4.7 \times 10^{17}$ | $1.04 \times 10^{19}$ |
| Effective density of states in valence band, $N_v$, (cm$^{-3}$) | $1.04 \times 10^{19}$ | $7.0 \times 10^{18}$ | $6.0 \times 10^{18}$ |
| Intrinsic carrier concentration (cm$^{-3}$) | $1.5 \times 10^{10}$ | $1.8 \times 10^{6}$ | $2.4 \times 10^{13}$ |

**B.4** *(concluded)*

| Property | Si | GaAs | Ge |
|---|---|---|---|
| Mobility (cm²/V-s) | | | |
| Electron, $\mu_n$ | 1350 | 8500 | 3900 |
| Hole, $\mu_p$ | 480 | 400 | 1900 |
| Effective mass, $\left(\dfrac{m^*}{m_0}\right)$ | | | |
| Electrons | $m_l^* = 0.98$ | 0.067 | 1.64 |
| | $m_t^* = 0.19$ | | 0.082 |
| Holes | $m_{lh}^* = 0.16$ | 0.082 | 0.044 |
| | $m_{hh}^* = 0.49$ | 0.45 | 0.28 |
| Effective mass (density of states) | | | |
| Electrons, $\left(\dfrac{m_n^*}{m_0}\right)$ | 1.08 | 0.067 | 0.55 |
| Holes, $\left(\dfrac{m_p^*}{m_0}\right)$ | 0.56 | 0.48 | 0.37 |

**B.5**  Other semiconductor parameters

| Material | $E_g$ (eV) | $a$ (Å) | $\epsilon_r$ | $\chi$ | $\bar{n}$ |
|---|---|---|---|---|---|
| Aluminum arsenide | 2.16 | 5.66 | 12.0 | 3.5 | 2.97 |
| Gallium phosphide | 2.26 | 5.45 | 10 | 4.3 | 3.37 |
| Aluminum phosphide | 2.43 | 5.46 | 9.8 | | 3.0 |
| Indium phosphide | 1.35 | 5.87 | 12.1 | 4.35 | 3.37 |

**B.6**  Properties of $SiO_2$ and $Si_3N_4$ ($T = 300°K$)

| Property | $SiO_2$ | $Si_3N_4$ |
|---|---|---|
| Crystal structure | [Amorphous for most integrated circuit applications] | |
| Atomic or molecular density (cm⁻³) | $2.2 \times 10^{22}$ | $1.48 \times 10^{22}$ |
| Density (g-cm⁻³) | 2.2 | 3.4 |
| Energy gap | $\approx 9$ eV | 4.7 eV |
| Dielectric constant | 3.9 | 7.5 |
| Melting point (°C) | $\approx 1700$ | $\approx 1900$ |

# THE PERIODIC TABLE

| Period | I a | I b | II a | II b | III a | III b | IV a | IV b | V a | V b | VI a | VI b | VII a | VII b | VIII a | VIII b |
|---|---|---|---|---|---|---|---|---|---|---|---|---|---|---|---|---|
| I | 1 H 1.0079 | | | | | | | | | | | | | | | 2 He 4.003 |
| II | 3 Li 6.94 | | 4 Be 9.02 | | | 5 B 10.82 | | 6 C 12.01 | | 7 N 14.01 | | 8 O 16.00 | | 9 F 19.00 | | 10 Ne 20.18 |
| III | 11 Na 22.99 | | 12 Mg 24.32 | | | 13 Al 26.97 | | 14 Si 28.06 | | 15 P 30.98 | | 16 S 32.06 | | 17 Cl 35.45 | | 18 Ar 39.94 |
| IV | 19 K 39.09 | 29 Cu 63.54 | 20 Ca 40.08 | 30 Zn 65.38 | 21 Sc 44.96 | 31 Ga 69.72 | 22 Ti 47.90 | 32 Ge 72.60 | 23 V 50.95 | 33 As 74.91 | 24 Cr 52.01 | 34 Se 78.96 | 25 Mn 54.93 | 35 Br 79.91 | 26 Fe 55.85  27 Co 58.94  28 Ni 58.69 | 36 Kr 83.7 |
| V | 37 Rb 85.48 | 47 Ag 107.88 | 38 Sr 87.63 | 48 Cd 112.41 | 39 Y 88.92 | 49 In 114.76 | 40 Zr 91.22 | 50 Sn 118.70 | 41 Nb 92.91 | 51 Sb 121.76 | 42 Mo 95.95 | 52 Te 127.61 | 43 Tc 99 | 53 I 126.92 | 44 Ru 101.7  45 Rh 102.91  46 Pd 106.4 | 54 Xe 131.3 |
| VI | 55 Cs 132.91 | 79 Au 197.2 | 56 Ba 137.36 | 80 Hg 200.61 | 57–71 Rare earths | 81 Tl 204.39 | 72 Hf 178.6 | 82 Pb 207.21 | 73 Ta 180.88 | 83 Bi 209.00 | 74 W 183.92 | 84 Po 210 | 75 Re 186.31 | 85 At 211 | 76 Os 190.2  77 Ir 193.1  78 Pt 195.2 | 86 Rn 222 |
| VII | 87 Fr 223 | | 88 Ra 226.05 | | 89 Ac 227 | | 90 Th 232.12 | | 91 Pa 231 | 93 Np 237 | 92 U 238.07 | 94 Pu 239 | 95 Am 241 | 96 Cm 242 | 97 Bk 246  98 Cf 249  99 Es 254  100 Fm 256 | 101 Md 256 |

### Rare Earths

| Period | | | | | | | | | | | | | | | |
|---|---|---|---|---|---|---|---|---|---|---|---|---|---|---|---|
| VI 57–71 | 57 La 138.92 | 58 Ce 140.13 | 59 Pr 140.92 | 60 Nd 144.27 | 61 Pm 147 | 62 Sm 150.43 | 63 Eu 152.0 | 64 Gd 156.9 | 65 Tb 159.2 | 66 Dy 162.46 | 67 Ho 164.90 | 68 Er 167.2 | 69 Tm 169.4 | 70 Yb 173.04 | 71 Lu 174.99 |

The numbers in front of the symbols of the elements denote the atomic numbers; the numbers underneath are the atomic weights.

# THE ERROR FUNCTION

$$erf(z) = \frac{2}{\sqrt{\pi}} \int_0^z e^{-t^2} dt$$

$$erf(0) = 0 \qquad erf(\infty) = 1$$

$$erfc(z) = 1 - erf(z)$$

| $z$ | $erf(z)$ | $z$ | $erf(z)$ |
|------|----------|------|----------|
| 0.00 | 0.00000 | 1.00 | 0.84270 |
| 0.05 | 0.05637 | 1.05 | 0.86244 |
| 0.10 | 0.11246 | 1.10 | 0.88021 |
| 0.15 | 0.16800 | 1.15 | 0.89612 |
| 0.20 | 0.22270 | 1.20 | 0.91031 |
| 0.25 | 0.27633 | 1.25 | 0.92290 |
| 0.30 | 0.32863 | 1.30 | 0.93401 |
| 0.35 | 0.37938 | 1.35 | 0.94376 |
| 0.40 | 0.42839 | 1.40 | 0.95229 |
| 0.45 | 0.47548 | 1.45 | 0.95970 |
| 0.50 | 0.52050 | 1.50 | 0.96611 |
| 0.55 | 0.56332 | 1.55 | 0.97162 |
| 0.60 | 0.60386 | 1.60 | 0.97635 |
| 0.65 | 0.64203 | 1.65 | 0.98038 |
| 0.70 | 0.67780 | 1.70 | 0.98379 |
| 0.75 | 0.71116 | 1.75 | 0.98667 |
| 0.80 | 0.74210 | 1.80 | 0.98909 |
| 0.85 | 0.77067 | 1.85 | 0.99111 |
| 0.90 | 0.79691 | 1.90 | 0.99279 |
| 0.95 | 0.82089 | 1.95 | 0.99418 |
| 1.00 | 0.84270 | 2.00 | 0.99532 |

# "DERIVATION" OF THE SCHRODINGER'S WAVE EQUATION

Schrodinger's wave equation was stated in Equation (2–6). The time-independent form of Schrodinger's wave equation was then developed and given by Equation (2–13). The time-independent Schrodinger's wave equation can also be developed from the classical wave equation. We may think of this development more in terms of a justification of the Schrodinger's time-independent wave equation rather than a strict derivation.

The time-independent classical wave equation, in terms of voltage, is given as

$$\frac{\partial^2 V(x)}{\partial x^2} + \left(\frac{\omega^2}{v_p^2}\right) V(x) = 0 \tag{E–1}$$

where $\omega$ is the radian frequency and $v_p$ is the phase velocity.

If we make a change of variable and let $\psi(x) = V(x)$, then we have

$$\frac{\partial^2 \psi(x)}{\partial x^2} + \left(\frac{\omega^2}{v_p^2}\right) \psi(x) = 0 \tag{E–2}$$

We can write that

$$\frac{\omega^2}{v_p^2} = \left(\frac{2\pi\nu}{v_p}\right)^2 = \left(\frac{2\pi}{\lambda}\right)^2 \tag{E–3}$$

where $\nu$ and $\lambda$ are the wave frequency and wavelength, respectively.

From the wave-particle duality principle, we can relate the wavelength and momentum as

$$\lambda = \frac{h}{p} \tag{E–4}$$

Then

$$\left(\frac{2\pi}{\lambda}\right)^2 = \left(\frac{2\pi}{h} \cdot p\right)^2 \tag{E–5}$$

and since $\hbar = \dfrac{h}{2\pi}$, we can write

$$\left(\frac{2\pi}{\lambda}\right)^2 = \left(\frac{p}{\hbar}\right)^2 = \frac{2m}{\hbar^2}\left(\frac{p^2}{2m}\right) \tag{E-6}$$

Now

$$\frac{p^2}{2m} = T = E - V \tag{E-7}$$

where $T$, $E$, and $V$ are the kinetic energy, total energy, and potential energy terms, respectively.

We can then write

$$\frac{\omega^2}{v_p^2} = \left(\frac{2\pi}{\lambda}\right)^2 = \frac{2m}{\hbar^2}\left(\frac{p^2}{2m}\right) = \frac{2m}{\hbar^2}(E - V) \tag{E-8}$$

Substituting Equation (E–8) into Equation (E–2), we have

$$\frac{\partial^2\psi(x)}{\partial x^2} + \frac{2m}{\hbar^2}(E - V)\psi(x) = 0 \tag{E-9}$$

which is the one-dimensional, time-independent Schrodinger's wave equation.

# UNIT OF ENERGY—THE ELECTRON-VOLT

The electron-volt (eV) is a unit of energy that is used constantly in the study of semiconductor physics and devices. This short discussion may help in "getting a feel" for the electron-volt.

Consider a parallel plate capacitor with an applied voltage as shown in Figure F–1. Assume that an electron is released at $x = 0$ at time $t = 0$. We may write

$$F = m_0 a = m_0 \frac{d^2x}{dt^2} = eE \qquad \text{(F–1)}$$

where $e$ is the magnitude of the electronic charge and E is the magnitude of the electric field as shown. Upon integrating, the velocity and distance versus time are given by

$$v = \frac{eEt}{m_0} \qquad \text{(F–2)}$$

and

$$x = \frac{eEt^2}{2m_0} \qquad \text{(F–3)}$$

where we have assumed that $v = 0$ at $t = 0$.

Assume that at $t = t_0$ the electron reaches the positive plate of the capacitor so that $x = d$. Then

$$d = \frac{eEt_0^2}{2m_0} \qquad \text{(F–4a)}$$

or

$$t_0 = \sqrt{\frac{2m_0 d}{eE}} \qquad \text{(F–4b)}$$

The velocity of the electron when it reaches the positive plate of the capacitor is

$$v(t_0) = \frac{eEt_0}{m_0} = \sqrt{\frac{2eEd}{m_0}} \qquad \text{(F–5)}$$

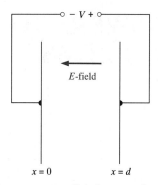

**Figure F–1**  Parallel plate capacitor.

The kinetic energy of the electron at this time is

$$T = \frac{1}{2} m_0 v(t_0)^2 = \frac{1}{2} m_0 \left( \frac{2eEd}{m_0} \right) = eEd \qquad \text{(F–6)}$$

The electric field is

$$E = \frac{V}{d} \qquad \text{(F–7)}$$

so that the energy is

$$T = e \cdot V \qquad \text{(F–8)}$$

If an electron is accelerated through a potential of 1 volt, then the energy is

$$T = e \cdot V = (1.6 \times 10^{-19})(1) = 1.6 \times 10^{-19} \text{ joule} \qquad \text{(F–9)}$$

The electron-volt (eV) unit of energy is defined as

$$\text{Electron-volt} = \frac{\text{joule}}{e} \qquad \text{(F–10)}$$

Then, the electron that is accelerated through a potential of 1 volt will have an energy of

$$T = 1.6 \times 10^{-19} \text{ J} = \frac{1.6 \times 10^{-19}}{1.6 \times 10^{-19}} \text{ (eV)} \qquad \text{(F–11)}$$

or 1 eV.

We may note that the magnitude of the potential (1 volt) and the magnitude of the electron energy (1 eV) are the same. However, it is important to keep in mind that the unit associated with each number is different.

# ANSWERS TO SELECTED PROBLEMS

## Chapter 1

1. *a.* 4 atoms, *b.* 2 atoms, *c.* 8 atoms
3. *a.* 74 percent, *b.* 68 percent, *c.* 34 percent
7. *a.* 4.50 Å, *b.* 3.18 Å, *c.* 2.60 Å
9. *a.* $6.78 \times 10^{14}$ cm$^{-2}$, *b.* $9.59 \times 10^{14}$ cm$^{-2}$, *c.* $7.83 \times 10^{14}$ cm$^{-2}$
11. $2 \times 10^{23}$ cm$^{-3}$
13. $2 \times 10^{-6}$ percent

## Chapter 2

3. $\lambda = 0.254$ $\mu$m (Gold), $\lambda = 0.654$ $\mu$m (Cesium)
5. $E = 10.3$ keV
7. *a.* $\Delta p = 1.054 \times 10^{-28}$, *b.* $\Delta E = 0.198$ eV
9. *a.* $\Delta p = 1.054 \times 10^{-24}$, *b.* $\Delta t = 6.6 \times 10^{-16}$ sec
11. $A = +1, -1, +j,$ or $-j$
13. *a.* $n = 1.35 \times 10^{28}$, *b.* $E = 0.010 + 1.48 \times 10^{-30}$ joule
17. *b.i.* $\Delta E = 3.85 \times 10^{-3}$ eV, *b.ii.* $\Delta E = 2.46 \times 10^{-17}$ eV
19. *a.* 0.118 percent, *b.* $1.9 \times 10^{-10}$ percent
21. $T = 3.06 \times 10^{-5}$
23. $T = \dfrac{4k_1 k_3}{(k_1 + k_3)^2}$
25. $-13.58$ eV, $-3.40$ eV, $-1.51$ eV, $-0.849$ eV

## Chapter 3

7. $m^*$ (curve B) $> m^*$ (curve A)
13. *a.* $E = E_F + 4.6$ $kT$, *b.* $f_F(E) \cong 0.01$
15. *a.* 0.304 percent, *b.* 14.96 percent, *c.* 99.7 percent, *d.* $f_F(E_F) = 0.5$
17. *a.* $f_F(E_1) = 9.3 \times 10^{-6}$, $1 - f_F(E_2) = 1.66 \times 10^{-19}$, *b.* $f_F(E_1) = 7.88 \times 10^{-18}$, $1 - f_F(E_2) = 1.96 \times 10^{-7}$
19. $T = 461°$K

## Chapter 4

3. $T = 531°$K
5. $E_{Fi} - E_{mid} = -0.0128$ eV (Si), $E_{Fi} - E_{mid} = -0.0077$ eV (Ge), $E_{Fi} - E_{mid} = +0.038$ eV (GaAs)
9. $n_o = (kT)K \exp\left[-(E_c - E_F)/kT\right]$
11. $r_1 = 15.4$ Å, $E = 0.029$ eV
13. $n_o = 5.59 \times 10^{15}$ cm$^{-3}$, $p_o = 9.57 \times 10^{10}$ cm$^{-3}$
15. $n_o = 5.19 \times 10^{14}$ cm$^{-3}$, $p_o = 2.08 \times 10^4$ cm$^{-3}$
17. $p_o = 7.04 \times 10^{18}$ cm$^{-3}$
19. *a.* $T = 149°$K, *b.* $T = 225°$K
21. *a.* $p_o = 2.95 \times 10^{13}$ cm$^{-3}$, $n_o = 1.95 \times 10^{13}$ cm$^{-3}$, *b.* $n_o = 5 \times 10^{15}$ cm$^{-3}$, $p_o = 1.15 \times 10^{11}$ cm$^{-3}$
23. *a.* n-type, *b.* $n_o = 9.24 \times 10^{10}$ cm$^{-3}$, $p_o = 2.44 \times 10^9$ cm$^{-3}$

25. Add $N_a = 6 \times 10^{15}$ cm$^{-3}$, $n_o = 6.5 \times 10^{-4}$ cm$^{-3}$

30. $N_d = 1.2 \times 10^{16}$ cm$^{-3}$

32. *a*. Add donors, *b*. $N_d = 5.33 \times 10^{14}$ cm$^{-3}$

## Chapter 5

1. *a*. $n_o = 10^{16}$ cm$^{-3}$, $p_o = 3.24 \times 10^{-4}$ cm$^{-3}$, *b*. $J = 136$ A/cm$^2$, *c*. $p_o = 10^{16}$ cm$^{-3}$, $n_o = 3.24 \times 10^{-4}$ cm$^{-3}$, $J = 6.4$ A/cm$^2$

3. $v_d = 6.25 \times 10^5$ cm/sec

5. *a*. $p_o = 7.8 \times 10^{16}$ cm$^{-3}$, $n_o = 4.15 \times 10^{-5}$ cm$^{-3}$, *b*. $n_o = 5.79 \times 10^{14}$ cm$^{-3}$, $p_o = 3.89 \times 10^5$ cm$^{-3}$

7. *a*. $J = 1.60$ A/cm$^2$, *b*. $T = 456°$K

9. *a*. $\mu_n = 2388$ cm$^2$/V-sec, *b*. $\mu_n = 844$ cm$^2$/V-sec

11. $n(0) = 0.25 \times 10^{14}$ cm$^{-3}$

13. $E = 14.5 - 26 \exp(x/18)$ V/cm

15. *a*. $n(x) = 6.5 \times 10^{15} - 3.24 \times 10^{15} \exp(-x/2.15 \times 10^{-3})$ cm$^{-3}$, *b*. $n(0) = 3.26 \times 10^{15}$ cm$^{-3}$, $n(50) = 6.18 \times 10^{15}$ cm$^{-3}$, *c*. $J_{dft}(50) = 94.9$ A/cm$^2$, $J_{dff} = 5.1$ A/cm$^2$

17. $V = -2.73$ mV

19. *a*. $V_H = 2.19$ mV, *b*. $E_H = 0.219$ V/cm

21. *a*. p-type, *b*. $p = 8.08 \times 10^{15}$ cm$^{-3}$, *c*. $\mu_p = 387$ cm$^2$/V-sec

23. *a*. n-type, *b*. $n = 8.68 \times 10^{14}$ cm$^{-3}$, *c*. $\mu_n = 8182$ cm$^2$/V-sec, *d*. $\rho = 0.88$ ohm-cm

## Chapter 6

1. $R' = 5 \times 10^{19}$ cm$^{-3}$ sec$^{-1}$

7. $\sigma = 8 + 0.114(1 - e^{-t/\tau_{po}})$ where $\tau_{po} = 10^{-7}$ sec

9. $I = (54 + 2.2e^{-t/\tau_{po}})$ mA where $\tau_{po} = 3 \times 10^{-7}$ sec

11. $\delta n = 10^{13} e^{-t/\tau_{no}}$ where $\tau_{no} = 0.1$ $\mu$sec

13. *b.i.* $\delta n = 0$, *b.ii.* $\delta n = 0.865 \times 10^{14}$ cm$^{-3}$, *b.iii.* $\delta n = 0$

15. *a*. $p_{po} = 10^{14}$ cm$^{-3}$, $n_{po} = 2.25 \times 10^6$ cm$^{-3}$, *b*. $\delta n(0) = -n_{po} = -2.25 \times 10^6$ cm$^{-3}$, *c*. $\delta n(x) = -n_{po}e^{-x/L_n}$

17. *a*. $\delta n(x) = \delta n(0)e^{-\alpha x}$, where $\alpha = 1/L_n\{L_n/2L' + \sqrt{(L_n/2L')^2 + 1}\}$ and where $L' = (kT/e) \cdot 1/E_o$

19. $\delta p = (G_o'/2D_p)(5L^2 - x^2)$ for $-L < x < L$, $\delta p = (G_o'L/D_p)(3L - x)$ for $L < x < 3L$, $\delta p = (G_o'L/D_p)(3L + x)$ for $-3L < x < -L$

22. *a*. $E_{Fi} - E_F = 0.3294$ eV, *b*. $E_{Fn} - E_{Fi} = 0.2697$ eV, $E_{Fi} - E_{Fp} = 0.3318$ eV

24. $E_{Fi} - E_{Fp} = 0.5814$ eV, $E_{Fn} - E_{Fi} = 0.4619$ eV

27. *a*. $R/\delta n = 10^7$ sec$^{-1}$, *b*. $R/\delta n = 1.67 \times 10^6$ sec$^{-1}$, *c*. $R/\delta n = 2 \times 10^6$ sec$^{-1}$

29. *a*. $\delta n = 10^{15} \cdot \dfrac{\sinh((W - x)/L_n)}{\sinh(W/L_n)}$ where $L_n = 35.4$ $\mu$m, *b*. $\delta n = 10^{15}(1 - x/W)$

31. $\delta n = (G_o'/2D_n)[2W^2 - 2Wx - x^2]$ for $-W < x < 0$, $\delta n = (G_o'W/D_n)(W - x)$ for $0 < x < W$

## Chapter 7

1. *A.a.* $V_{bi} = 0.575$ volt, *b*. 0.635 volt, *c*. 0.695 volt, *d*. 0.754 volt

3. *a*. $V_{bi} = 0.713$ volt, *b*. $x_n = 4.8 \times 10^{-7}$ cm, $x_p = 0.960 \times 10^{-4}$ cm, $W = 0.965 \times 10^{-4}$ cm, *c*. $E_{max} = 1.48 \times 10^4$ V/cm

5. $T = 490°$K

7. *a*. $V_{bi} = 0.671$ volt, *b*. $W = 0.691$ $\mu$m ($V_R = 0$), $W = 2.48$ $\mu$m ($V_R = 8$), *c*. $E_{max} = 1.94 \times 10^4$ V/cm, ($V_R = 0$), $E_{max} = 7.0 \times 10^4$ V/cm ($V_R = 8$)

9. *a*. $N_a = 4.28 \times 10^{15}$ cm$^{-3}$, $N_d = 2.14 \times 10^{17}$ cm$^{-3}$, *b*. $x_p = 1.80$ $\mu$m, *c*. $C' = 5.74 \times 10^{-9}$ F/cm$^2$

11. *a*. $f_o = 1.67$ MHz, *b*. $f_o = 2.63$ MHz

13. *a.* $N_a = 9.92 \times 10^{14}$ cm$^{-3}$,
    *b.* $N_d = 8.93 \times 10^{15}$ cm$^{-3}$,
    *c.* $V_{bi} = 0.632$ volt

15. *a.* $V_R = 193$ volts, *b.* $x_n = 0.5$ $\mu$m, *c.* $E_{max} = 7.72 \times 10^4$ V/cm

17. *a.* $N_d = 5 \times 10^{16}$ cm$^{-3}$, *b.* $N_a = 8.82 \times 10^{17}$ cm$^{-3}$, *c.* $E_{max} = 2.71 \times 10^5$ V/cm

## Chapter 8

1. *a.* $\Delta V = 59.6$ mV, *b.* $\Delta V = 119.3$ mV

3. $V = -59.6$ mV

5. $J_n/J_p = 0.0104$

7. *a.* $V_{bi} = 0.723$ volt, *b.* $n_p(0) = 1.12 \times 10^{14}$ cm$^{-3}$, $p_n(0) = 6.28 \times 10^{13}$ cm$^{-3}$, *c.* $J_n(0) = 0.106$ A/cm$^2$, $J_p(0) = 0.354$ A/cm$^2$

9. *a.* $\delta p_n = 3.81 \times 10^{14}$ exp $[-x/2.83 \times 10^{-4}]$ cm$^{-3}$, *b.* $J_p = 0.597$ A/cm$^2$, *c.* $J_n = 1.39$ A/cm$^2$

11. $E_g = 0.769$ eV

13. $T = 519°$K

15. $Y = 0.0772 + j\omega(3.86 \times 10^{-8})$

17. *a.* $I_D = 0.518$ mA, *b.* $V_a = 0.594$ volt, *c.* $r_d = 50\ \Omega$

19. $V_a = 0.443$ volt

21. $J_s = 8.57 \times 10^{-18}$ A/cm$^2$, $J_{gen} = 1.93 \times 10^{-9}$ A/cm$^2$

23. $V_a = 0.477$ volt

25. $J_{gen} = 1.5 \times 10^{-3}$ A/cm$^2$

27. *a.* $E_{crit} \cong 3 \times 10^5$ V/cm, $4.5 \times 10^5$ V/cm, $6.2 \times 10^5$ V/cm, *b.* $V_B \cong 291$ volts, 65.5 volts, 12.4 volts

29. $V_B \approx 15$ volts

31. *a.* $I_R/I_F = 1.11$, *b.* $t_2/\tau_{po} = 0.65$

## Chapter 9

1. *c.* $\phi_{B0} = 0.27$ volt, $x_d = 0.091\ \mu$m, $E_{max} = 1.41 \times 10^4$ V/cm, *d.* $\phi_{B0} \cong 0.55$ volt, $x_d = 0.211\ \mu$m, $E_{max} = 3.26 \times 10^4$ V/cm

3. *a.* $\phi_{B0} = 1.03$ volts, *b.* $\phi_n = 0.058$ volt, *c.* $V_{bi} = 0.972$ volt, *d.* $x_d =$

$0.416\ \mu$m, *e.* $E_{max} = 2.87 \times 10^5$ V/cm

5. *a.* $C = 4.75$ pF, *b.* $C = 15$ pF

7. *a.* $V_{bi} = 0.344$ volt, $x_d = 0.211$ $\mu$m, $E_{max} = 3.26 \times 10^4$ V/cm, *b.* $\Delta\phi = 20.0$ mV, $x_m = 0.306 \times 10^{-6}$ cm, *c.* $\Delta\phi = 37.8$ mV, $x_m = 0.163 \times 10^{-6}$ cm

9. *a.* $V_{bi} = 0.812$ volt, $x_d = 0.153$ $\mu$m, $E_{max} = 1.06 \times 10^5$ V/cm, *b.* $V_R = 7.47$ volts

11. $D_{it} = 4.97 \times 10^{11}$ eV$^{-1}$ cm$^{-2}$

13. *a.* $V_a = 0.278$ volt, *b.* $V_a = 0.125$ volt

15. *a.* $I_R = 15.2$ nA, *b.* $I_R = 18.6$ nA

19. Area $= 1.62 \times 10^{-3}$ cm$^2$

21. *a.* $V_a = 0.691$ volt (pn junction), $V_a = 0.445$ volt (Schottky diode), *b.* $I = 120$ mA (pn junction), $I = 53.7$ mA (Schottky diode)

23. *b.* $N_d = 1.24 \times 10^{16}$ cm$^{-3}$, *c.* 0.20 volt

## Chapter 10

3. $V_{BE} = 0.635$ volt, *b.* $p_n(0) = 10^{14}$ cm$^{-3}$, *c.* $V_{BE} = 0.375$ volt, $p_n(0) = 10^{14}$ cm$^{-3}$

9. *a.* $V_{CB} = 0.70$ volt, *b.* $V_{EC}$(sat) $= 0.05$ volt, *c.* $Q_B = 5.45 \times 10^{-8}$ C/cm$^2$, *d.* $8.82 \times 10^{13}$ electrons/cm$^2$

11. $V_{CB} = 0.48$ volt

13. *a.* $\alpha = 0.787$, *b.* $\gamma = 0.923$, *c.* $\alpha_T = 0.983$, *d.* $\delta = 0.867$, *e.* $\beta = 3.69$

19. *a.* $x_B = 0.742\ \mu$m, *b.* $\delta = 0.9999994$

21. *a.* $N_B = 3.19 \times 10^{16}$ cm$^{-3}$, *b.* $V_A \cong 95$ volts

23. *a.* $N_B = 1.83 \times 10^{15}$ cm$^{-3}$, *b.* $N_B = 4.02 \times 10^{16}$ cm$^{-3}$

25. $S = 2.45\ \mu$m

27. *a.* $BV_{CBO} \cong 90$ volts, *b.* $BV_{CEO} = 18$ volts, *c.* $BV_{BE} \cong 12$ volts

29. $V_{pt} = 18.3$ volts, $BV_{(expected)} \approx 25$ volts

31. $I_B = 0.105$ mA

35. *a.* $\tau_e = 41.4$ ps, $\tau_b = 98$ ps, $\tau_c = 4.8$ ps, $\tau_d = 20$ ps, *b.* $f_T = 970$ MHz, $f_\beta = 19.4$ MHz

37. $x_B = 1.26$ $\mu$m

## Chapter 11

3. *a.* $V_{p0} = 5.79$ volts, $V_p = 4.91$ volts, *b.i.* 0.215 $\mu$m, *b.ii.* 0.0653 $\mu$m, *b.iii.* zero

5. *a.* $V_{p0} = 3.79$ volts, $V_p = -2.90$ volts, *b.i.* 0.102 $\mu$m, *b.ii.* 0.044 $\mu$m, *b.iii.* zero

7. *a.* $I_{p1} = 1.03$ mA, *b.* $V_{DS}(\text{sat}) = 1.06$ volts, 0.795 volt, 0.53 volt, 0.265 volt, *c.* $I_{D1}(\text{sat}) = 0.258$ mA, 0.140 mA, 0.061 mA, 0.0145 mA

9. *a.* $G_{o1} = 2.69 \times 10^{-3}$, *b.* $V_{DS}(\text{sat}) = +0.35$ volt, $+0.175$ volt, *c.* $I_{D1}(\text{sat}) = 0.0504$ mA, 0.0123 mA

11. *a.* $g_{ms}(\text{max}) = 0.295$ mS, *b.* $g_{ms}(\text{max}) = 1.48$ mS

13. *a.* $N_d = 8.1 \times 10^{15}$ cm$^{-3}$, *b.* $V_T = +0.051$ volt

15. *a.* $a = 0.26$ $\mu$m, $V_T = 0.092$ volt, *b.* $V_{DS}(\text{sat}) = 0.258$ volt

17. *a.* $k_n = 2.51$ mA/V$^2$, *b.* $V_{DS}(\text{sat}) = 0.06$ volt, 0.12 volt, *c.* $I_{D1}(\text{sat}) = 9.04$ $\mu$A, 36.1 $\mu$A

19. $L = 1.54$ $\mu$m

21. *a.* $V_{DS} = 2$ volts, *b.* $h_{\text{sat}} = 0.306$ $\mu$m, *c.* $I_{D1}(\text{sat}) = 3.72$ mA, *d.* $I_{D1}(\text{sat}) = 9.08$ mA

23. *a.* 0.39 pA, *b.* 0.42 pA, *c.* 0.50 pA

25. $f_T = 9.76$ GHz

26. *a.* $V_{\text{off}} = -2.07$ volts, *b.* $n_s = 3.25 \times 10^{12}$ cm$^{-2}$

28. $d_d = 251$ A$^\circ$

## Chapter 12

1. *a.*  (Si): $x_{dT} = 0.30$ $\mu$m, $|Q'_{SD}(\text{max})| = 4.8 \times 10^{-8}$ coul/cm$^2$
 (GaAs): $x_{dT} = 0.410$ $\mu$m, $|Q'_{SD}(\text{max})| = 6.56 \times 10^{-8}$ coul/cm$^2$
 (Ge): $x_{dT} = 0.235$ $\mu$m, $|Q'_{SD}(\text{max})| = 3.76 \times 10^{-8}$ coul/cm$^2$

3. *a.* $\phi_{ms} = -0.926$ volt, *b.* $-0.876$ volt, *c.* $+0.244$ volt

5. *a.i.* $V_{FB} = -0.295$ volt, *a.ii.* $-0.504$ volt, *a.iii.* $-1.43$ volts

7. *a.* $V_T = -0.528$ volt, *b.* $-0.478$ volt, *c.* $+0.642$ volt

9. $N_a = 1.71 \times 10^{16}$ cm$^{-3}$

11. $Q'_{ss} = 6.23 \times 10^{11}$ cm$^{-2}$

13. *b.* $\phi_{ms} = -1.11$ volts, *c.* $V_{TN} = +0.557$ volt

15. *a.* $C_{\text{ox}} = C'(\text{inv}) = 3.45 \times 10^{-8}$ F/cm$^2$, $C'_{FB} = 2.14 \times 10^{-8}$ F/cm$^2$, $C'_{\text{min}} = 0.70 \times 10^{-8}$ F/cm$^2$, *b.* $C_{\text{ox}} = 3.45 \times 10^{-8}$ F/cm$^2$, $C'_{FB} = 2.14 \times 10^{-8}$ F/cm$^2$, $C'_{\text{min}} = C'(\text{inv}) = 0.70 \times 10^{-8}$ F/cm$^2$, *c.* $V_{FB} = -0.34$ volt, $V_{TP} = -1.15$ volts

17. *a.* $\Delta V_{FB} = -1.74$ volts, *b.* $\Delta V_{FB} = -0.869$ volt, *c.* $\Delta V_{FB} = -1.16$ volts

23. $V_T = 0.6$ volt, $\mu_n = 650$ cm$^2$/V-sec

25. *a.* $g_{mL} = 0.333$ mS, *b.* $g_{ms} = 2.13$ mS

29. *a.* $r_s = 198$ $\Omega$, *b.* 12 percent reduction

31. *a.* $f_T = 3.18$ GHz, *b.* $f_T = 0.83$ GHz

## Chapter 13

1. *a.* $L = 0.606$ $\mu$m, *b.* $L = 3.77$ $\mu$m

5. $\Delta V_T = -0.137$ volt

9. $W = 1.44$ $\mu$m

11. $t_{\text{ox}} = 1000$ Å

13. $V_{DS} = 2.08$ volts

15. $L = 1.08$ $\mu$m

17. *a.* $V_T = -3.74$ volts, *b.* $D_I =$ $9.32 \times 10^{11}$ cm$^{-2}$

19. $\Delta V_T = -2.09$ volts

## Chapter 14

1. *a.* $\lambda = 1.88$ $\mu$m(Ge), $\lambda = 1.11$ $\mu$m(Si), $\lambda = 0.873$ $\mu$m(GaAs), *b.* $E = 2.18$ eV ($\lambda = 570$ nm), $E = 1.77$ eV ($\lambda = 700$ nm)

3. $g' = 1.44 \times 10^{19}$ cm$^{-3}$ sec$^{-1}$, $\delta n = 1.44 \times 10^{13}$ cm$^{-3}$

5. *a.* $x = 1.98$ $\mu$m, *b.* $x = 0.41$ $\mu$m

7. $\delta p_n = \dfrac{\alpha \Phi_0 \tau_{p0}}{\alpha^2 L_p^2 - 1} \left\{ \left[ \dfrac{s + \alpha D_p}{s + \dfrac{D_p}{L_p}} \right] e^{-x/L_p} - e^{-\alpha x} \right\}$

11. $P_m = 276$ mW, Increased by factor of 11.1

14. *a.* $\delta p = 10^{13}$ cm$^{-3}$, *b.* $\Delta \sigma = 2.29 \times 10^{-3}$, *c.* $I_L = 1.15$ mA, *d.* $\Gamma_{ph} = 7.15$

16. $I_L = 0.131$ $\mu$A

20. $x = 230$ $\mu$m

22. *a.* $E_g = 1.85$ eV, *b.* $\lambda = 0.670$ $\mu$m

26. $\Delta \lambda = 5.08 \times 10^{-3}$ $\mu$m

## Chapter 15

1. *a.* $C_0 = 0.069$ pF, *b.* $\phi_{s0} = 10.2$ volts, *c.* $x_{d0} = 3.63$ $\mu$m, *d.* $Q = -1.16 \times 10^{-13}$ coul

3. *a.* ($V_G = 10$ volts) $\phi_{s0} = 11.5$ volts, $x_{d0} = 5.46$ $\mu$m ($V_G = 20$ volts) $\phi_{s0} = 22.1$ volts, $x_{d0} = 7.56$ $\mu$m, *b.* ($V_G = 10$ volts) $\phi_{s2} = 8.53$ volts, $x_{d0} = 4.7$ $\mu$m ($V_G = 20$ volts) $\phi_{s2} = 19.2$ volts, $x_{d0} = 7.05$ $\mu$m

5. *a.* 1 millisec, *b.* $0.90 \times 10^6$ electrons

11. *a.* $n_u/n_L = 2.61 \times 10^{-4}$, *b.* $T = 1043°$K

13. *a.* 0.25 watt, *b.* 2.08 watts

15. *a.* $L = 0.875$ $\mu$m, *b.* ($+Q$) $= 0.25$ pC

# INDEX

## A

Abrupt junction approximation; *see* pn junction

Absorption
  coefficient, 610–613, 624, 657
  optical, 610–615

a-c Equivalent circuit
  BJT, 413–417
  JFET, 466–468
  MOSFET, 541–545

Acceptor
  action in semiconductors, 119, 120
  atoms, 120, 123–125, 128, 150
  doping profile, 257
  ionization energy, 123, 124
  level, 123, 133
  traps, 217, 445

Accumulation layer; *see* MOS capacitor

Admittance, small-signal, 285–293

AlGaAs, 3, 20, 351, 356, 639, 640, 647, 648

Alpha; *see* Common base current gain

Aluminum, 123, 325, 490, 501, 503, 509–512, 516

Ambipolar
  diffusion coefficient, 201, 202, 228
  mobility, 201, 202, 228
  transport, 199–214, 228, 281
  transport equation, 200–202, 204, 205, 207, 209, 228, 274

Amorphous silicon, 623, 624

Amorphous solids, 3

Amphoteric, 124

Amplification, 410

Amplifier, 410, 413, 466, 478, 541, 672, 673

Anisotype junction, 363

Annealing, 17

Anode
  GUNN diode, 693
  pnpn diode, 678–680, 682

Arrays, solar cell, 623, 625

Atomic arrangement; *see* Crystal structures

Atomic bonding; *see* Bonds

Atomic structure; *see* Periodic table of the elements

Atomic weight, 723

Au; *see* Gold (Au)

Auger recombination, 637, 638, 657

Avalanche breakdown, 315
  BJT, 407–410
  diode, 301–305
  doping dependence, 304, 305
  electron and hole ionization rates, 302, 303
  mechanism, 301–305
  multiplication factor, 302, 408, 585, 586
  pnpn structure, 680, 681
  voltage, 303

Avalanche photodiode, 634

Avalanche transit time diode; *see* IMPATT diode

Avogadro's number, 720

Axes, crystal, 4, 5

## B

Balance of drift and diffusion, 241, 271

Band bending, 242, 243, 491–496

Band diagram; *see* Energy bands

Bandgap energy
  definition, 63, 64
  dependence on alloy composition, 639, 640
  of common semiconductors, 721
  relation to intrinsic carrier concentration, 114

Band-to-band transitions, 192, 196, 637, 639

Band discontinuities, 351–353

Bandwidth, 634

Barrier
  to injection, 242, 270, 271
  tunneling, 44

Base (BJT)
  contact, 372
  current components, 384, 385, 388
  drift in, 404, 405
  minority carrier concentration, 378–380, 383
  narrowing, 395–398, 405
  region, 372, 377–380, 392
  resistance, 402, 413
  transit time, 417, 418, 421, 429
  transport factor, 386–389, 392, 394, 429

# C

# N

## Q

## R

## U

## V

## W

## X

## Z